NEURAL TRANSPLANTS

Development and Function

NEURAL TRANSPLANTS

Development and Function

Edited by

John R. Sladek, Jr.
and
Don M. Gash

University of Rochester
School of Medicine and Dentistry
Rochester, New York

SPRINGER SCIENCE+BUSINESS MEDIA, LLC

Library of Congress Cataloging in Publication Data

Main entry under title:

Neural transplants.

Includes bibliographical references and index.
1. Nerve tissue — Transplantation. 2. Surgery, Experimental. I. Sladek, John R. II. Gash, Don M., Date–
RD124.N49   1984   599′.0188   84-8219
ISBN 978-1-4684-4687-6

ISBN 978-1-4684-4687-6   ISBN 978-1-4684-4685-2 (eBook)
DOI 10.1007/978-1-4684-4685-2

© 1984 Springer Science+Business Media New York
Originally published by Plenum Press, New York in 1984
Softcover reprint of the hardcover 1st edition 1984

We wish to dedicate
this book to our wives,
CELIA D. SLADEK AND SHARON LEE GASH,
in appreciation of their
support and encouragement.

Contributors

GARY W. ARENDASH, Department of Biological Sciences, University of South Florida, Tampa, Florida 33620

ANDERS BJÖRKLUND, Departments of Histology and Ophthalmology, University of Lund, Lund, Sweden

HÅKAN BJÖRKLUND, Department of Histology, Karolinska Institute, Stockholm, Sweden

MILTON W. BRIGHTMAN, Laboratory of Neuropathology and Neuroanatomical Sciences, National Institutes of Health, Bethesda, Maryland 20205

MARIANNE BRONNER-FRASER, Department of Physiology and Biophysics, University of California, Irvine, California 92717

CARL W. COTMAN, Department of Psychobiology, University of California, Irvine, California 92717

WILLIAM J. FREED, Preclinical Neurosciences Section, Adult Psychiatry Branch, National Institute of Mental Health, Saint Elizabeths Hospital, Washington, D.C. 20032

DON M. GASH, Department of Anatomy, University of Rochester School of Medicine and Dentistry, Rochester, New York 14642

MARIE J. GIBSON, Division of Endocrinology, Mount Sinai School of Medicine, New York, New York 10029

ROGER A. GORSKI, Department of Anatomy and Laboratory of Neuroendocrinology of the Brain Research Institute, University of California School of Medicine, Los Angeles, California 90024

P. P. C. GRAZIADEI, Department of Biology, Florida State University, Tallahassee, Florida 32306

WILLIAM A. HARRIS, Department of Biology, University of California, San Diego, La Jolla, California 92093

BARRY J. HOFFER, Department of Pharmacology, University of Colorado Medical Center, Denver, Colorado 80262

GLORIA E. HOFFMAN, Department of Anatomy, University of Rochester School of Medicine and Dentistry, Rochester, New York 14642

H. KIMURA, Kinsmen Laboratory of Neurological Research, Department of Psychiatry, University of British Columbia, Vancouver, British Columbia V6T 1W5, Canada

DOROTHY T. KRIEGER, Division of Endocrinology, Mount Sinai School of Medicine, New York, New York 10029

LAWRENCE F. KROMER, Department of Anatomy and Neurobiology, College of Medicine, University of Vermont, Burlington, Vermont 05405

BRUCE S. MCEWEN, Laboratory of Neuroendocrinology, The Rockefeller University, New York, New York 10021

E. G. MCGEER, Kinsmen Laboratory of Neurological Research, Department of Psychiatry, University of British Columbia, Vancouver, British Columbia V6T 1W5, Canada

P. L. MCGEER, Kinsmen Laboratory of Neurological Research, Department of Psychiatry, University of British Columbia, Vancouver, British Columbia V6T 1W5, Canada

LINDA K. MCLOON, Departments of Anatomy and Ophthalmology, University of Minnesota, Minneapolis, Minnesota 55455

STEVEN C. MCLOON, Departments of Anatomy and Ophthalmology, University of Minnesota, Minneapolis, Minnesota 55455

G. A. MONTI GRAZIADEI, Department of Biology, Florida State University, Tallahassee, Florida 32306

HOWARD NORNES, Department of Anatomy, Colorado State University, Fort Collins, Colorado 80523

LARS OLSON, Department of Histology, Karolinska Institute, Stockholm, Sweden

CHARLES M. PADEN, Department of Biology, Montana State University, Bozeman, Montana 59717

JOSEPH ROGERS, Department of Neurology, University of Massachusetts Medical School, Worcester, Massachusetts 01605

JEFFREY M. ROSENSTEIN, Department of Anatomy, George Washington University Medical Center, Washington, D.C. 20037

ANN-JUDITH SILVERMAN, Department of Anatomy, Columbia University, College of Physicains and Surgeons,New York, New York 10027

JOHN R. SLADEK, JR., Department of Anatomy, University of Rochester School of Medicine and Dentistry, Rochester, New York 14642

ULF STENEVI, Departments of Histology and Ophthalmology, University of Lund, Lund, Sweden

WYLIE W. VALE, Peptide Biology Laboratory, The Salk Institute, La Jolla, California 92138

RICHARD JED WYATT, Preclinical Neurosciences Section, Adult Psychiatry Branch, National Institute of Mental Health, Saint Elizabeths Hospital, Washington, D.C. 20032

STEVEN F. ZORNETZER, Department of Pharmacology, University of California, Irvine, California 92668

Preface

The story of mammalian neural transplantation really begins eighty-one years ago. In Chicago in December of 1903, a 34-year-old physician, Elizabeth Hopkins Dunn, working as a research assistant in neurology, initiated a series of experiments to examine the ability of neonatal rat cerebral tissue to survive transplantation into the brain of matched littermates. Out of 46 attempts, four clearly successful grafts were identified. The publication of Dunn's results in 1917, the first credible report to demonstrate the feasibility of mammalian CNS transplants, generated little interest. In fact, the next significant experiment in this field did not appear until 1930. The field continued to grow slowly and quietly as investigators gradually realized the value of neural transplantation to study problems of development and plasticity in the mammalian nervous system. With the discovery in 1979 that grafted neurons were capable of appropriate and functional interactions with the host brain, interest in neural transplantation escalated sharply. The extraordinary opportunities created by using functional neural transplants in investigating basic issues in neurobiology, as well as the clinical implications, excited both scientists and the public. The popularity of neural transplantation has been growing rapidly in the past five years and shows no signs of abating.

The present volume was designed to meet two needs created by the rapid growth of this subdiscipline of neurobiology. The first was to provide a thorough review of the experimental foundations of neural transplantation. The second was to document the state of the art of brain cell transplantation as a field of scientific inquiry at an important juncture in its development. Accordingly the first chapter examines the historical development of mammalian neural grafting studies and discusses the conceptual basis of present-day experimentation on transplant development and function. The next two chapters are devoted to analyzing the extensive body of literature on neural transplants in nonmammalian vertebrates. Such work on the lower vertebrates has greatly influenced our understanding of neural development and is very relevant to interpreting results from transplants in mammals. The remaining fifteen chapters survey the present domain of mammalian neural transplantation. They describe the model systems that are being used to explore the properties and principles of brain cell transplantation and delineate the significant scientific advances that have been made. We believe each chapter reflects not only the significant progress that has been made but also the great challenges that await us.

We wish to thank the authors for their meaningful contributions. We also wish to thank Kirk Jensen of Plenum Press for his initiative and advice on this project, and our wives, Celia Sladek and Sharon Gash, for their patience and understanding.

John R. Sladek, Jr.
Don Marshall Gash

Contents

Chapter 3

Neural Transplants in Lower Vertebrates

WILLIAM A. HARRIS

Chapter 4

Transplantation of the Developing Mammalian Visual System

STEVEN C. McLOON AND LINDA K. McLOON

Chapter 5

Camera Bulbi Anterior: New Vistas on a Classical Locus for Neural Tissue Transplantation

LARS OLSON, HÅKAN BJÖRKLUND, AND BARRY J. HOFFER

Chapter 6

The Olfactory Organ: Neural Transplantation

G. A. MONTI GRAZIADEI AND P. P. C. GRAZIADEI

Chapter 7

Correction of Genetic Gonadotropic Hormone-Releasing Hormone
Deficiency by Preoptic Area Transplants

DOROTHY T. KRIEGER AND MARIE J. GIBSON

Contents

Chapter 8

Hypothalamic Grafts and Neuroendocrine Cascade Theories of
Aging: Immunocytochemical Viability of Preoptic Hypothalamic
Transplants from Fetal to Reproductively Senescent Female Rats

JOSEPH ROGERS, GLORIA E. HOFFMAN, STEVEN F. ZORNETZER, AND
WYLIE W. VALE

Chapter 9

Brain Tissue Transplants and Reproductive Function: Implications
for the Sexual Differentiation of the Brain

GARY W. ARENDASH AND ROGER A. GORSKI

Chapter 10

Morphological and Functional Properties of Transplanted
Vasopressin Neurons

JOHN R. SLADEK, JR., AND DON M. GASH

Chapter 11

The Use of Fetal Hypothalamic Transplants in Developmental
Neuroendocrinology

CHARLES M. PADEN, ANN-JUDITH SILVERMAN, ULF STENEVI, AND
BRUCE S. MCEWEN

Chapter 12

Specificity of Termination Fields Formed in the Developing Hippocampus by Fibers from Transplants

CARL W. COTMAN

Chapter 13

Use of CNS Implants to Promote Regeneration of Central Axons across Denervating Lesions in the Adult Rat Brain

ULF STENEVI, ANDERS BJÖRKLUND, AND LAWRENCE F. KROMER

Chapter 14

Intracephalic Embryonic Transplants: A New Experimental
Preparation for Developmental Neurobiology

LAWRENCE F. KROMER

Chapter 15

Transplantation of Newborn Brain Tissue into Adult Kainic-Acid-
Lesioned Neostriatum

P. L. McGEER, H. KIMURA, AND E. G. McGEER

Chapter 16

Transplantation of Catecholamine-Containing Tissues to Restore
the Functional Capacity of the Damaged Nigrostriatal System

WILLIAM J. FREED, BARRY J. HOFFER, LARS OLSON, AND
RICHARD JED WYATT

Chapter 17

Transplantation Strategies in Spinal Cord Regeneration

Howard Nornes, Anders Björklund, and Ulf Stenevi

Chapter 18

Some Consequences of Grafting Autonomic Ganglia to Brain Surfaces

Jeffrey M. Rosenstein and Milton W. Brightman

1

Neural Transplants in Mammals
A Historical Overview

Don M. Gash

A scholar must review the old so as to find out what is new.
Confucius, *The Analects*
c. 500 B.C.

We have inherited the wisdom of the past, in Institutions and in Libraries. Most times we have chosen to disregard it and start afresh.
George E. Brown, D-U.S. Congress
In a speech at the Winter Conference on Brain Research; January 23, 1983

1. Introduction

The basic techniques of neural transplantation have long been within the repertoire of the neuroscientist and have been productively employed to address many of the major questions in neurobiology. With some notable exceptions, the animal models for grafting studies have come from the amphibian and avian classes. Only within the last 6 years have experiments examining the properties of transplanted mammalian central nervous system (CNS) tissue become widespread. It is illustrative that more papers on mammalian CNS grafts were published in 1980 alone than were published in the entire first 60 years (1890–1950) of work in this area (see Fig. 1). The present volume surveys the state of the art of neural transplantation in vertebrates. In the present chapter, we trace

Don M. Gash ● Department of Anatomy, University of Rochester School of Medicine and Dentistry, Rochester, New York 14642.

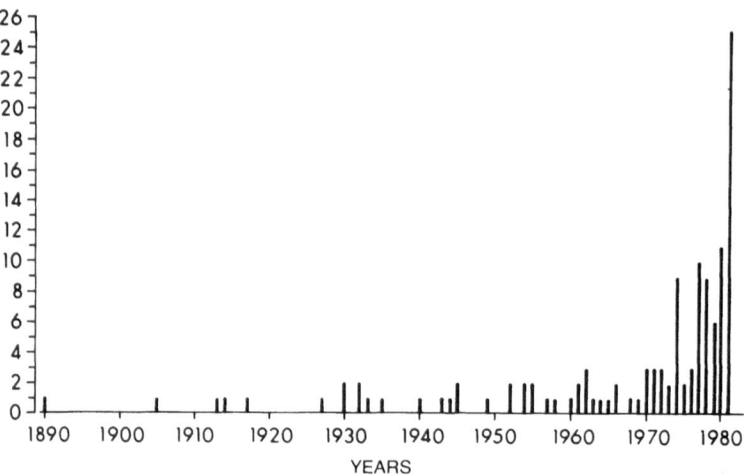

FIGURE 1. This graph shows the number of papers published each year from 1890 to 1980 on experiments utilizing transplantation procedures involving CNS neurons of mammals. Compiled from *Index Medicus* and other sources.

the slow development of the field of mammalian brain cell transplantation and examine the reasons why, until recently, there has been a relative dearth of studies on mammals as compared to other vertebrate classes.

2. The First Fifty Years: 1890–1940

Saltykow[1] in 1905 presented the first detailed report on transplanting mammalian brain tissue. His experiments, which began at the University of Basel in 1903, involved excising and rapidly reimplanting slabs of cerebral cortex tissue in young rabbits ranging in age from 6 weeks to several months. Histological examination of the grafts recovered at intervals ranging from 20 min up to 233 days after reimplantation revealed good survival of neurons up to 8 days following transplantation. From day 8 on, neurons showed progressive signs of degeneration with protoplasmic swelling and increases in nuclear size. Glial cells proliferated with large numbers of mitotic figures seen, especially on days 7 through 20. Endothelial and perivascular cells also showed abundant mitotic activity with blood vessels rapidly revascularizing the reimplanted tissue.

After reviewing the literature, Saltykow believed his to be the first report on "transplantation of healthy brain tissue." While he had indeed conducted a well-designed and thoroughly documented study, an American, W. Gilman Thompson, professor of physiology at the New York University Medical College, 15 years earlier had published[2] a brief cryptic description of his attempts to graft mammalian cortical tissue. Thompson made several transplants of mature feline cerebral cortex into cortical cavities in adult dogs. No neurons could be identified in the grafts which were examined histologically 6 weeks following transplantation.

Saltykow had conducted the type of experiment which should logically provide for the optimal survival of grafted neurons; that is, rapid reimplantation of brain tissue into the site from which it was excised. His results demonstrated some of the pitfalls to be encountered in transplanting mammalian brain tissue and have formed the cornerstone of the concept that fully differentiated CNS neurons cannot survive grafting. Two reports from Italy, D'Abundo[3] in 1913 and Altobelli[4] in 1914, did, however, suggest prolonged survival of transplanted adult neurons. The first study entailed transplants of feline spinal cord, either intraperitoneally or subcutaneously, into other adult cats while the latter employed cerebral cortical grafts between rabbits. D'Abundo confined his experiments to grafts with 12- to 21-day survival periods and reported that while in most instances neurons were observed in various states of degeneration, a few histologically normal anterior horn neurons could be identified in spinal cord grafts. Whether these cells were present in both 12- and 21-day grafts was not stated nor was the number of surviving neurons per transplant given. Altobelli claimed that grafted neurons were present in his long-term cortical grafts of 102-day duration and that neurites from both transplant and host passed through the glial scar. Altobelli's interpretation of his data must be viewed with caution. His transplants taken 8 days after grafting showed extensive necrosis, lymphocytic infiltration, and provided no evidence for the survival of grafted neurons. He identified the boundaries of the grafts recovered after 30 days or longer based on differences in staining intensity and notes that the transplants had become fully incorporated into the brain. Without adequate criteria for distinguishing graft versus host tissue, neurons in the host cortex injured by the transplantation procedures may have been mistakenly identified as grafted cells.

The first study to convincingly demonstrate survival of transplanted CNS neurons came from experiments conducted by Elizabeth Hopkins Dunn[5] while working as a research assistant in H. H. Donaldson's laboratory at the University of Chicago. Dunn initiated her experiments in December 1903 and published her results in 1917 in the *Journal of Comparative Neurology*. Her procedures involved transplanting cerebral cortical tissue from 9- to 10-day-old rats into cortical cavities in littermate recipients. In every successful case, the graft had come into contact with the choroid plexus in the lateral ventricles in the course of the grafting procedures and had developed an adequate blood supply. Dunn described neurons in the graft as being myelinated and organized into relatively normal cytoarchitectonic patterns. She saw no evidence that axons from the transplant crossed through the glial scar to connect with host neurons. While the survival times were not clearly stated for her successful transplants, data in her report indicated that several of these grafts had gone for 66 days. The important contributions from Dunn's work were: (1) demonstration that immature neural tissue could be successfully grafted and (2) providing the first experimental evidence that transplantation onto a rich vascular bed was important for graft survival. Dunn did not follow up on this study and, in fact, waited 10 years after it was completed to publish the results. While known to the scientific community, her report did not immediately inspire others to attempt transplantations, and 13 years was to pass until the next significant report on mammalian neural transplants appeared.

In 1930, Raoul M. May,[6] Harvard educated and then working at the Laboratory of Comparative Histology at the College of France, published the first of what was to be an extensive series of experiments on transplants into the anterior chamber of the eye. His initial study involved grafting cerebral tissue from 0- to 2-day-old neonatal rats into the eyes of adult albino rats. The grafts became vascularized by blood vessels from the iris and remained intact for the duration of the experiment; a period close to 6 months for some grafts. Golgi-staining revealed neurons dispersed throughout the transplant. While not displaying normal patterns of cytoarchitectural organization, the neurons did appear to possess the features of normal cells in that their nuclei and immediate projections were not different from the corresponding structures of the latter. Fibers, identified by Golgi-staining in the transplant, stayed within the boundaries of the graft and were not observed to pass through the graft–host interface. Similarly, fibers of host origin within the iris were not observed to innervate the graft. May's experiments, as he pointed out, supported and extended Dunn's observations. Again, the immature nature of the donor tissue seemed to be an important factor for successful transplantation. An equally important concept to emerge from May's study was that neural explants could survive for prolonged periods (in this case, 6 months) in sites that, from other studies, were beginning to be characterized as immunologically privileged.[7]

Confirmation of both Dunn's and May's work came in 1940, in a study conducted at the Universiy of Oxford by Le Gros Clark.[8] Using 15- to 20-day fetal rabbits for tissue donors, portions of the fetal cortex were freed from their mesodermal coverings and implanted as isolated fragments into the lateral cerebral ventricles of 6-week-old host rabbits. Successful grafts, examined histologically 4 weeks after implantation, showed evidence of continued growth and differentiation. The neurons were organized into clumps, whorls, and occasional laminar formations. While clearly abnormal, the cytoarchitectural arrangements of the grafted cells were reminiscent of organizational patterns seen in the normal brain. As had May earlier, Le Gros Clark reported the transplanted fetal neurons had attained the features of typical mature cortical neurons, indicating the former had retained their capacity for normal differentiation.

Le Gros Clark did not pursue this line of investigation or at least he was not to publish again on mammalian brain cell transplantation. Scientists, along with the other citizens of the world, had their attention drawn elsewhere; namely, to the consuming conflagration of World War II.

3. 1941–1970: Sporadic Progress

The only scientist to maintain a sustained program of investigation focused on neural transplantation in the period from 1941 to 1970 was Raoul May.[9–15] We shall examine his work in detail later. Other individuals and groups did contribute occasional important papers to the literature, but their papers, because of the general lack of interest in neural transplantation at the time of publication, were not followed up and now seem to be virtually unknown.

A study that does deserve attention, primarily as a graphic example of a premature and ill-founded human study, was an attempt to conduct a neural transplant in man.[16] Citing Le Gros Clark's study as part of their rationale for proceeding, in July 1942, a surgical group in St. Louis grafted a spinal cord into a 16-year-old boy with a spinal cord injury. The transplanted piece had been removed at autopsy 2 weeks earlier, fixed in 10% formalin, rinsed in distilled water, and kept in 70% ethanol. The patient was operated on under ether anesthesia with the spinal cord exposed by a laminectomy, and the spinal cord transected at the third to fourth thoracic level. A cellophane bed was prepared in the vertebral canal and a 3-inch section of the cadaver cord glued into the gap between the cut ends of the patient's cord. About 4 months later the patient died. No positive effects of the transplantation were reported.

Following this bizarre episode, a study of fundamental importance involving human neural tissue was published. Harry Greene and Hildegerde Arnold,[17] from the Yale University School of Medicine, conducted an extensive series of experiments on transplants from human, rabbit, and guinea pig sources into the anterior chamber of the eye of the rabbit and guinea pig. Normal adult brain tissue failed to grow when grafted, while embryonic and neoplastic brain tissue survived cross-species transplantation. Homologous transplants of normal embryonic rabbit brain into adult rabbit eyes were successful in 42 of 46 attempts. Several of the grafts were followed for time periods up to 2½ years without demonstrating signs of regressive changes. When examined histologically, the transplants contained numerous healthy ganglion cells in addition to glial elements. Transplants of rabbit and human embryonic brain tissue into the anterior chamber of the guinea pig eye were successful in 41 of 43 rabbit and 50 of 55 human brain grafts. Guinea pigs bearing human brain cell grafts were followed for more than 2 years up to the termination of the experiment without the appearance of regressive changes. Histologically, the heterologous transplants appeared to be quite similar to homologous transplants and consisted of well-formed ganglion cells and glial elements. The Greene and Arnold study remains to the present as the only documented investigation on transplanted human embryonic brain tissue. In considering the possible future clinical applications of neural transplants, it is indeed important to note that in this initial study human brain grafts behaved in an identical manner to grafts from other mammalian species.

In 1945, studies by Raoul M. May again appear in the literature. Indeed, no review of neural transplantation would be complete without a discussion of May's contributions. Born in Hermosillo, Mexico, educated at Stanford, Harvard, and the University of Paris, May was a gifted linguist who began his scientific career in 1928 by translating Ramón y Cajal's book *Degeneration and Regeneration of the Nervous System* from the original Spanish into French and English editions. May maintained a lifelong interest in development, regeneration, and transplantation. While he worked with both mammalian and nonmammalian systems and many types of embryonic tissue, we shall limit our review here to his experiments involving the transplantation of nerve cells in mammals. In 1930, as has already been noted, May reported that neonatal rat cerebral tissue survived transplantation into the anterior chamber of the adult rat eye. Fifteen years passed before he returned to this experimental paradigm (this time using the adult mouse as a host animal) and in 1945 published[9] the first in a series of papers on double intraocular grafts. He initially demonstrated that the simultaneous transplantation of neonatal mouse cerebral tissue and the sciatic nerve apparently induced the regeneration of cerebral cell fibers. These cerebral fibers invaded the degenerated nerve and once within the nerve sheath coursed through in a manner similar to normal peripheral axons. Next[10] he implanted cerebral tissue from a newborn mouse along with a small piece of muscle from the thigh of the same donor. Cerebral neurons innervated the cografted muscle within several days. Entire tracts of fibers left the cerebral transplant and enveloped the striated muscle fibers. Sixteen days after the double graft, the innervation pattern of the muscle tissue showed similarities to that in a newborn mouse with nerve endings of a primitive type formed by a simple terminal enlargement of the nerve fiber on the striated muscle fiber.

Some of the suggestions raised by these studies were (1) would the cerebral grafts innervate structures which are normally sparsely innervated in the adult animal and (2) are neurotrophic factors important in inducing fiber outgrowth from the cerebral grafts? May[11] addressed these questions, the first directly and the second indirectly, by grafting neonatal cerebral tissue into the anterior chamber near either living or killed thymus tissue. In 112 cases, the thymus fragment was from the same donor as the cerebral tissue, while in 40 host animals the thymus fragment had been taken from a newborn mouse, fixed in ethanol and washed in saline before transplantation. Fragments of the thymus grafted in the living state were invaded by fibers from cerebral cells within 18 days after implantation. These axons crossed the thymus from one end to the other forming a rich plexus of fibers by their divisions and intermingling. The fragments of dead thymus were completely restructured by the immigration of lymphoctyes, fibroblasts, and other unidentified cells by 12 days after implantation. These remodeled fragments were then invaded by fibers from cerebral cells just as in the case of fragments of living thymus. Clearly the grafted neural tissue exhibits a propensity for innervating cografted tissue. The results from grafting killed thymus fragments are difficult to interpret. On the basis of our current knowledge, it could be postulated either that the killed tissue contained sufficient neurotrophic substance(s) to induce fiber outgrowth from the cerebral graft or that growth factors from the fibroblasts and other invading cells contributed to the observed results. Indeed, May, in the terminology of his day, listed both of these as possible explanations and noted that they are not mutually exclusive hypotheses.

One additional study conducted during this period merits further discussion. Paris E. Royo and Wilbur B. Quay[18] examined fetal retinal transplants into the adult rat eye and made some unexpected and possibly quite important observations. Their experimental paradigm involved transplanting the retina from a 16-day fetus into the anterior chamber of its mother's eye, using pigmented rats of the Long–Evans strain. The retinal grafts were followed for up to 50 days and, in 40 of 45 cases examined, were found to have survived. By 5 days after transplantation, grafted cells had moved through the pupil and were attached to the posterior surface of the host's iris. By 9 days, small ganglion cells, fibroblasts, and other grafted retinal cells were beginning to migrate around the lens, through the hyaloid canal and toward the optic nerve. The transplanted ganglion cells were observed to send many of their axons into the host's optic nerve by 50 days after grafting. The grafted tissue showed normal patterns and rates of development with, for example, transplanted rods exhibiting the same growth rate as rods in intact developing eyes. The transplants apparently induced the regrowth of the fetal hyaloid artery in that by 5 days after surgery, a vascular extension from the central artery of the host's optic disc entered the vitreous humor and continued throughout the course of the experiment to grow along the hyaloid canal. Some negative effects of the transplants were noted with fibroblasts invading the host's lens and degenerative changes occurring at points in the host's photoreceptive layer, especially near the optic disc. Royo and Quay left their study at this point. Royo went on to medical school and became a practicing ophthalmologist, while Quay had other pressing experiments to pursue. No one picked up on the exciting leads they had uncovered.

4. Summary of Progress from 1890 to 1970

Though relatively few studies were conducted in the first 80 years of neural transplantation in mammals, the pioneering experiments of Dunn, Le Gros Clark, May, and Gross had established some basic principles. Developing neural tissue from fetal and neonatal donors could survive when placed in the CNS of a host animal near an adequate blood supply. The grafted neurons continued normal patterns of cytodifferentiation and were often organized in cytoarchitectural patterns reminiscent of their site of origin. Transplants, at least in cocultures in the anterior chamber of the eyes, could innervate adjacent tissues. Some evidence for the survival of cross-species transplants and the prolonged survival of grafted tissue had also been presented. As interesting as these data seem in retrospect, they had not yet excited the general interest of the neuroscience community. It was not until the decade of the 1970s that intensive research efforts were to begin in this field.

5. 1970–1980: A Quickening Tempo

The pace of research on neural transplantation began to pick up in 1970 and, by the end of the decade, was proceeding at a feverish rate. This decade of research is, in great measure, covered in the other chapters of this book. Accordingly, of the numerous important papers published in this period, only those landmark studies that served as prognosticators of future developments will be discussed.

Olson and Malmfors,[19] in 1970, inaugurated the study of chemically identified neurons in grafts by employing fluorescence, histochemical, and autoradiographic techniques to identify adrenergic neurons grafted into the anterior chamber of the eye. Gopal Das and Joseph Altman[20] (1971) confirmed and extended the earlier studies of Dunn and Le Gros Clark by showing that cerebellar tissue could be grafted between 7-day neonatal rats. From grafts juxtaposed to the host cerebellum, donor neuroblasts, labeled with [^{3}H]thymidine, apparently migrated into the host cerebellar cortex and differentiated into basket and granule cells. Hoffer and co-workers[21] in 1974 published their initial electrophysiological study on neural transplants in oculo and showed that grafted neurons retained normal electrical properties. It is clear that an increasing number of research groups were becoming interested in brain cell transplantation and were utilizing state-of-the-art techniques to investigate the properties of grafted neurons.

Two particularly important papers were published in 1976. Stenevi and co-workers[22] provided the first systematic study of factors important for obtaining good and consistent survival of transplanted CNS tissue. They also reported the first successful transplantation of central monoaminergic neurons into the brain. Their experiments examined the suitability of three different sites for transplant survival: within the brain parenchyma of the diencephalon and caudate nucleus, onto the dorsal surface of the caudate nucleus, and into a transplantation cavity created by aspirating out cortical tissue so as to expose a portion of the choroid plexus. In the latter of these three sites, the best survival of grafted fetal central catecholamine neurons was observed in terms of the number of grafts surviving (22/25) and the number of viable neurons within the transplants (up to 500 monoaminergic neurons in a graft). The few surviving grafts on the surface of the caudate and in the parenchyma of the caudal diencephalon (3 out of a total of 44 grafts) contained only a few monoaminergic neurons. The success of the transplantation cavity as transplantation site seemed to be due, in large measure, to the choroid plexus which provided a rich blood supply to the grafts.

While the study by Stenevi and co-workers started to transform transplantation from a phenomenological to an analytical science, Lund and Hauschka,[23] also in 1976, published a paper that illustrated how powerful transplantation techniques could be when applied to basic studies on the developing mammalian nervous system. They grafted strips of tissue from the superior colliculus of fetal rats into the superior colliculus region of newborn rats. Four to six weeks after transplantation, considerable numbers of axons were found running between the host brain and the grafted tissue. Retrograde degeneration following removal of the contralateral eye showed that some of these fibers were visual afferents. Others were thought to be efferents from the transplant to the host. An important observation was that anatomically normal-appearing synapses could be identified in the transplant by electron microscopy.

Given that transplants survive for extended periods, show anatomically normal features, and send fibers into the host brain, the issue arises as to whether grafted neurons can interact with the host in an appropriate and functional manner. A question of obvious fundamental importance, it excited the interest of a number of investigators. Perlow et al.[24] in 1979 were the first to report a successful study. Their experimental model entailed unilaterally destroying the substantia nigra in adult rats by 6-hydroxydopamine injections. Because of the resulting loss of dopaminergic innervation to the caudate nucleus, the rats showed deficits in motor control that were exacerbated by apomorphine treatment (the animals showed pronounced rotations contralateral to the side of the lesion). Grafts of fetal substantia nigra tissue reversed, to some degree, these motor abnormalities and the recovery was associated with the growth of dopaminergic neurons and fibers from the graft into the host caudate nucleus.

A second model system for studying functional development of grafted neurons was described a year later by Gash and co-workers[25,26] (also see Chapter 10). In these studies, vasopressin neurosecretory neurons were transplanted from normal donors into Brattleboro rats which congenitally lack vasopressin-producing neurons. Vasopressin is the antidiuretic hormone and in its absence, mammals exhibit a pronounced diabetes insipidus, that is, a polydipsia and polyurea. Vasopressin was identified in the grafts by both immunocytochemistry and radioimmunoassay. The symptoms of diabetes insipidus were significantly ameliorated in 9 of 40 Brattleboro rats receiving transplants of hypothalamic tissue.

Neural grafting studies had thus entered a new era. Investigators were becoming increasingly aware that transplantation technology offered potential avenues to correct neurological dysfunctions resulting from the loss of circumscribed populations of neurons. The initial explorations into this developing field, with its extraordinary potential, are described in the remaining chapters of this volume.

6. Analysis

Two major questions concerning the history of neurotransplantation remain to be examined. Why did the field develop so slowly until the early 1970s? What has led to the recent exponential growth of studies on neural grafts in mammals?

First, it should be recognized that there has been a general upsurge of interest, beginning around 1960, by the scientific community in neurobiology. Neuroscience, in the opinion of many,[27] has become the new frontier of biology. One illustration of this increasing attention is that the number of papers published on the nervous system increased by an order of magnitude from 1950 to 1980 (see Fig. 2). The American Society of Neuroscience has similarly grown 10-fold in the past few years. In part then, the increased interest in neural transplantation is a reflection of a much broader scientific movement toward studies on the nervous system. Certainly, this is not the whole story.

Was the scientific community not aware of the successful experiments with CNS grafts before 1970? The answer seems to be no. Dunn, who published the first credible report of transplant survival in 1917, conducted her studies at the University of Chicago at a time when it was a leading American center in the neurosciences.[28] Stephen W. Ranson, an internationally recognized neuroscientist, was working in the same laboratory with Dunn and indeed conducted his own studies on peripheral nervous system grafts.[29] The preeminent neurophysiologist C. S. Sherrington visited Dunn while her experiments were in progress and even advised her on surgical techniques.[5] May and

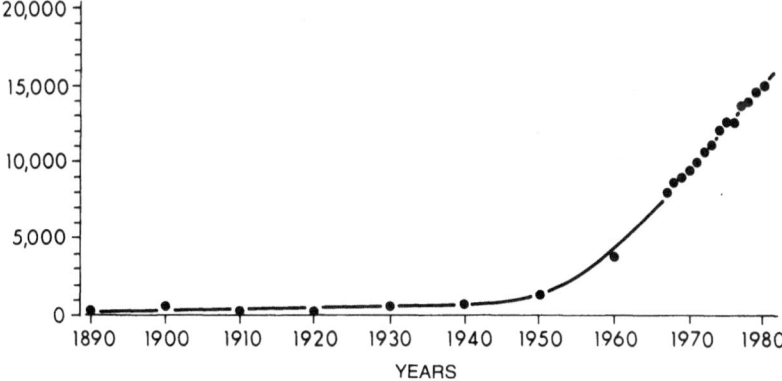

FIGURE 2. The number of papers indexed under the "Nervous System" and related headings in *Index Medicus* are shown for the first year in each decade from 1890 to 1960. Similar data, in this case from papers listed in the "Nervous System" (A8) category of *Index Medicus,* are shown for each year from 1967 to 1980.

Le Gros Clark were internationally prominent scientists whose papers were published in first-line journals. One must assume that the basic concepts of neural transplantation were known to the neuroscience community and were simply not exploited.

The most likely explanation is that a sufficient rationale for conducting extended studies on neural transplants in mammals did not exist until the late 1960s. Prior to that time, experimental issues involving transplants could be more logically and conveniently conducted on amphibian or avian species. Studies of fundamental importance in developmental neurobiology utilizing neural transplantation techniques were conducted in the 1920s and earlier on nonmammalian vertebrates[30,31] (also see Chapters 2 and 3). Even May, who was the most productive early investigator of mammalian CNS transplants, worked extensively on amphibian model systems.[32]

Until the 1960s the only reason for studying neural grafts in mammals was to examine their anatomy. Concepts of the brain until that time essentially negated the consideration of function transplants. J. Z. Young[33] perhaps most succinctly stated the prevailing viewpoint, when arguing against the possibility of replacing destroyed nerve cells:

> On general biological grounds, it's perhaps hardly to be expected that the intricate morphogenetic processes necessary to produce the finer details of the higher nervous centers could be reproduced in the adult.

While Young was surely aware of Le Gross Clark's work on embryonic neural transplants published 2 years before this was written (both men were at Oxford), he did not mention it as a possible factor in the functional repair of nervous tissue.

Our concepts of the nervous system have undergone radical revisions in the last 20 years. Development of the Flack–Hillarp technique in 1962,[34,35] which made possible the histological study of monoaminergic neurons, and along with methodological advances in neurochemistry, led to a new way of looking at the brain; essentially viewing the nervous system in terms of its chemical anatomy. The power of this approach was quickly demonstrated by showing that Parkinson's disease was the result of the loss of dopaminergic neurons in the substantia nigra, which in turn led to a significant depletion of dopamine levels in the striatum.[36] Greatly improved methods of treating Parkinson's disease through the use of drugs to replenish brain dopamine, followed these discoveries.[36]

Now it was apparent that the loss of circumscribed populations of neurons could lead to quite specific behavioral dysfunctions. Since drug treatment could at least partially reverse these symptoms, one could also anticipate that discrete grafts of nervous tissue, even without making all of the connections, could have a functional role in the host brain. A rationale for conducting CNS transplants in mammals now existed. A new technology began to develop.

ACKNOWLEDGMENTS. I am indebted to Leslie B. Dick, Susan Connor, and Michael Aschner for their assistance with translating manuscripts and library research; Janet B. Berk and Charles King provided the advice and assistance on various phases of the historical research. I also wish to thank Joyce Goodberlet and Mary Beth Horan for their expert typing of the manuscript.

References

1. Saltykow, S., 1905, Versuche über Gehirnreplantation, zugleich ein Beitrag zur Kenntniss reactiver Vorgänge an den zelligen Gehirnelementen, *Arch. Psychiatr.* **40**:329.
2. Thompson, W. G., 1890, Successful brain grafting, *N.Y. Med. J.* June 28, 701.
3. D'Abundo, G., 1913, Sulle manifestazioni di vitalita nei trapianti del tessuto nervoso, *Riv. Ital. Neuropat. Psichiat. Elettroter* **6**:145.
4. Altobelli, R., 1914, Innesti cerebrali, *Gazz. Int. Med. Chir.* **17**:25.
5. Dunn, E. H., 1917, Primary and secondary findings in a series of attempts to transplant cerebral cortex in the albino rat, *J. Comp. Neurol.* **27**:565.
6. May, R. M., 1930, La greffe dans l'oeil de rat blanc adulte du tissu cerebral de rat nouveau-ne, *Arch. Anat. Microsc. Morphol. Exp.,* **26**:433.
7. Murphy, J. B., and Sturm, E., 1923, Conditions determining the transplantability of tissues in the brain, *J. Exp. Med.* **38**:183.
8. Le Gros Clark, W. E., 1940, Neuronal differentiation in implanted foetal cortical tissue *J. Neurol. Psychiatry* **3**:263.
9. May, R. M., 1945, Régénération cérébrale provoquée par la greffe intraoculaire simultanée de tissu cérébral de nouveau-né et de nerf sciatique chez la souris, *Bull. Biol. Fr. Belg.* **79**:151.
10. May, R. M., 1949, Connexions entre des cellules cérébrales et des muscles de la cuisse dans leur greffe brephoplastique intra-oculaire simultanée chez la souris, *Arch. Anat. Microsc. Morphol. Exp.* **38**:145.
11. May, R. M., 1952, La greffe brephoplastique intra-oculaire simultanée de tissu cérébral et de thymus vivant ou mort chez la souris, *Arch. Anat. Microsc. Morphol. Exp.* **41**:237.
12. May, R. M, 1954, La greffe brephoblastique intraoculaire du cervelet chez la souris, *Arch. Anat. Microsc. Morphol. Exp.* **43**:42.
13. May, R. M., 1955, Cerebral transplantation in mammals, *Transplant. Bull.* **2**:62.
14. May, R. M., 1957, The possibilities of brephoplastic transplants, *Ann. N.Y. Acad. Sci.* **64**:937.
15. May, R. M., and Barres, M. C., 1962, La greffe brephoplastique de tissu cérébral sous la capsule du rein chez la souris, *C.R. Acad. Sci.* **254**:2839.
16. Woolsey, D., Minckler, J., Rezende, N., and Klemme, R., 1944, Human spinal cord transplant, *Exp. Med. Surg.* **2**:93.
17. Greene, H. S. N., and Arnold, H., 1945, The homologous and heterologous transplantation of brain and brain tumors, *J. Neurosurg.* **2**: 315.
18. Royo, P. E., and Quay, W. B., 1959, Retinal transplantation from fetal to maternal mammalian eye, *Growth* **23**:313.
19. Olson, L., and Malmfors, T., 1970, Growth characteristics of adrenergic nerves in the adult rat: Fluorescence, histochemical and ^{3}H-noradrenaline uptake studies using tissue transplantations to the anterior chamber of the eye, *Acta. Physiol. Scand. Suppl.* **348**:1.
20. Das, G. D., and Altman, J., 1971, Transplanted precursors of nerve cells: Their fate in the cerebellums of young rats, *Science* **173**:637.
21. Hoffer, B., Seiger, A., Ljungberg, T., and Olson, L., 1974, Electrophysiological and cytological studies of brain homografts in the anterior chamber of the eye: Maturation of cerebellar cortex *in oculo, Brain Res.* **79**:165.
22. Stenevi, U., Björklund, A., and Svendgaard, N. A., 1976, Transplantation of central and peripheral monamine neurons to the adult rat brain: Techniques and conditions for survival, *Brain Res.* **114**:1.
23. Lund, R. D., and Hauschka, S. D., 1976, Transplanted neural tissue develops connections with host rat brain, *Science* **193**:582.
24. Perlow, M. J., Freed, W. J., Hoffer, B. J., Seiger, A. Olson, L., and Wyatt, R. J., 1979, Brain grafts reduce motor abnormalities produced by destruction of nigrostriatal dopamine system, *Science* **204**:643.
25. Gash, D., Sladek, J. R., Jr., and Sladek, C. D., 1980, Functional development of grafted vasopressin neurons, *Science* **210**:1367.
26. Gash, D., Sladek, C. D. and Sladek, J. R., Jr., 1980, A model system for analyzing functional development of transplanted peptidergic neurons, *Peptides* **1**(Suppl. 1):125.
27. Anonymous, 1981, A *Nature* survey of the neurosciences, *Nature (London)* **293**:515.
28. DeJong, R. N., 1982, *A History of American Neurology*, pp. 59–61, Raven Press, New York.

29. Ranson, S. W., 1914, Transplantation of the spinal ganglion with observations on the significance of the complex types of spinal ganglion cells, *J. Comp. Neurol.* **24**:547.

30. Spemann, H., 1921, Die erzeugung tierischer chimären durch heteroplastische embryonale transplantation zwischen, *Triton cristatus* und *taeniatus, Arch. Entwicklungsmech. Org.* **48**:533.

31. Spemann, H., and Mangold, H., 1924, Über induktion von embryonalanlagen durch implantation artfremder organisatoren, *Arch. Mikrosk. Anat. Entwicklungsmech.* **100**:599–638.

32. May, R. M., 1932, Répercussions de la transplantation nerveuse chez le porte-greffe, *Encephale* **27**:885.

33. Young, J. Z., 1942, The functional repair of nervous tissue, *Physiol. Rev.* **22**:318.

34. Carlsson, A., Falck, B., and Hillarp, N. Å., 1962, Cellular localization of brain monamines, *Acta Physiol. Scand.* **56**(Suppl. 196):1.

35. Falck, B., and Owman, C., 1965, A detailed methodological description of the fluorescence method for the cellular demonstration of biogenic monamines, *Acta Univ. Lund. Sect. 2* **1965**:1.

36. Yahr, M. D., 1981, Introduction, in: *Research Progress in Parkinson's Disease* (F. C. Rose and R. Capildeo, eds.), pp. 3–8, Pitman Medical, London.

2

Transplantation by Microinjection

Heterospecific Analysis of Avian Neural Crest
Migration and Differentiation

MARIANNE BRONNER-FRASER

1. Introduction

The vertebrate neural crest is an embryonic structure which gives rise to cells of the peripheral nervous system as well as many other cell types. Because of the diversity of derivatives and the extensive migratory ability of neural crest cells, this structure serves as an important model system for exploring questions of cell migration and differentiation.

Neural crest migration is a phenomenon ordered in both space and time. The neural crest arises on the dorsal side of the avian embryo during neurulation, the process that initiates formation of the nervous system. The neural folds first close to form the neural tube in the anterior (head) regions; the process of neural tube closure then progresses posteriorly. As the neural folds close, neural crest cells depart from the dorsal neural tube. Just as tube closure proceeds as an anterior to posterior wave, the onset of crest migration starts in the head and progresses caudally. Therefore, several stages of neural crest migration exist within a single embryo (Fig. 1). While crest migration in the anterior regions of the neural axis may be well underway, migration at the posterior level is just beginning.

Along the embryonic axis, several populations of crest cells exist that differ in their migratory pathways as well as in their range of derivatives. These populations are referred to as the cranial, vagal, trunk, and lumbosacral neural crest (Fig. 2). The axial level through which the cells migrate appears to determine, in large measure, the final

---

MARIANNE BRONNER-FRASER ● Department of Physiology and Biophysics, University of California, Irvine, California 92717.

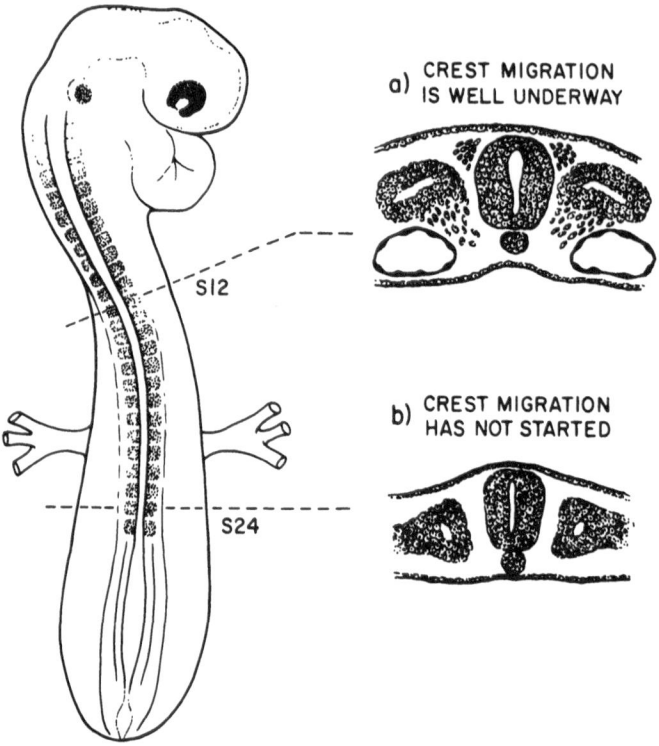

FIGURE 1. Diagram illustrating selected stages of migration of neural crest cells in a stage 14 to 15 embryo. At anterior levels (somite 12) of the neural axis, crest cell migration is well underway while at posterior levels (somite 24) crest migration has not yet started. From Ref. 24 with permission.

localization pattern of the crest cells.[1] The "cranial" neural crest arises in the head region. Initiation of cranial crest migration occurs prior to fusion of the neural folds.[2,3] The cells first enter a cell-free space, and then distribute according to the brain level of origin.[3] The cranial crest gives rise to skeletal and connective tissue of the head and face as well as neural and supportive elements of the cranial ganglia.[2,3] The cells from the "vagal" neural crest arise from axial levels corresponding to somites 1 to 7. The vagal crest cells undergo some of the most extensive migrations of any embryonic cell type. These cells enter the gut in the anterior regions and migrate progressively posteriorly as far as the umbilicus and rectum. Vagal crest cells form the parasympathetic ganglia in this region.[1,4] The "lumbosacral" neural crest, which arises caudal to the 28th somite, also contributes some supportive cell types to the parasympathetic ganglia of the gut.

The crest cells derived from the "trunk" levels (somites 8 to 28) give rise to melanocytes, the neurons and supportive cells of the autonomic and sensory ganglia, and chromaffin cells of the adrenal medulla (Table I). Neural crest cells from the trunk region follow migratory routes distinct from both the cranial and the vagal crest. These cells depart from the dorsal aspect of the neural tube following neural tube closure. Similar to the cranial crest, the trunk neural crest cells enter a cell-free space adjacent to the neural tube. The crest cells then migrate away from the neural tube along two routes (Fig. 3). Some cells migrate between the neural tube and the adjacent somites along the "ventral" pathway of migration. Cells following this ventral route settle in three main regions: (1) the sensory ganglion site adjacent to the neural tube; (2) the sympathetic ganglion site; and (3) around the dorsal aorta where the cells eventually coalesce to form the adrenal medulla. Other cells follow the "dorsal" pathway of migration. These cells migrate just under the ectoderm, settle in the dermatome, and there give rise to melanocytes.

The initial stages of crest migration take place through a small cell-free space that contains extracellular matrix molecules produced by neighboring tissues. Studies are

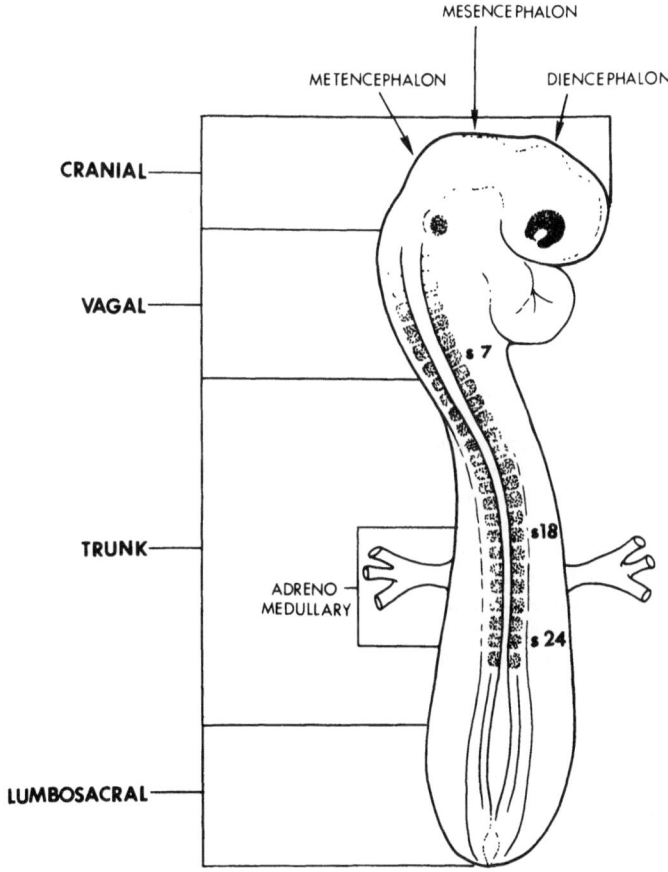

FIGURE 2. Diagram illustrating regions along the neural axis that differ in their range of neural crest derivatives and in crest migratory routes. The cranial crest emerges from the levels of the neural tube above the otic vesicles. The vagal crest arises from the neural tube between somitic levels 1 and 7. The trunk crest emerges from axial levels between somites 8 and 28, with those crest cells that contribute to the adrenal gland arise from somitic levels 18 to 24. The lumbosacral crest emerges from axial levels beyond somite 28. From Ref. 40 with permission.

currently under way in many laboratories to elucidate the nature of this matrix, and the role the extracellular matrix molecules may play in neural crest migration. The extracellular matrix through which crest cells migrate includes glycosaminoglycans, fibronectin, collagen, as well as other proteins and glycoproteins. The neural crest migratory routes are rich in fibronectin at the onset of crest migration.[5,6] Although crest cells do not synthesize fibronectin, they migrate readily on a fibronectin matrix.[7] Hyaluronic acid is also present in high concentrations along both the dorsal and the ventral route.[8,9] In contrast, chondroitin sulfate appears to be present predominantly on the dorsal pathway. Chondroitin sulfate may retard crest cell migration.[7] It has been observed that the time of onset of migration of neural crest cells on the dorsal and ventral pathways is somewhat different. In quail–chick chimeras, the timing of initial emigration along the dorsal pathway is delayed with respect to initiation of migration on the ventral route.[10] The presence of chondroitin sulfate on the dorsal pathway may account for this unequal distribution of migrating cells along the dorsal and ventral pathways.[7] In tissue culture experiments, crest cells themselves have been found to produce some glycosaminoglycans and glycoproteins.[11] Therefore, as crest cells enter the migratory routes, they may contribute to or modify the extracellular matrix through which they move.

In order to study neural crest cells in the absence of other cell types, the neural crest can be examined in tissue culture. After explantation *in vitro*, neural crest cells

TABLE I
List of Neural Crest Derivatives

Neuronal derivatives
  Sympathetic neurons
  Parasympathetic neurons
  Sensory neurons
    Spinal dorsal root ganglia
    Parts of facial, trigeminal, glossopharyngeal, and vagal ganglia
Neurosecretory cells
  Adrenomedullary cells
  Calcitonin-producing cells
  Carotid body type I and II cells
Supportive cells of the peripheral nervous system
  Glia
  Sheath cells of Schwann
  Satellite cells
  Pigment cells
  Skeletal and connective tissues (derived from the cranial crest)
Bones and cartilage of visceral and facial skeleton
Muscle
  Ciliary muscle
  Some muscles of the facial and visceral regions
  Smooth muscle of cranial vasculature
Mesenchyme
  Dermis of the face and ventral neck
  Corneal fibroblasts and endothelium
  Connective tissues

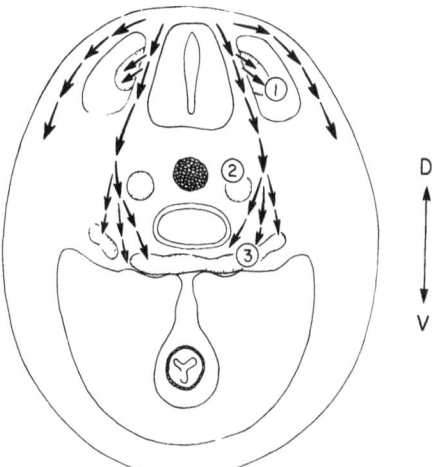

FIGURE 3. Diagram illustrating the normal routes of neural crest migration in the trunk region of the avian embryo. Those cells choosing the ventral (V) pathway localize in three main areas: (1) the sensory ganglia; (2) the primary sympathetic chain; and (3) the adrenal gland, aortic plexuses, and some cells of the metanephrogenic mesenchyme. Crest cells following the dorsolateral (D) pathway migrate under the ectoderm and become skin melanocytes. From Ref. 23 with permission.

will continue to migrate out of the dorsal aspect of the neural tube. A short time after explanting neural tubes onto petri plates, the first crest cells can already be observed leaving the tube.[12] By 1 day after plating, a monolayer of cells is present. At this time, the neural tube is removed, leaving a relatively pure population of crest cells. When these cells are grown in a nutrient medium containing horse serum and embryo extract, several phenotypes differentiate *in vitro*. These include melanocytes, adrenergic neurons, cholinergic neurons, and other as yet unidentified derivatives.[12,13] A major advantage of the culture method is that single neural crest cells can be isolated and cloned in order to study questions of neural crest cell fate.[14-16] Thus, tissue culture can be used as a tool to produce neural crest cells of homogeneous origin.

Elaborate grafting experiments have been used to map the routes of neural crest migration and to categorize the derivatives arising from the neural crest (see Table I).[2,3,17,18] Recently, information about the macromolecular composition of the regions through which crest cells migrate has been accumulating. However, surprisingly little is known about the *mechanisms* that direct neural crest migration or govern cell lineage decisions in this population of cells.

2. Neural Crest Cell Marking Techniques

Because neural crest cells first migrate into a cell-free space, it is possible to distinguish neural crest cells in the initial stages of migration. As development proceeds, however, it becomes increasingly difficult to separate crest cells from other cell types. In order to map the crest migratory pathways and to determine the number of crest-derived cell types, it is necessary to label crest cells in order to distinguish them from the tissues through which they migrate.

Cell marking techniques were first applied to avian embryos by Weston[17] who labeled the neural tube containing the crest cells with [³H]thymidine. The labeled neural tubes were then exchanged for a similar length of neural tube in an unlabeled host chicken embryo. The grafts healed and neural crest cells appeared to emigrate from the donor neural tubes in a normal manner. The host embryos were fixed at various times after the transplantation procedure; slides of the serially sectioned embryo were then dipped in photographic emulsion. Using this approach, it was possible to autora-diographically follow crest movement. This technique was, however, only useful for studying the relatively early stages of migration because crest cells are a rapidly dividing population. With progressive cell division, the [^{3}H]thymidine marker was diluted out.

The problem of dilution of marker was overcome by the advent of the heterospecific chimera, which combines tissues from quail and chicken embryos.[18] Quail neural tubes can be isolated and transplanted into host chicken embryos, thereby producing a chimeric embryo. Though the two species appear to develop similarly, quail cells can be distinguished from chicken cells because of the nucleolus-associated heterochromatin unique to the quail. By staining for DNA, it is possible to distinguish donor (i.e., quail neural crest and neural tube cells) from host chicken cells. This marker is stable through cell division and therefore enables long-term identification of the implanted cells.

The transplantation techniques have been extremely useful for determining the pathways of crest migration and for establishing the list of crest-derived cell types. These methods do, however, have some inherent shortcomings. Such grafts tend to be traumatic and cause a delay in the onset of crest migration.[17] The transplantations can only be performed successfully at early developmental stages prior to extensive vascularization; at later stages, hemorrhaging following these operations can lead to death or abnormal development.[19] The grafts entail extirpating and replacing the host chicken neural tube with a donor quail tube. This surgery disrupts the extracellular matrix around the neural tube. In recent years, it has become obvious that these matrices are important substrates for crest cell migration and localization. Although the neural tube grafts apparently heal and the grafted embryos look grossly normal, it is impossible to assess the effects of the disrupted extracellular matrix on the migration of the grafted cells. Some additional complications of the quail–chick chimera arise from the fact that although the quail and chick are indeed similar species, some species differences do exist that may alter the response of the grafted crest cells. In tissue culture, for example, quail and chick cells from gastrulating tissues exhibit marked differences in adhesive and inva-sive properties.[20]

3. The Injection Technique

Alternative methods for introducing labeled cells into embryos at the time of neural crest migration have recently been devised. These include implanting pellets of cells next to the neural tube[21,22] or microinjecting cells into the desired regions of the embryo.[15,23,24] In the trunk region, neural crest cells appear to migrate as individual cells rather than as tight aggregates.[25] Therefore, it seems advantageous to implant cells in suspension to minimize experimentally produced cell contacts and to maximize interaction of cells with neural crest pathways. The injection technique, which is described in detail below, uses cells or particles in suspension, thus avoiding the potential confounding effects incurred by using pelleted cells with abundant intercellular adhesions.

The injection technique for implanting cells into embryos is adapted from electro-physiological methods for injecting substances intracellularly. Briefly, the technique involves placing cells or inert particles into a micropipette with an opening on the order of 50 μm. The micropipette is mounted onto a micromanipulator and connected to a pressure injection apparatus. The micromanipulator is used to accurately maneuver the injection needle to the surface of the embryo above the chosen implantation site. By slowly lowering the pipette using the micromanipulator, the micropipette punctures the ectoderm. The needle is carefully lowered to the desired region. The contents in the tip of the pipette are then expelled into the embryo with a pulse of pressure.

In order to study questions of cell migration, it is necessary to place an implant in a site that is identifiable, localized, and reproducible. Because the injection technique uses a cell suspension, the force of the pressure injection tends to disperse the cells. It is, therefore, necessary to introduce the suspension into a natural cavity in the embryo. For this purpose, the embryonic somites in the posterior (trunk) region were chosen as the injection site. At the time of injection, the somites are epithelial sacs with a cavity suitable for initially retaining the injected cell suspension (Fig. 4A). In the course of normal somitic development, the cells of the ventromedial wall of the somite (adjacent to the neural tube) disperse. The cells contained within the cavity of the somite are, therefore, released along one of the normal neural crest migratory pathways in the trunk—the ventral pathway (see Fig. 3). Because the implant is placed directly into the somite with no incision adjacent to the neural tube, the extracellular matrix around the neural tube is left intact. The dorsal aorta lies immediately below the somite at the time of injection; therefore, embryonic bleeding serves as a useful assay for an improperly placed implant. The somitic injection site fulfills all the requirements for a proper implantation site since the cells are initially contained in a reproducible location and are subsequently exposed to a normal neural crest migratory route.

The injection technique makes it possible to implant many different cell types onto a neural crest pathway. Using this method, it has been possible: (1) to return crest cells that have already differentiated back to a younger embryonic environment; (2) to place neural crest and other cell types onto a pathway that they would not normally encounter; and (3) to implant inert substances with specific surface components, to serve as inert probes of the crest migratory routes. The utility of the technique lies largely in its flexibility.

In addition, the injection technique has several important advantages over previous methodology. In contrast to transplantation techniques, the injection technique is far less deleterious to the host. After withdrawal of the micropipette, only a small puncture wound remains in the roof of the somite and the overlying ectoderm (Fig. 4A). Using the injection technique, it is possible to implant neural crest cells grown *in vitro*, includ-ing cells of clonal origin. The latter is particularly significant in examining questions of neural crest cell potentiality, where it is important to study cells of homogeneous origin (see Section 5). An additional advantage of the injection technique is that it allows the investigator to work with small numbers of cells. The results obtained with this tech-nique are described in the following sections.

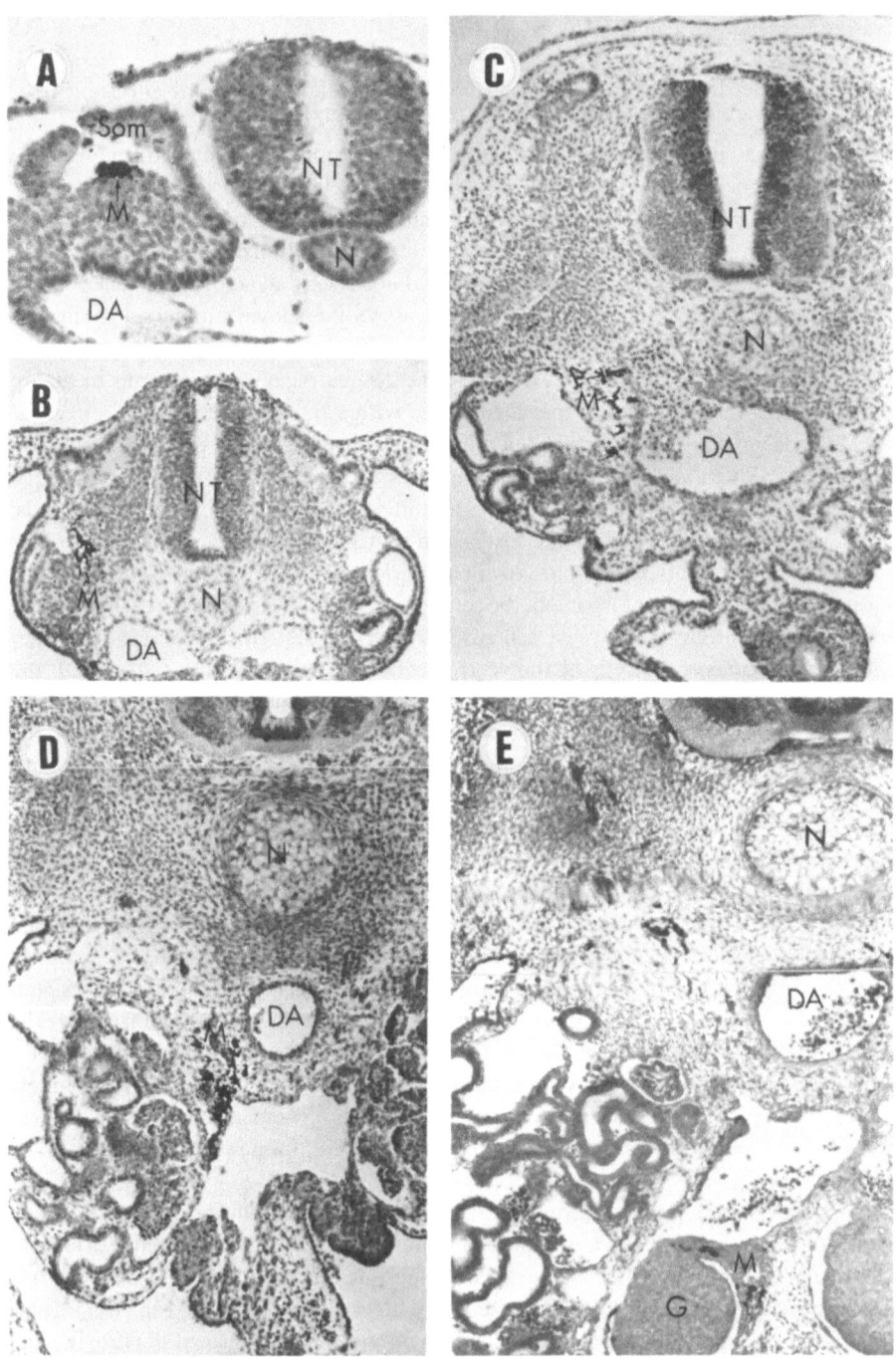

FIGURE 4. Light photomicrographs of transverse sections demonstrating the ventral migration of cloned quail melanocytes (M) injected into the newly formed somites (Som) of 2½-day chick embryos and fixed at progressively older stages of development. (A) Immediately after injection, the dark black melanocytes are restricted to the somitic cavity. (B) One day after injection, melanocytes have migrated to the region of the sensory ganglion. Note the single melanocyte adjacent to the dorsal aorta. (C) Two days after injection,

4. Migration of Embryonic Cells along the Ventral Crest Pathway

All of the pigmented cells of the body, with the exception of retinal pigment epithelial cells, are derived from the neural crest. In their undifferentiated form, avian neural crest cells are highly migratory; they are, however, no longer motile upon differentiation into the pigment phenotype in both the intact embryo[26] as well as in tissue culture.[23]

In the tissue culture dish, neural crest cells readily differentiate into melanocytes after a few days. The initial experiments using the injection technique employed clonally derived quail melanocytes. In this way, it was possible to examine the effects of the embryonic environment on already differentiated crest cells. The cloned melanocytes were injected into the trunk somites of 2½-day-old chicken embryos at a region of the neural axis where crest migration was beginning. The host embryos were fixed at various times after injection, serially sectioned, and stained. The implanted cells were originally contained in the lumen of the somite (Fig. 4A). Following release from the somitic lumen, the melanocytes distributed along the ventral neural crest pathway normally followed by endogenous precursors to neuronal and supportive cells (see Fig. 3). The injected melanocytes were found in progressively ventral sites along the pathway with time (Figs. 4B–E). One day after injection, some cells were in areas corresponding to the future site of the sensory ganglia; the majority had moved to the region where the sympathetic ganglia and adrenal gland form. Two and three days after injection, the pigment cells were found still further along the ventral pathway. The cells generally settled in regions where endogenous trunk neural crest cells localize—in the vicinity of the dorsal aorta where the adrenal chromaffin cells and adrenergic cells of the aortic plexuses differentiate. In some cases, the pigment cells invaded the gut and were found in the region where parasympathetic ganglia (derived from vagal neural crest) normally differentiate.[23] The duration of time spent in clonal culture did not affect the route or time course of migration, with melanocytes grown from 8 to 24 days all exhibiting the same behavior. These results demonstrated that the neural crest-derived melanocytes (which were nonmotile prior to injection) were capable of ventral translocation when introduced into the proper embryonic environment.

In analyzing the factors that may influence the previously nonmotile pigment cells to resume migration along the ventral pathway, it is essential to eliminate potential artifacts inherent in the experimental paradigm. The possible complicating factors include (1) the potential effects of long-term culture on the melanocytes and (2) potential species incompatibilities between the quail and the chick. The melanocytes used in these experiments were derived from the quail and were cultured for many days prior to injection. In order to rule out any confounding effects of long-term culture on the migration of

---

melanocytes are found further ventrally in the region of the sympathetic ganglia and lateral to the dorsal aorta.(D)Three days after injection, melanocytes have migrated beyond the dorsal aorta and can be identified by their intense black cell bodies and processes next to the developing adrenal gland, metanephrogenic mesenchyme, and aortic plexuses.(E) Four days after injection, melanocytes are present further ventrally in the vicinity of the gonads (G) and mesentery of the gut. NT, neural tube; N, notochord; DA, dorsal aorta. From Ref. 40 with permission.

pigment cells, melanocytes were freshly isolated from the skin of 9- to 11-day-old quail embryos; these cells were injected in the same manner as were the cultured melanocytes. The movement of these skin-derived melanocytes, which were not exposed to the culture environment, was indistinguishable from that of the cloned melanocytes described previously.[24] Therefore, prolonged time *in vitro* does not account for the migration of the pigment cells. Another factor that may contribute to the ventral distribution of the quail melanocytes might be species incompatibility between the quail and the chick. A tacit assumption in the interpretation of the results from experiments involving quail–chick chimeras is that the two species are developmentally homologous. The quail and chick do differ in gestation time, as well as in the size of the embryo. In addition, under some culture conditions, quail cells do appear to be more adhesive and invasive than do chick cells.[20] To examine the possibility that chick and quail cells may differ, crest cells were cloned from a pigmented species of chicken and were cultured for 8 or more days by which time they contained high levels of melanin. The melanin marker permits distinction of the injected cells from host cells since the host embryos were devoid of pigment cells at the times considered. The chick melanocytes distributed analogously to the quail pigment cells.[24] Thus, the resumption of migratory behavior does not reflect species differences based on the quail–chick chimera.

Why did the injected cells follow the ventral pathway normally followed by precursors to neuronal and supportive cells as opposed to the dorsal route normally taken by presumptive melanocytes? The melanocytes were released onto the ventral pathway by the dispersion of the sclerotomal cells of the somite. The opening of the somite, therefore, biased the injected cells onto the ventral route. In addition, the dermomyotomal remnant of the somite (which matures from the dorsal aspect of the somitic tissue) is epithelial and probably excluded the melanocytes from the dorsal pathway. As discussed in the previous section, the dorsal route is invaded by crest cells at a later time than is the ventral pathway.[10] This delay in time of initial migration may also be a contributing factor.

These studies showed that melanocytes injected into the posterior trunk somites at the onset of crest migration, moved progressively ventral along a normal neural crest pathway. The cells settled in ventral sites in the vicinity of the adrenal gland, sympathetic ganglia, and aortic plexuses. How, then, do crest cells become localized in the more dorsal sites such as the dorsal root ganglion site in the trunk region? Previous investigators have proposed that environmental alterations occurring with progressive embryonic age ultimately limit crest dispersion along the ventral route.[19] Using the [3H]thymidine marker, Weston and Butler[19] transplanted labeled neural tubes from "young" donor embryos into progressively "older" unlabeled hosts. They found a decreasing number of labeled cells in the more ventral regions with increasing age of the host axial level. These results indicate that the distal migration of crest cells is limited with increasing developmental age of the host. Their conclusions were, however, weakened by the limited success of the grafts especially with the oldest animals; damage inflicted during grafting caused death or abnormal development in many cases. Thus, the restricted migration observed could well be a consequence of the greater trauma caused to the older host by the neural tube transplantation.

Using the injection technique, it was possible to reexamine this problem in a far less traumatic way. Melanocytes can be injected into embryos at later stages of neural crest migration. The cells were injected into the somitic cavity of anterior somites in regions of the embryo that were typically 10 to 12 hr more developed (with respect to neural crest migration) than the posterior somites used in the experiments described previously. Even after the cells of the ventromedial wall of the somite began to disperse, a somitic cavity remained that was capable of containing the injected cells (Fig. 5A). When melanocytes were implanted into the anterior somites, their ventral migration was considerably less than that of melanocytes introduced into the posterior somites (Fig. 5B). In a single embryo, melanocytes derived from the same clone can be injected into both anterior and posterior somites, since several stages of neural crest migration exist simultaneously in a single embryo (Fig. 1). Because the cells were of homogeneous origin, any differences in migratory ability between melanocytes placed in anterior versus posterior sites must be attributed to the environment through which they moved, and not to the cells themselves. In all the embryos examined, the melanocytes introduced into the posterior somites migrated further ventrally than those injected into the anterior somites. Thus, it appears that axial level and not differences between injected clones was responsible for restriction of migration. It was important to ensure that the restricted migration was due to the degree of maturation at the level of the neural axis and not to inherent regional differences between anterior and posterior somites. Therefore, embryos were allowed to develop for 10 to 12 hr past the times used previously for injections. The melanocytes were then placed into the "older" posterior somites; the maturation level of these "older" posterior somites was comparable to that of the anterior somites described above. Under these conditions, the pigment cells were again restricted in their degree of migration (Fig. 5C). These results confirm the findings of Weston and Butler and demonstrate that changes in the environment that occur as a function of developmental age limit the extent of ventral translocation.

After injection into the somites, neural crest-derived cells distributed along the ventral neural crest pathway similarly to endogenous crest cells. Is this phenomenon unique to cells of neural crest origin, or will any cell type placed onto this pathway distribute similarly? To determine whether any cell type exposed to the somitic milieu would migrate ventrally, non-neural crest-derived cells were injected into the posterior somites at the onset of host crest migration. Two cell types, somite cells and fibroblasts, were chosen for these experiments. Fibroblast cells have been the object of much study because of their migratory behavior *in vitro;* this migratory behavior probably reflects their *in situ* involvement in wound healing. Somite cells on the ventromedial wall of the somite disperse during the course of normal somitic development; some somite cells are, therefore, motile during a portion of their development. Cell suspensions of both the somite and the fibroblast cells were prepared from quail embryonic tissue. The somite or fibroblast cells were injected into the trunk somites of chicken embryos at a time when host crest migration was beginning. The donor quail cells were distinguishable from host cells because of the heterochromatin marker unique to the quail. Unlike the neural crest-derived melanocytes, neither the fibroblast nor the somite cells moved ventrally along the neural crest pathway. The quail somite cells were found associated with the somitic mesenchyme of the host (Figs. 6A, B). The quail fibroblasts were usually found in the dermomyotomal region of the host tissue (Figs. 6C, D). Thus, neither of these cell types

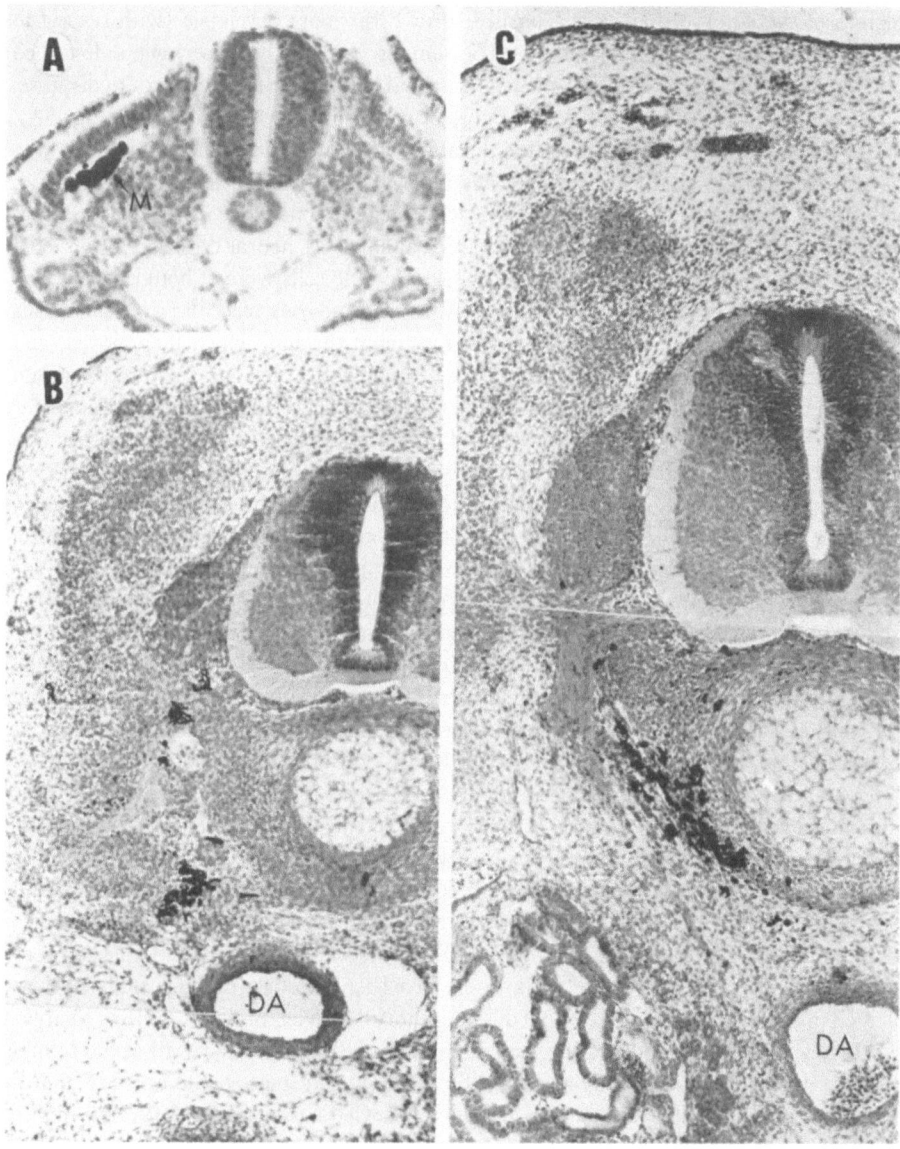

FIGURE 5. Transverse sections illustrating restricted migration of melanocytes injected into somites at developmentally older axial levels. (A) Injected melanocytes (M) are contained within the older somite. (B) Four days after quail melanocytes were injected into an anterior level (somite 15) of a 25-somite embryo, the jet-black melanocytes migrate no further than the dorsal aorta (DA). (C) Four days after quail melanocytes were injected into posterior levels (somite 21) of a 32-somite embryo, the melanocytes migrate no further than the dorsal aorta. From Ref. 24 with permission.

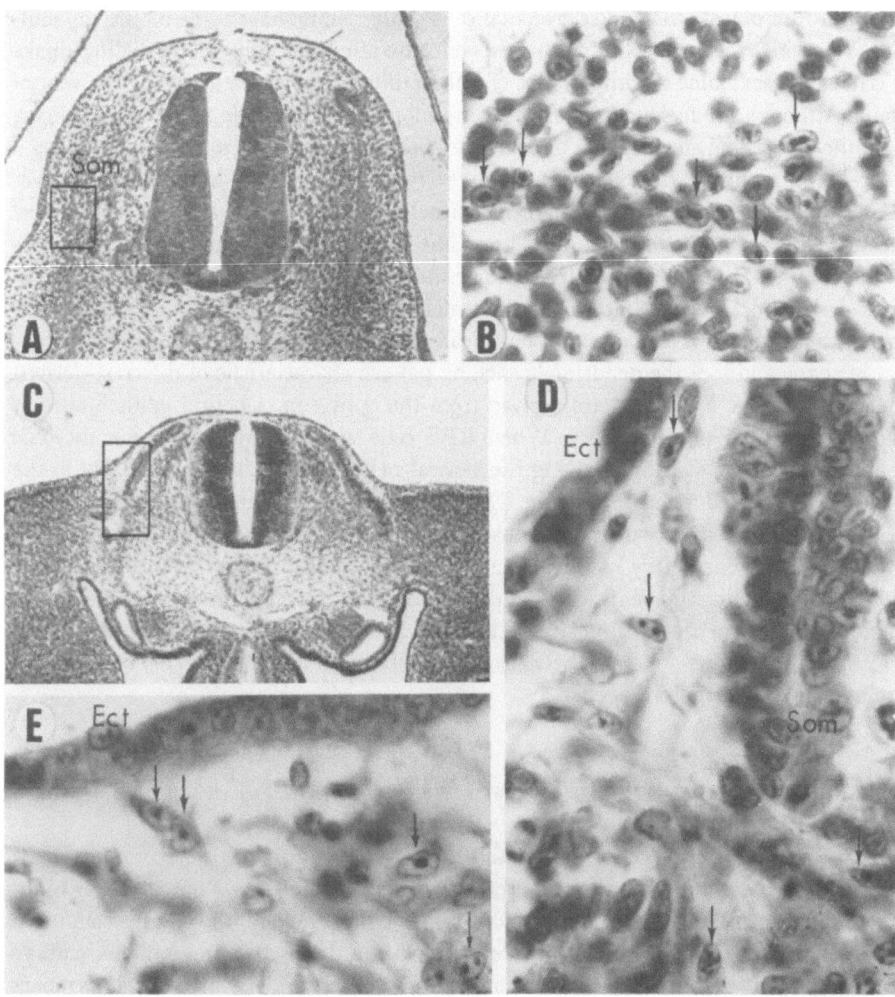

FIGURE 6. Light photomicrographs of transverse sections through a chick embryo fixed after injection of quail somite or fibroblast cells into the posterior somites. Quail cells are identified by a Feulgen-positive nucleolar mass. Arrows indicate cells in which the quail marker is most visible. (A) Chicken host 2 days after quail somite cells were injected into the somite (Som). (B) At higher magnification of boxed area in (A), it is seen that the quail somite cells remain associated with the host somitic mesenchyme. (C) Embryo fixed 1 day after injection of quail fibroblast cells into the posterior somites. (D) At higher magnification of boxed area in (C), the quail fibroblasts are found at the base of the somite and between the somite and the ectoderm (Ect). (E) Another embryo 1 day after injection with quail fibroblasts located under the ectoderm and closer to the neural tube. From Ref. 24 with permission.

migrated along the ventral neural crest pathway after injection into the somites. These results demonstrate that the embryonic environment along the ventral migratory route is selective. Certain cell types, such as somite and fibroblast cells, were apparently excluded from the pathway whereas neural crest-derived melanocytes distributed along the ventral route similarly to endogenous crest cells.

Both fibroblast cells and sclerotomal cells of the somite have some motile capabilities. Yet neither of these cell types moved ventrally after injection. These findings make it critical to reexamine the importance of cell motility on translocation along the ventral route. Although the melanocytes used for injection were nonmotile at the time of injection, the embryonic environment appeared to "induce" these cells to translocate along a neural crest pathway. In order to examine the role of cell motility in this ventral movement, another cell type, retinal pigment epithelial (RPE) cells, was injected into host chicken embryos at the onset of migration. The RPE cells are the only pigmented cells of the body that are not derived from the neural crest. These cells are similar to neural crest-derived melanocytes in terms of their melanin metabolism. However, neural crest melanocytes and RPE cells are morphologically quite different: the RPE cells are cuboidal in shape and lack the dendritic branching pattern characteristic of the crest-derived melanocytes. The RPE cells are derived from the optic cup, a neural epithelium that invaginates but is not migratory. When RPE cells were injected into the embryonic somites, these cells distributed along the ventral neural crest pathway similarly to the neural crest-derived melanocytes (Fig. 7). Like the crest-derived pigment cells, the RPE cells settled in the vicinity of endogenous trunk crest sites. The RPE cells did not, however, invade the mesentery of the gut, as had some quail melanocytes in the previous study. These findings make it clear that the neural crest ventral pathway supports movement of not only neural crest-derived cells but other cell types as well. Thus, the pathway appears to show some selectivity (excluding fibroblast and somite cells) but does not exclude all non-neural crest cells.

5. Clonal Analysis of Neural Crest Cell Differentiation

Avian neural crest cells can be studied in tissue culture. *In vitro* studies of the neural crest have the advantage of probing crest cell differentiation under controlled conditions and in the absence of other cell types. Unfortunately, the culture conditions used for avian cells are typically ill-defined—the nutrient medium contains embryo extract and horse serum. Cells depend on medium-derived growth factors and hormones (many of which remain unidentified) for their survival and differentiation into many phenotypes. In culture, a subset of neural crest derivatives will differentiate. The conditions are not, however, permissive for all the known derivatives of the neural crest to be expressed. By using the injection technique, neural crest cells that were isolated in tissue culture can subsequently be returned to the embryonic milieu. The embryo can thus be used as an *in situ* culture system that provides the nutritive environment necessary for a more complete range of differentiation.

In tissue culture, clones of neural crest cells can be isolated. The cloning procedure utilizes cells from primary crest cultures (described in Section 1). These are removed from the culture dish, replated at sparse densities, and screened for single cells.[14] The colonies derived from the continued proliferation of these single cells are referred to as clones. The cloned cells were returned to the embryonic milieu by injecting the quail cells into host chicken embryos. In the host, the quail cells were allowed to differentiate in a natural and nutritive environment. By studying differentiation of cloned cells, it is

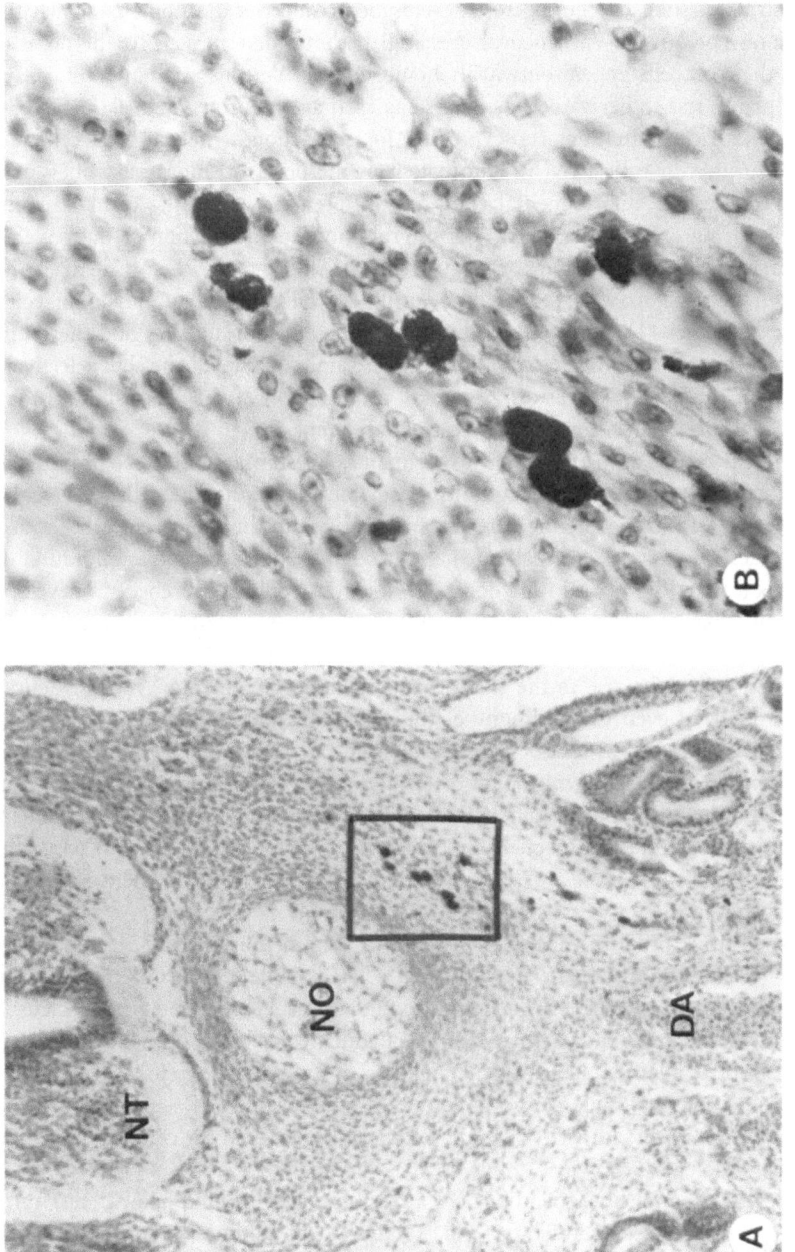

FIGURE 7. Light photomicrographs of a transverse section through a chick embryo 3 days after retinal pigment epithelial (RPE) cells were injected into the posterior somites. (A) At lower magnification, the RPE cells are found on the ventral pathway below the notochord and extending adjacent to the dorsal aorta. RPE cells in adjacent sections are predominantly adjacent to the aorta. (B) High magnification of boxed area in (A) demonstrates that the RPE cells maintain their cuboidal morphology after injection. NT, neural tube; NO, notochord; DA, dorsal aorta. From Ref. 31 with permission.

possible to ask specific questions about the fate of the descendants of a single neural crest cell and thereby to probe the cell lineage decisions of individual neural crest cells.

A central question of neural crest development concerns the point at which the developmental potential of each neural crest cell becomes determined. At the onset of migration, the crest cells appear outwardly homogeneous. Yet this population of cells is capable of giving rise to derivatives as diverse as cartilage, neurons, and pigment cells. Several alternative explanations could account for restriction of neural crest cell fate. First, the neural crest cells may be multipotent at the onset of migration. They may migrate in response to environmental cues, and differentiate according to interactions experienced during or at the conclusion of migration. Thus, restriction of the possible fates might result from environmental interactions. Alternatively, before leaving the neural tube, the crest may consist of predetermined cells or subpopulations of cells. These cells can either (1) migrate preferentially to the appropriate sites and there differentiate according to their fate or (2) migrate indiscriminately to all sites but selectively survive only at the appropriate sites. One or a combination of these possibilities may explain the cell lineage decisions in the crest population.

Using neural tube transplantation experiments, Weston and Butler[19] probed the question of neural crest cell pluripotentiality. They found that the first cells to leave the neural tube do, in fact, migrate the furthest and give rise to the more ventrally located neural crest derivatives (e.g., sympathetic ganglia). This extensive ventral migration of the first cells leaving the neural tube does not, however, reflect differences in developmental potential between early and late migrating cells. When neural tubes from "older" neural tube axial levels were transplanted into "young" regions of the embryo, the emerging crest cells gave rise to the same range of derivatives as did "young" neural tubes. This occurred even though many of the crest cells had already migrated away from the "old" donor neural tubes. These results indicate that either (1) the crest cells are pluripotent at the onset of migration or (2) the crest cells are predetermined prior to migration but leave the neural tube at random times (such that the full complement of cell types is represented at any one time). Because the population of cells leaving the transplanted neural tubes is *heterogeneous,* Weston and Butler's experiment could not distinguish between these two possibilities.

Clonal cultures produce cells of *homogeneous* origin, thus making it possible to distinguish between the above alternatives. The isolated single cells proliferate to form clones consisting of sister cells derived from a common precursor. Three types of neural crest clones arise after 4 to 5 days in culture: all pigmented cells, all unpigmented cells, and a mixture of pigmented and unpigmented cells.[14] By employing clonal techniques to isolate the progeny of single cells and combining this technology with the injection technique for introducing cultured cells to the embryo, clonal analysis can be adapted to the *in vivo* situation.

In order to examine the potential derivatives arising from one neural crest cell, clones were grown in tissue culture for 7 to 10 days. At these times, pigment cells had differentiated but no neuronal or other cell types were identified. Like the cloned melanocytes described previously, unpigmented clones or clones of mixed character (containing both pigmented and unpigmented cells) were injected into the posterior somites of 2½-day-old host chicken embryos. The cultured cells were always of quail origin and could, therefore, be distinguished from host cells. The unpigmented cells, like those from the pure melanocyte colonies, moved along the ventral neural crest pathway after injection. The distribution of quail cells was similar for cells cultured from 7 to 10 days. The unpigmented cells localized within the sympathetic ganglion, along the ventral nerve cord, as well as around the dorsal aorta (Fig. 8). These are areas where endogenous trunk neural crest cells settle and give rise to neurons and supportive cells of the sympathetic ganglia, Schwann cells around the nerve cord, and chromaffin cells of the adrenal medulla, respectively. Unlike the pigmented sister cells, the unpigmented cells did not invade the mesentery of the gut, but were specifically located in sites normally occupied by endogenous trunk crest cells. Even when the pigmented and unpigmented cells were derived from the same clone, this difference in distribution was observed. These results demonstrate that crest cells derived from a common precursor can express differences in migratory behavior. It should be noted that the injected quail cells were not found in the sensory ganglion site (adjacent to the neural tube) when either pigmented or unpigmented cells were injected into the posterior somites. This may stem from the fact that the cells were injected at times when endogenous crest migration was just beginning; perhaps, only the more ventral sites are filled by cells departing at this time whereas more dorsal sites (i.e., the sensory ganglion sites) may be filled by the later migrating cells.[19,24] Alternatively, the fact that no cells settled in the sensory ganglion site might reflect some *restriction* in possible fates of the cloned cells.

Because many of the unpigmented quail cells settled in areas of potential adrenergic differentiation (such as the sympathetic ganglion and adrenal medullary region), we examined whether the unpigmented donor cells in fact became catecholamine-producing neuroblasts. Adrenergic neurons can be histochemically identified by the techniques of formaldehyde-induced fluorescence,[27] by which catecholamine-containing cells become fluorescent. In the host chicken embryo, the adrenergic cells are normally observed in the sympathetic and adrenal sites after 3 to 5 days of incubation.[28] Host embryos were examined for catecholamine histofluorescence several days after injection of mixed quail clones into the host somites. At the times of observation, the sympathetic and adrenergic sites in the host already contained catecholamine-producing cells. The embryos were prepared for formaldehyde-induced fluorescence and subsequently serially sectioned. After photographically recording those areas in the embryo that contained fluorescent cells, the same sections were stained for DNA using the Feulgen–Rossenbeck method[29]; this stain permits the distinction of quail and chick cells. The same sections were again photographed, and the fluorescence and light photomicrographs were matched. By double staining for catecholamine fluorescence and for the quail marker, it was possible to determine if any of the quail neural crest cells had differentiated into adrenergic neurons. A large percentage of the injected cells had become adrenergic after several days in the host environment (Fig. 9); the quail cells contributed to sympathetic ganglia, adrenomedullary cells, and neurons of the aortic plexus. Thus, some of the cells from mixed clones (containing both unpigmented and pigmented cells) differentiated into neurons. In the same animal, cells derived from the same clone expressed the pigment phenotype as well. Therefore, cells from the same clone can differentiate into both melanin-containing and catecholaminergic cell types. Other clonally derived cells either remained undifferentiated or gave rise to some as yet unidentified derivative.

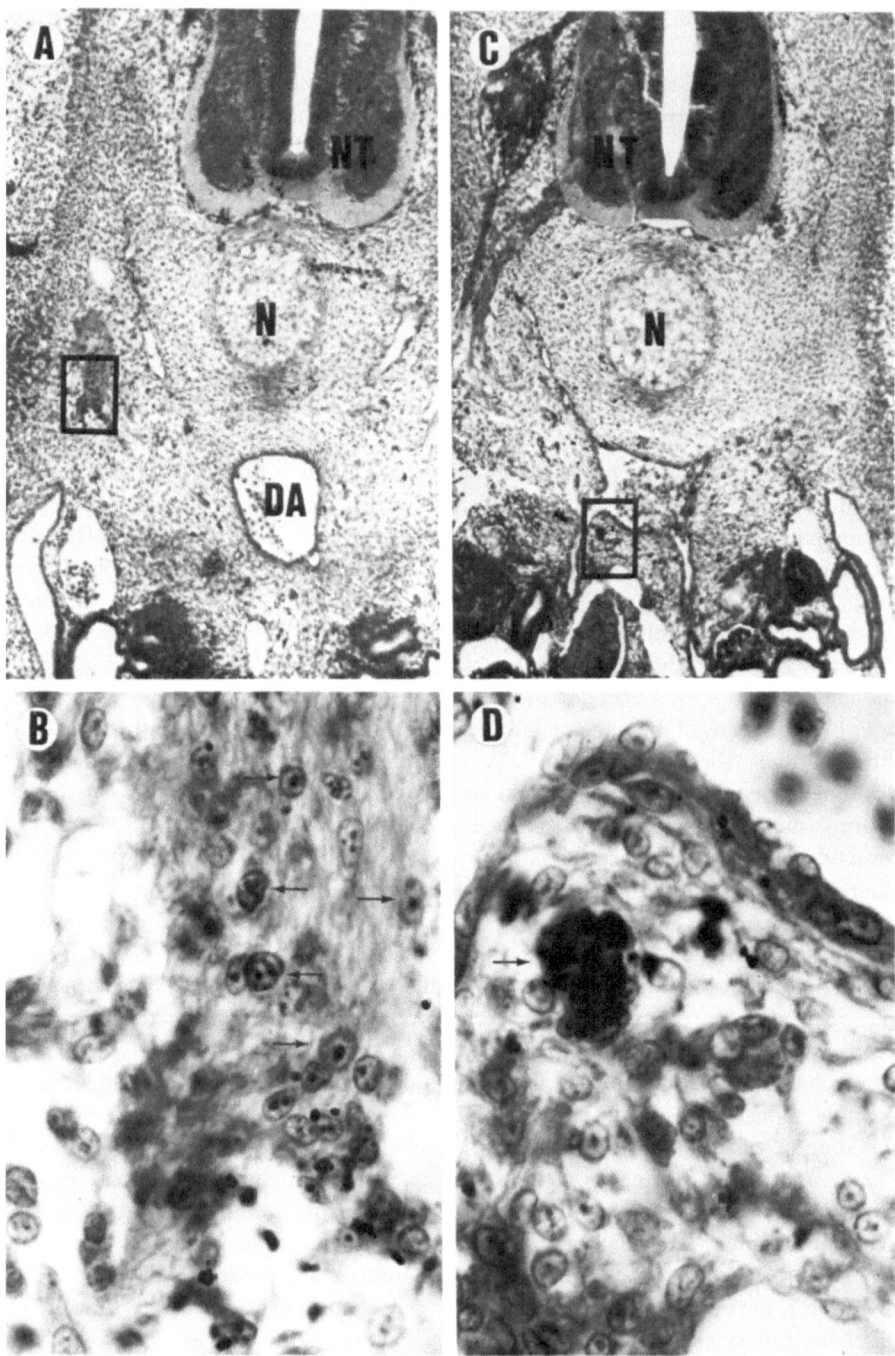

FIGURE 8. Light photomicrograph of transverse sections demonstrating the ventral migration and localization of unpigmented quail neural crest cells derived from mixed clones and injected into the newly formed somites of 2½-day chick hosts. (A) Three days after injection, some quail cells are present in the sympathetic ganglion. (C) Higher magnification of boxed area in (A) shows many quail cells within the ganglion. (B) Another embryo 3 days after injection in which quail neural crest cells have migrated to the level of the

By cloning cells and injecting them into the host, it was found that the progeny of a single cell can give rise to derivatives as diverse as melanoctyes and adrenergic neurons. These experiments, together with the *in vitro* results of Sieber-Blum and Cohen,[16] are the first demonstration that a single neural crest cell can differentiate into multiple phenotypes. Thus, at least a subpopulation of the premigratory neural crest in the trunk region is *bipotent* and probably pluripotent. The crest cells probably become determined in accordance with interactions with cells of the environment and with other crest cells. Clonal analysis adapted *in vivo* provides a methodology for looking at cell lineages that is not possible using heterogeneous populations of crest cells. Future experiments will extend this technology to probe the more complete repertoire of single crest cells and to elucidate the role of the embryonic environment in the induction of certain phenotypes.

6. Latex Beads as Inert Probes of Neural Crest Migration

The injection technique together with cell implantation methods have demonstrated that the embryonic environment can promote movement of already differentiated, post-migratory neural crest cells.[21,23,24,30] The neural crest ventral pathway may be partially selective for crest-derived cells since some cell types (somitic and fibroblast cells) did not migrate after implantation. However, other non-neural crest cell types did move along the ventral pathway after implantation. RPE cells, which are neither crest-derived nor migratory, moved along the ventral neural crest pathway after injection onto this route.[31] In addition, sarcoma 180 cells were capable of migrating after implantation.[22] Those cells capable of crestlike movement along the ventral route tended to localize near endogenous neural crest sites. These experiments suggest that the embryonic environment (1) plays some role in determining the pattern of crest cell distribution and (2) influences the selection of cell types that translocate along the crest pathways.

Interactions between cell surface components and the extracellular matrix along the migratory routes may contribute to crest cell movement and localization. Several laboratories are currently studying the macromolecular composition of the pathways along which neural crest cells migrate in order to gain understanding of such interactions. During the initial phase of migration, the ventral neural crest pathway is largely cell-free and contains many extracellular matrix components including hyaluronate, fibronectin, and collagen. Crest cells themselves synthesize glycoproteins and glycosaminoglycans.[11] They do not, however, synthesize fibronectin,[5] though they avidly migrate on a fibronectin substrate.[7] Fibronectin has recently received much attention with respect to cell motility. Cell migration *in vitro* is stimulated by addition of fibronectin,[32] which seems to alter both cell-to-cell and cell-to-substrate adhesivity. The motility of neural crest cells may be enhanced by the presence of exogenous fibronectin precisely because these cells lack fibronectin on their cell surface.

---

adrenal gland. (D) Higher magnification of boxed area in (B) shows the quail nucleolar marker within adrenal structures. NT, neural tube; N, notochord; DA, dorsal aorta. Arrows indicate cells in which the quail marker is most visible. From Ref. 15 with permission.

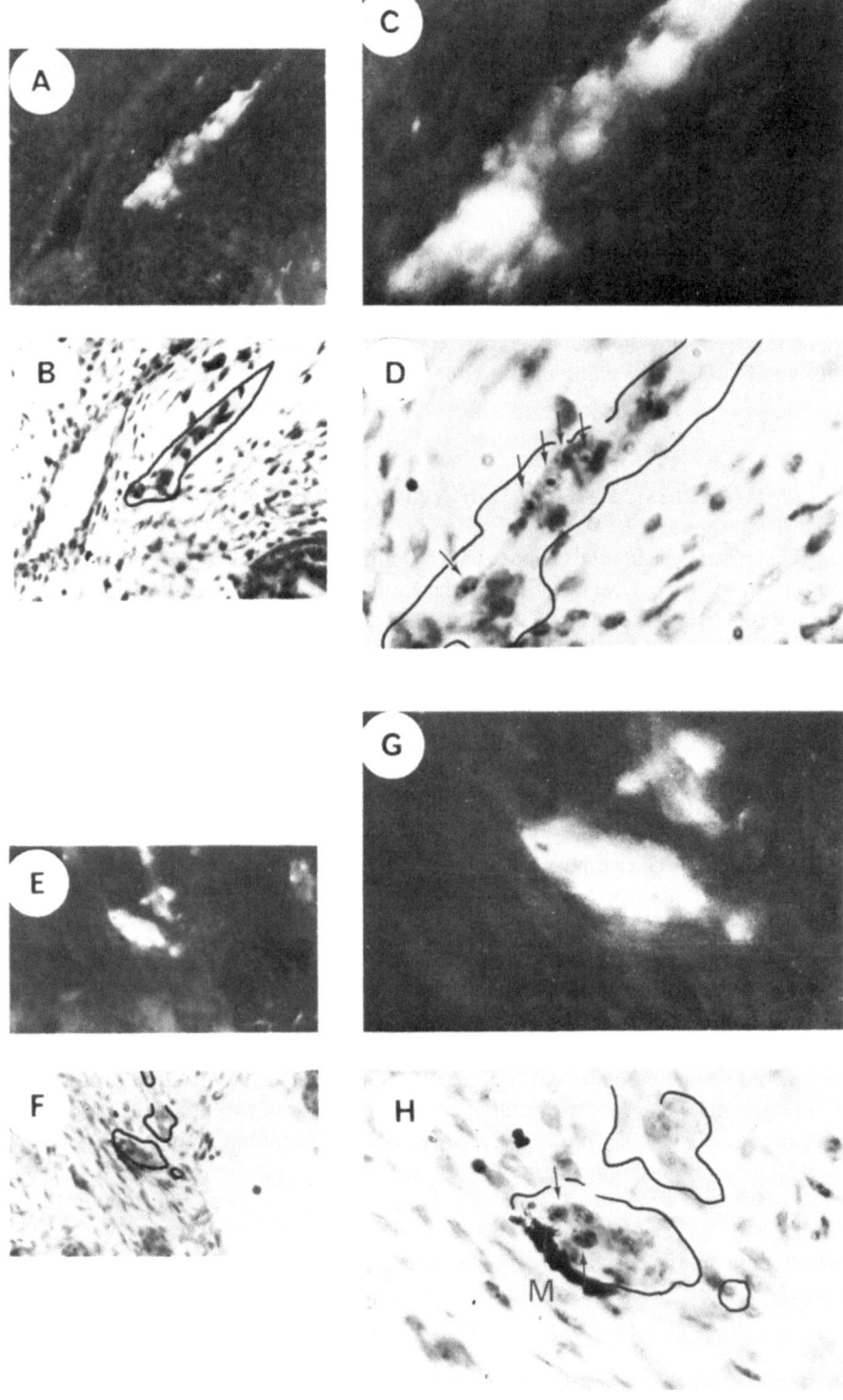

The ability to distribute along the ventral route does *not* appear to correlate with migratory ability: both fibroblast and somite cells (which did not translocate ventrally) are inherently migratory whereas RPE cells are not. The RPE cells, in fact, retained their cuboidal morphology even several days after injection into the host; they failed to extend pseudopods, which are often morphologically correlated with migratory cells. This observation raises the possibility that factors other than active migration may affect localization of the injected RPE cells, as well as of endogenous crest cells.

To test this possibility, latex polystyrene particles, which are inert and clearly non-motile, were placed onto the ventral neural crest pathway as probes of the embryonic environment. The particles were chosen to be approximately the same size as endogenous neural crest cells. The bead size did not, however, affect the distribution of the latex particles since beads of many different sizes behaved identically. Using the injection technique, the bead suspension was implanted onto the ventral neural crest pathway. In this way, it was possible to examine events independent of cell motility that may be involved in crest cell distribution and localization.

The latex beads were injected into the posterior somites of chicken embryos in regions along the neural axis where crest migration was just beginning. Like the neural crest-derived pigment cells described previously, the latex beads were initially contained within the somitic cavity (Fig. 10). In embryos fixed and sectioned at progressively older stages after injection, the latex beads distributed ventrally along the neural crest pathway with a time course remarkably similar to that of endogenous crest cells. The beads localized in the vicinity of host neural crest sites; i.e., near the sympathetic ganglion, ventral nerve cord, or adjacent to the dorsal aorta in the region where the adrenal chromaffin cells coalesce (Fig. 11). Thus, even an inert particle can distribute along the ventral neural crest pathway after injection onto the pathway.

A possible explanation for the observed "migration" of the latex particles is that the beads may have adhered nonspecifically to host neural crest cells. To minimize this adhesion, the beads were coated with bovine serum albumin (BSA); in another cell system, BSA-coated beads exhibited reduced binding to cells compared to uncoated latex beads.[33] In addition, by using a protein like BSA that is not normally present along the crest pathway, it was possible to assess whether coating the bead surface with any protein would alter the distribution of the latex beads after injection. After the bead surface was coated with the protein, fluorescence-labeled antibodies to BSA were used to assay

FIGURE 9. Light and fluorescence photomicrographs of transverse sections demonstrating catecholamine histofluorescene in quail neural crest cells 3 days after injection of mixed clones into chick embryos. (A) Catecholamine histofluorescence and (B) light photomicrograph of the same section with the fluorescent area outlined around quail cells adjacent to the dorsal aorta. At higher magnification, histofluorescence (C) and light photomicrograph (D) demonstrate that the injected quail cells contain catecholamines; small arrows indicate cells in which the quail marker is most visible; the larger arrow indicates quail cells in which the heterochromatin marker is out of the plane of focus. (E–H) Another embryo with catecholamine histofluorescence in a more ventral site. (E) Catecholamine histofluorescence and (F) light photomicrograph of the same section with the fluorescent area outlined around a group of fluorescent cells lateral to the inferior vena cava; (G) histofluorescence of same section at higher power; (H) light photomicrograph at higher power of the same section shows that the injected quail cells are positive for catecholamines; note the quail melanocyte adjacent to the fluorescent cluster of cells. From Ref. 15 with permission.

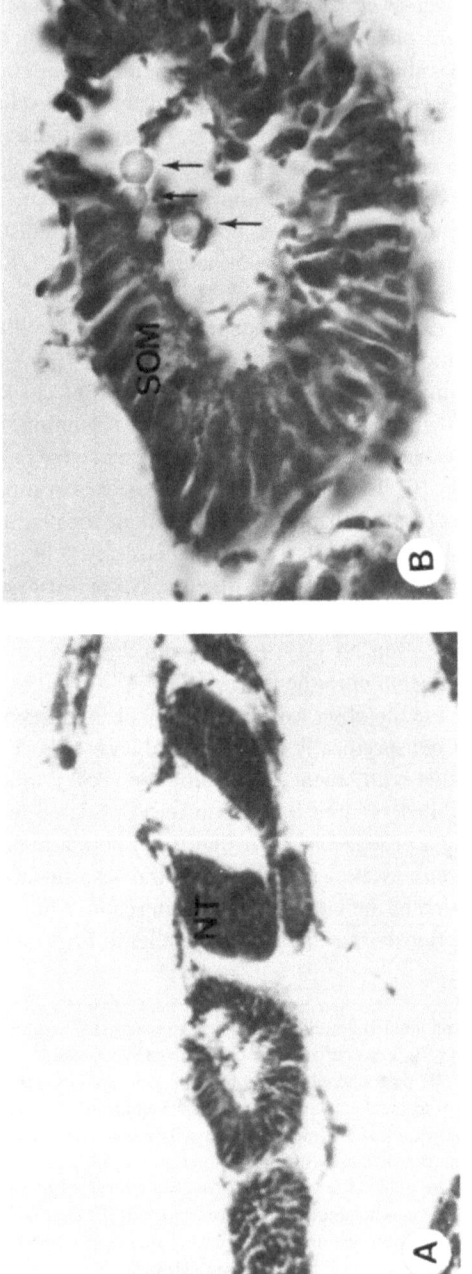

FIGURE 10. A transverse section through an embryo fixed immediately after latex polystyrene beads were injected into the somite. The beads are contained within the somitic cavity. Arrows indicate the positions of the beads. NT, neural tube; SOM, somite. From Ref. 31 with permission.

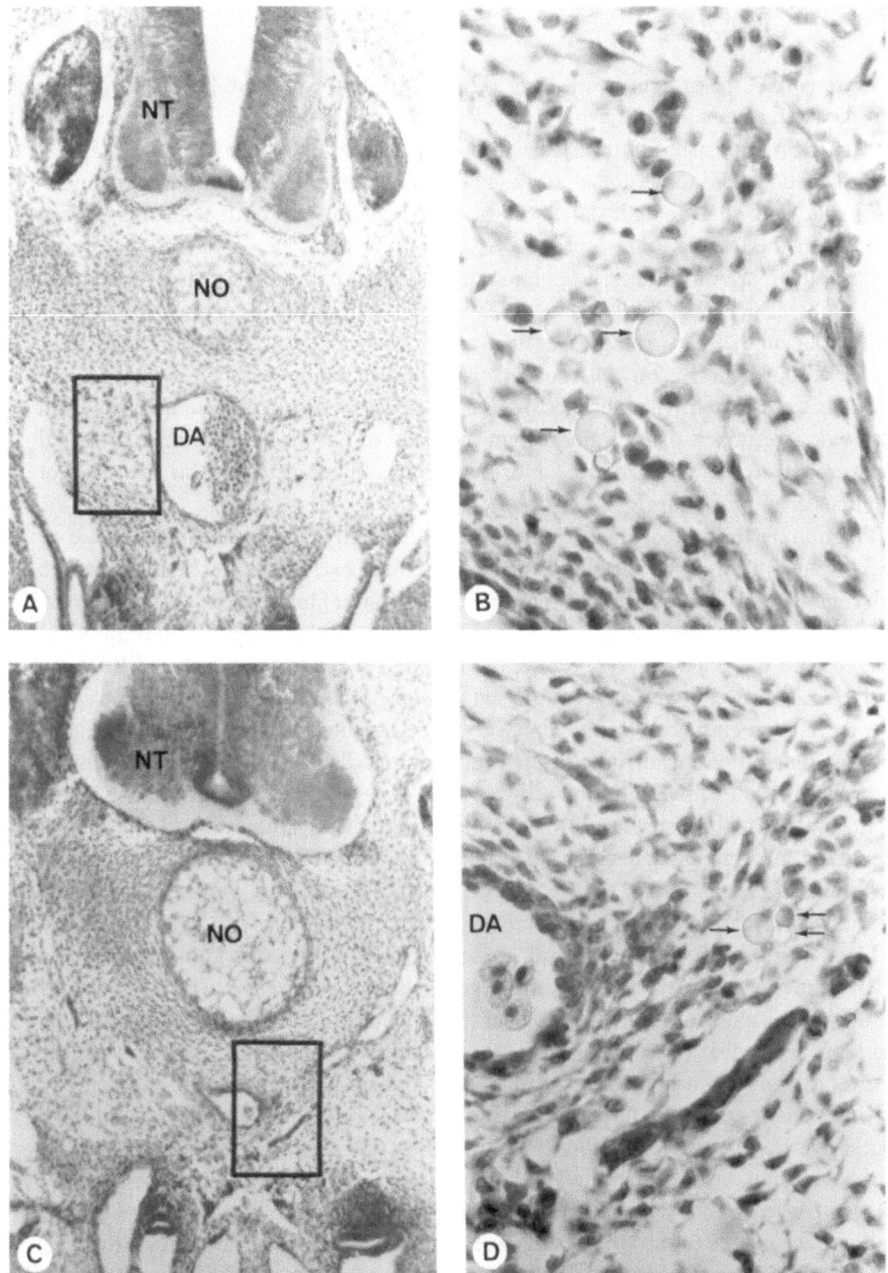

FIGURE 11. Transverse sections through chick embryos after uncoated beads were injected into the posterior somites. (A) An embryo fixed 2 days after injection. (B) At higher magnification of boxed area in (A), the latex particles are clearly visible adjacent to the dorsal aorta in close proximity to endogenous crest cells, (C) An embryo fixed 3 days after injection. (D) At higher magnification of boxed area in (C), the beads can be seen adjacent to the dorsal aorta and close to adrenomedullary crest cells. Arrows indicate the positions of the beads. NT, neural tube; NO, notochord; DA, dorsal aorta. From Ref. 31 with permission.

for the presence of the protein on the bead. The localization pattern of BSA-coated beads was observed in host embryos fixed at consecutive days after injection. Like the uncoated beads, the BSA-coated beads translocated along the ventral pathway and settled near the sympathetic ganglia and dorsal aorta (Fig. 12). The distribution pattern was indistinguishable from that of plain latex beads. Thus, coating the bead with this protein does not alter the translocation process.

Why, then, are certain cell types like somite and fibroblast cells restricted from the ventral pathway whereas inert beads are capable of ventral displacement? The difference might lie on the cell surface. One surface component that somite and fibroblast cells produce,[34,35] but neural crest cells do not,[5] is cell surface fibronectin. For this reason, latex beads were coated with fibronectin. The presence of the glycoprotein on the bead surface was ascertained using a fluorescent antibody to fibronectin. Fibronectin-coated beads were subsequently introduced into the somites of host embryos at the onset of endogenous crest migration.

The fibronectin-coated beads behaved quite differently from either the uncoated or the BSA-coated beads. The fibronectin beads remained exclusively at the injection site (the dermomyotomal or sclerotomal remnant of the somite) and did *not* distribute ventrally along the neural crest pathway (Fig. 13). Occasionally, the fibronectin beads were associated with the ectoderm. Therefore, the fibronectin beads were excluded from the ventral neural crest pathway as were the somite and fibroblast cells.

A correlation exists between the presence of cell surface fibronectin and the *inability* to move along the ventral route. Embryonic somite cells and fibroblasts both synthesize fibronectin.[34-36] Conversely, the cell and bead types capable of translocation along the ventral pathway lack surface fibronectin. For example, melanocytes and other crest-derived cells, sarcoma 180 cells (a transformed fibroblastic cell line that has lost fibronectin), and RPE cells all are deficient in cell surface fibronectin; yet they distributed along the ventral pathway after injection.

These results suggest that neural crest migration is, in part, controlled by the embryonic environment. Although neural crest cells are inherently highly migratory, much of their localization must be under environmental control. Both plain and BSA-coated latex beads distributed along the ventral pathway and localized near host neural crest sites; it follows that localization of the beads and crest cells must, in part, be influenced by an environmentally imparted driving force. Coating the beads with an adhesive molecule like fibronectin was sufficient to overcome this driving force. Therefore, cell surface fibronectin, or other proteins with similar adhesive properties, may be a "stop" signal that prevents classes of cells from entering this route. In this way, differences on the cell surface may account for selection at the entrance of the neural crest pathway.

The mechanisms that cause the movement and localization of the beads are still not clear. A possible driving force could be haptotaxis[37,38] in response to general adhesiveness of certain extracellular matrix components along the crest pathway. In culture, Wiseman[39] has shown that solid tissue aggregates are able to move cell-sized spheres of metal, plastic, and glass from the surface to the interior of the aggregate. The movement of the inert spheres in tissue masses may be analogous to the translocation of inert latex beads along the ventral crest pathway. Small movements of cells surrounding the beads could mix them and allow them to move in response to their adhesive qualities. All of

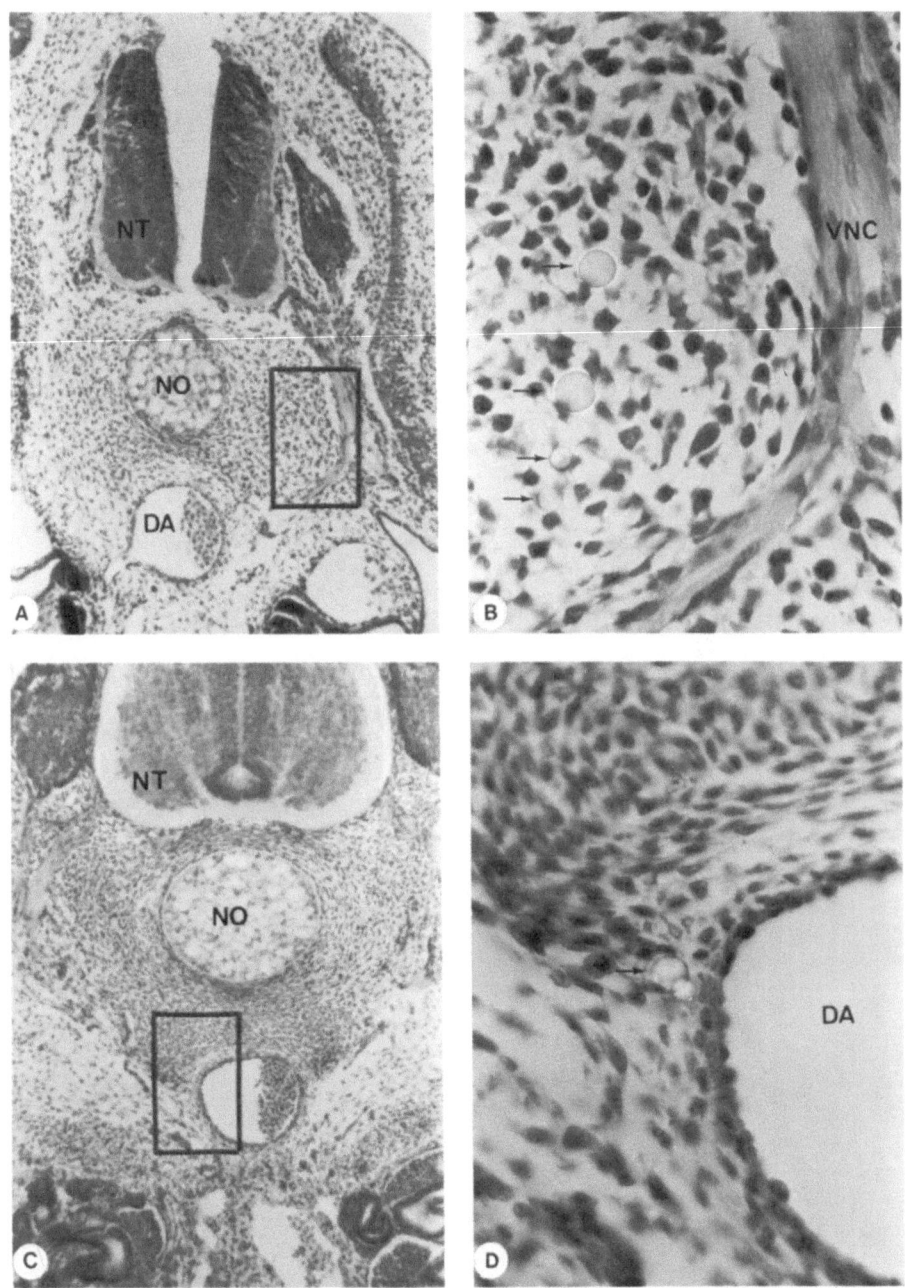

FIGURE 12. Transverse sections through chick embryos after BSA-coated beads were injected into the posterior somites. (A) An embryo fixed 2 days after injection. (B) Higher magnification of boxed area in (A) shows BSA-coated beads are found adjacent to the sympathetic ganglion and ventral nerve cord. (C) An embryo fixed 3 days after injection. (D) Higher magnification of boxed area in (C) demonstrates that the BSA-coated beads are localized adjacent to the dorsal aorta. Arrows indicate the positions of the beads. NT, neural tube; NO, notochord; DA, dorsal aorta; VNC, ventral nerve cord. From Ref. 31 with permission.

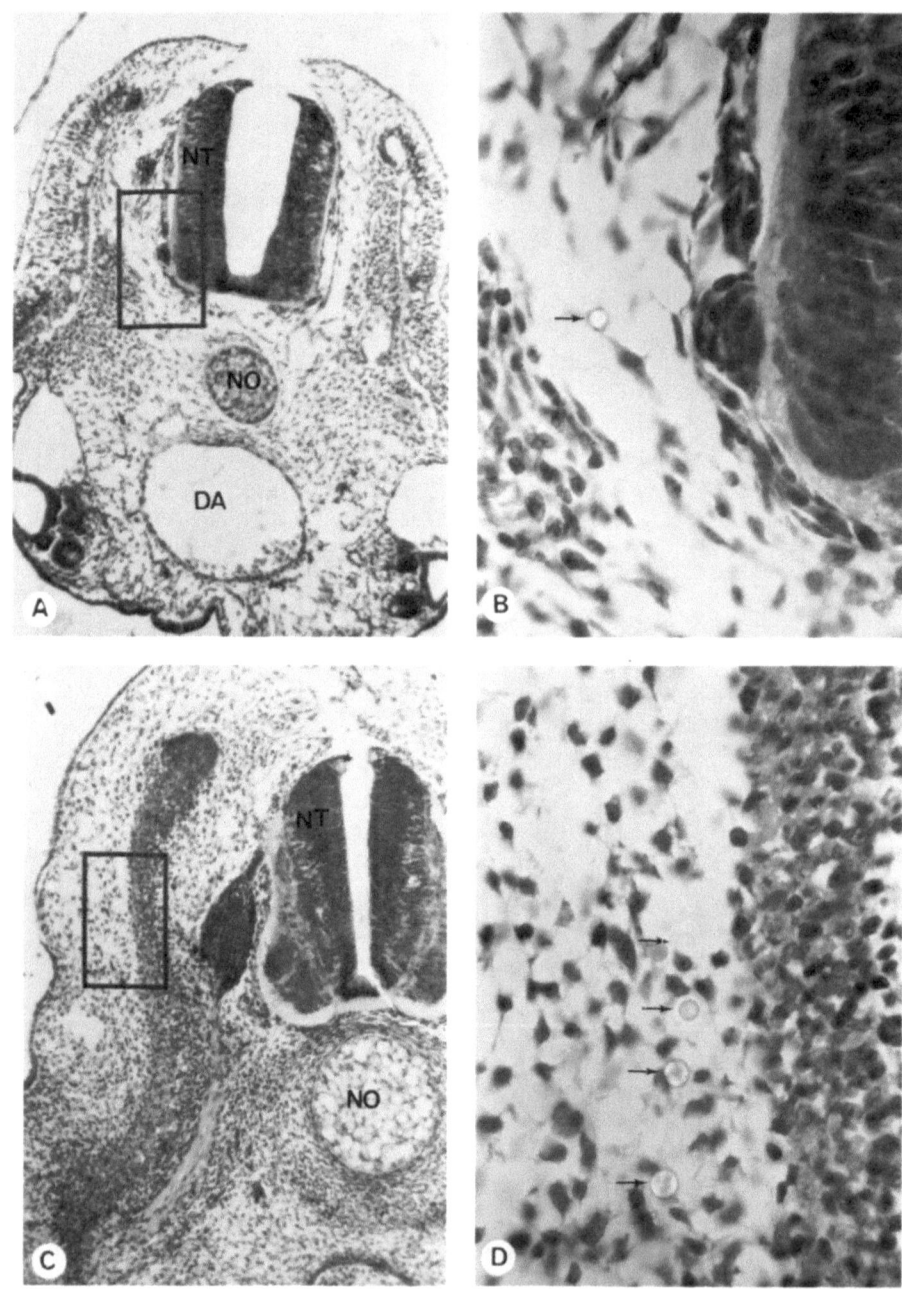

FIGURE 13. Transverse sections through chick embryos after fibronectin-coated beads were injected into the posterior somite. (A) An embryo fixed 2 days after injection. (B) Higher magnification of boxed area in (A) has a fibronectin-coated bead adjacent to the neural tube and just below the dorsal root ganglion. (C) An embryo fixed 3 days after injection with fibronectin-coated beads. (D) At higher magnification of boxed area in (C), the fibronectin-coated beads are found associated with the dermomyotome. Arrows indicate the positions of the fibronectin-coated beads. NT, neural tube; NO, notochord; DA, dorsal aorta. From Ref. 31 with permission.

the bead types could then be moved according to their surface properties; for example, the adhesiveness of the fibronectin-coated beads may be sufficient to keep them at the implantation site, whereas less-adhesive BSA beads or uncoated beads may be moved to more ventral sites. The similar pattern of distribution of beads, RPE cells, neural crest cells, and melanocytes implies a common localization mechanism for all of these cells or particles.

The use of latex beads as an environmental probe offers a clear-cut methodology for studying the embryonic environment with the potential to uncover the mechanisms directing neural crest cell localization. The results demonstrate that, in addition to the inherent motility of neural crest cells, an important component independent of active migration may be involved in both selection of cells entering the crest pathway and final localization of crest cells. We are presently in the process of exploring the role of other extracellular matrix molecules on the ventral migratory pathway with the hopes of learning more about cell surface and extracellular matrix interactions, and their role in crest cell localization.

7. Conclusion

The experiments reviewed in this chapter utilize a microinjection technique that enables the introduction of cultured cells or inert particles onto the trunk ventral pathway of neural crest migration. By placing beads or heterospecifically marked cells onto this migratory route, we hope to elucidate the mechanisms controlling neural crest migration and to establish the cell lineage decisions of this population. The injection technique has the advantage of being much less deleterious than previous methods used to study these questions. An additional advantage of this methodology is that it allows one to place small numbers of cells or latex particles into the embryo.

Using this technique, neural crest-derived pigment cells have been implanted onto a pathway normally followed by precursors to neuronal and supportive cells. The pigment cells distributed along the ventral route and localized adjacent to endogenous crest cells even though they had already differentiated and were no longer motile. The ventral pathway seems to be somewhat selective, because somite and fibroblast cells, two embryonic cell types known to be migratory, did not migrate ventrally after injection. Another non-neural crest-derived cell type, the retinal pigment epithelium, which is not inherently migratory, was able to translocate ventrally after injection. The restriction of the somite and fibroblast cells from the ventral route may be caused by the presence of cell surface fibronectin. When latex beads were coated with fibronectin and subsequently injected into the somites, the beads did not move along the ventral pathway but remained at the injection site. This is in contrast to the uncoated beads or beads coated with BSA. The latter two bead types both distributed along the ventral pathway following injection. It thus seems that an adhesive molecule like fibronectin on the surface of cells may serve as a recognition mechanism that restricts those cell types from entering the pathway. Once cells or beads enter the ventral path, they progress ventrally and generally settle adjacent to endogenous neural crest sites. This suggests that the factors responsible for crest cell localization are common for both endogenous and implanted cells. Thus, the localization of crest cells may be independent of cell motility. These results point out the necessity of critically reexamining the findings of some previous experiments. Several investigators have implanted differentiated cells of neural crest origin onto the ventral pathway.[21,23,30] The implanted cells distributed along the pathway similarly to the host crest cells. It was concluded that these cells retained the capacity to migrate to appropriate neural crest sites even after differentiation; the environment along the ventral route was thought to "induce" the appropriate migratory behavior in these cells. In light of the fact that latex beads are capable of remarkably similar ventral translocation, it must be concluded that the ventral movement of the implanted neural crest-derived cells is probably *not* a specific phenomenon; rather, the cell distribution pattern may reflect a more general sorting phenomenon resulting from interactions between the cell surface of the implanted cells and the extracellular matrix along the embryonic pathway. With increasing age, some changes occur in the embryonic environment that restrict the degree of ventral displacement of implanted cells. These age-related changes may, to some extent, account for the localization of crest cells in the sensory ganglion site.

The results obtained by injecting mixed clones of neural crest cells indicate that some cells of the premigratory neural crest are bipotent and become determined in accordance with interactions between crest cells and other cell types or among crest cells themselves. Prior to injection, the mixed clones contained pigmented cells and undifferentiated neural crest cells. After injection, some of the cells became adrenergic neurons. The embryo thus may serve as an *in vivo* culture system that allows the investigator to study the potential of the progeny of a single cell. Since the descendants of one crest cell can form both melanin-containing and adrenergic phenotypes, at least a subpopulation of the neural crest has multiple potential.

The mechanisms underlying neural crest migration and differentiation are still poorly understood. The experimental paradigm outlined in this chapter aspires to at least partially elucidate some of the controlling factors behind these intriguing and complicated phenomena. The injection technique represents a new technology for introducing exogenous cells or substances onto the crest migratory route in order either (1) to characterize the properties of the pathway or (2) to characterize the injected cells themselves. In the future, we will continue to explore the cell lineage decisions that lead to final commitment in neural crest cells and to examine the role of cell surface and extracellular matrix interactions in neural crest cell migration and localization.

ACKNOWLEDGMENTS. I would like to thank Dr. Scott Fraser, James Coulombe, and Marthe Howard for their careful reading and helpful criticism of the manuscript. Parts of the research reviewed in this chapter were supported by U.S. Public Health Service Grant HD-15527-01 and by Basil O'Connor Starter Research Grant 5-312 from the March of Dimes Birth Defects Foundation.

References

1. LeDouarin, N. M., and Teillet, M. A., 1974, Experimental analysis of the migration and dfferentiation of neuroblasts of the autonomic nervous system and of neuroectodermal mesenchymal derivatives using a biological cell marking technique, *Dev. Biol.* **41**:162.

2. Johnston, M. C., 1966, A radioautographic study of migration and fate of cranial neural crest cells in the chick embryo, *Anat. Rec.* **156**:143.

3. Noden, D. M., 1975, An analysis of the migratory behavior of avian cephalic neural crest cells, *Dev. Biol.* **42**:106.

4. Allan, I. J., and Newgreen, D. F., 1980, The origin and differentiation of enteric neurons of the intestines of the fowl embryo, *Am. J. Anat.* **157**:137.

5. Newgreen, D. F., and Thiery, J.-P., 1980, Fibronectin in early avian embryos: Synthesis and distribution along the migration pathways of neural crest cells, *Cell Tissue Res.* **211**:269.

6. Mayer, B. W., Hay, E. D., and Hynes, R. O., 1981, Immunocytochemical localization of fibronectin in embryonic chick trunk and area vasculosa, *Dev. Biol.* **82**:267.

7. Newgreen, D. F., Gibbins, I. L., Sauter, J., Wallenfels, B., and Wutz, R., 1982, Ultrastructural and tissue-culture studies on the role of fibronectin, collagen and glycosaminoglycans in the migration of neural crest cells in the fowl embryo, *Cell Tissue Res.* **221**:521.

8. Pintar, J., 1978, Distribution and synthesis of glycosaminoglycans during quail neural crest morphogenesis, *Dev. Biol.* **67**:444.

9. Derby, M. A., 1978, Analysis of glycosaminoglycans within the extracellular environments encountered by migrating neural crest cells, *Dev. Biol.* **66**:321.

10. Teillet, M. A., and LeDouarin, N. M., 1970, La migration de cellules pigmentaires etudiee par la methode des greffes heterospecifiques de tube nerveux chez l'embryon d'Oiseau, *C. R. Acad. Sci. Ser. D* **270**:3095.

11. Manasek, F. J., and Cohen, A. M., 1977, Anionic glycopeptides and glycosaminoglycans synthesized by embryonic neural tube and neural crest. *Proc. Natl. Acad. Sci. USA* **74**:1057.

12. Cohen, A. M., 1977, Independent expression of the adrenergic phenotype by neural crest in vitro, *Proc. Natl. Acad. Sci. USA* **74**:2899.

13. Kahn, C. R., Coyle, J. T., and Cohen, A. M., 1980, Head and trunk neural crest in vitro: Autonomic neuron differentiation, *Dev. Biol.* **77**:340.

14. Cohen, A. M., and Konigsberg, I. R., 1975, A clonal approach to the problem of neural crest determination, *Dev. Biol.* **46**:462.

15. Bronner-Fraser, M. E., Sieber-Blum, M., and Cohen, A. M., 1980, Clonal analysis of the avian neural crest: Migration and maturation of mixed neural crest clones injected into host chicken embryos, *J. Comp. Neurol.* **193**:423.

16. Sieber-Blum, M., and Cohen, A. M., 1980, Clonal analysis of quail neural crest cells: They are pluripotent and differentiate in vitro in the absence of noncrest cells, *Dev. Biol.* **80**:96.

17. Weston, J.A., 1963, A radioautographic analysis of the migration and localization of trunk neural crest cells in the chick, *Dev. Biol.* **6**:279.

18. LeDouarin, N. M., 1973, A biological cell labelling technique and its use in experimental embryology, *Dev. Biol.* **30**:217.

19. Weston, J. A., and Butler, S. L., 1966, Temporal factors affecting localization of neural crest cells in the chicken embryo, *Dev. Biol.* **14**:246.

20. Bellairs, R., Ireland, G. W., Sanders, E. J., and Stern, C. D., 1981, The behavior of embryonic chick and quail tissues in culture, *J. Embryol. Exp. Morphol.* **61**:15.

21. LeDouarin, N. M., Teillet, M. A., Ziller, C., and Smith, J., 1978, Adrenergic differentiation of cells of the cholinergic ciliary and Remak's ganglia in avian embryos after in vivo transplantation, *Proc. Natl. Acad. Sci. USA* **75**:2030.

22. Erickson, C. A., Tosney, K. W., and Weston, J. A., 1980, Analysis of migratory behavior of neural crest and fibroblastic cells in embryonic tissues, *Dev. Biol.* **77**:142.

23. Bronner, M. E., and Cohen, A. M., 1979, Migratory patterns of cloned neural crest melanocytes injected into host chicken embryos, *Proc. Natl. Acad. Sci. USA* **76**:1843.

24. Bronner-Fraser, M. E., and Cohen, A. M., 1980, Analysis of the neural crest ventral pathway using injected tracer cells, *Dev. Biol.* **77**:130.

25. Tosney, K., 1978, The early migration of neural crest cells in the trunk region of the avian embryo: An electron microscopic study, *Dev. Biol.* **62**:317.

26. Hamilton, H. L., 1940, A study of the physiological properties of melanocytes with special reference to their role in feather coloration, *Anat. Rec.* **78**:525.

27. Falck, B., Hillarp, N.-Å, Thieme, G., and Torp, A., 1962, Fluorescence of catecholamines and related compounds condensed with formaldehyde, *J. Histochem. Cytochem.* **10**:348.

28. Enemar, A., Falck, B., and Hakanjon, R., 1965, Observations on the appearance of norepinephrine in the sympathetic nervous system of the chick embryo, *Dev. Biol.* **11**:268.

29. Feulgen, R., and Rossenbeck, H., 1924, Mikroskopisch-Chemischer Nachweis einer Nucleinsaure von Typus der Thymonucleinsaure und die Darauf Beruchende Elektive Farburg von Zellkernen in Mikroskopischen Praperaten, *Hoppe-Seylers Z. Physiol. Chem.* **135**:203.

30. LeLievre, C. S., Schweizer, G. G., Ziller, C. M., and LeDouarin, N. M., 1980, Restrictions of developmental capabilities in neural crest derivatives as tested by *in vivo* transplantation, *Dev. Biol.* **77**:362.

31. Bronner-Fraser, M. E., 1982, Distribution of latex beads and retinal pigment epithelial cells along the ventral neural crest pathway, *Dev. Biol.* **91**:50.

32. Ali, I. U., and Hynes, R. O., 1978, The effects of LETS glycoprotein on cell motility, *Cell* **14**:439.

33. Grinnell, F., 1980, Fibroblast receptors for cell–substratum adhesion: Studies on the interaction of baby hamster kidney cells with latex beads coated by cold insoluble globulin (plasma fibronectin), *J. Cell Biol.* **86**:104.

34. Loring, J., Erickson, C. A., and Weston, J. A., 1977, Surface proteins of neural crest, crest-derived, and somite cells *in vivo*, *J. Cell Biol.* **75**:71a.

35. Engvall, E., and Ruoslahti, E., 1977, Binding of soluble form of fibroblast surface protein, fibronectin to collagen, *Int. J. Cancer* **20**:1.

36. Yamada, K. M., and Olden, K., 1978, Fibronectin—Adhesive glycoproteins of cell surface and blood, *Nature (London)* **275**:179.

37. Carter, S. B., 1965, Principles of cell motility: The direction of cell movement and cancer invasion, *Nature (London)* **208**:1183.

38. Harris, A., 1973, Behavior of cultured cells on substrata of variable adhesiveness, *Exp. Cell Res.* **77**:285.

39. Wiseman, L. L., 1977, Contact inhibition and the movement of metal, glass, and plastic beads within solid tissues, *Experientia* **33**:734.

40. Bronner-Fraser, M. E., and Cohen, A. M., 1980, The neural crest: What can it tell us about cell migration and determination? *Curr. Top. Dev. Biol.* **15**:1.

3

Neural Transplants in Lower Vertebrates

WILLIAM A. HARRIS

1. Introduction

Taken as a developmental problem, the nervous system in higher metazoans represents the most complex example of pattern formation in all of biology. An understanding of the mechanisms regulating neural ontogeny cannot simply come from descriptive anatomy or physiology. As the early experimental embryologists realized, hypotheses about development may be born of observations on normal embryos, but experimental manipulations are necessary to prove them.[1-3] The manipulations often take the classical form of tissue transplantation, and in the nervous system the results are analyzed with modern techniques of electrophysiology and neuroanatomical tracing. The power of combining these approaches has been enormously successful in generating basic rules of causal neurogenesis, even to the point of suggesting molecular mechanisms. Transplantation studies cannot, however, discover the molecules involved. The remarkable development of the topographic retinotectal projection as revealed by experimental manipulations *in vivo* has attracted many groups interested in the molecular biology of the formation of specific neural connections.[4-9] If, as soon seems likely, theories derived from transplantation experiments are substantiated biochemically, it will constitute a major breakthrough in our understanding of the brain.

This review will cover studies in amphibians and fish in such a way that issues in neural development rather than types of transplantation are dealt with systematically.

WILLIAM A. HARRIS ● Department of Biology, University of California, San Diego, La Jolla, California 92093.

1.1. The Organisms

Of all the classes of animals, none compares to Amphibia for studies on the transplantation of embryonic tissue. The amphibian embryo develops outside the mother inside of a soft jelly case, at ambient temperature, in pond water (or the laboratory equivalent). Even after drastic surgery, such as the transplantation of entire heads, the tissues may heal completely, allowing development to proceed within minutes. Amphibians allow transplantation at all stages of development, and because these operations can be done at very early embryonic times, they share many of the advantages of the "genetic transplants" caused by homeotic mutations (mutations that transform one body part into another) in insects, without the conceptual problems associated with mutants. In addition, all neural tissues, even those deep within the CNS, are available for transplantation at most stages.

Fish have also been used for many transplantation studies because of the ability of the adult nervous system to heal and regenerate. These studies do not ask questions directly about development, but assume that regeneration and development are similar in many ways. Certainly, however, there are differences. For example, in the development of the retinal projection in lower vertebrates, a small number of retinal ganglion cells send axons directly and in topographic order to their targets,[10,11] whereas in regeneration a large number of cells regrow axons that are initially disordered and travel along abnormal pathways to their targets.[12]

Avian embryos are also used for similar experiments although not as extensively as amphibians, as it is more difficult to do embryonic transplantations with chicks. It is more difficult still to do embryonic surgery on mammals. Birds and mammals are covered in the other chapters of this volume. Of the invertebrates, the study of homeotic mutations in *Drosophila* and the surgical transplantations of tissues in other insects have also been extremely informative with regard to neural development. Reviews of such studies with insects are available elsewhere.[13–15]

1.2. The Transplantations: Definitions

Rotations: Parts of the brain have been rotated at different times in development to find out when the orientation of the tissue becomes determined, or to find out if the rotated piece contains a fixed set of positional markers. *Translocations:* Tissue may be moved from one place to another to find out whether there is a time when the fate of the tissue is determined or whether determined tissues develop particular abnormalities as a result of translocation. *Supernumeraries:* Extra tissue is sometimes transplanted to an animal to examine the response of surrounding tissues or connecting structures. *Duplications:* Duplicating some fragments of the nervous system at the expense of others has been done to ask questions about the specificity of innervation, by examining the targets of the duplicated pieces. *Heteroplastic transplants:* Transplantations done between phenotypically different animals, including between species, can reveal whether certain genetically regulated differences are important in determining the pattern of connections in the nervous system. *Temporal or heterochronic:* Some types of transplantation experiments can serve to disrupt the normal temporal sequence of development.

Labeling: Still other transplantations are performed merely to label cells of specific origins, such that their migrations or the growth of their axons can be monitored.

1.3. The Neural Structures

Of the many structures that comprise the nervous system, relatively few have been exploited for transplantation purposes (Fig. 1). Among those that have, are the visual, auditory, and olfactory structures, the limb, the spinal cord, and the giant Mauthner cells. A feature that these systems share is their accessibility to surgical transplantation. The retina has most often been used because it is a two-dimensional sheet of neural

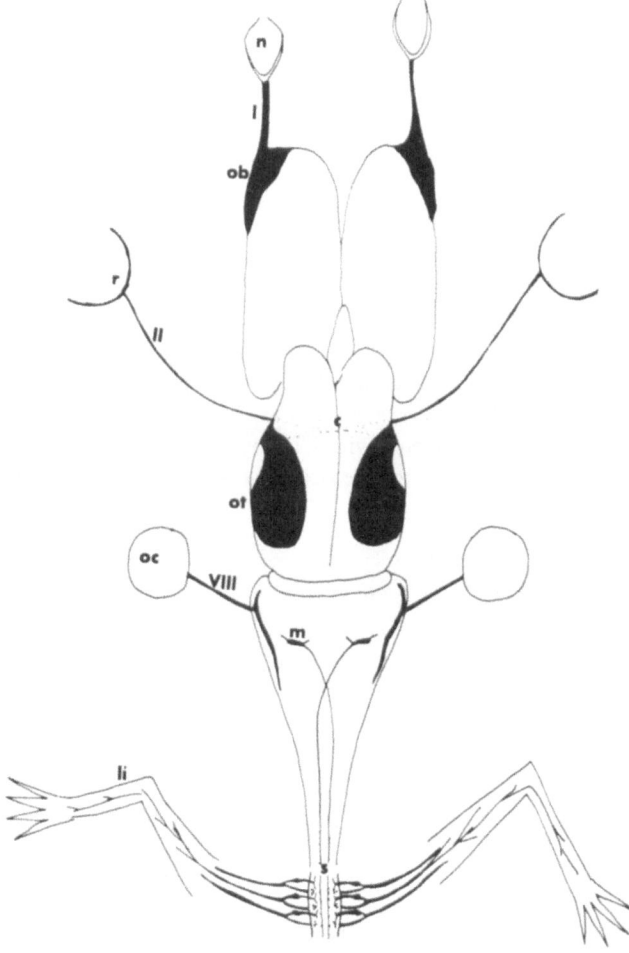

FIGURE 1. The structures commonly transplanted in the amphibian to study neurogenesis, as seen from a dorsal view of the adult brain. n, nasal epithelium; ob, olfactory bulb; r, retina; ot, optic tectum; oc, otic capsule, m, Mauthner cells; s, spinal cord; li, limb, I, II, VIII, cranial nerves; c, optic chiasm.

tissue that projects topographically to target structures in the diencephalon and mesencephalon. Both the retina and these visual centers in the brain can be studied electrophysiologically. The retina is located peripherally making it accessible for manipulation with minimal damage. It can easily be labeled anatomically. Finally, the normal neuroanatomy and physiology of the visual system are fairly well known.

The tectum, the main mesencephalic target of the retina, has also been extensively transplanted for some of the above reasons, and because results from retinal studies bring up issues that can be resolved only by the transplantation of the tectum. For instance, the question of whether retinal axons form their "map" by seeking out intrinsic tectal labels can best be approached with tectal transplants.

There have been a number of transplantation studies using the olfactory and auditory systems, both of which share only some advantages with the visual system. The giant Mauthner cells in the medulla of larval amphibians are studied because they are large identifiable neurons with identifiable major dendritic arbors and axonal projections. The limb and its innervation have been transplanted to pursue questions of nerve branching and navigation, muscular development, cell death, and origin of motor patterns.

2. The Fate and Orientation of the Neural Tissues

2.1. The Brain

At early neural plate or preneural stages, it seems that the committed primordial CNS is highly regulative. Deleted pieces of the neural plate are replaced, translocated pieces change fate, and rotated pieces develop as though they were oriented normally. For example, forebrain tissue labeled with [^{3}H]thymidine and transplanted to the presumptive medulla has been shown to give rise to Mauthner cells[16] (Fig. 2 and 3). These

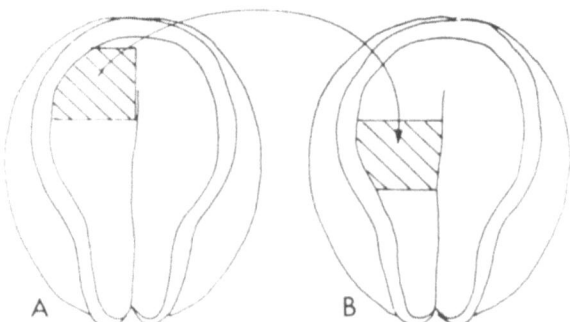

FIGURE 2. Diagram of the unilateral substitution of prospective forebrain–midbrain for prospective hindbrain in the neural plate of an axolotl neurula. The neuroepithelium of the cross-hatched area (sometimes with the underlying mesoderm) in embryo A was excised and implanted in normal orientation into the crosshatched area in embryo B from which the neuroepithelium (sometimes with the underlying mesoderm) had earlier been removed. From Ref. 16.

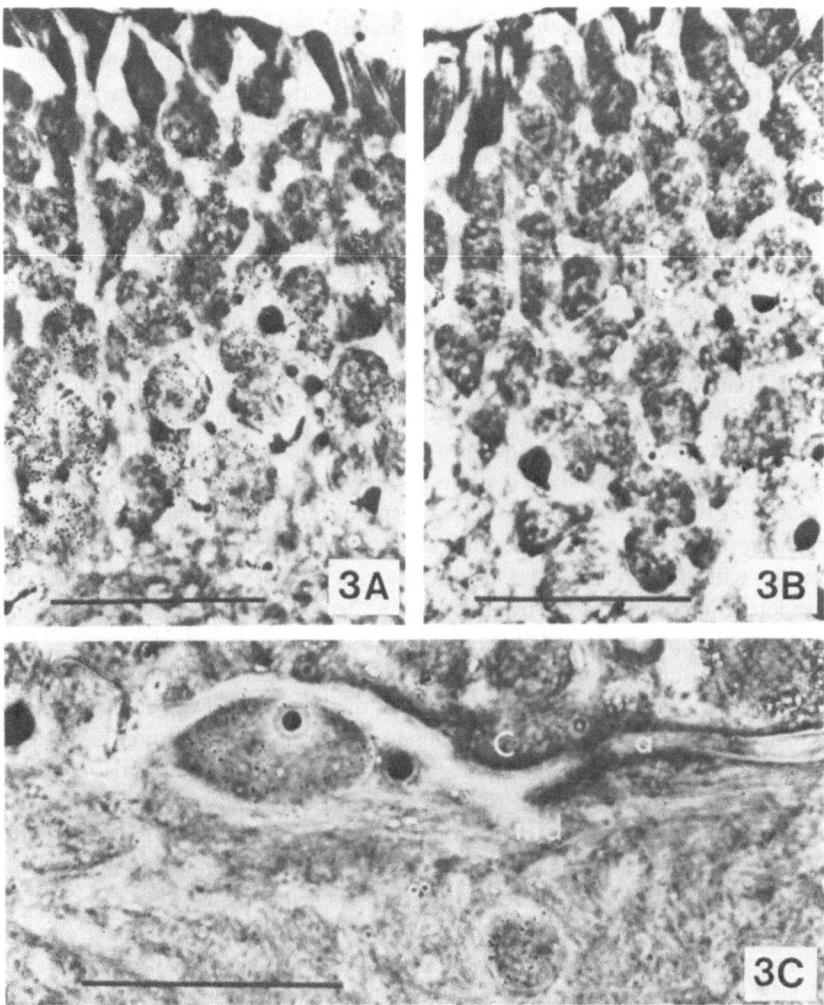

FIGURE 3. Phase micrographs of autoradiographs of Epon-embedded, 4-μm-thick cross-sections through medullae of 21-mm larvae that had undergone unilateral substitution of [³H]thymidine-labeled prospective forebrain–midbrain for prospective hindbrain in otherwise unlabeled host brains. Focus is at the level of the autoradiographic silver grains. (A) Section through the gray matter of the grafted half of a medulla. Virtually all of the nuclei are labeled. (B) Section through the gray matter of the control half of the medulla shown in (A). The number of silver grains over nuclei approximates that of background. (C) Section through a Mauthner cell in the grafted half of the medulla of another animal. The Mauthner cell nucleus is labeled. md, medial dendrite; a, axon; C, axon cap. Scale: 50 μm. From Ref. 16.

cells normally originate from hindbrain tissue. Later, the neural plate becomes mosaically determined, which means that translocated or rotated tissues will not change their fates or orientations. Thus, for example, rotation of presumptive medulla leads to the reversed polarity of Mauthner cells and cranial nerve nuclei.[17] Reversal of the diencephalon and midbrain from late neural plate stage shows autonomous axial development in terms of both morphology and rostrocaudal pattern of cellular proliferation.[18] These statements may oversimplify the situation as reports from different laboratories do not always agree on the stage at which fates and orientations in the nervous system are fixed. Perhaps the conflicting results are due to differences in the sizes of the grafts, the rapidity of healing, or the integration of the transplants.

2.2. The Retina

Primordial retinae lie together at the anterior extreme of the neural plate. As the neural tube closes, the anlagen separate, and shortly thereafter, optic bulges begin to evaginate from the ventral walls of the presumptive diencephalon. To find out when the axes of the eye are laid down, investigators have rotated eye primordia at different stages of development and looked in the adult at the resulting orientation of the visual projection on the tectum behaviorally and electrophysiologically. The first reports on the rotation of eyes at early tailbud stages showed respecification of the retinal axes, such that the rotated eye mapped onto the tectum according to its new orientation.[19,20] These studies conflict with more recent experiments in which careful examination of the rotated eyes and genetic marking of the transplants have demonstrated that the axes of primordial eyes appear to be polarized at least as soon as the earliest tailbud stages.[21,22] There have been numerous explanations for the apparent respecification found in the earlier reports. These include reversal of the rotation of the eye,[23] differences in operating solutions which cause differential healing,[22] abnormalities in the orientation of fascicles of retinal ganglion cell axons in the retina,[24] the amount of surrounding tissue that was rotated with the retina,[25] and cell movement from the untransplanted optic stalk.[10,26] This last explanation is perhaps the most likely. If the primordial eye is excised, labeled with [³H]thymidine in vitro, and replaced, the eye that develops is labeled dorsally and unlabeled ventrally[10,26] (Fig. 4). Such in vitro labeling experiments show that the ventral retina arises from cells migrating out of the optic stalk. The ventral retina grows much faster than the dorsal during metamorphosis in Xenopus,[27] so a small group of ventral retina cells give rise to a much larger fraction of the adult retina than the same sized group of dorsal cells. The sector of the retina that projects to the part of the tectum most easily mapped is virtually all of nonrotated stalk origin (ventral retina). If transplants done early in development do not include the stalk (presumptive ventral retina), then eye rotations may show an apparent repolarization of the ventral retina (Figs. 5 and 6). This explanation fits with data from studies of albino-marked transplants,[21] from which it is clear that whatever bits of ventral retina arise from the host always map normally and whatever retina arises from the rotated transplant always maps in a reversed orientation.

2.3. The Ear and Limb

The time of polarization of the ear ectoderm has also been studied in Ambystoma by rotation. The ear is determined by late gastrula or early neurula stages.[28] The ante-

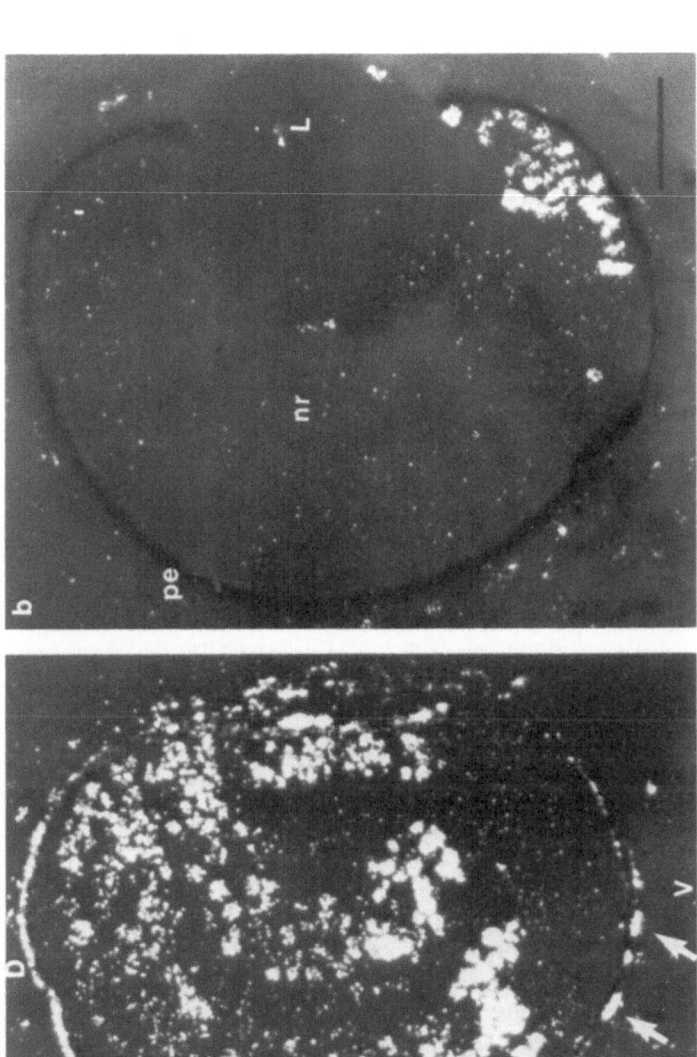

FIGURE 4. Dark-field autoradiographs of coronal sections of eyes (stage 41/2) following exposure of (a) the eye vesicle and (b) the optic stalk to [³H]-TdR. (a) The eye vesicle was incubated in label at stage 22; the section shown is nasal to the ventral fissure. The ventral neural retina is unlabeled whereas the rest of the retina is labeled. The densely labeled nuclei immediately dorsal to the unlabeled ventral region probably became postmitotic shortly after [³H]-TdR incorporation. Cells in the more dorsal retina are more weakly labeled, indicating mitotic dilution of the label. Arrows indicate labeled pigment epithelial cells adjacent to the unlabeled ventral neural retina. (b) The embryo, minus the eye vesicle, was incubated in [³H]-TdR at stage 25/6. Labeled cells are confined to the ventralmost retina. The notch in the pigment epithelium marks the position where the optic nerve leaves the eye. pe, pigment epithelium; nr, neural retina; L, lens; D, dorsal; V, ventral. Scale; 50 μ m. From Ref. 10.

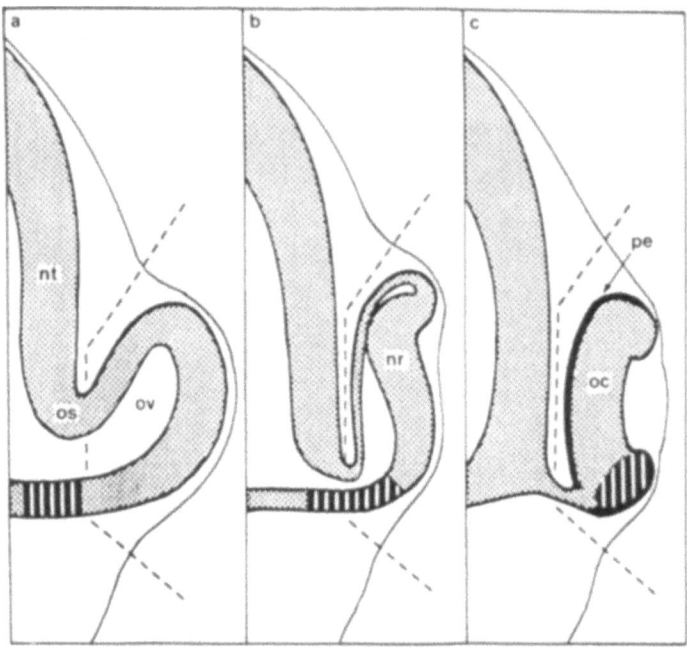

FIGURE 5. Schematic representation of eye development showing 180° rotation surgery (dashed lines) in suggested relation to the ventral fissure (VF)-forming cells (striped region). (a) The optic vesicle stage (before stage 28 in *Xenopus* and stage 12 in chick). The rotation surgery does not include the VF-forming cells. (b) Invaginating eye vesicle around the time of VF formation (stage 27–31 in *Xenopus* and stage 12 in chick). The surgery splits the population of VF-forming cells in two, rotating one half into 'a dorsal position. (c) The newly invaginated eye cup (after stage 31/2 in *Xenopus* and stage 12 in chick). The surgery rotates the whole population of VF cells dorsally. nt, neural tube; os, optic stalk; ov, optic vesicle; nr, neural retina; oc, optic cup; pe, pigment epithelium. From Ref. 10.

rior–posterior axis appears to be fixed at neural folding. Slightly later, the dorsal–ventral axis becomes fixed. If the anterior–posterior axis is reversed during critical stages of its polarization, the ear ectoderm often adjusts only partially to its new environment. An enantiomorphic twin ear then results. Such twinning occurs along a transverse plane of symmetry and produces a duplicated ventral half with mirror-image sets of semicircular canals, or a duplicated dorsal half with two saccules[28-31] (Fig. 7).

Rotation of the presumptive limb region on the flank of a tailbud embryo reveals that here too there is an apparent sequence of axis specification. As with the ear, the anterior–posterior axis is fixed first, and the dorsal–ventral next.[32-34] As with the retina, studies of cell movements with radioactively labeled or genetically marked transplants might also provide mechanistic insight into the concept of "axis specification" in the ear and limb.

In many of these systems, there appear to be steps in the pathway leading to a particular fate. What events actually cause these changes? How are tissues committed to particular fates, and what are the mechanisms by which polarity is induced in the differentiating structure? Is incomplete rotation of the primordium really the explanation for the phenomenon of repolarization in the retina, and could it account for the apparent respecification reported in other systems?

3. Axonal Outgrowth

As neurons differentiate, they send out axons and dendrites. The correct shaping and orienting of these neural processes is, of course, crucial to the development of specific neural connections. We have learned much about axonal outgrowth from neural transplantations. Experiments with Mauthner cells in *Ambystoma* and *Xenopus* are the most illustrative. Wherever a transplant results in changing the polarity of a piece of medulla containing the Mauthner cells, these cells develop their primary neuritic processes as though rotated.[17,35,36] The axon will always start out medially and ventrally with respect to the axis of the transplanted tissue, even though the host may influence the axon's course when it grows into host-derived tissue. Thus, local or intrinsic cues are important in the primary outgrowth of processes. Similarly, transplantation of olfactory placodes[37-41] and eye primordia to ectopic locations has shown that axonal outgrowth will proceed normally even in the absence of contact with the CNS.[42,43]

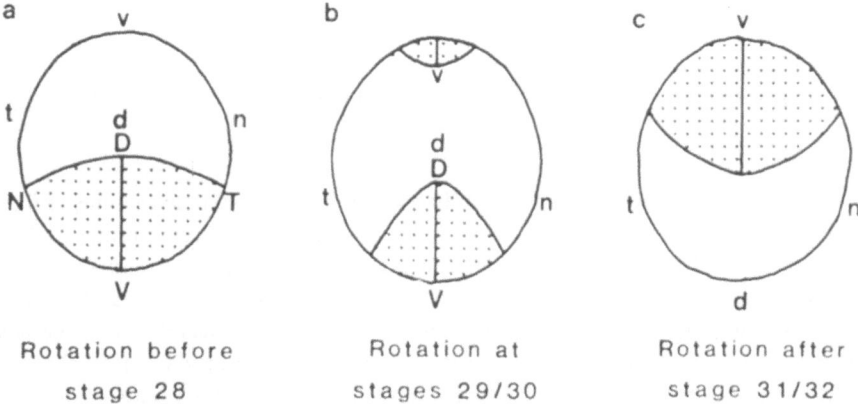

FIGURE 6. Schematic representation of the position of the ventral retina- and fissure-forming cells (stippled region) in the adult eye following 180° rotation of the optic vesicle around the time of eye cup formation. Capital letters indicate the *unrotated* axes of the late arriving ventral retinal cells, and lower-case letters indicate the *rotated* axes. (a), (b) and (c) show the suggested outcome of the rotation surgery illustrated in Fig. 5. (a) shows the ventral retina oriented normally and the dorsal part rotated following eye rotation operations made before stage 28. (b) shows a wedge of unrotated ventral retina bounded on either side by rotated nasal (n) and temporal (t) retina as a result of eye inversions made at stage 29/30. (c) shows the whole eye rotated following inversions made after stage 31/32. D, d, dorsal. V, v, ventral; T, t, temporal; N, n, nasal. Notice that rotations done before stage 28 would yield tectal maps (from ventral retina) that would appear respecified in both axes, rotations done around stages 29/30 would yield maps that appear respecified in the DV axis, and rotations done after stage 31/32 would yield nonrotated maps. From Ref. 10.

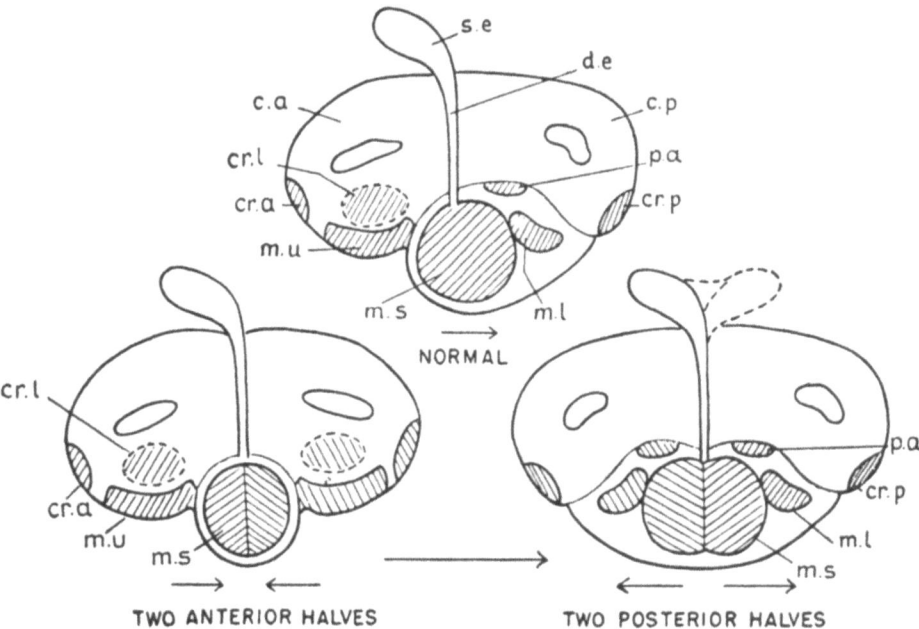

FIGURE 7. Normal right labyrinth and two types of enantiomorphic twins, medial surface. The large arrow points to the posterior end of the embryo; the small arrows point posteriorly with respect to the labyrinth or half-labyrinth to which they refer. c.a, anterior semicircular canal; c.p, posterior canal; cr.a and cr.p, the corresponding sensory cristae; cr.a, lateral crista, projected from the ventrolateral wall but not visible in this view; d.e, endolymphatic duct; m.1, macula of lagena; m.s, macula of sacculus; m.u, macula of utriculus; p.a, papilla amphibiorum; s.e, endolymphatic sac. From Ref. 32.

Classical studies with the spinal cord and limb bud showed that the innervation of the limb is of central origin. When limb buds are transplanted from normal to CNS-deprived host embryos, which are kept alive through parabiosis with normal embryos, nerveless limbs develop.[44] In the reciprocal experiment in which a CNS-deprived embryo's limb is grafted to a normal embryo, the previously nerveless limb becomes innervated by host fibers.[44] Isolated pieces of spinal cord transplanted to abnormal regions send out motor and sensory axons toward nearby targets.[1] Extirpation of the neural crest and dorsal plate reveals that motoneuron axons can grow out in the absence of sheath cells and sensory neurons.[45,46]

Transplantations between embryos of different species indicate that the timing of nerve fiber outgrowth from the CNS is independent of the developmental stage of the target tissue. Thus, a fast-maturing piece of neural tube prematurely innervates the limbs of a slower developing host. In the reverse situation, innervation of the limb is delayed.[47]

From these studies we see that there is probably a great deal of autonomy in the initial outgrowth of processes in differentiating nerve cells. The rules for initial short-range axonal outgrowth, however, seem to be rather different from those governing the long-range trajectory of these axons to their targets.

4. Axonal Navigation

4.1. Reaching the Target

Developing axons may grow hundreds of cell diameters to reach their targets. Because adjacent neurons in one nucleus may innervate different targets spatially separated in the brain, there is not likely to be some simple mechanical explanation for target-directed growth. Transplantation studies have convincingly ruled out purely mechanical models.

Transplantation of the embryonic retina to abnormal locations on the body, followed by anatomical studies of the transplanted optic projection, has shown that for optic fibers to enter the CNS, there must be a neuroepithelial continuity from the presumptive retina to the primordial brain. Experiments in which *Rana, Xenopus,* or *Ambystoma* optic cups without optic stalks are transplanted to abnormal locations show a lack of central innervation; axons grow out of the developing eye and end in a "blind neuroma"[42,43,48] (and Hibbard, personal communication). When the optic stalk is transplanted along with the retina, and the stalk is integrated into the underlying neuroepithelium of the CNS, the fraction of transplants with a successful central innervation increases dramatically.

When transplanted optic nerves enter the main neural axis, they are able to find their normal targets starting from a variety of abnormal entry points. Thus, whether the transplanted optic nerve enters near the olfactory nerve at the rostral end of the telencephalon (Constantine-Paton and Hibbard, personal communications) or enters near the caudal mesencephalon,[49] these fibers can grow to, and make appropriate connections with, the normal primary visual centers of the brain. This is so even when the host for such transplantation studies is genetically eyeless, and therefore lacks optic tracts of host origin[50] (Fig. 8). Thus, axons are not restricted to follow a particular pathway in order to reach their destinations.

In similarly conceived studies in salamanders and frogs, olfactory and otic placodes have been transplanted to ectopic regions in host embryos. In normal development, there is no olfactory stalk joining presumptive nasal epithelium to the forebrain; they are merely closely apposed. For these experiments a wound need not be made in the host neuroepithelium to ensure central innervation. These nasal transplants are also able to grow nerves that innervate the host's olfactory bulb.[41,51,52] Otic placodes from which ears arise have also been transplanted such that the eighth nerve that grows out of the graft penetrates the medulla at abnormal locations.[53] These axons too are able to navigate to their targets along novel pathways.[53]

When *Ambystoma* forelimbs are grafted a few segments rostral or caudal to their normal position, the brachial spinal nerves that innervate these limbs travel in unusual rostral or caudal pathways to innervate them. Similarly, isolated spinal brachial segments, transplanted as isolated pieces to a region near a limb, can form nerves that travel along abnormal courses to their appropriate targets.[1] All these situations indicate that long-range cues may be used by axons for navigation. Neurite outgrowth-promoting factors recently identified *in vitro* in several laboratories may, in fact, be these cues.[54–58]

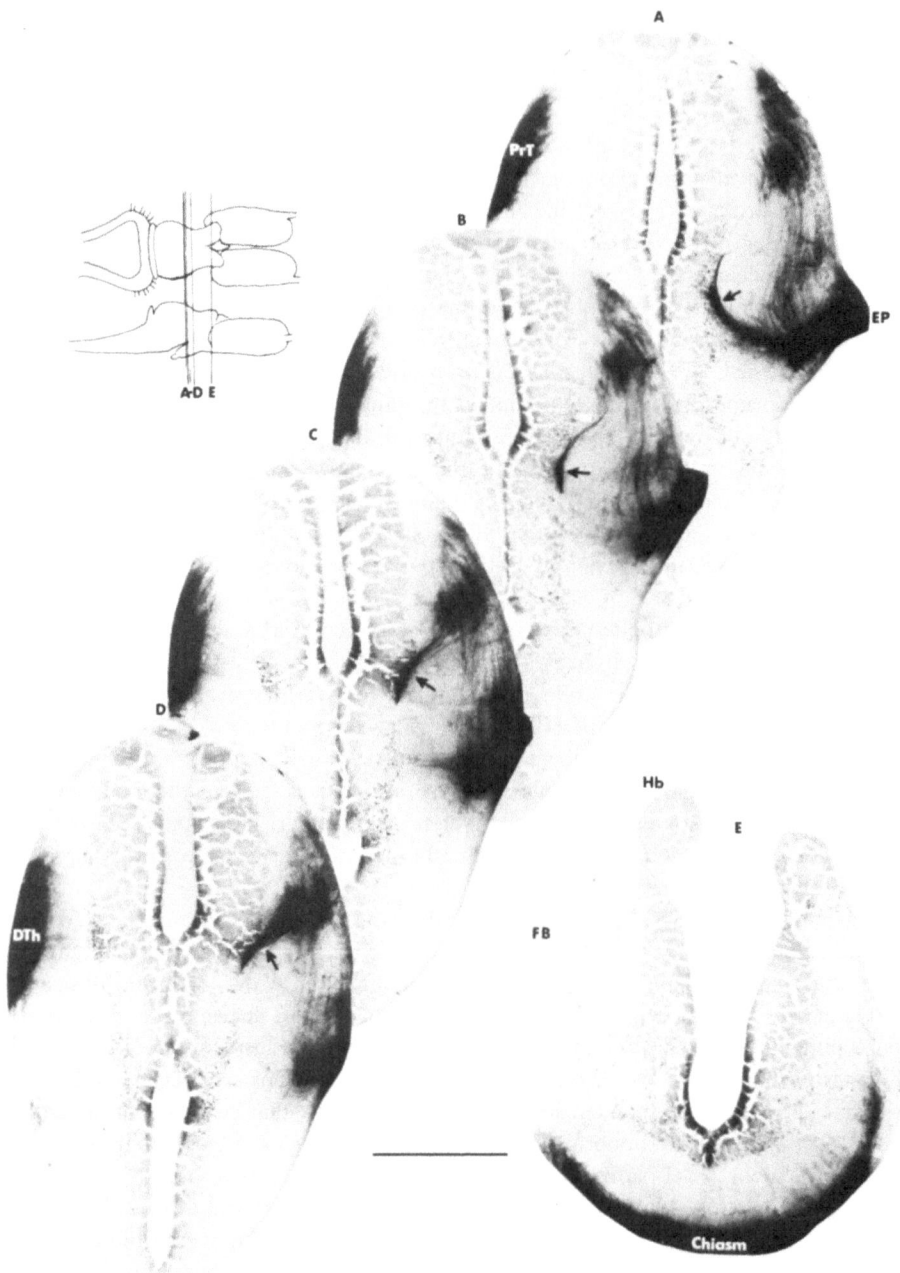

FIGURE 8. Example of bilateral projection from an eye transplanted to an eyeless mutant axolotl and labeled with HRP. EP indicates the entry point in section A near the beginning of the mesencephalon. A, B, C, and D are serial sections showing the internal trajectory (arrows) of the nerve fibers as they curve toward their final targets in the dorsal thalamus (DTh) and pretectum (PrT). In section E, the chiasm or commissure by which the labeled axons traverse from one side of the brain to the other is shown. Hb, habenula; FB, forebrain. The inset shows the approximate location of the sections. Scale: 500 μm. From Ref. 50.

4.2. Secondary Targets

If optic nerves from transplanted embryonic eyes penetrate the spinal cord, the medulla, or even the posterior mesencephalon, their axons, instead of reaching the tectum, travel along particular dorsolateral tracts at the surface of the spinal cord[48-50,59-61] (Fig. 9). It may be significant that (1) these tracts carry other sensory axons[60] and (2) most of the primary visual centers in the diencephalon and mesencephalon are dorsolateral and superficial.[59]

When ear placodes are put in place of eye primordia, ears develop where eyes would have. The auditory nerve that develops from these transplanted ears enters the CNS abnormally through the diencephalon. While the final destination of the axons is not known, the initial branching pattern of these fibers in the diencephalon closely resembles the branching pattern that the eighth nerve normally makes when it enters the medulla (Constantine-Patton, personal communication). Thus, either these axons are intrinsically programmed to branch in a particular ways regardless of the nature of the tissue they innervate, or perhaps more likely, the cues in the CNS that lead to eighth nerve fiber branching in the medulla are repeated in the diencephalon even though these fibers normally never pass through the latter. Since optic[49] or olfactory[41] axons do not show similar branching patterns nor do they replicate the other's pattern when they are made to enter the same region of the diencephalon, each set of axons may follow different cues.

In many cases, transplanted olfactory placodes send out axons that preferentially innervate the dorsal diencephalon,[41,62] and this is especially true when the transplant is located more caudally on the head. More caudally still, the transplanted axons tend to innervate the myelencephalon.[41]

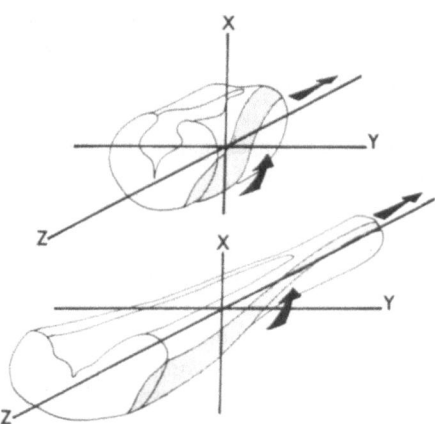

FIGURE 9. The relationship between optic tract growth and the three axes of the CNS. (Top) Schematic diagram of the anuran diencephalon with the medial–lateral Y axis, the dorsal–ventral X axis, and the longitudinal Z axis superimposed. The darkened area represents the growth trajectory of the normal optic tract subsequent to its decussation at the optic chiasm. (Bottom) Schematic diagram of the anuran medulla oblongata and the anterior spinal cord with the same axes superimposed. The darkened area represents the growth trajectory of the transplanted optic tract. From Ref. 48.

If forelimb buds are transplanted more than about four segments caudally, brachial spinal nerves will not find them but rather will innervate nearby axial musculature.[1] Isolated spinal lengths not transplanted sufficiently near to target limbs will fail to innervate them.[1] In all of these cases, we see axons choosing secondary targets, targets that they innervate only when the primary targets are inaccessible, usually as a result of being too far displaced. Similar results have been found for extra wings in homeotic mutants of *Drosophila*. Also, in mammals, when the primary visual centers are removed from neonatal hamster brains, the eyes usually form projections to auditory centers, targets of secondary preference.[63]

4.3. Direction of Growth

The Mauthner cells are dealt with separately here because results from these transplants seem easier to interpret in terms of the direction and choice of the axonal pathway rather than target seeking. If a salamander or frog bilateral medullary segment containing both Mauthner cells is rotated 180°, left and right Mauthner cells are reversed, and as was mentioned above, their axons start out rostromedially instead of caudomedially.[17,35,36] In many cases, however, when these axons exit the rostral end of the transplanted tissue, they make an "about face" and head back toward the graft[17,35,36] (Fig. 10). Thus, while within the graft they are traveling caudally with respect to the tissue that surrounds them, when they leave the graft they turn and again head caudally, this time with respect to the host tissue that they have entered. But when they then enter the grafted region once more, they do not again change direction. They travel through the graft toward the tail, therefore heading rostrally with respect to the grafted tissue. The simple interpretation[64] of this phenomenon is that Mauthner cell axons usually follow a rostrocaudal gradient. When the axons are first forming in the graft, the prevailing gradient is determined by grafted tissue, but as they leave this tissue they encounter the nonrotated gradient of the rest of the nervous system. This gradient is thought eventually to overpower that in the graft so that when the axons reenter this tissue they can travel without interruption through it to the tail. If this were so, then one would not expect Mauthner axons to be able to travel against the prevailing gradient for very long before being forced to turn around. To test this idea, Janus-like embryos were created by fusing the transected spinal cords of two embryos[65] (Fig. 11). Mauthner cells from each head start toward the junction, and once they cross this junction, they encounter the supposed reversed gradient from the other embryo. If the rostrocaudal axis of the spinal cord is critical to the direction of Mauthner cell growth, the axons should reverse directions and head back toward the junction, and perhaps terminate there. But this is not what happens. The Mauthner cell axons simply ignore the junction and grow rostrally in the upper spinal cord of the other half-embryo.[65] Throughout the entire double cord, two Mauthner cell axons from one head are traveling in one direction and the two from the other head are traveling in basically the same tract but in the reverse direction (Fig. 12). From this experiment, it is clear that Mauthner axons are able to grow for long distances in a reversed direction.

This result is reminiscent of those from eye transplant experiments, in that when

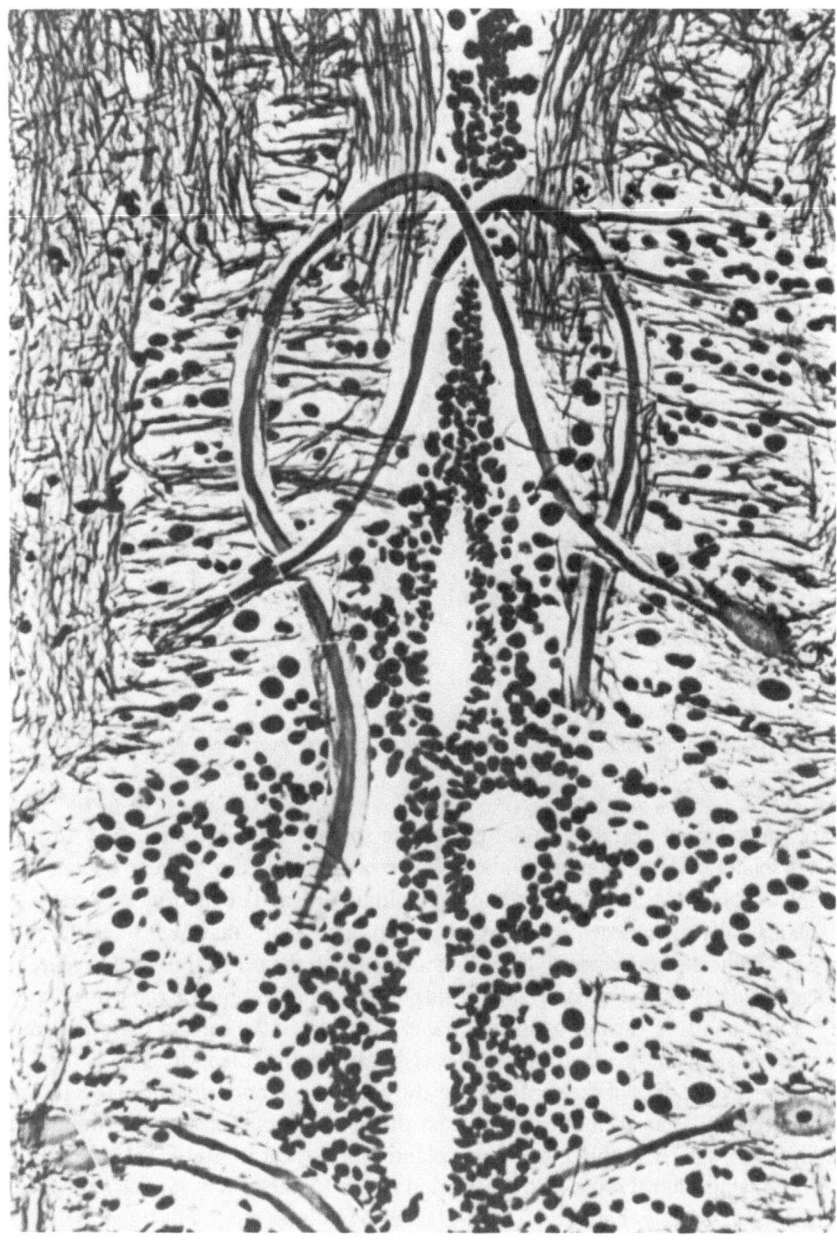

FIGURE 10. A composite photograph combining successive serial sections to show the recurving Mauthner cell axons in an animal having a graft in reversed anteroposterior orientation. The point at which the axons recurve corresponds approximately to the orignal posterior end of the grafted segment. From Ref. 36.

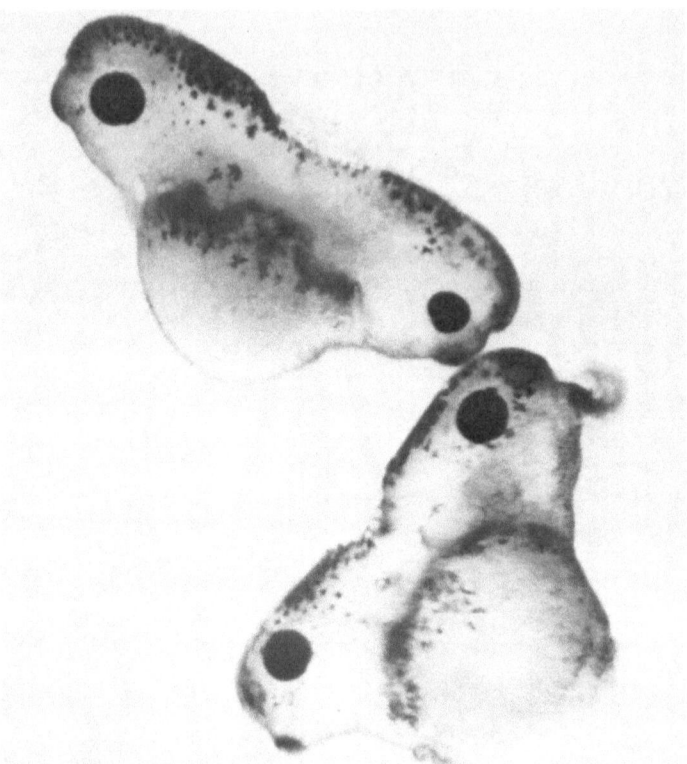

FIGURE 11. Living Janus telobionts, lateral aspect, 3 days postoperatively. From Ref. 65.

eyes are transplanted near the caudal end of the spinal cord, optic axons run rostrally in the dorsolateral tracts mentioned above.[60] If eyes are grafted to the ear region, however, the transplanted axons grow into the spinal cord and then travel caudally.[42,48] Transplants halfway between the two give rise to optic axons that may travel in either direction.[59] Some of these rostrally traveling axons reach the cerebellum, traverse it, and then head caudally down the contralateral side of the spinal cord in the same dorsolateral tract.[59] Thus, axons of various sorts may travel up or down the cord in preferred tracts. Similar conclusions have come from studies of homeotic mutants in insects.[13,14]

The transverse axes of the cord, such as the dorsolateral and superficial–deep placement of the tract, seem to matter more than the rostrocaudal axis in determining the direction of growth. Mauthner cells transplanted to the cord have axons that tend to grow in the same ventral cross-sectional region of the spinal cord as do normal Mauthner cells.[66] Occasionally, however, a transplanted Mauthner axon will travel in a more dorsal pathway.[66] The fact that transplanted optic and Mauthner cell axons tend to travel repeatedly in particular pathways has led some to think that there might be only a few embryonically determined substrate pathways in the CNS into which axons are somehow channeled.[67]

What then is the explanation for the Janus results? One possibility is that there is indeed a rostrocaudal gradient of an adhesivity or growth-promoting factor in the tracts that usually carry the Mauthner cell axons.[65] When axons first exit the cell body, they head medially and, because of the rostrocaudal gradient, they bend caudally as they decussate. In the Janus experiment, unlike the case of the rotated medulla, the axons are oriented along the direction of the gradient when they encounter the transplant border. Since no component of their growth is perpendicular to the gradient at this point,

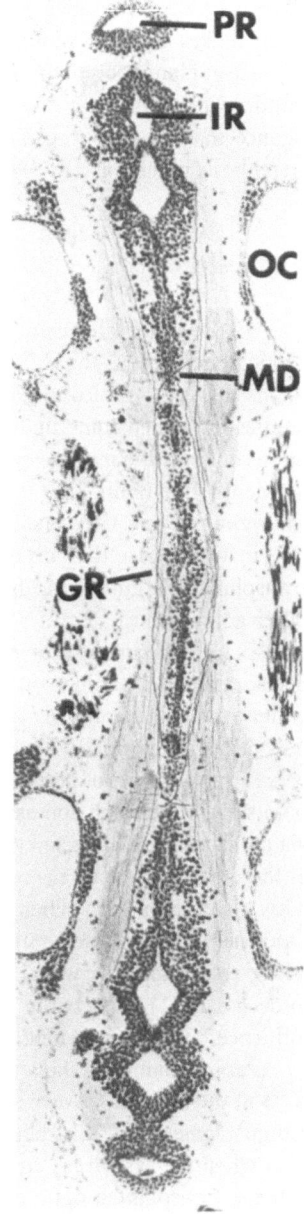

FIGURE 12. Composite photograph of Bodian preparation showing Mauthner axons in a Janus telobiont. Fixed 8 days postoperatively. PR, preoptic recess, IR, infundibular recess, OC, otic capsule, MD, Mauthner decussation; GR, graft. From Ref. 65.

the axons cannot bend, and so they grow straight but in the wrong direction. This hypothesis predicts that the growing axons advance more slowly in the reverse than in the normal direction.

Other ideas on Mauthner cell axonal navigation come from experiments with unilateral rotations of the presumptive medulla.[35] These studies show a high incidence of irregular Mauthner axon growth in the rotated sectors. Often these axons run rostrally all the way to the forebrain. A possible explanation for these results is that the pathway chosen by the Mauthner cell may depend on the origin of the fibers with which this axon makes the first contact. Mauthner cell axons, according to this idea, are using contact guidance for pathway formation. Unilateral rotations cause greater tissue disruption medially, near the site of Mauthner cell axon outgrowth, and increase the chance that an inappropriate fascicle of fibers will guide the developing axon. This hypothesis is reminiscent of that proposed for pioneer fibers in insects.[68]

4.4. Decussation

In the vertebrate brain, a number of axons and axon tracts decussate at the midline. Optic fibers cross at the chiasm, and Mauthner cell axons cross in the medulla. How do axons make the choice to go to one or the other side of the midline? Transplantation studies have shed some light on this process. The majority, but not all Mauthner cell axons do cross the midline even when (1) they are transplanted in an isolated medullary segment to the flank,[69] (2) they are rotated 180° rostrocaudally so that the left cell is on the right side and vice versa,[35,36] and (3) their axons grow rostrally into the diencephalon, in which case they decussate in the optic chiasm or posterior commisure.[36] Some transplanted Mauthner cells bifurcate with one branch remaining isilateral and the other going contralateral.[35,36] These data are comparable to results obtained from eye transplant experiments; for example, in amphibians there is normally almost complete decussation of optic fibers at the chiasm. Eyes transplanted such that their optic nerves either join the optic nerves of a nearby host eye or enter the brain in the telencephalon will decussate normally at the chiasm, even if they are right eyes on left sides or vice versa[49] (and Hibbard, personal communication). The entry of optic nerves at abnormal points in the diencephalon usually leads to bilateral projections (contralateral component via the chiasm). This is so even when the transplanted eye comes from the opposite side of the donor.[49,50] Optic nerves that penetrate the midbrain tend to project ipsilaterally. This also occurs even when the transplanted eye is taken from the opposite side.[49,50] Optic nerve fibers growing up the spinal cord may decussate across the cerebellar ridge.[59] These results together with those from using Mauthner cells suggest that the propensity to decussate (1) is probably intrinsic to the neurons, (2) is greater than any left–right influence, (3) is greatly modified by local tissues, and (4) is not absolute.

Albino mammals show a defect in optic fiber decussation such that too many fibers cross at the chiasm.[70] Every species of mammal with genetic variants leading to reduced retinal pigment has this defect, yet it is not found in albino amphibians.[71] Moreover, even when the system is "stressed" using albino eyes transplanted to albino hosts such that the transplanted optic nerve entry point is mesencephalic, the projection remains totally ipsilateral, as in normal transplants.[72] Thus, there seems to be some difference between mammals and amphibians such that the absence of pigment causes decussation aberrations in one class but not the other.

4.5. Physiological Activity

The role of impulse activity in axonal navigation has been examined by transplantations from the Mexican axolotl to the California newt.[43,73] The axolot's nervous system is sensitive to the neurotoxin tetrodotoxin (TTX), which blocks voltage-dependent sodium channels. In contrast, the newt manufactures TTX while being relatively insensitive to it.[43,73] Axolotl-to-newt eye transplantations (Fig. 13) result in optic fibers that never carry normal TTX-sensitive action potentials and yet can navigate to their proper targets in the newt host[43] (Fig. 14). Thus, sodium action potentials appear not to be involved in this process.

Calcium action potentials, however, which are found in some embryonic neurons at initial stages of axon outgrowth in vitro,[74] and which recent work has shown probably also occur in embryonic Xenopus retinal ganglion cells (Taylor, personal communication), may in fact play a role in axonal navigation. Paramecia use Ca^{2+} action potentials for chemotaxis,[75] Ca^{2+} influxes correlate with the direction of pseudopodial extension in amoebae,[76] and current influxes have recently been shown to correlate with the direction of filopodial extension in retinal growth cones.[77] Some regenerating axons,[78] axons in culture,[79] and nerve-growth-factor-transformed PC-12 cells[80] have Ca^{2+} action potentials at their growth cones. Furthermore, some mammalian sensory axons in vitro turn toward localized sources of Ca^{2+} in the presence of A23187, a Ca^{2+} ionophore.[81] A23187 also accelerates optic nerve regeneration in fish.[82] It would be interesting, therefore, to block Ca^{2+} action potentials in parts of the developing nervous system to see the effect on axonal navigation.

4.6. Common Cues

The axolotl–newt transplantation experiment demonstrates another point, which is that axons from one species of amphibian can find appropriate targets in another

FIGURE 13. Method of transplanting axolotl (A) eyes to newts (T). (Top) The eye cup is excised along with the entire optic stalk. (Middle) Wounds are made in the host's epidermis and neuroepithelium, dorsal and posterior to the eye on the same side as the one being transplanted. (Bottom) The transplant is put into place. Underneath each embryo is a schematic cross-section of the operation. From Ref. 43.

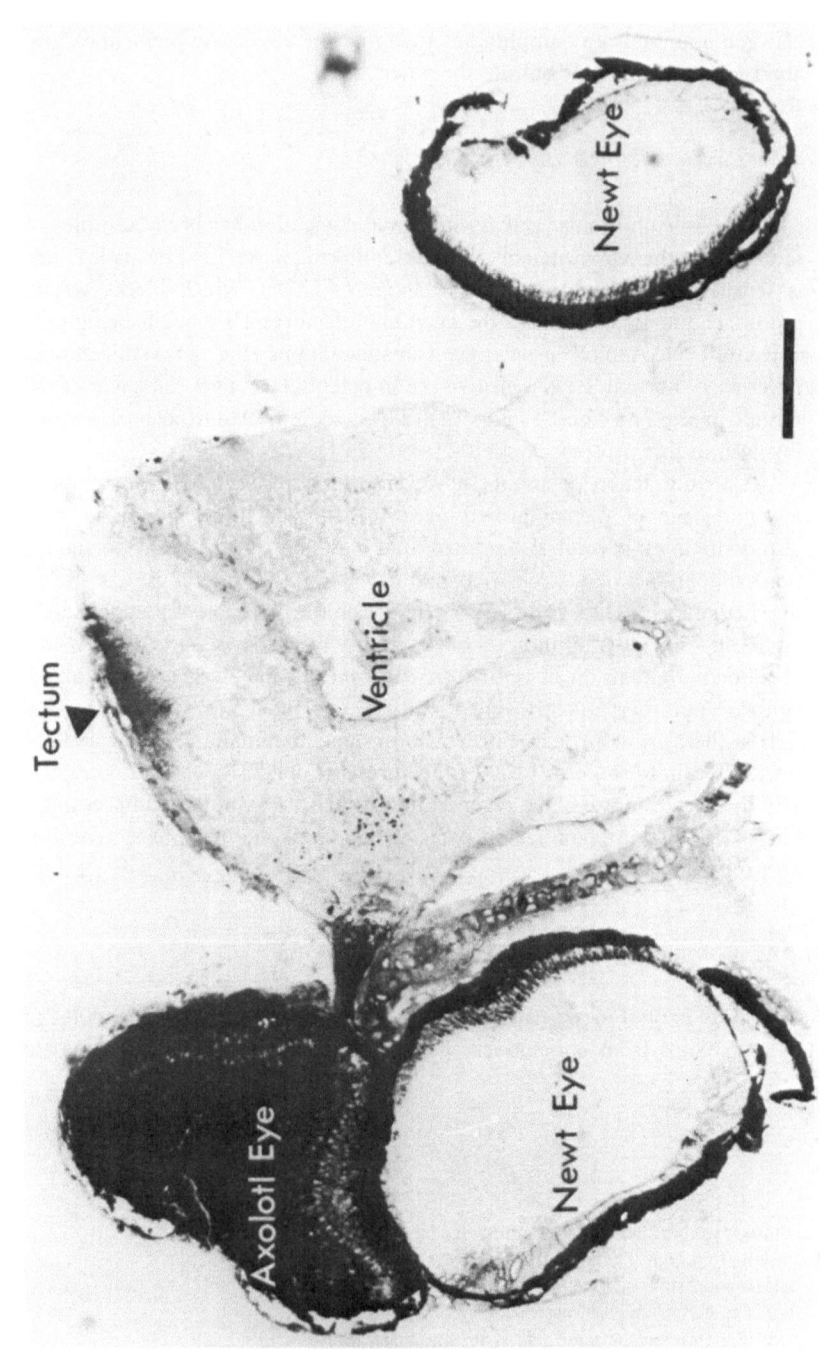

FIGURE 14. Axolotl eye projecting to the tectum of a 1.3-cm newt. All three eyes are visible in this cross-section autoradiograph of the brain, but only the

species.[43] *Xenopus*-to-*Rana* retinal transplants are also successful in this way (Ferguson and Holt, personal communication). Cross-species transplantation followed by correct axonal navigation has also been shown for the spinal-limb system. Such results indicate that whatever cues are used to guide axons to their targets, they are not necessarily species specific. Until we discover what these cues actually are, it will be worthwhile to examine how large a species divergence will allow appropriate axonal navigation to a target.

5. The Formation of Specific Connections

5.1. Pathways and Fiber Interactions

The topography of the retinotectal projection is not dependent on the pathway that retinal fibers use to find the target. This is known from electrophysiological and anatomical studies of animals with eyes transplanted to abnormal regions[49,50,83] No matter where the transplanted optic nerve penetrates the brain, if a projection is made to the tectum, the retinotectal map is normally topographic. The same conclusion has been reached through regeneration studies (see below).

It has been suggested that the relative ordering of optic fibers in the optic nerve and tract is largely responsible for the topographic mapping of the retinotectal projection.[84] Although ordering probably helps to establish tectal topography in normal development, transplantation experiments have shown that it is not crucial. Compound eyes can be created with two ventral, nasal, or temporal halves. When the course of axons from such compound eyes is followed, fibers arising from a particular quadrant of the compound retina are found to consistently enter the same quadrant of the optic nerve head irrespective of their origin.[85] Thus, fibers from any retinal sector can be forced into abnormal positions in the optic nerve. Yet when these fibers innervate the tectum, they do so according to their original and not their transplanted retinal locations. In fact, by the time the transplanted axons have reached the optic tracts, they are distributed in the appropriate brachia for their final terminations.[85] Thus, fibers from a double ventral compound eye will approach the tectum through the medial brachium, and at first terminate only in the medial half of the tectum. These fibers have obviously received cues somewhere along their course from the eye to the optic tract that allow them to choose the most direct route to the proper terminations. Ectopic transplantation studies (see above) and regeneration studies (see below) have shown that axons need not even be in the appropriate brachium to find their normal position in the target.

5.2. Timing

From the above results, it is clear that the position of an optic fiber in the retina, the optic nerve, or the optic tract up to the anterior tectum does not determine its eventual site of termination. The argument has often been made, however, that timing as well as position is important for the formation of topographic maps.[84] Indeed, in certain arthropods, delaying the maturation of parts of the retina causes abnormalities in the development of the retinal projection.[86,87]

In one experiment on the role of timing in the development of the *Xenopus* retinal projection, an embryo was enucleated, resulting in an animal with one tectum that had never received optic fiers, and in larval life the remaining retinal projection was deflected to this visually "virgin" tectum.[88] This affected the delayed innervation of the tectum by regenerating ipsilateral fibers. The result of the experiment was a normal topographic retinotectal projection. There are certain obvious difficulties in using the above result to rule out a role for temporal factors in the development of topography: (1) regeneration is not development, and (2) temporal relationships within the retina were not disturbed in this experiment.

An experiment that is now under way will more directly address the role that timing of axonal outgrowth plays. This study involves replacing a particular retinal sector from an older embryo with one from a younger embryo. At the time of transplantation, neither retina had begun axonal outgrowth. Preliminary results (Holt, personal communication) show than the transplanted sector indeed is delayed with respect to outgrowth of axons and innervation of the target, yet makes an appropriate topographic projection.

5.3. Activity

Activity in the form of TTX-sensitive action potentials has been ruled out as a cue for axonal navigation to a target, and can also be ruled out for the formation of topographical connections. First, axolotl eyes transplanted to newts do form topographic retinotectal connections in the absence of impulse activity.[43] Second, experiments in which *Xenopus* are raised in a stroboscopic illumination, which is likely to synchronize the activity of retinal fibers, show that these fibers develop a normal topographic projection.[89]

This is not to say that activity plays no role in the formation of neuronal connections. The stroboscopic experiment revealed abnormalities in the layering, if not the topography, of visual responses in the tectum,[89] and the axolotl-to-newt transplant assessed the retinotectal map only roughly, with anatomical techniques.[43] Recent results show that the order of the retinotectal map is less precise if the eye is injected with TTX during optic nerve regeneration.[90,91] In addition, application of α-bungarotoxin, a blocker of the ACh receptor, to a patch of the tectal surface causes retinal terminals to leave the treated area,[92,93] whereas if TTX is simultaneously injected into the eye, this shift does not take place.[94] Some studies have claimed that compression of the retinotectal projection (see Section 7) does not occur in animals deprived of darkness,[95] although others have not confirmed this result.[96] It has also been shown that activity is necessary for the segregation of eye-specific termination bands both in the tectum of lower vertebrates[97,98] and in the visual cortex of mammals.[99]

5.4. Secondary Selection

It is possible that selective innervation of the target is a result of an initially random formation of connections throughout the entire target area followed by retraction of inappropriate connections. Indeed, in developing mammalian[100–103] and avian[104] brains, there is evidence for an overproduction of connections followed by secondary selection of only those that conform to the normal adult pattern. Exuberance in embryonic projections does not imply randomness. For instance, the visual callosal projection in mammals is exuberant and may also be topographic (Shatz and LeVay, personal communication). Yet, in other systems there is evidence that the early exuberant projection also contains mistakes or random connections.[100] The same studies indicate that the random projections may be a minor component of an at least grossly topographic set of connections.[104]

Since none of these investigations has examined the very first projection to the target, the possibility remains that the initial projection is indeed random, and that by the time of these studies, some secondary selection has taken place. The problem in addressing this hypothesis lies in the difficulty associated with studying early embryonic projections anatomically and physiologically. Recently, however, a technique has been developed (Figs. 15 and 16) by which embryonic tissue can be excised, labeled *in vitro*

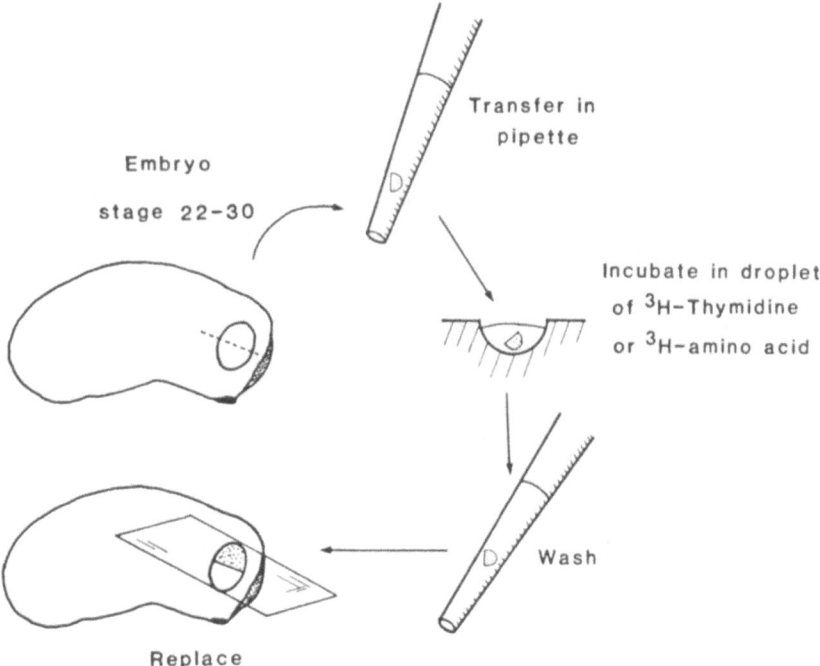

FIGURE 15. Diagram illustrating the procedure for incubating embryonic eye tissue in isotopes. From Ref. 10.

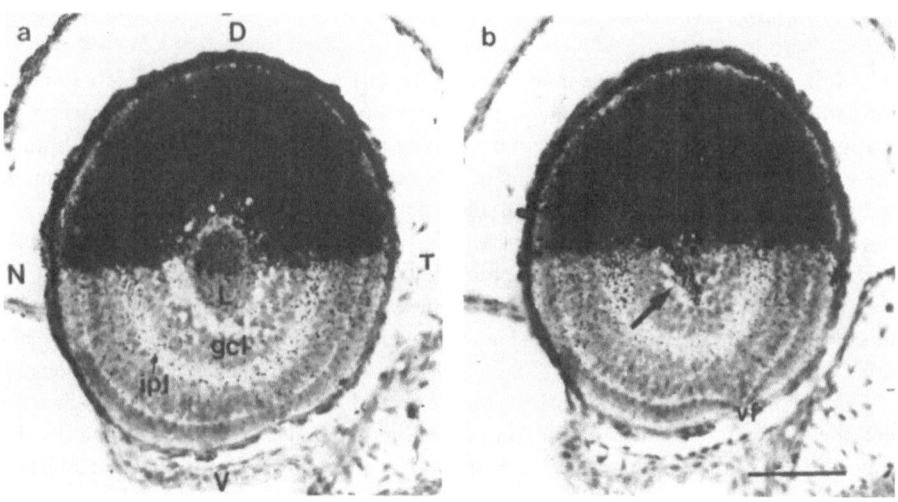

FIGURE 16. Light-field autoradiographs of an eye (stage 43) following [³H]proline incubation of the dorsal half at stage 27/8. (a) Sagittal section through the retina and lens. The label is confined to the dorsal half of the retina and lens with a sharp boundary between labeled and unlabeled cells. Some label is seen in the inner plexiform layer extending ventrally from the labeled boundary. (b) Sagittal section behind the lens through the ganglion cell axonal layer; arrow indicates labeled axons coursing toward the ventral fissure and optic nerve head. gcl, ganglion cell layer; ipl, inner plexiform layer; L, lens; vf, position of ventral fissure; D, dorsal; V, ventral. Scale: 100 μm. From Ref. 10.

with [³H]proline, and then replaced in the embryo before the time of initial axonal outgrowth.[10,11] The result of this procedure is that axons are radioactively labeled as they first grow out. Such experiments can be done easily in amphibians. *In vitro* labeling of sectors of the embryonic *Xenopus* retina clearly established that the initial projection to the tectum was, in fact, topographically ordered[10,11] (Fig. 17). Earlier electrophysiological experiments failed to reveal this order, but a recent electrophysiological investigation using more suitable stimulus presentation confirmed the *in vitro* labeling experiment.[11] Thus, the development of the topographic retinotectal projection is not the result of exuberant random termination followed by secondary selection. The retraction of inappropriate connections might nevertheless be involved in the refinement of the adult map.

5.5 Stripes

A surprising discovery was made when third eyes were embryonically transplanted in frogs. Tecta, whether ipsilateral or contralateral to the transplant, that were innervated by both the host eye and the extra eye showed a pattern of stripes, not unlike the ocular dominance columns demonstrated for the geniculocorcortical projection in the visual system of mammals[105] (Fig. 18). The host and transplanted retinal projections divided the primary visual neuropil of the tectum into alternating stripes of about 250 μm in width and running rostrocaudally.[83,105] The topographies of the two projections were normal.[83] Subsequently, it has been shown that stripes form in salamanders with dual tectal innervation (personal observation) (Fig. 19), in frogs that have dual tectal innervation due to either disruption of the optic chiasm[106] or removal of one tectum,[107] and in fish due to translocation of retinal fibers or unilateral tectal excision.[108] Indeed, it seems that whenever two eyes innervate a single tectal lobe, eye-specific stripes of terminals are formed. What causes them to develop? Is it that the two eyes have different biochemical labels such that fibers with similar topographic labels can still be distinguished? The answer seems to be no. First, two left eyes from embryos of the same spawning produce stripes, indicating that a left–right difference is not responsible[83] (and personal observation). Second, compound eyes (e.g., double ventral eyes) produce stripes, indicating that retinal ganglion cells even within one globe can somehow be distinguished[109] (Fig. 20). Third, duplicated eyes that arise from embryonic transection of a developing primordium can produce stripes[110] (Fig. 21). This rules out different early lineages as the mechanism for distinction.

Stripes do not develop initially. Instead, early dual innervation is overlapping and stripes segregate with time (Constantine-Paton, personal communication). This too

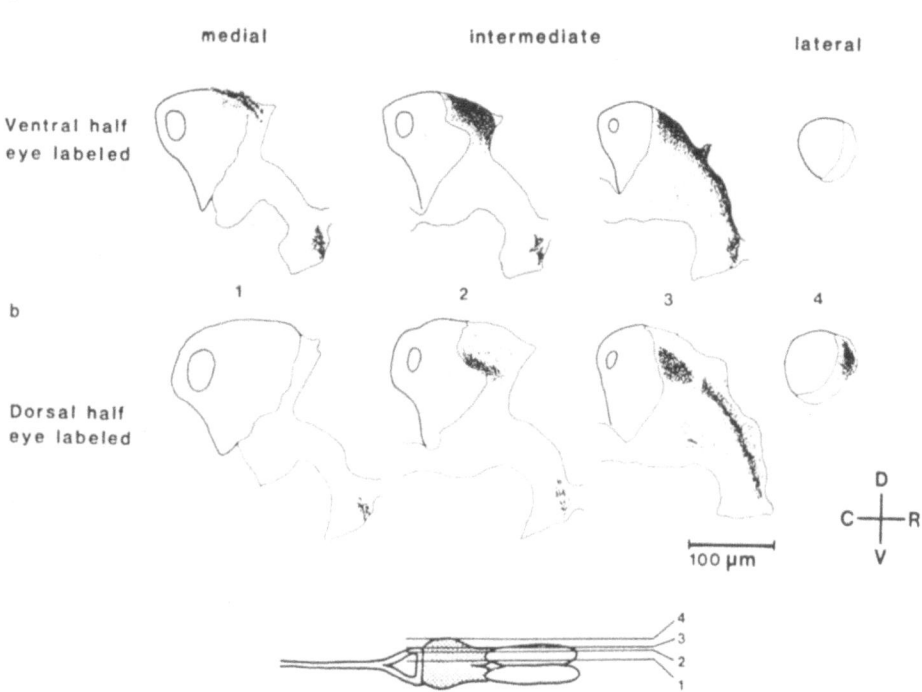

FIGURE 17. Tracings of (a) dorsally and (b) ventrally labeled optic projections at stages 40 (b) and 41 (a). Sections 1, 2, 3, and 4 are at the levels indicated in the whole brain outline. The ventral half of the retina projects to the medial tectal neuropil and the dorsal half to the lateral tectal neuropil. In the optic tract (sections 3), dorsal fibers are located ventrally and ventral fibers dorsally. From Ref. 10.

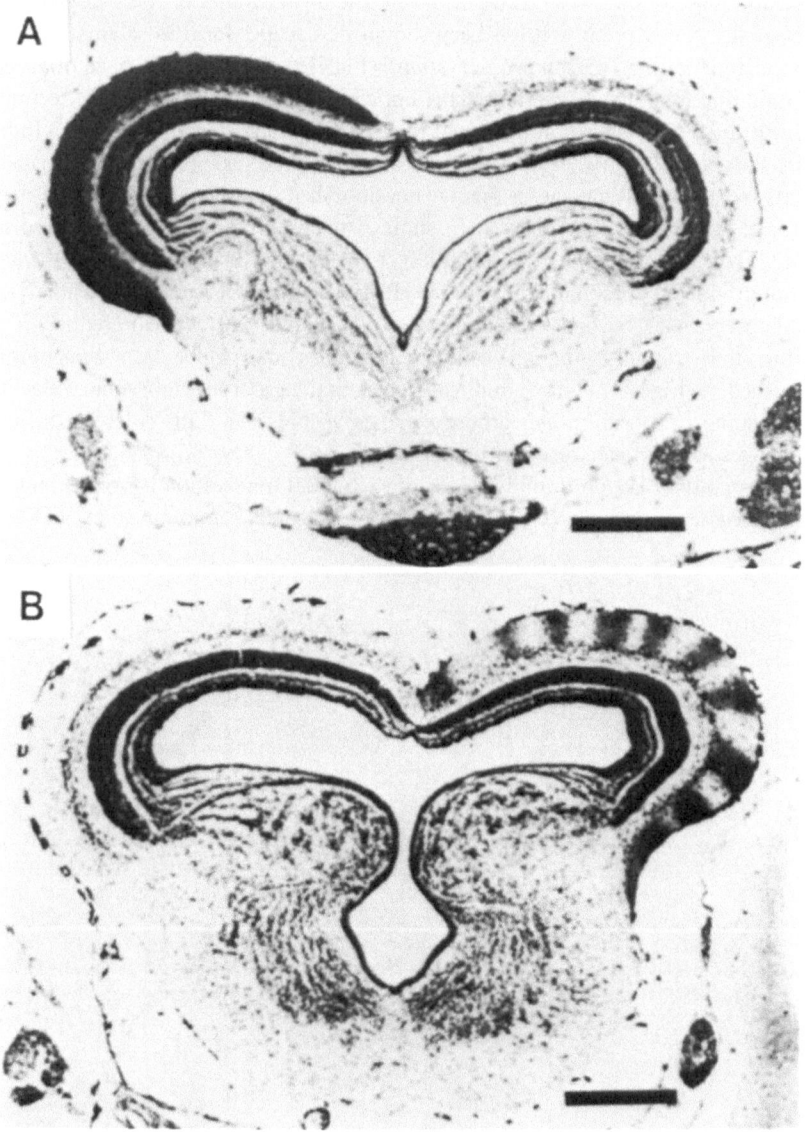

FIGURE 18. (A) Autoradiograph of a coronal section through the midbrain of a normal frog. This animal's eye was injected with 10 μCi of [^{3}H]proline 3 days before it was killed. The entire superficial neuropil of the right optic lobe (left side of the picture) is filled with developed silver grains indicating the region occupied by synaptic terminals from the labeled (contralateral) eye. (B) Autoradiograph of a coronal section through the midbrain of a three-eyed frog following intraocular injection of 10 μCi of [^{3}H]proline into its normal right eye. The left optic lobe of this animal was innervated by the labeled eye plus the supernumerary eye. The normally continuous retinotectal synaptic zone of the contralateral eye has consequently been divided into regularly spaced bands of terminal neuropil. Bar: 400 μm. From Ref. 176.

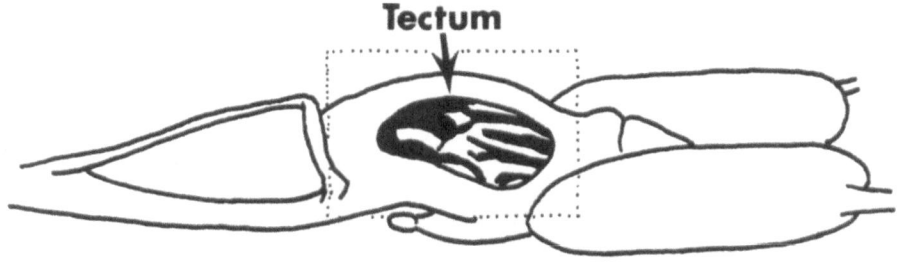

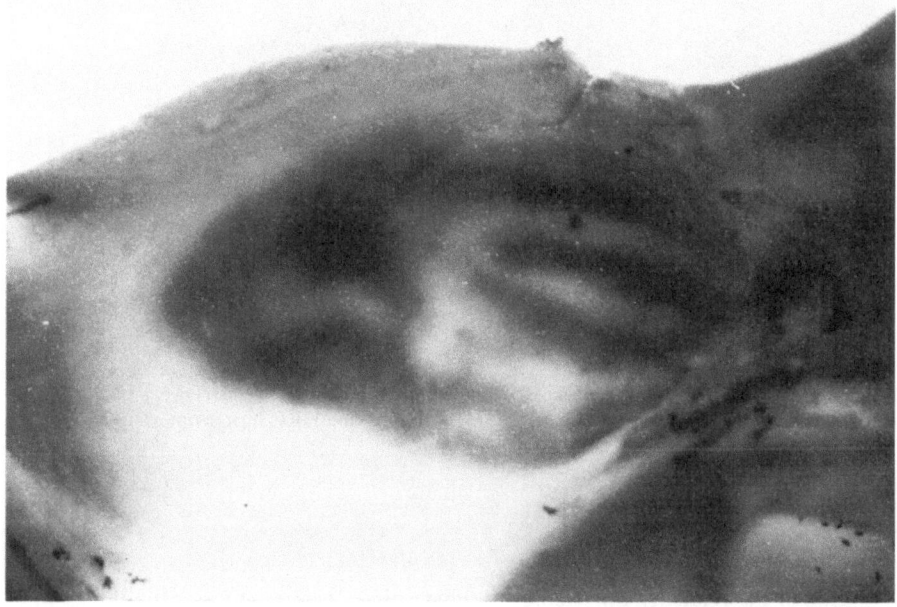

FIGURE 19. Stripes on a three-eyed axolotl's tectum. Whole mount procedure uing HRP. Drawing shows location of the labeled tectum.

mimics the development of ocular dominance columns in the mammalian cortex.[101,111] The late shifting of terminals suggests mechanisms similar to those responsible for the shifting of terminals in an α-bungarotoxin-treated area. These, as was argued earlier, might be activity dependent. Indeed, ocular dominance columns in the kitten cortex do not develop when retinal activity is blocked with TTX injected binocularly.[99] Similar experiments in fish also indicate that activity is necessary for stripe formation[97,98] (Fig. 22). Thus, it is possible that fibers with particular retinal coordinates tend to map to particular tectal coordinates, but their tendency to do so may be overwhelmed by the even stronger tendency of fibers with synchronous activity to innervate the same area of the tectum.

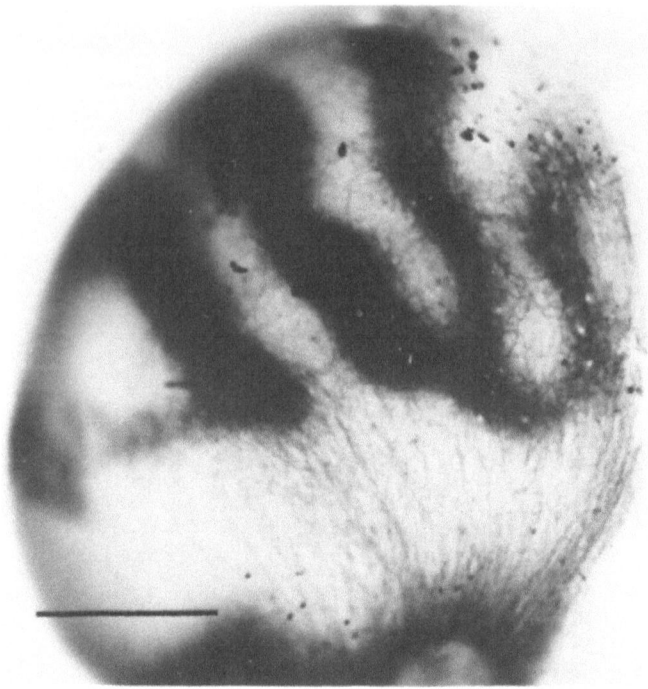

FIGURE 20. Dorsal view of a whole mount of a tectum innervated by a double nasal eye: medial tectum to the right, caudal to the top. The donor half of the eye was filled with HRP. Stripes run rostrocaudally: the fibers running to them are unstriped. Scale: 200 μm.

5.6. Induction of Polarity

One of the relativity few studies in which pieces of the central visual areas were transplanted in development has provided one of the most intriguing findings about the development of the retinotectal projection. The primordial *Xenopus* dorsal mesencephalon, diencephalon, or both were removed, rotated 180°, and replaced. The results indicate that the dorsal diencephalon somehow induces the polarity of the retinotectal projection[18,112,113] (Fig. 23). If the tectal primordium is rotated alone, electrophysiological data show that the retinotectal map remains unrotated. If, however, the primordial diencephalon is rotated along with the tectum, the diencephalon develops caudal to the mesencephalon, and the retinotectal map is reversed. Additional evidence for the role of the diencephalon in inducing the polarity of the retinotectal map comes from those instances in which the rotation splits the diencephalon. In this case, part of the diencephalon remains rostral and part develops caudal to the tectum. Electrophysiological studies on such animals indicate the presence of two tectal maps of opposing orientations.

How is it that the diencephalon can specify the polarity of the retinotectal map? Could it be that retinal axions pick up a map location while passing through the diencephalon on their way to the tectum? That this is unlikely can be seen by tracing retinal

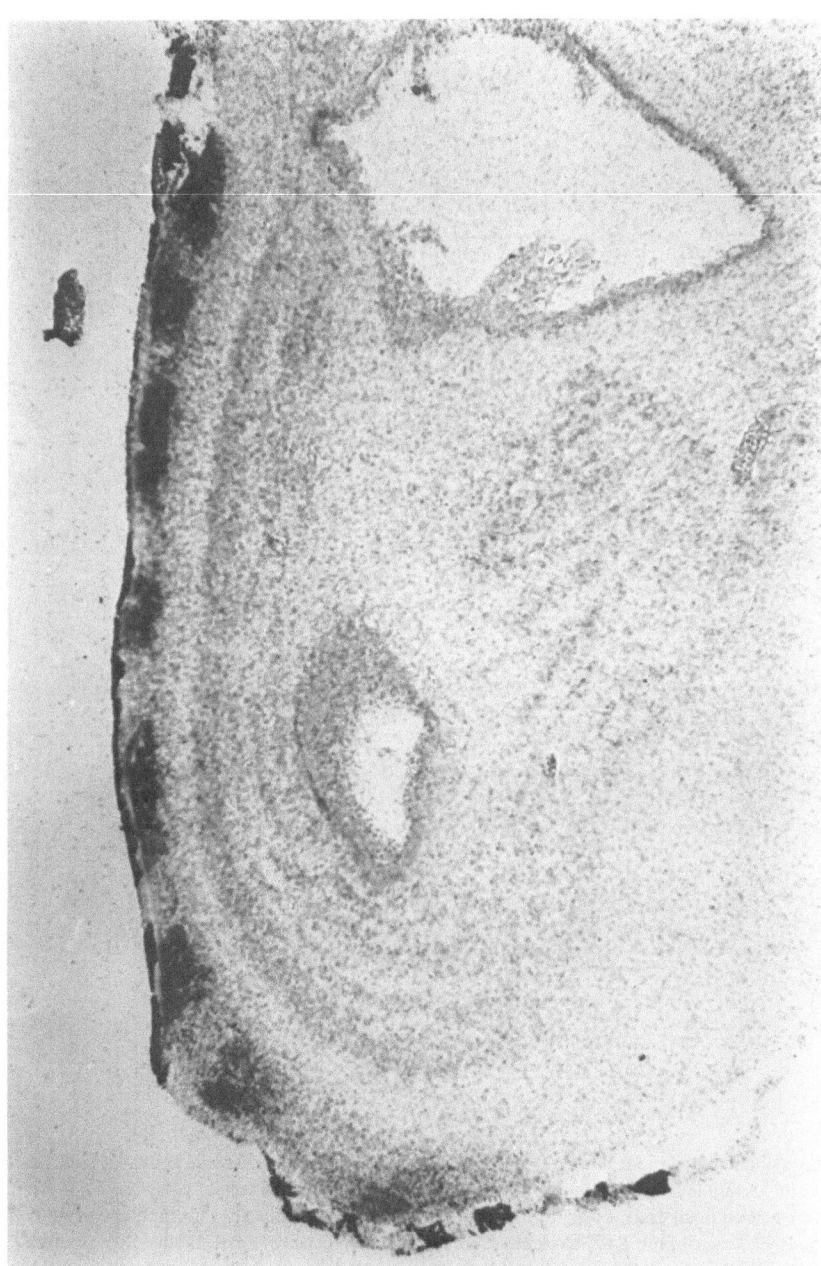

FIGURE 21. Projection pattern of retinal ganglion cells in the optic tectum assayed by autoradiographic techniques. Half-labeling the "temporal" half-retina of an isogenic double nasal (nasal 1/3) eye demonstrates label covering nearly the entire extent of the tectum, but distributed in a discontinuous manner. The clusters of the label are "stripes" cut in cross-section.

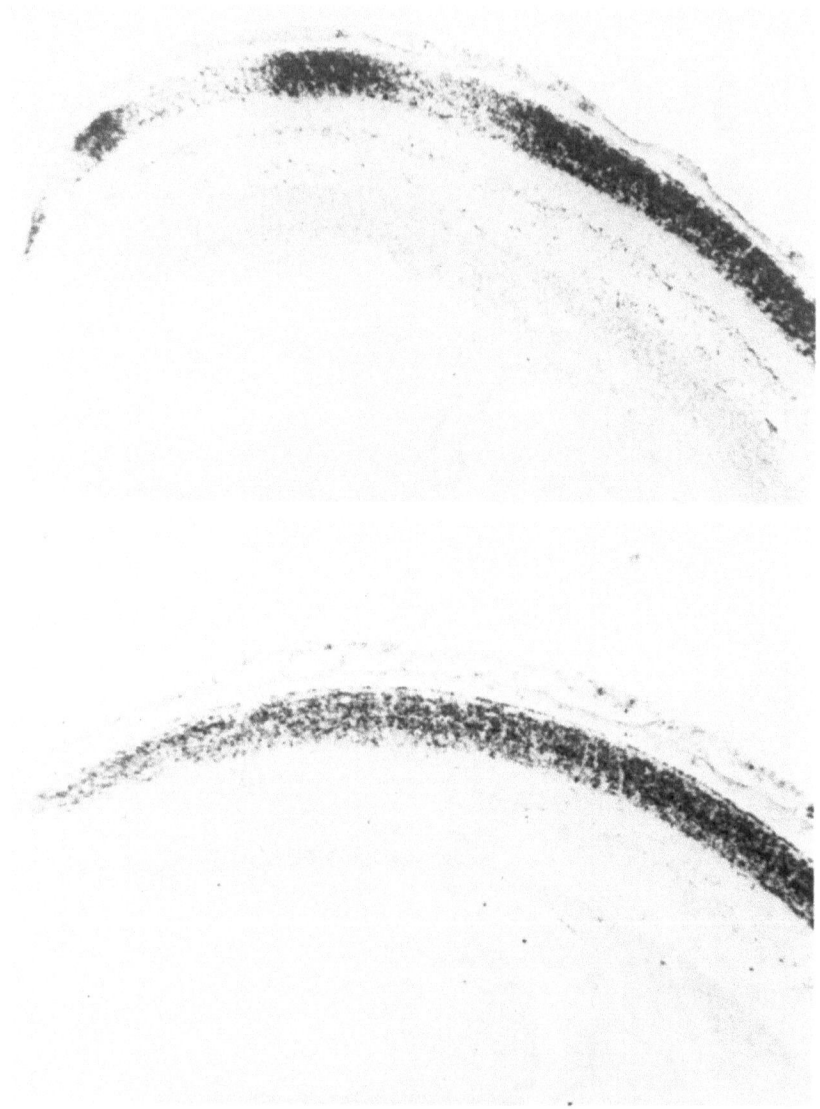

FIGURE 22. Frontal autoradiographs through the dorsal posterior host tectum. Dorsal is at the top and the medial edge of the tectum is at the extreme left. Most of the lateral half of the tectum is not shown. Host optic fibers were labeled in all cases. (Top) 61 days, subthreshold TTX. Notice the gaps in the main optic lamina. (Bottom) 90 days, chronic TTX blockade. Note the decreased density of medial label. Apparently, the label, which would have been confined to discrete columns of normal density, is now spread over a greater area of the tectum in lower density. From Ref. 97.

fibers from transplanted eyes injected with HRP. If an eye is transplanted to a genetically eyeless axolotl such that the optic nerve enters the middle or caudal mesencephalon, and if only a few axons are labeled, they can be followed as they enter the CNS and traverse directly to their tectal termination without having passed through the diencephalon (personal observation). Unless these fibers sent transient processes to the diencephalon, a normal retinotectal map was formed without retinal fibers having been instructed by direct diencephalic interactions. Experiments in which innervation of the tectum is delayed until larval development[88] also argue against the necessity of an optic fiber-diencephalon interaction. The mechanism by which this transfer of information from diencephalon to tectum takes place remains a mystery.

Embryonic hindbrain has been transplanted to a region where the tectum would normally develop (Sterling and Cooke, personal communication) (Fig. 24). If retinal fibers merely sought out appropriate CNS coordinates, or if the diencephalon could induce this abnormally located hindbrain to accept an orderly retinal map, this tissue should become visually innervated. Instead, what happens is that retinal fibers, after coursing through the diencephalon, turn ventrally to avoid the trasplanted medulla. If a piece of tectum lies caudal to the transplant, the retinal fibers find and innervate it.

5.7. Other Systems

Roles for timing and mechanical guidance have also been ruled out in the innervation of the Mauthner cell from eighth nerve fibers. Exchange of the ear-forming

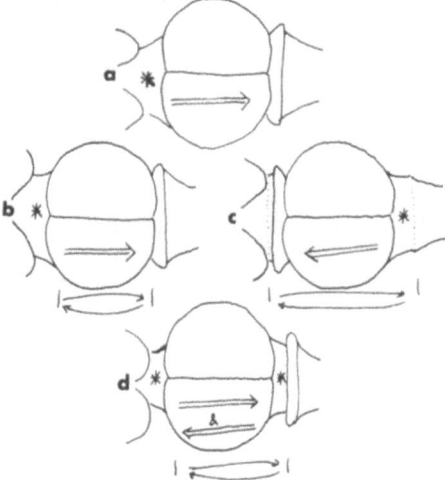

FIGURE 23. (a) Normal axis of a tectal map. Double arrow indicates an anteroposterior progression in the visual field. (b) The tectum has been embryonically rotated (see arrows underneath), but the diencephalon (*) has not. The map is normal. (c) Both tectum and diencephalon have been rotated. The map is reversed. (d) The tectum has been totally rotated but only the posterior part of the diencephalon has been rotated. A double map ensues. After Ref. 113.

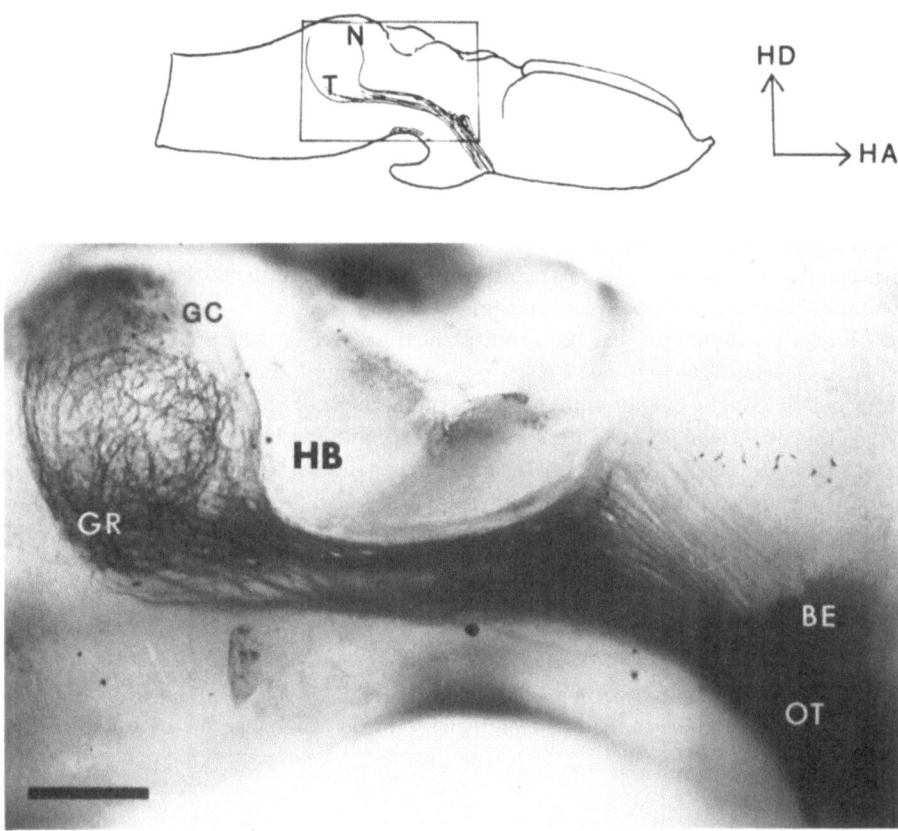

FIGURE 24. Lateral view of the brain of tectum-rotated animal. The visual map of the contralateral eye was recorded electrophysiologically before its optic nerve was filled with cobalt. T and N mark the positions to which the temporal and nasal retina projected. In normal animals, the temporal retina projects to the rostral tectum where the optic fibers enter. The graft of tectum and hindbrain has been rotated 120° so that the rostral tectum is oriented at an angle to the host brain axes, and its hindbrain (HB) lies between the tectum and the host diencephalon. The optic fibers course up the unrotated host optic tract (OT) and supply the host neuropil of Bellonci (BE); they then sweep laterally to "avoid" the grafted hindbrain to enter the grafted tectum at its rostral pole (GR), fanning out to reach its caudal pole (GC). Scale: 200 μm. Courtesy of V. Sterling.

region from older to younger embryos results in precocious synapse formation on the dendrites of the Mauthner axon, which receives vestibular nerve input.[114] Thus, the transplanted axons are able to find their appropriate targets in spite of abnormally timed axonal outgrowth. If otic placodes are shifted slightly so that their nerves grow into the brain at abnormal locations, ultrastructural studies demonstrate the existence of appropriate synapses on the Mauthner cell dendrites from the transplanted organ.[53] Furthermore, vestibular axons can find their way to synapse on experimentally displaced Mauthner cells,[115] or Mauthner cells of abnormal forebrain origin.[116]

Transplantation experiments indicate that development of specificity in avain neuromuscular connections shares many features with the formation of the retinotectal map. Motoneurons from particular motor columns of the spinal cord may find their appropriate muscles, either when the limb is rotated or when the spinal segments containing these neurons are rotated.[117] This indicates that the motoneurons are specified to innervate distinguishable targets. Such experiments along with others involving slight shifts of the limb bud, also show that axons can travel along abnormal pathways to reach their targets.[117] In addition, as with retinal fibers, motor axons seem to grow directly to their targets, taking appropriate nerve branches along the way. Observations of fibers labeled during axonal outgrowth in chick embryos rule out random outgrowth followed by secondary selection as a mechanism for the formation of specific connections.[117] Furthermore, neuromuscular synaptic activity is not involved in the specificity of these connections, as they can develop in the presence of cholinergic blockers.[118] Activity may, however, be involved in the refinement of these connections, for neuromuscular synaptic function allows the death of overproduced motoneurons, a few of which may be making inappropriate projections.[119]

Grafting limbs from a quickly developing to a slower developing species can lead to substantial differences in the timing of axonal outgrowth with respect to target maturation (see above). Limbs from slowly developing species become innervated and begin moving precociously, but appropriately, when grafted onto a more rapidly developing *Ambystoma* species.[47] In the reverse case, i.e., when the limb is from a rapidly developing animal, it is retarded in its innervation and initial movements. When movements do begin, however, they are normal.[47] These studies give a third line of evidence against a major role of timing in the developing of specific neural connections.

5.8. Peripheral Specification

Early studies on grafting an extra limb to the side of the embryo in reversed orientation, next to one in normal orientation, showed that the two limbs moved with synchronous flexions and extensions.[120-122] This was first interpreted to mean that motoneurons that would not usually innervate leg muscles not only were doing so, but were also somehow achieving functional precision. The explanation given was that each innervated muscle "informed" the innervating motoneurons of its identity. The "informed" neuron would then be able to develop the appropriate central connections for normal movements. This is the theory of myotypic specification.[122] Skin transplants gave results that led to similar conclusions for sensory axons. Here, frog belly skin was exchanged with back skin. After healing, if one irritated the frog on its back, in the region of the transplanted belly skin, it would wipe its belly. Conversely, when irritated on the belly, it would wipe its back.[123,124] This result was interpreted to mean that the skin transplants could inform the innervating sensory axons of their region of origin. The "informed" sensory axons could then make appropriate central connections for the reflex wiping response.

Both ideas of myotypic and sensory specifications were demolished as a result of later studies. The case against myotypic specification comes most strongly from an experiment that originally was thought to argue in its favor.[125] Following an operation in which flexor and extensor nerves were exchanged, limbs eventually regained their normal movements, yet the exchanged nerves appeared in place. Thus, it seemed as though the muscles had respecified the motoneurons. However, more careful examination of limb nerves using electrical stimulation revealed that instead of being respecified by the muslces, the motor axons had actually recrossed to their normal targets. A similar study of the exchanged back and belly skin showed that sensory axons, instead of being respecified, simply found their way back to their normal but ectopic targets.[126] Grafts of third limbs onto chicks also give evidence against myotypic specification. The populations of motoneurons that innervate particular muscles in the extra appendage usually come from the same or closely related motor columns as those that innervate these muscles in the normal limb.[127]

5.9. Ectopic Terminations

Olfactory axons in one sense seem to obey different rules from optic, auditory, or motor fibers. Transplanted olfactory placodes develop projections to several different parts of the brain, but on contact with the CNS, instead of growing toward their normal targets, they often simply terminate at their entry points. The normal olfactory ending is a glomerulus, in which many different afferent fibers synapse in a characteristic way on the dendrites of a single mitral cell, the principal cell of the olfactory bulb. When transplanted nasal epithelial axons make contact with the CNS at abnormal locations (say the dorsal diencephalon or hindbrain), where there are no mitral cells, characteristic olfactory glomeruli are made at these ectopic sites[41] (Figs. 25 and 26). Is it the case that CNS cells near the site of abnormal contact are induced to have mitral dendritic morphology, or do the transplanted axons make glomeruli autonomously, without postsynaptic participation? Studies in mammals have demonstrated actual synapse formation in ectopic glomeruli,[128] so the former possibility seems more likely.

5.10. Interacting Topographies

The tectum or superior colliculus of many species of vertebrates receives not only topographic visual but also topographic somatosensory information. In salamanders, immediately subjacent to the superficial tectal neuropil containing retinal terminals is another neuropil layer containing somatosensory terminals.[129,130] Electrophysiological studies of this region show a map of the body surface in which anterior body parts are represented in the rostral tectum, posterior body parts in the caudal tectum, dorsal parts in the medial tectum, and ventral parts in the lateral tectum.[129] The somatosensory map is thus in register with the visual one; for example, the anterior world, both visual and somatosensory, maps to the rostral tectum (Fig. 27). This registration of axes holds true for all species that have been examined in which different sensory modalities project to the dorsal midbrain. What factors are responsible for the lamination of the tectal neuropil and what controls the registration of the tectal maps? Embryonic eye transplantations have shown that early visual innervation influences the lamination but not the

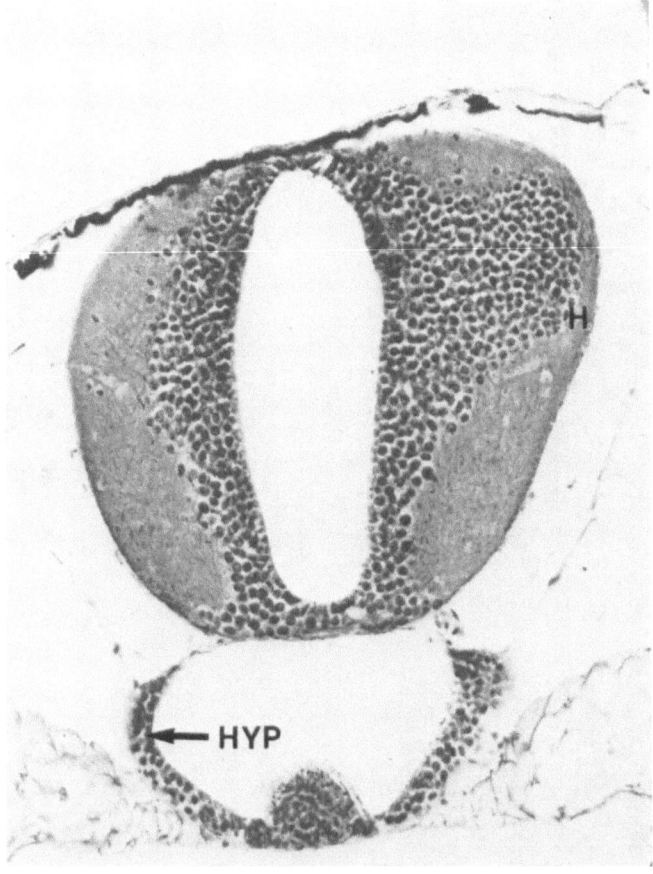

FIGURE 25. Coronal section showing hyperplasia (H) of the dorsal thalamus, following penetration of the supranumerary, ectopic olfactory nerve into this region. The transplant was performed on a stage 29/30 donor–host pair and the host was killed at stage 48. HYP, hypothalamus. From Ref. 41.

topography of the somatosensory input to the tectum. In genetically eyeless or embryonically enucleated salamanders, somatosensory innervation of the tectum is superficial rather than deep, whereas if eyes are transplanted to genetically eyeless embryos, the somatosensory innervation develops in its normally deep neuropil.[130] Similar experiments of either adding or subtracting retinal input, done with larval or adult rather than embryonic salamanders, show that changes in the visual innervation of the tectum do not produce significant changes in the distribution of somatosensory terminals.[131] Thus, only embryonic innervation affects the lamination of the somatosensory tectal input.

Transplantation of eyes in rotated orientations to genetically eyeless salamanders results in animals that have rotated visual maps in the tectum from the earliest stages of development.[50] In these situations, the somatosensory projection to the tectum does

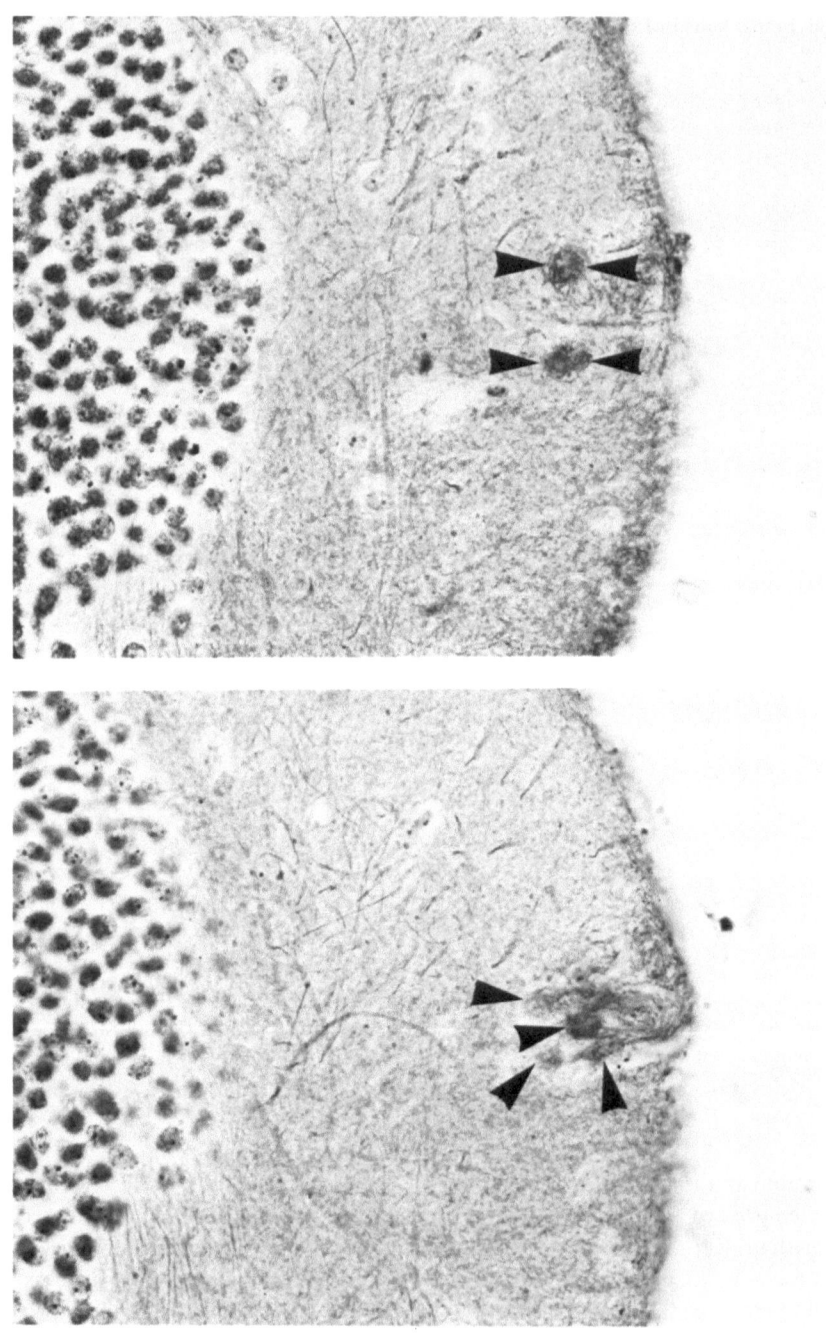

FIGURE 26. (Top) Glomeruler structures (arrowheads) in the diencephalon of a host. Just 12 μm rostral to this section, the supranumerary olfactory nerve penetrated into the diencephalon. The operation was performed on a donor–host pair at stage 32 and the host was killed at stage 49. (Bottom) Glomerular structures (arrowheads) in the diencephalon of a host. The supranumerary olfactory nerve was seen penetrating the diencephalon in the section cranial to the one illustrated. The operation was performed on a donor–host pair at stage 32 and the host was killed at stage 49. From Ref. 41.

not rotate in order to be in register with the overlying visual one[50] (Fig. 28). Instead, the axes of the somatosensory tectal map are unaffected in such animals, and the visual–somatosensory mismatch is stable. The two systems, therefore, do not functionally interact to assure registration.

This lack of functionally mediated topographic interaction between the visual and the somatosensory system can be contrasted to the situation in which normal visual activity is responsible for the functional recovery seen with artificially mismatched tectotectal projections created by the rotation of one eye in a normal larval frog. In these studies, binocular tectal fields are misaligned by the rotation of an eye.[132] Frogs raised postoperatively in ordinary visual environments regain normal binocularity, while those visually deprived by being kept in darkness do not.[133] The rewiring that is responsible for the neural realignment of the binocular fields is in the isthmotectal rather than the tec-

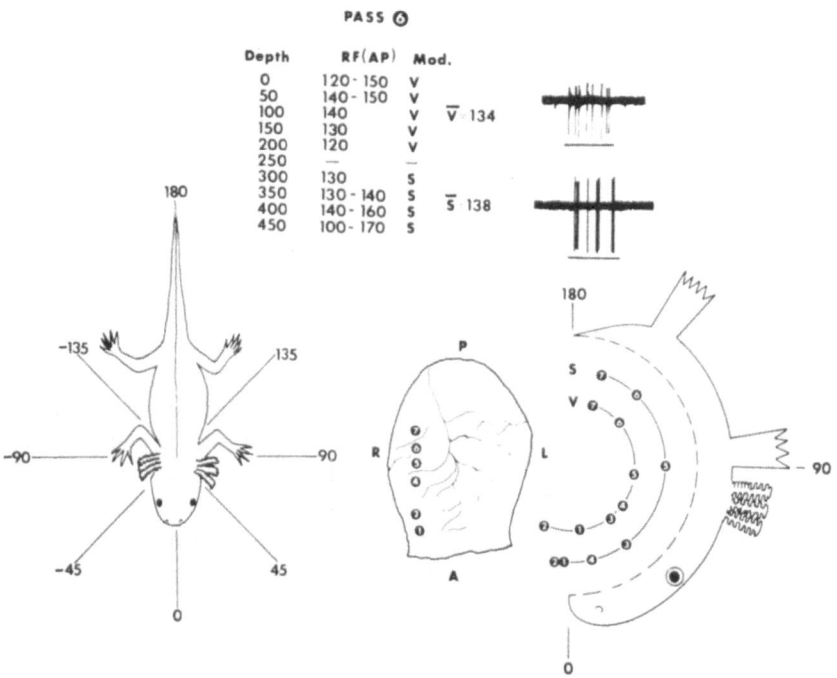

FIGURE 27. Visual and somatosensory data from a normal axolotl's tectum. The bottom left represents an axolotl of normal shape as it appears positioned above the polar grid. The bottom center shows a diagram of a tectal surface (redrawn from a Polaroid photograph) with the superficial electrode penetration sites labeled. A, P, R, and L refer to anterior, posterior, right, and left. At the bottom right are plotted the visual (V) and somatosensory (S) average receptive field locations for each electrode pass. Only anteroposterior data are shown. . The diagrammatic half-salamander shows the approximate distortion involved in plotting the somatosensory data in polar coordinates. (The dashed line is not meant to indicate the backbone.) The top left shows actual data from electrode pass 6. Depth is given in micrometers; anteroposterior receptive fields, RF(AP), are in degrees. Modality (Mod.) also is indicated at each recording depth. Finally, the average field centers for each modality are calculated. The top right shows the response of a visual unit (upper) and a somatosensory unit (lower) to stimulus presentations (solid line). From Ref. 50.

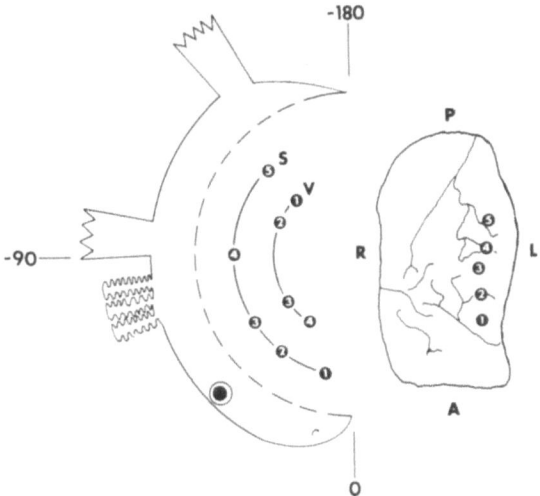

FIGURE 28. Example of a visual–somatosensory mismatched map recorded from the tectum of an eyeless mutant that had received a rotated eye transplant. Conventions are as in Fig. 27. From Ref. 50.

toisthmic leg of the tectotectal pathway.[134] Isthmotectal axons take abnormal routes (Fig. 29) to the tectum to achieve this match.[135]

6. Hypoplasia and Hyperplasia

In general, an increased amount of target causes hyperplasia of the neuronal population projecting to it, whereas a decreased amount of target leads to hypoplasia in the projecting neurons. Conversely, if the neuronal population projecting to a particular target is increased, the result is hyperplasia of the target, whereas a decreased projection population leads to hypoplasia in that target. The specific experiments that lead to these general conclusions use the same systems as have been discussed previously.

6.1. Changes in Size of Projecting Populations

Embryonic enucleation of a frog eye leads to fewer than the normal number of cells in the deprived tectum[136–138] (Fig. 30a). It has been questioned whether the decreased number of cell divisions in the tectum due to this lack of visual input is restricted to the glial or neuronal population.[137] Recent evidence favors effects on both populations.[138] Specific markers for the separate populations of dividing cells are needed to settle this issue finally, but are not yet available in ampibians. Transplantation of an extra eye of the same species, or substitution of an eye by a larger one from a different species, leads to tectal hyperplasia[139] (Fig. 30b). Whether this is due to increased cell division or decreased cell death is not yet known. Transplantation of extra or bigger olfactory placodes to amphibian embryos leads to hyperplasia of the innervated targets even when these are ectopic ones such as the dorsal diencephalon.[51,52] Unusually high indices of mitotic figures are seen at these sites, suggesting hyperplasia by cell division. Removal of the nasal placode results in the underdevelopment of the anterior telencephalon and its olfactory bulb.[140]

6.2. Changes in Size of Target

Eyes or olfactory placodes from embryos of a small *Ambystoma* species have been substituted with those of embryos of a larger species. The targets become hypotrophic, because of the reduced projecting population. But what of the size of the transplanted organs? Do they become larger due to their larger than normal targets? No cell counts or measurements have yet been done on such size-disparate transplants.

Extra eyes transplanted to ectopic locations sometimes share the tectum with the normal eyes. In these cases, there are fewer than the normal number of retinal ganglion cells[141] possibly as a result of excessive cell death.

Survival and death in cells projecting to enlarged or reduced targets have been most extensively studies with the motoneuron-limb system. Normally, almost 50% of the motoneurons that initially innervate the limb die at a particular developmental stage.[118,127] If the limb bud is removed, virtually all of the limb motoneurons die.[142] Conversely, in birds and amphibians, an extra limb allows a substantial fraction, but not all, of the neurons that would have normally died, to survive.[143,144] Adding an extra limb, however, is not like doubling the size of the normal limb. Because it cannot be

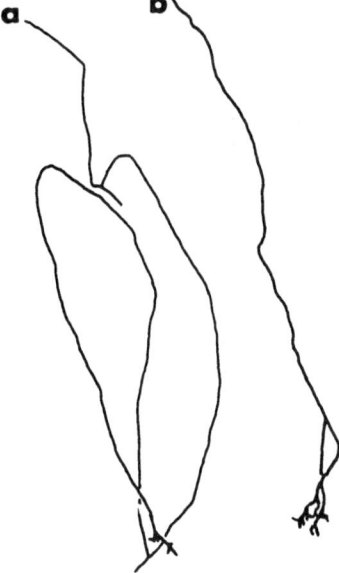

FIGURE 29. HRP-filled isthmotectal axons in normal *Xenopus* (b) and in *Xenopus* in which one eye had been rotated *without optic nerve transection* during midlarval life (a). HRP was injected into the right nucleus isthmi; after 3 days, the left tectum was exposed, removed, and flattened between a coverslip and slide. After glutaraldehyde fixation and buffer rinse, the tecta were reacted to visualize the HRP. In normal *Xenopus,* most isthmotectal axons take fairly straight routes from the point where they enter the tectum to the point where they terminate. However, in *Xenopus* that have had abnormal visual input due to early eye rotation, many axons take highly abnormal, circuitous routes. These abnormal trajectories result from the initial mismatch in the orientation of the left and right eyes' visual maps. Courtesy of S. Udin.

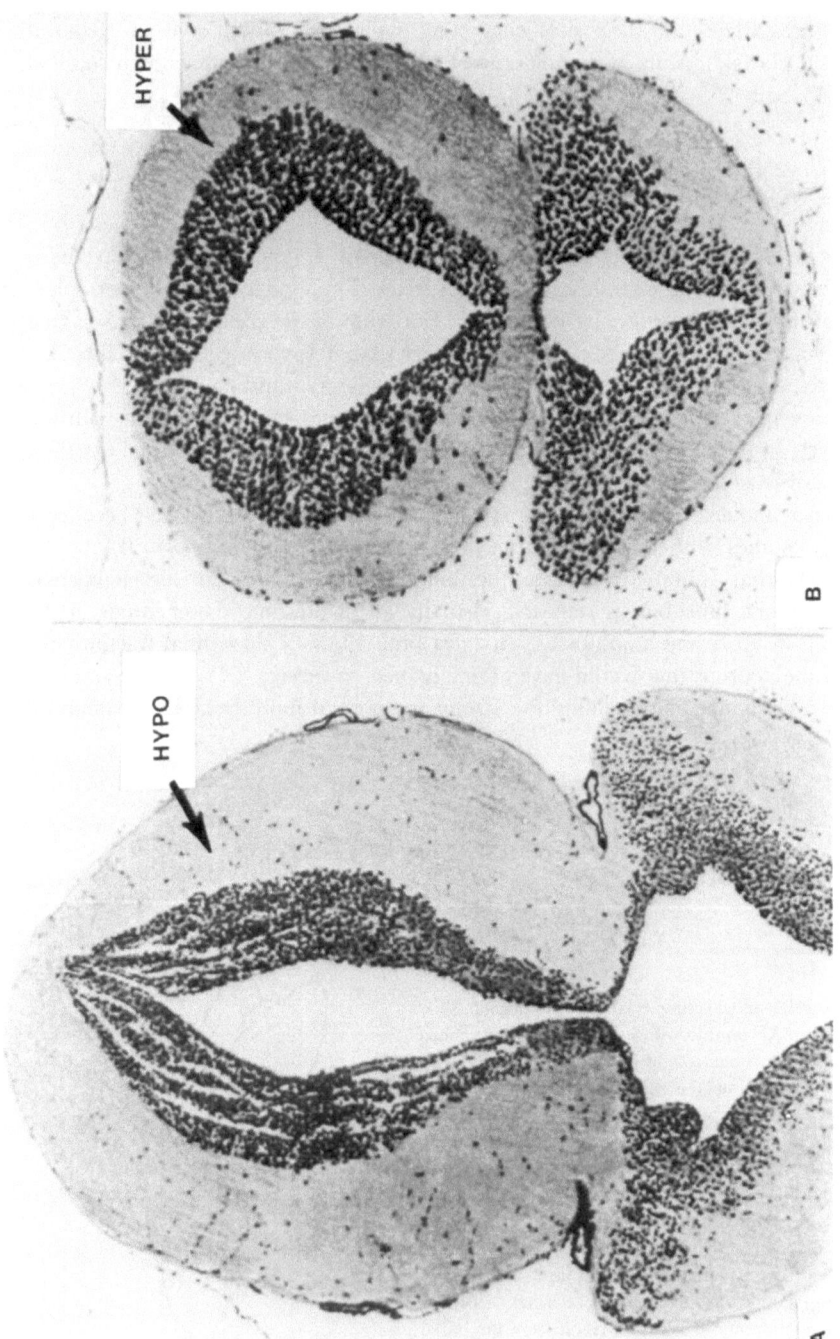

FIGURE 30. (a) Example of unilateral hypoplasia as a result of innervation of an *A. trigrinum* tectum with an *A. punctatum* eye. (b) Example of unilateral hyperplasia as a result of an eye transplantation from *A. tigrinum* to *A. punctatum*. From Ref. 132.

transplanted to exactly the same location as the normal limb, the transplanted limb usually receives a smaller than normal spinal projection, composed of neurons from normal and abnormal motor pools.[145] For these reasons, an exact algebraic relationship between the size of the projecting population and the size of the target has yet to be established, especially for increases in target size. If one limb bud is removed and its nerves forced to share the contralateral limb, only about half of each motor pool dies.[146] This is a strange result for peripheral deletion experiments suggest a more strictly proportional relationship. The matching of the size of the motor pool to that of the target seems to be functionally regulated. Drugs that block neuromuscular transmission can delay normal motoneuron death until after recovery from the blockade.[118]

7. Regeneration

7.1. Pathway Selection and the Formation of Connections

The development of the nervous system has been assumed to be similar in many ways to the process of neural regeneration. The results discussed below demonstrate that this assumption is at least partly correct. Regeneration and development, however, are distinct processes. This distinction is especially clear in the case of the retinotectal system where the developing and regenerating projections form in apparently different ways. Comparison of the results described in this section to those discussed earlier on development, reveals many of these differences.

The mature nervous system of a lower vertebrate after having developed its specific connections, retains the capacity to restore damaged functional pathways. In young *Xenopus*, Mauthner axons can regenerate across spinal transections to make appropriate connections caudally.[147] Cut or crushed optic nerve fibers in fish and amphibians regrow to their specific tectal locations.[148,149] They do so even when forced to enter the brain at ectopic locations.[150] The pathways they take to their final destinations can be through tissue that previously had no contact with optic nerve fibers, as is the case for developing connections.[150] Regenerating optic fibers may even find their way to targets that have been considerably displaced from their normal position, e.g., a tectal lobe transplanted to an enucleated eye socket.[151] Regenerating fibers, unlike developing ones, seem to take abnormal courses through the target neuropil to their particular destinations, arriving there by a series of approximations[12] (Fig. 31).

A compound eye forced to regenerate to an ipsilateral tectum that is innervated by a normal eye connects to its appropriate restricted tectal sites.[152] If the normal eye is then removed, however, the restricted map from the compound eye quickly expands to fill the tectum.[152] If a compound eye is forced to regenerate to an ipsilateral "virgin" tectum, which has never been innervated by retinal fibers, a double temporal eye is initially restricted in its projection, whereas a double nasal one is not.[152] These last somewhat confusing results open the question of whether there exist markers intrinsic to the tectum that guide appropriate optic fibers to them, or whether retinal terminals simply sort themselves out as they innervate the tectum.

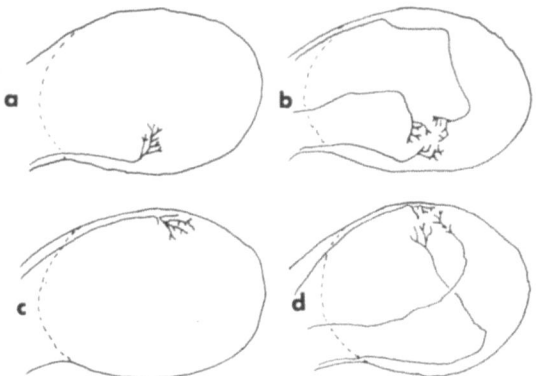

FIGURE 31. A semischematic representation of the trajectories of normal retinal axons (a, c) and regenerating retinal axons at the 10th week of regeneration (b, d) within the optic tract and tectum. These trajectories were traced using a camera lucida in the specimens of the dorsal (a, b), ventral (c, d) retinal lesions. After Ref. 12.

7.2. Tectal Markers

Two types of experiments in goldfish and frogs have been used to address the problem of whether the tectum does in fact have its own set of markers. The first type of experiment suggests a role for retinal innervation. These are the *size disparity* studies in which half the retina or half the tectum is removed.[94] As a result, the half-retina "expands" and projects to the whole tectum in the former case[153] (Fig. 32a). In the latter, the whole retina "compresses" into the half-tectum[154] (Fig. 32b). If fibers from a normal eye are surgically deflected onto an expanded tectal map, only half of them innervate most of the tectum[155] (Fig. 32d). If a half-eye that has an expanded projection is deflected onto a normal map, it immediately and topographically innervates the entire tectum[155] (Fig. 32c). If the tectum remains visually uninnervated for 6 months before optic nerve regeneration, a half-retina immediately forms an expanded projection onto a whole tectum[94] (Fig. 32f), and a whole retina immediately forms a compressed projection, onto a half-tectum[156] (Fig. 32e). These experiments all suggest that tectal markers seem somehow to depend on retinal innervation.

A second type of experiment, using *tectal transplantation,* suggests that the adult tectum has a stable set of intrinsic markers. If pieces of the adult tectum are rotated, inverted, or both, regenerating retinal axons will recognize the transplanted piece and map according to its new orientation.[157] If two pieces of tectal tissue are interchanged (e.g., a square of rostral tectum exchanged with one of caudal tectum) or interchanged and rotated, the regenerated retinotectal projections show not only corresponding reciprocal transpositions along the nasotemporal axis but also a localized 180° rotation in the order of receptive fields within the reimplanted tissue.[158] Thus, pieces of tectum retain their topographic labels when transplanted.

The two types of experiments, size disparity and tectal transplantation, have been combined.[159] The tectum was halved and a piece of the remaining half rotated. There

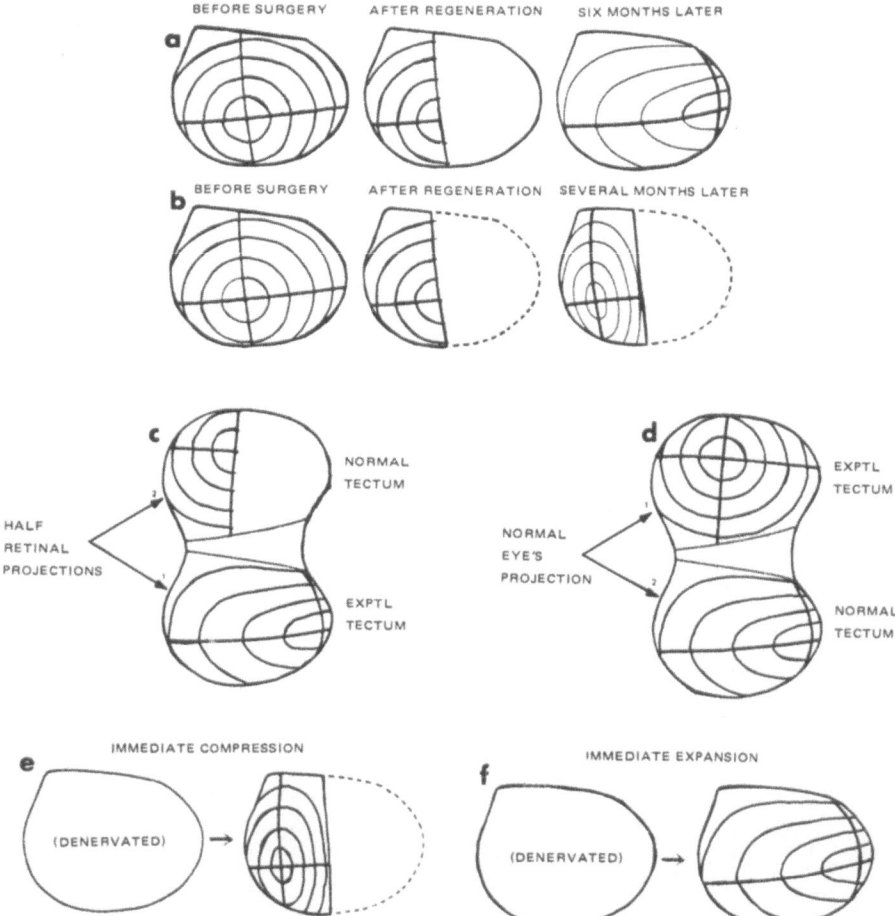

FIGURE 32. (a, b) Schematic drawings showing the steps (from left to right) in half-retinal expansion (a) and half-tectal compression (b). The tectal surface is shown in each case along with representation of the retinotopic projection. The dark horizontal and vertical lines represent the projections of the horizontal and vertical meridians of the eye and the circles represent retinal contours at 20° intervals from the center. For the half-retinal case (a), regeneration restores an appropriate half-projection, leaving the other half of the tectum denervated, but the half-projection eventually expands over the tectal surface. For the half-tectal case (b), only the appropriate half of the projection is initially regenerated, but the other half later shares the remaining tectum in an orderly map. (c, d) Schematic diagrams of the tests for changes in the half-retinal (c) and tectal (d) markers following expansion. In (c), the half-retina that had expanded its projection (1) subsequently innervated the opposite normal tectum (2), which was still innervated by a normal eye. It always retained its selectivity for the appropriate half of a normal tectum. In (d), a normal eye's fibers were deflected to innervate a tectum receiving an expanded projection, and the normal eye also formed an expanded projection there, indicating altered tectal markers. (e, f) Results of innervation following long-term denervation. (e) Immediate compression is produced by a normal eye innervating half of a denervated tectum. (f) Immediate expansion occurs when a half-retina innervated the denervated tectum. Note that both cases skip the intermediate step of an appropriate half-projection. From Ref. 94.

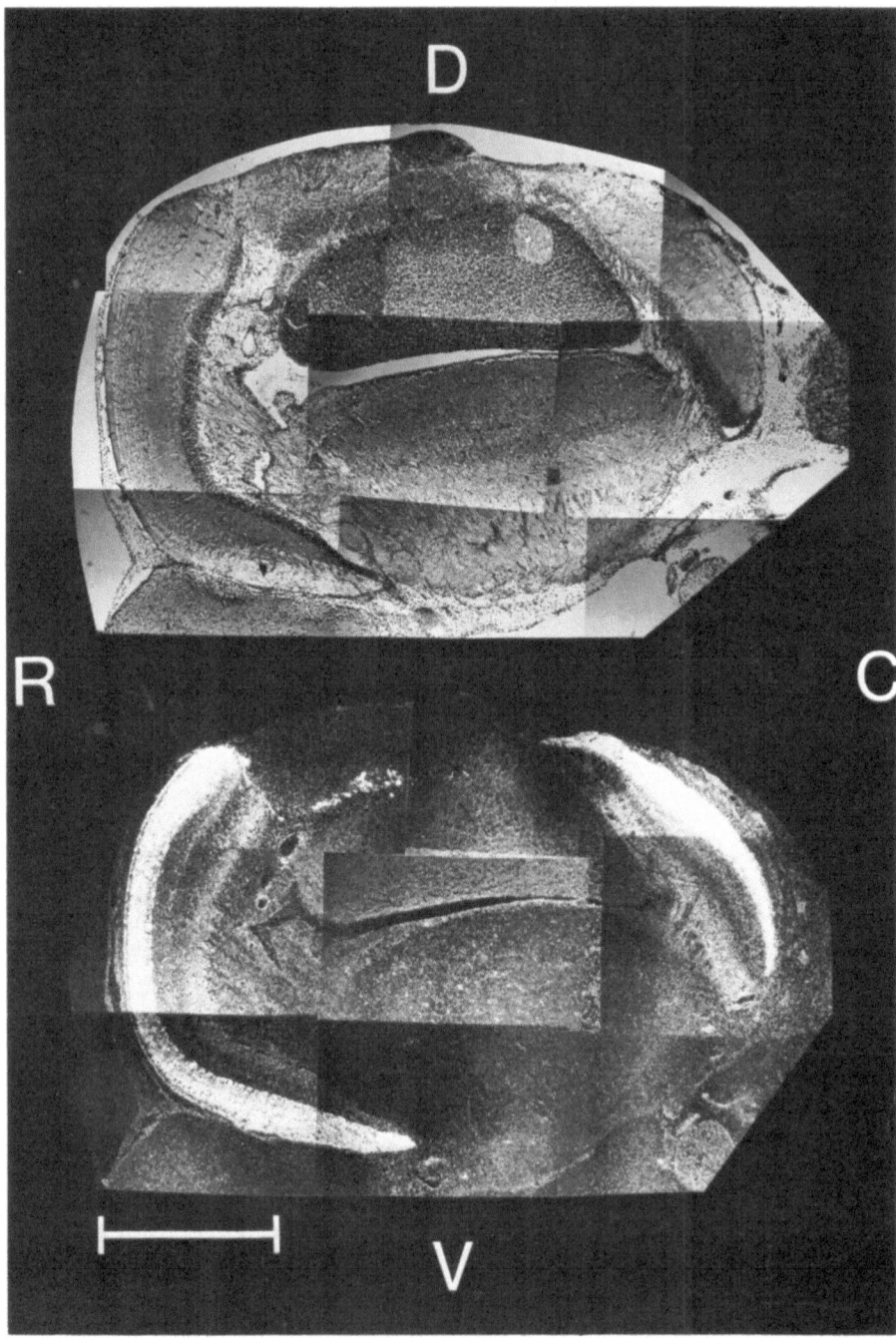

FIGURE 33. Micrographs of parasagittal sections of the operated tectum, which contains the 180° rotated cerebellar transplant in the middle. The brain was fixed days after the reciprocal transplantation, embedded in paraffin, and serially sectioned at 10 μ. The fish received an intraocular injection of L-[^{3}H]proline 11 days before the fixation. The upper section was stained by a modified Bodian protargol impregnation method. It shows the 180° rotated cerebellar graft, interposed in the middle between the rostral and the

formed, upon regeneration, a compressed map with an inverted square.[159] These results argue for the presence of tectal markers that can in certain situations be modified by retinal innervation. Cell adhesion assays are being used to look into the molecular basis of tectal labels. Some potentially relevant molecules have been discovered and are being characterized.[7]

If a square of the tectum is replaced with a square of cerebellar tissue,[160] regenerating retinal fibers confronting the foreign tissue turn ventrally and run along the ventral border of the implant. They do not innervate it (Fig. 33). Instead, the entire retina projects in a compressed fashion onto the remaining tectum. Thus, compression is not simply a result of some measure of the tectal perimeter. Another conclusion from this experiment is that retinal fibers can recognize the difference between cerebellum and tectum even when the nontarget neural tissue is positioned in the middle of the target. This result is analogous to that of transplanting a piece of hindbrain to the tectum embryonically.

7.3. The Pigment Epithelium

Urodele retinae are able to regenerate after removal. The pigment epithelium gives rise to the new retina, which maps in a normal topographic way onto the tectum. Rotations of pieces of the pigment epithelium at the time of retinal removal, reveal that positional information is transmitted from the pigment epithelium to the retina. This is demonstrated by showing that only the piece of regenerated retina overlying the rotated piece of pigment epithelium projects in a rotated sense onto the tectum.[161]

7.4. Motoneurons and Foreign Innervation

Regenerating motoneurons in amphibians can regrow terminals to the original synaptic endings in a denervated muscle.[162] They do this with ultrastructural precision.[162] In mammals, when foreign nerves are transplanted to an innervated muscle, they do not form ectopic synapses on it unless the original nerve is cut.[163] Exclusion of foreign innervation is under control of muscle activity, for a denervated but electrically stimulated muscle can also repel foreign innervation.[164] When a foreign nerve does make functional neuromuscular contacts, these are not restricted to the original sites but can be at various other locations on the muscle fibers.[163]

Original nerves seem to have an advantage over foreign nerves in the innervation of lower vertebrate muscle. As an original nerve grows back to a muscle innervated by

caudal parts of the tectum. The dark-field micrograph of an autoradiograph was obtained from another parasagittal section, taken from about 50 μm medial to the above section in the same brain. It shows that the stratum opticum and the stratum fibrosum et griseum superficiale in both the rostral and the caudal parts of the operated tectum are intensely labeled. In contrast, the cerebellar tissue, interposed in the middle, does not show any noticeable labeling. Scale: 0.5 mm. R, rostral; C, caudal, D, dorsal; V, ventral. From Ref. 160.

a foreign nerve, it may "drive off" much of the foreign innervation.[165,166] Some of the foreign innervation will usually remain, and a few fibers may show a stable dual innervation by both foreign and original motoneurons.[167] It was initially proposed that when the orignal nerve regrew to its muscle with foreign innervation, it functionally suppressed the anatomically stable foreign synapses.[165] It now appears more likey that foreign terminals actually retract from the muscle as the original nerve reestablishes contact.[168]

7.5. Transfer of Memory and Behavior

In exotic experiments, entire adult amphibian brains or telencephela have been exchanged between animals or species.[169-172] In others, brains were sliced up and reinserted into the cranium in jumbled configurations.[173] Investigators who performed such studies not only reported regeneration of connections, but claimed to have demonstrated recovery of memory traces in "shuffled" brains, and exchange of memories or species-specific behaviors from donors to hosts in transplant experiments. One cannot help thinking there might be something a bit fanciful in the conception of these experiments and the reports of the results. Still, until other laboratories try to repeat these studies, they stand unchallenged.

8. Discussion

The transplantations reported here have provided the information that various hypotheses of neural development have tried to unify. They have also, however, provided the facts that eventually demolished many such hypotheses. Finally, they provided the phenomena that must eventually be explained in cellular and molecular terms if we are to understand neural development. The time for such transplantation studies is by no means over; the recent work is perhaps the most informative, and many questions will be answered by future transplants. Insofar as other animals have been used to ask similar questions, the studies with fish and amphibians have often yielded complementary rather than contradictory results. The mechanisms of such things as axonal navigation, and topographic map formation, may therefore be rather general.

Transplantation studies have forced us to acknowedge the existence of cellular interactions, long-range forces, and certain formal rules of neural development. They have not told us much about the molecular biology of neural development; yet, perhaps they can. Transplants between species or studies using genetic variants can allow one to focus on the actual cellular differences between the donor and the host. This sort of analysis has begun, for example, on two mutants of the axolotl that affect neural development. They are *spastic,* which causes intrinsic derangement of the cerebellum, and *eyeless,* in which a factor from the mutant epidermis seems to inhibit the formation of optic vesicles (see Refs. 14, 174, 175 for reviews). The correlation of processes, in cases where developmental events are the *same* between species as shown by cross-species transplantations, might also reveal pertinent, cellular mechanisms of neural development. Ultrastructural examination of the interaction of labeled transplanted and unlabeled host tissue may also provide considerable insight into the cellular biology of neural development. Finally, transplantations coupled with biochemical perturbations of development, for instance by exposing pieces of transplanted tissue to particular drugs or antibodies developed from *in vitro* studies, might help us discover or test the *in vivo* roles of certain molecules in neural development.

9. Summary

1. Lower vertebrates offer many advantages for the study of pattern formation in the nervous system. This is largely due to the fact that neural tissue can be easily transplanted throughout the development.

2. For many neural structures, there is a developmental stage (early neural plate) when the fate of cells can be changed by transplantation. Later, these fates become irreversibly determined. Some studies were thought to show that after determination, the polarity of certain structures was still modifiable. More recent studies using the retina have not supported this as a separate event in pattern formation.

3. The initial stages of axonal and dendritic outgrowth seem to be governed by mechanisms intrinsic to neurons.

4. Axonal navigation to a target is not dependent on mechanical guidance.

5. There appears to be a hierarchical order of targets in the CNS. If the target of choice is unavailable or too distant, axons will choose a structure of secondary preference.

6. Gradients of diffusible or substrate-bound materials released by the target may help guide axons toward the appropriate destination, from a distance.

7. Some axons have an intrinsic tendency to decussate.

8. Voltage-sensitive Na^+ channels and action potentials are not necessary for axonal navigation, although Ca^{2+} channels might be.

9. The cues governing axonal navigation are shared between many, if not all, amphibian species, indicating that the mechanisms involved might be evolutionarily conserved.

10. Topographic projections are not dependent on pathways or fiber–fiber interactions.

11. Correct timing of axonal outgrowth or of arrival at a target is not necessary for the formation of topographic projections.

12. Impulse activity is not necessary for the development of gross topography in the nervous system, although it seems to play a role in refining the topography.

13. In amphibians, at least, the initial projection to a particular target in the CNS is orderly. Thus, the formation of a topographic map in the brain is not the result of random innervation followed by the removal of inappropriate connections.

14. When two similar projecting populations are forced to innervate the same central target, they do so topographically, yet they segregate from each other. This results when two similar retinal fields project to the tectum in the formation of eye-specific stripes. The segregation appears to be dependent on impulse activity.

15. In early development, the diencephalon may induce axial polarization of the tectum with respect to the retinal projection. It probably does not do this by direct interaction with retinal fibers.

16. In spite of earlier studies, peripherally projecting neurons do not generally achieve their particular identities through their postsynaptic partners. Rather their identities are intrinsically or centrally specified, and central projections are not respecified to accommodate abnormal peripheral projections.

17. Registration of somatosensory and visual topographic maps in the tectum is coincidental, and mismatched maps cannot be realigned by experience. The primary retinotectal and the tectotectal visual maps, however, can be aligned as a result of visual experience.

18. Increased target leads to hyperplasia of the neural projection to it, and an increased neural projection leads to hyperplasia in the target. Similarly, hypoplasias are noted with decreased projections and targets.

19. In lower vertebrates, regeneration of some central pathways is possible. Regenerating projections have several features in common with developing ones, such as independence of pathways, of fiber–fiber interactions, and of timing, and a possible dependence on chemical gradients.

20. Translocation and rotation experiments in the tectum indicate that this target is somehow labeled such that specific regenerating retinal cells will only connect with specific tectal cells. This projection can be either expanded or compressed as a result of hemisection of the retina or tectum, indicating the target labeling is an interactive wholefield phenomenon.

21. In the neuromuscular system, a regenerating normal projection is favored over a foreign one. The host nerve may even drive off functioning foreign synapses.

22. It is possible that memory may be transferred by brain transplantation.

ACKNOWLEDGMENTS. Substantial revision of this manuscript was due to Jeff Hall's spilling of suntan oil all over the first draft. I thank John Lovell Bixby, James Fawcett, Betty Alice Ferguson, Fred Samuel Harris, and Christine Holt for other help with the manuscript. Support was from NIH (Grant HD 14490), the McKnight Foundation (Scholar's Award), and the March of Dimes (Basil O'Connor Grant).

References

1. Detwiler, S. R., 1936, *Neuroembryology,* Hafner, New York.
2. Huxley, J. S., and DeBeer, G. R., 1934, *The Elements of Experimental Embryology,* Cambridge University Press, London.
3. Spemann, H., 1938, *Development and Induction,* Yale University Press, New Haven, Conn.
4. Balsamo, J., McDonough, J., and Lilien, J., 1976, Retinal-tectal connections in the embryonic chick: Evidence for regionally specific cell surface components which mimic the pattern of innervation, *Dev. Biol.* **44**:338.
5. Barbera, A. J., 1975, Adhesive recognition between developing retinal cells and optic tecta of the chick embryo, *Dev. Biol.* **47**:167.
6. Letourneau, P. C., 1982, Nerve fiber growth and its regulation by extrinsic factors, in: *Neuronal Development* (N. C. Spitzer, ed.), pp. 213-254, Plenum Press, New York.

7. Gottlieb, D. I., and Glaser, L., 1980, Cellular recognition during neural development, *Annu. Rev. Neurosci.* **3**:303.

8. Marchase, R. B., 1977, Biochemical investigations of retinotectal adhesive specificity, *J. Cell Biol.* **75**:237.

9. Trisler, C. D., Scheider, M. D., and Niremberg, M., 1981, A topographic gradient of molecules in retina can be used to identify neuron position, *Proc. Natl. Acad. Sci. USA* **78**:2143.

10. Holt, C. E., 1982, The development of the eye and its central conections, Ph.D. thesis, King's College, London University. Biophysics.

11. Holt, C. E., and Harris, W. A., Order in the initial retinotectal map in *Xenopus:* A new technique for labelling growing nerve fibers, *Nature* **301**:150.

12. Fujisawa, H., Tani, H., Watanabe, K., and Ibata, Y., 1982, Branching of regenerating retinal axons and preferential selection of appropriate branches for specific neuronal connections in the newt, *Dev. Biol.* **90**:43.

13. Anderson, H., Edwards, J. S., and Palka, J., 1980, Developmental neurobiology of invertebrates, *Annu. Rev. Neurosci.* **3**:97.

14. Palka, J., 1982, Genetic manipulation of sensory pathways in *Drosophila,* in: *Neuronal Development* (N. C. Spitzer, ed.), pp. 121–170, Plenum Press, New York.

15. Hall, J. C., Greenspan, R. J., and Harris, W. A., 1982, *Genetic Neurobiology,* MIT Press, Cambridge, Mass.

16. Model, P. G., 1982, Prospective forebrain-midbrain from axolotl neurulae can be reprogrammed to differentiate as Mauthner cell-containing medulla, *Dev. Brain Res.* **3**:109.

17. Jacobson, C. O., 1964, Motor nuclei, cranial nerve roots, and fiber pattern in the medulla oblongata after reversal experiments on the neural plate of axolotl larvae. I. Bilateral operations, *Zool. Bidr. Uppsala* **36**:73.

18. Cooke, J., 1980, Early organization of the central nervous system: Form and pattern, *Curr. Top. Dev. Biol.* **15**:373.

19. Stone, L. S., 1960, Polarization of the retina and development of vision, *J. Exp. Zool.* **145**:85.

20. Hunt, R. K., and Jacobson, M., 1972, Development and stability of positional information in *Xenopus* retinal ganglion cells, *Proc. Natl. Acad. Sci. USA* **69**:780.

21. Gaze, R. M., Feldman, J. D., Cooke, J., and Chung, S.-H., 1979, The orientation of the visual map in *Xenopus:* Developmental aspects, *J. Embryol. Exp. Morphol.* **53**:39.

22. Sharma, S. C., and Hollyfield, J. G., 1980, Specification of retinotectal connections during the development of the toad *Xenopus laevis, J. Embryol. Exp. Morphol.* **55**:77.

23. Goldberg, S., 1976, Polarization of the avian retina: Ocular transplantation studies, *J. Comp. Neurol.* **168**:379.

24. Grant, P., and Rubin, E., 1980, Disruption of optic fiber growth following eye rotation in *Xenopus laevis* embryos, *Nature (London)* **287**:845.

25. Katz, M. J., and Silver, J., 1981, Inverted *Xenopus* eye primordia develop into anatomically inverted eyes, *Dev. Biol.* **86**:510.

26. Holt, C. E., 1980, Cell movements in *Xenopus* eye development, *Nature (London)* **287**:850.

27. Beach, D. A., and Jacobson, M., 1979, Patterns of cell proliferation in the retina of the clawed frog during development, *J. Comp. Neurol.* **183**:603.

28. Yntema, C. L., 1955, Ear and nose, in *Analysis of Development* (B. H. Willier, P. Weiss, and V. Hamburger, eds.), pp. 415–427, Saunders, Philadelphia.

29. Harrison, R. G., 1935, Factors concerned in the development of the ear in *Amblystoma punctatum, Anat. Rec.* **64**(Suppl. 11):38.

30. Harrison, R. G., 1936a, Relations of symmetry in the developing ear of *Amblystoma punctatum, Proc. Natl. Acad. Sci. USA* **22**:238.

31. Harrison, R. G., 1936, Relations of symmetry in the developing embryo, *Coll. Net.* **11**:217.

32. Harrison, R. G., 1945, Relations of symmetry in the developing embryo, *Trans. Conn. Acad. Arts Sci.* **36**:277.

33. Harrison, R. G., 1921, On relations of symmetry in transplanted limbs, *J. Exp. Zool.* **32**:1.

34. Harrison, R. G., 1925, The effect of reversing the medio-lateral or transverse axis of the forelimb bud in the salamander embryo (*Amblystoma punctatum* Linn.), *Arch. Entwicklungsmech. Org.* **104**:469.

35. Jacobson, C. O., 1976, Motor nuclei, cranial nerve roots, and fiber pattern in the medulla oblongata after reversal experiments on the neural plate of axolotl larvae. II. Unilateral operation, *Z.O.O.N.* **4**:87.

36. Hibbard, E., 1965, Orientation and directed growth of Mauthner's cell axons from duplicated vestibular nerve roots, *Exp. Neurol.* **13**:289.

37. Luna, E., 1915, Ricerche sperimentali gulla morfologia dell'organo dell'olfatto negli amfibi, *Arch. Ital. Anat. Embriol.* **14**:609.

38. Zwilling, E., 1940, An experimental analysis of the development of the anuran olfactory organ, *J. Exp. Zool.* **84**:291.

39. Bell, E. T., 1907, Some experiments on the development and regeneration of the eye and the nasal organ in frog embryos, *Arch. Entwicklungsmech. Org.* **23**:457.

40. Burr, H. S., 1916a, The effects of removal of the nasal pits in *Amblystoma* embryos, *J. Exp. Zool.* **20**:27.

41. Stout, R. P., and Graziadei, P. P. C., 1980, Influence of the olfactory placode on the development of the brain in *Xenopus laevis* (Daudin). I. Axonal growth and connections of the transplanted olfactory placode, *Neuroscience* **5**:2175.

42. Constantine-Paton, M., and Capranica, R. R., 1976, Axonal guidance of developing optic nerves in the frog. I. Anatomy of the projection from the transplanted eye primordia, *J. Comp. Neurol.* **170**:17.

43. Harris, W. A., 1980, The effects of eliminating impulse activity on the development of the retinotectal projection in salamanders, *J. Comp. Neurol.* **194**:303.

44. Harrison, R. G., 1907, Experiments in transplanting limbs and their bearing upon the problems of the development of nerves, *J. Exp. Zool.* **4**:239.

45. Harrison, R. G., 1904, Neue versuche und beobachtungen über die entwicklung fer peripheren nerven der wirbeltiere, *Sitzungsber. Neiderrh. Ges. Natur. Heilkunde. Bonn. Sitzung* **11**:1.

46. Harrison, R. G., 1924, Neuroblast versus sheath cell in development of peripheral nerves, *J. Comp. Neurol.* **37:123.**

47. Harrison, R. G., 1924, Some unexpected results of heteroplastic transplantation of limbs, *Proc. Natl. Acad. Sci. USA* **10**:69.

48. Constantine-Paton, M., and Capranica, R. R., 1975, Central projection of optic tract from translocated eyes in the leopard frog *(Rana pipiens)*, *Science* **189**:480.

49. Harris, W. A., 1980, Regions of the brain influencing the projection of developing optic tracts in salamander, *J. Comp. Neurol.* **194**:319.

50. Harris, W. A., 1982, The transplantation of eyes to genetically eyeless salamanders: Visual projections and somatosensory interactions, *J. Neurosci.* **2**:339.

51. Burr, H. S., 1924, Some experiments on the transplantation of the olfactory placode in *Amblystoma.* I. An experimentally produced aberrant cranial nerve, *J. Comp. Neurol.* **37**:455.

52. Burr, H. S., 1930, Hyperplasia in the brain of *Amblystoma, J. Exp. Zool.* **55**:171.

53. David, W. S., and Model, P. G., 1982, Synapse localization is unaffected by displacement of a major afferent projection, *Neurosci. Abstr.* **8**:436.

54. Adler, R., Manthorpe, M., Skaper, S., and Varon, S., 1981, Polyornithine-attached neurite-promoting factors (PNPFs) culture sources and responsive neurons, *Brain Res.* **206**:129.

55. Collins, F., 1980, Neurite outgrowth induced by the substrate associated material from nonneuronal cells *Dev. Biol.* **79**:247.

56. Coughlin, M. D., Bloom, E. M., and Black, I. B., 1981, Characterization of a neuronal growth factor from mouse heart cell-conditioned medium, *Dev. Biol.* **82**:56.

57. Lander, A. D., Fujii, D. K., Gospodarowicz, D., and Reichardt, L. F., 1982, Characterization of a factor that promotes neurite outgrowths: evidence linking activity to a heparin sulfate proteoglycan, *J. Cell. Biol.* **94**:574.

58. Turner, J. E., Schwab, M. E., and Thoenen, H., 1982, Nerve growth factor stimulates neurite outgrowth from goldfish retinal explants: The influence of a prior lesion, *Dev. Brain Res.* **4**:59.

59. Constatine-Paton, M., 1978, Central projections of anuran optic nerves penetrating hindbrain or spinal cord regions of the neural tube, *Brain Res.* **158**:31.

60. Katz, M. J., and Lasek, R. L., 1978, Eyes transplanted to tadpole tails send axons rostrally in two spinal cord tracts, *Science* **199**:204.

61. Giorgi, P. P., and Van der Loos, H., 1978, Axons from eyes grafted in *Xenopus* can grow into the spinal cord and reach the optic tectum, *Nature (London)* **275:**746.

62. Burr, H. S., 1932, An electrodynamic theory of development suggested by studies of proliferation rates in the brain of *Amblystoma, J. Comp. Neurol.* **56:**347.

63. Frost, D. O., 1981, Orderly anomalous retinal projections to the medial geniculate, ventrobasal, and lateral posterior nuclei of the hamster, *J. Comp. Neurol.* **203:**227.

64. Jacobson, M., 1978, *Developmental Neurobiology,* Plenum Press, New York.

65. Swisher, J. E., and Hibbard, E., 1967, The course of Mauthner axons in Janus-headed *Xenopus* embryos, *J. Exp. Zool.* **165:**433.

66. Katz, M. J., and Lasek, R. J., 1981, Substrate pathways demonstrated by transplanted Mauthner axons, *J. Comp. Neurol.* **195:**627.

67. Katz, M. J., Lasek, R. J., and Nauta, H. J. W., 1980, Ontogeny of substrate pathways and the origin of the neural circuit pattern, *Neuroscience* **5:**821.

68. Bate, C. M., 1976, Pioneer neurones in an insect embryo, *Nature (London)* **260:**54.

69. Piatt, J., 1944, Experiments on the decussation and course of Mauthner's fibers in *Amblystoma punctatum, J. Comp. Neurol.* **80:**335.

70. Guillery, R. W., 1974, Visual pathways in albinos, *Sci. Am.* **230:**44.

71. Guillery, R. W., and Updyke, B. V., 1976, Retinofugal pathways in normal and albino axolotls, *Brain Res.* **109:**235.

72. Cole, J., Dolin, R., Fahrner, K., Gallenson, N., Hall, J., and Harris, W., 1982, Retinofugal pathways from albino eyes embryonically transplanted to normal and albino axolotls, *Dev. Brain Res.* **5:**346.

73. Twitty, V. C., 1937, Experiments on the phenomenon of paralysis produced by a toxin occurring in *Triturus* embryos, *J. Exp. Zool.* **76:**67.

74. Spitzer, N. C., 1979, Ion channels in development. *Annu. Rev. Neurosci.* **2:**363.

75. Kung, C., 1979, Biology and genetics of *Paramecium* behavior, in: *Neurogenetics* (X. O. Breakfield, ed.), pp. 1–26, Elsevier/North-Holland, Amsterdam.

76. Nuccitelli, R., Poo, M.-M., and Jaffee, L. F., 1977, Relations between amoeboid movement and membrane-controlled electrical currents, *J. Gen. Physiol.* **69:**743.

77. Freeman, J. A., Mayes, B., Snipes, G. J., and Wiskwo, J. P., 1982, Real-time measurements of minute neuronal currents with a circularly vibrating microprobe, *Neurosci. Abstr.* **8:**302.

78. Meiri, H., Spira, M. E, and Parnas, I., 1981, Membrane conductance and action potential of a regnerating axonal tip, *Science* **211:**709.

79. Grinvald, A., and Farber, I. C., 1981, Optical recording of calcium action potentials from growth cones of cultured neurons with a laser microbeam. *Science* **212:**1164.

80. Huttner, S. L., and O'Lague, P. H., 1982, Excitability of growth cones of mutinucleate PC-12 cells, *Neurosci. Abstr.* **8:**125.

81. Gunderson, R. W., and Barrett, J. N., 1980, Characterization of the turning response of dorsal root neurites toward nerve growth factor, *J. Cell Biol.* **87:**546.

82. Grafstein, B., and Meiri, H., 1980, Increased rate of regeneration in goldfish optic axons resulting from intraocular injection of calcium ionophore A23187, *Neurosci. Abstr.* **6:**387.

83. Law, M. I., and Constantine-Paton, M., 1981, Anatomy and physiology of experimentally produced striped tecta, *J. Neurosci.* **1:**741.

84. Horder, T., and Martin, K. A. C., 1978, Morphogenetics as an alternative to chemospecificity in the formation of nerve connections, in: *Cell–Cell Recognition* (A. S. G. Curtis, ed.), *J. Embryol. Exp. Morphol.* **46:**147.

85. Fawcett, J. W., and Gaze, R. M., 1982, The retinotectal fiber pathways from normal and compound eyes in *Xenopus, J. Embryol. Exp. Morphol.* **72:**19.

86. Anderson, H., 1978, Postembryonic development of the visual system of the locust *Schistocerca gregaria.* II. An experimental investigation of the formation of the retina-lamina projection, *J. Embryol. Exp. Morphol.* **46:**147.

87. Macagno, E. R., 1981, Cellular interactions and pattern formation in the development of the visual system in *Daphnia magna* (Crustacea, Branchiopoda). II. Induced retardation of optic axon ingrowth results in a delay in laminar neuron differentiation, *J. Neurosci.* **1:**945.

88. Feldman, J. D., Gaze, R. M., and Keating, M. J., 1971, Delayed innervation of the optic tectum during development in *Xenopus laevis, Exp. Brain Res.* **14:**16.

89. Chung, S.-H., Gaze, R. M., and Sterling, R. V., 1973, Abnormal visual function in *Xenopus* following stroboscopic illumination, *Nature New Biol.* **246**:186.
90. Schmidt, J. T., and Edwards, D. L., 1982, Activity sharpens the map during the regeneration of the retinotectal projection in goldfish, *Neurosci. Abstr.* **8**:668.
91. Meyer, R. L., 1982, Tetrodotoxin inhibits the formation of refined retinotopography in goldfish, *Dev. Brain Res.* **6**:293.
92. Schmidt, J. T., 1979, Movement of optic terminals in goldfish tectum after local pre or postsynaptic blockade of transmission, *Neurosci. Abstr.* **5**:635.
93. Freeman, J. A., 1977, Possible regulatory function of acetylcholine receptor in maintenance of retinotectal synapses, *Nature (London)* **269**:218.
94. Schmidt, J. T., 1982, The formation of retinotectal projections, *Trends in Neuro Sci.* **4**:111.
95. Yoon, M. G., 1975, Effects of post-operative visual environments on reorganization of retinotectal projection in goldfish, *J. Physiol. (London)* **246**:673.
96. Meyer, R. L., and Scott, M. Y., 1977, Failure of continuous light to inhibit compression of retinotectal projection in goldfish, *Brain Res.* **128**:153.
97. Meyer, R. L., 1982, Tetrodotoxin blocks the formation of ocular dominance columns in goldfish, *Science* **218**:589.
98. Boss, V., and Schmidt, J. T., 1982, Tests for a role of activity in the formation of ocular dominance patches, *Neurosci. Abstr.* **8**:668.
99. Stryker, M. P., and Harris, W. A., Impulse activity is necessary for the segregation of ocular dominance columns in the cat's visual cortex, submitted for publication.
100. Innocenti, C. M., Fiore, L., and Cominiti, R., 1977, Exuberant projections into the corpus callosum from the visual cortex of newborn cats, *Neurosci. Lett.* **4**:237.
101. Hubel, D. H., Wiesel, T. N., and LeVay, S., 1977, Plasticity of ocular dominance columns in monkey striate cortex, *Philos. Trans. R. Soc. London Ser. B* **287**:377.
102. Rakic, P., 1979, Genesis of visual connections in the rhesus monkey, in: *Developmental Neurobiology of Vision* (R. D. Freeman, ed.), pp. 249–279, Plenum Press, New York.
103. Land, P. W., and Lund, K. R., 1979, Development of the rats uncrossed retinotectal pathway and its relation to plastic studies, *Science,* **205**:698.
104. McLoon, S. C., 1982, Alterations in precision of the crossed retinotectal projection during chick development, *Science* **215**:1418.
105. Constantine-Paton, M., and Law, M. I., 1978, Eye-specific termination bands in tecta of three-eyed frogs, *Science* **202**:639.
106. Straznicky, C., Tay, D., and Hiscock, J., 1982, Segregation of optic fiber projections into eye-specific bands in dually innervated tecta in *Xenopus, Neurosci. Lett.* **19**:131.
107. Law, M. I., and Constatine-Paton, M., 1980, Right and left eye bands in frogs with unilateral tectal ablations, *Proc. Natl. Acad. Sci. USA* **77**:2314.
108. Meyer, R. L., 1979, Extra optic fibers exclude normal fibers from tectal regions in goldfish, *J. Comp. Neurol.* **183**:883.
109. Fawcett, J. W., and Willshaw, D. J., 1982, Compound eyes project stripes on the optic tectum in *Xenopus, Nature (London)* **296**:350.
110. Ide, C. F., Fraser, S. E., and Meyer, R. L., Eye dominance columns from an isogenic double nasal frog eye, *Science* **221**:293.
111. LeVay, S., and Stryker, M. P., 1979, The development of ocular dominance columns in the cat, *Soc. Neurosci. Symp.* **4**:83.
112. Chung, S.-H., and Cooke, J., 1975, Polarity and structure of ordered nerve connections in the developing amphibian brain, *Nature (London)* **258**:126.
113. Chung, S.-H., and Cooke, J., 1978, Observations on the formation of the brain and of nerve connections following embryonic manipulation of the amphibian neural tube, *Proc.R. Soc. Lond. B.* **201**:335.
114. Leber, S. M., and Model, P. G., 1982, Normal localization of synapses on the amphibian Mauthner cell despite precocious synaptogenesis, *Neurosci. Abstr.* **8**:436.
115. Model, P. G., 1978, Aspects of Mauthner cell differentiation in the axolotl, *Ambystoma mexicanum, Am. Zool.* **18**:253.
116. Model, P. G., and Wurzelmann, S., 1982, Vestibular axons form synapses on abnormally derived Mauthner cells, *Dev. Brain Res.* **3**:123.

117. Landmesser, L., 1981, Pathway selection by embryonic neurons, in: *Studies in Developmental Neurobiology* (W. M. Cowan, ed.), pp. 53–73, Oxford University Press, London.
118. Oppenheim, R. W., 1981, Neuronal cell death and some related regressive phenomena during neurogenesis: A selective historical review and progress report, in: *Studies in Developmental Neurobiology* (W. M. Cowan, ed.), pp. 74–133, Oxford University Press, London.
119. Lamb, A. H., 1979, Evidence that some developing limb motoneurons die for reasons other than peripheral competition, *Dev. Biol.* **71**:8.
120. Weiss, P., 1922, Die funktion transplantierter amphibien extremitäten: Aufstellung einer resonanz theorie der motorischen nerventätigkeit auf grund abgestimmter endorgane, *Arch. Mikrosk. Anat. Entwicklungsmech.* **102**:635.
121. Weiss, P., 1928, Erregungsspezifität und erregungsresonanz: Grundzüge einer theorie der motorischen nerventätigkeit auf grund spezifischer zuordnung ("Abstimmung") zwischen zentraler und peripherer erregung (Nach experimentellen Ergabnissen), *Ergeb. Biol.* **3**:1.
122. Weiss, P., 1936, Selectivity controlling the central–peripheral relations in the nervous system, *Biol. Rev.* **11**:494.
123. Miner, N., 1956, Integumental specification of sensory fibers in the development of cutaneous local sign, *J. Comp. Neurol.* **105**:161.
124. Jacobson, M., and Baker, R. E., 1968, Development of neuronal connections with skin grafts in the frog. *Science* **160**:543.
125. Grimm, L. M., 1971, An evaluation of myotypic respecification in axolotls, *J. Exp. Zool.* **178**:479.
126. Heidemann, M. K., 1977, Neurophysiological and behavioral evidence for selective reinnervation in skin-grafted *Rana pipiens, Proc. Natl. Acad. Sci. USA* **74**:5749.
127. Hollyday, M., and Grobstein, P., 1981, Of limbs and eyes and neuronal connectivity, in: *Studies in Developmental Neurobiology* (W. M. Cowan, ed.), pp. 188–217, Oxford University Press, London.
128. Graziadei, P. P. C., Levine, R. R., and Monti Graziadei, G. A., 1978, Regeneration of olfactory axons and synapse formation in the forebrain after bulbectomy in neonatal mouse, *Proc. Natl. Acad. Sci. USA* **75**:5230.
129. Gruberg, E. R., and Solish, S. P., 1978, The relationship of a monoamine fiber system to a somatosensory tectal projection in the salamander *Ambystoma tigrinum, J. Morphol.* **157**:137.
130. Gruberg, E. R., and Harris, W. A., 1981, The serotonergic somatosensory projection to the tectum of normal and eyeless salamanders, *J. Morphol.* **170**:55.
131. Harris, W. A., 1982, Differences between embryos and adults in the plasticity of somatosensory afferents to the axolotl tectum, *Dev. Brain Res.* **7**:245.
132. Keating, M. J., 1974, The role of visual function in the patterning of binocular visual connexions, *Br. Med. Bull* **30**:145.
133. Keating, M. J., and Feldman, J. D., 1975, Visual deprivation and intertectal neuronal connections in *Xenopus laevis, Proc. R. Soc. London Ser. B* **191**:467.
134. Udin, S. B., and Keating, M. J., 1981, Plasticity in a central nervous pathway in *Xenopus:* Anatomical changes in the isthmotectal projection after larval eye rotation, *J. Comp. Neurol.* **203**:575.
135. Udin, S. B., 1982, Abnormal visual input during development leads to abnormal trajectories of isthmotectal axons in *Xenopus* frog, *Neurosci. Abstr.* **8**:437.
136. Kollros, J. J., 1953, The development of the optic lobes in the frog. I. The effects of unilateral enucleation in embryonic stages, *J. Exp. Zool.* **123**:153.
137. Currie, J., and Cowan, W. M., 1974, Some observations on the early development of the optic tectum in the frog *(Rana pipiens),* with special reference to the effects of early eye removal on mitotic activity in the larval tectum, *J. Comp. Neurol.* **156**:123.
138. Kollros, J. J., 1982, Peripheral control of midbrain mitotic activity in the frog, *J. Comp. Neurol.* **205**:171.
139. Twitty, V. C., 1932, Influence of the eye on the growth of its associated structures, studied by means of heteroplastic transplantation, *J. Exp. Zool.* **61**:333.
140. Burr, H. W., 1916, The effects of the removal of the nasal pits in *Amblystoma* embryos, *J. Exp. Zool.* **20**:27.
141. Law, M. I., and Constantine-Paton, M., 1981, Morphometric examinaton of competing retinal projections in three-eyed frogs, *Neurosci. Abstr.* **7**:405.
142. Lamb, A. H., 1982, Target dependency of developing motoneurons in *Xenopus laevis, J. Comp. Neurol.* **203**:157.

143. Hollyday, M., and Hamburger, V., 1976, Reduction of the naturally occurring motor neuron loss by enlargement of the periphery, *J. Comp. Neurol.* **170:**311.

144. Lamb, A. H., 1979, Ventral horn cell counts in *Xenopus* with naturally occurring supernumerary hindlimbs, *J. Embryol. Exp. Morphol.* **49:**13.

145. Hollyday, M., 1980, Motoneuron histogenesis and the development of limb innervation, *Curr. Top. Dev. Biol.* **15:**181.

146. Lamb, A. H., 1980, Motoneurone counts in *Xenopus* frogs reared with one bilaterally-innervated hindlimb, *Nature (London)* **284:**347.

147. Lee, M. T., 1982, Regeneration and functional reconnection of an identified vertebrate central neuron, Ph.D. thesis, University of California, San Diego.

148. Sperry, R. W., 1944, Optic nerve regeneration with return of vision in anurans, *J. Neurophys.* **7:**57.

149. Attardi, D. G., and Sperry, R. W., 1963, Preferential selection of central pathways by regenerating optic fibers, *Exp. Neurol.* **7:**46.

150. Hibbard, E., 1967, Visual recovery following regeneration of the optic nerve through the oculomotor root in *Xenopus, Exp. Neurol.* **19:**350.

151. Levine, R. L., 1982, Widespread regeneration of central axons through the central nervous system of the goldfish, *Dev. Brain Res.* **9:**416.

152. Gaze, R. M., and Fawcett, J. W., 1982, Pathways of *Xenopus* optic fibers regenerating from normal and compound eyes under various conditions, *J. Emb. Exp. Morp.* **73:**17.

153. Schmidt, J. T., Cicerone, C. M., and Easter, S. S., 1978, Expansion of the half-retinal projection to the tectum in goldfish: An electrophysiological and anatomical study, *J. Comp. Neurol.* **177:**257.

154. Gaze, R. M., and Sharma, S. C., 1970, Axial differences in the reinnervation of the goldfish optic tectum by regenerating optic nerve fibers, *Exp. Brain Res.* **10:**171.

155. Schmidt, J. T., 1978, Retinal fibers alter tectal positional markers during the expansion of the half retinal projection in goldfish, *J. Comp. Neurol.* **177:**279.

156. Sharma, S. C., and Romeskie, 1977, Immediate "compression" of the goldfish retinal projection to a tectum devoid of degenerating debris, *Brain Res.* **134:**1.

157. Yoon, M. G., 1975, Readjustment of retinotectal projection following reimplantation of a rotated or inverted tectal tissue in adult goldfish, *J. Physiol. (London)* **252:**137.

158. Yoon, M. G., 1980, Retention of topographic addresses by reciprocally translocated tectal reimplants in adult goldfish, *J. Physiol. (London)* **308:**197.

159. Yoon, M. G., 1977, Induction of compression in the re-established visual projections onto a rotated tectal reimplant that retains its original topographic polarity within the halved optic tectum of adult goldfish, *J. Physiol (London)* **264:**379.

160. Yoon, M. G., 1979, Reciprocal transplantations between the optic tectum and the cerebellum in adult goldfish, *J. Physiol. (London)* **288:**211.

161. Levine, R. L., 1979, Involvement of pigment epithelium in the genesis of locus specificities during retinal regeneration in the newt, *Brain Res.* **162:**154.

162. Letinsky, M. S., Fischbeck, K. H., and McMahan, U. J., 1976, Precision of reinnervation of original postsynaptic sites in frog muscle after nerve crush, *J. Neurocytol.* **5:**691.

163. Frank, E., Jansen, J. K. S., Lømo, T., and Westgaard, R., 1975, The interaction between foreign and original motor nerves innervating the soleus muscle of rats, *J. Physiol. (London)* **247:**725.

164. Jansen, J. K. S., Lømo, T., Nicolaysen, K., and Westgaard, R. H., 1973, Hyperinnervation of skeletal muscle fibers: Dependence on muscle activity, *Science* **181:**559.

165. Mark, R. F., Marotte, L. R., and Mart, P. E., 1972, The mechanism of selective reinnervation of fish eye muscles. IV. Identification of repressed synapses, *Brain Res.* **46:**149.

166. Yip, J. W., and Dennis, M. J., 1976, Suppression of transmission at foreign synapses in adult newt muscle involves reduction in quantal content, *Nature (London)* **260:**350.

167. Scott, S. A., 1977, Maintained function of foreign and appropriate junctions on reinnervated goldfish extraocular muscles, *J. Physiol. (London)* **168:**87.

168. Dennis, M. J., and Yip, J. W., 1978, Formation and elimination of foreign synapses on adult salamander muscle, *J. Physiol. (London)* **274:**299.

169. Mellinger, M. W., 1975, Reconstruction of the cerebral hemispheres in *Ambystoma mexicanum* following bilateral ablation and ablation/transplantation, *Anat. Rec.* **75:**424.

170. Hershkowitz, M., Segal, M., and Samuel, D., 1972, The acquisition of dark avoidance by transplantation of the forebrain of trained newts *(Pleurodeles waltl)*, *Brain Res.* **48**:366.

171. Kirsche, K., and Kirsche, W., 1968, Ueber homotransplantation eines endhirndrittels von *Ambystoma mexicanum*, *Z. Mikrosk. Anat. Forsch.* **79**:223.

172. Pietsch, P., and Schneider, C. W., 1969, Brain transplantation in salamanders: An approach to memory transfer, *Brain Res.* **14**: 707.

173. Pietsch, P., 1981, *Shufflebrain,* Houghton Mifflin, Boston.

174. Harrison, R. G., 1934, Heteroplastic grafting in embryology, *Harvey Lect.* **29**:116.

175. Harris, W. A., 1979, Amphibian chimeras and the nervous system, *Soc. Neurosci. Symp.* **4**:228.

176. Constantine-Paton, M., 1981, Induced ocular dominance zones in tectal cortex, in: *The Organization of the Cerebral Cortex* (F. O. Schmitt, F. Worden, and S. Dennis, eds.), pp. 47–67, MIT Press, Cambridge, Mass.

4

Transplantation of the Developing Mammalian Visual System

STEVEN C. McLOON AND LINDA K. McLOON

1. Introduction

Processing of the visual image in the brain requires an orderly relay of information between the various visual centers. This is accomplished by having these visual centers interconnect in very precise patterns. A good example of this is the projection of the retinal ganglion cells to the brain. The axons of the ganglion cells course along a well-defined tract, enter only very specific nuclei in the brain, and within these nuclei terminate in a retinotopic fashion such that neighboring retinal ganglion cells terminate in neighboring areas of the central visual nuclei. A major problem in developmental neurobiology is to determine how these patterns of connections develop. The transplantation technique affords one method by which to study this problem.

A feature acquired by a given neuronal population during development is either a manifestation of the environment in which that tissue develops or it reflects a developmental program intrinsic to that tissue. By studying a neuronal population challenged with a novel context in which to develop, it may be possible to determine which features are specifically programmed into the developmental scheme for that population versus those features that are imposed by the environment. Numerous studies over the past 70 years have shown that neuronal tissue from fetal mammals will survive when placed in a host brain (see Chapter 1 for a review). Several studies have taken advantage of this technique to study the development of various visual centers transplanted to novel locations of the brain. These studies have addressed questions pertaining to the mechanisms that guide growing axons, determine which nuclei the axons will enter, and determine where within those nuclei an axon will form a synapse. This chapter summarizes the

STEVEN C. McLOON AND LINDA K. McLOON ● Departments of Anatomy and Ophthalmology, University of Minnesota, Minneapolis, Minnesota 55455.

most salient results of these studies and attempts to draw some preliminary conclusions from these results.

Three visual centers have been transplanted in the studies described herein, the retina, superior colliculus (or tectum), and occipital cortex. All three structures have been well characterized morphologically. The retina has three cell layers, a photoreceptor cell layer, an inner nuclear layer, and a ganglion cell layer. The photoreceptor cells are light sensitive and transduce the light into the electrochemical signals carried by neurons. This information is relayed via synapses to the neurons of the inner nuclear layer, bipolar cells, horizontal cells, and amacrine cells, which integrate and relay the information to the neurons of the ganglion cell layer. The ganglion cell layer contains amacrine cells and ganglion cells. The axons of the ganglion cells constitute the sole neuronal connections from the retina to the brain. The ganglion cell axons terminate in certain specific nuclei of the brain including the lateral geniculate nucleus of the thalamus and the superior colliculus of the midbrain. The lateral geniculate nucleus relays visual information to the visual cortex in the occiptal lobe of the cerebral hemispheres. Six cell layers have been described in the visual cortex, each with unique afferents and efferents. At no time does the cortex receive visual input directly from the retina. The superior colliculus is also a laminated structure with cells in the most superficial layers receiving synapses from retinal ganglion cell axons. The cortex also projects to the superior colliculus. The superior colliculus is important in mediating various motor functions, particularly those related to eye control. It is the rules governing the development of these connections that the following transplantation studies address.

2. The Transplantation Technique

The experiments described in this chapter involved transplanting fetal retina, tectum, or cortex adjacent to the superior colliculus of newborn host rats (see Ref. 18 for a complete description of the transplantation technique). Donor tissue was obtained from time-mated female rats of the same strain as the hosts. Embryos on gestational day 14 to 20 were removed from the uterus of the mother and placed in ice-cold tissue culture media. The retina, rostral portion of the tectum, or caudal portion of the telencephalic vesicle was dissected from the head. The vasculature and most of the pigment epithelium were stripped from the retinae, and the meninges were removed from the tecta and cortices. One to twelve hours after birth, host rats were anesthetized with ether. The head of a host animal was transilluminated to reveal internal landmarks, and a small slit was made in the skin and skull lateral to the right superior colliculus. The tissue to be transplanted was aspirated into a glass pipet (I.D. 0.5 mm). The pipet was inserted through the slit, the tip was positioned over the left superior colliculus, and the tissue was ejected along with 2–3 μl of culture media. Half of the animals receiving retinal transplants also had a unilateral enucleation contralateral to the transplant. After receiving the transplant, the host rats were returned to their mothers and allowed to survive 1–3 months, at which time the histology and connections of the transplants were studied as described in subsequent sections of this chapter.

Variations of this general technique were used in several experiments. In several instances, the donor tissue was transferred to the exposed surface of the host midbrain as a flat sheet carried on the meniscus of a hair loop.[19] This was done so as to better preserve the cytoarchitecture or orientation of the tissue. For several experiments, the donor tissue was held in culture for up to 2 weeks[29] or was dissociated and reaggregated prior to transplantation.[28] In other cases, the tissue was placed in locations other than the midbrain, particularly to the cerebral cortex. We have also transplanted fetal tissue to the brain of adult rats by aspirating out the cerebral cortex over the host superior colliculus and placing the tissue adjacent to the cortex and midbrain.[35]

3. General Aspects of Transplant Development in Neonates

Transplants of embryonic tectum, cortex, or retina survived well when placed adjacent to the superior colliculus of newborn host rats. Of all animals that received transplants, approximately 80% had identifiable transplants after 1 or more months of posttransplantation survival. This percentage did not vary substantially among the three types of tissue transplanted. Also, there was no apparent correlation between the proportion of transplants that survived and the posttransplantation survival time. Particularly for the retinal and cortical transplants, there was a reduction in transplant viability by using older donor tissue, whereas viability was not significantly affected for any of the three tissues by altering the age of the host at the time of transplantation. Although inbred rat strains were not used nor was any effort made to match histocompatibility markers of the donor and recipients, there was never evidence of transplant rejection even in those animals studied after a short transplant survival. Clearly the high transplant survival rate seen in these studies shows that, if any antigenic reaction occurs, it has little or no effect on transplant viability.

Transplants were most often found positioned over the inferior colliculus or rostral cerebellum 1 month after transplantation. A developmental series where the host animals were sacrificed at 2-day intervals revealed a correlation between the caudal growth of the cerebral hemispheres and the caudal displacement of the transplants from their initial position over the superior colliculus (Fig. 1). Presumably, this displacement was the result of mechanical factors, either a push from the growing hemispheres or changes in the relative position of the meninges to which the transplants were attached. In a few cases though, the transplants were embedded in the host midbrain.

Histological examination revealed numerous blood vessels coursing within the transplant and between the transplant and host. The majority of the vessels ran between the transplant and host superior colliculus, but vessels were also found in most cases running between the transplant and meninges. It was less common for vessels to be contributed by the host cortex to the transplant.

At the time of transplantation, each of the three tissue types was rather undifferentiated, presenting a simple pseudostratified neuroepithelium (Fig. 2A). The transplants exhibited considerable posttransplantation growth, which thymidine labeling showed was the result of continued cell division in the transplants. At least for retinal and cortical transplants, neurogenesis continues on the same schedule as if the tissue had remained in its natural position.[13,26] By 1 month posttransplantation, the histological

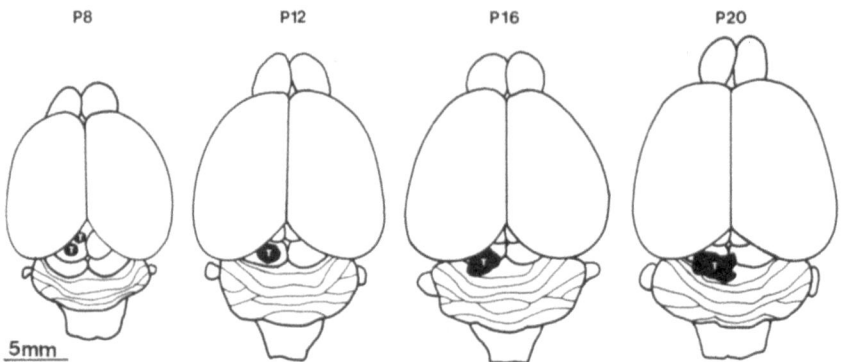

FIGURE 1. Camera lucida drawings of the dorsal surface of brains with retinal transplants (T). This shows a series of brains from 8 days after transplantation (P8) to 20 days after transplantation (P20). Notice that the caudal displacement of the transplants correlates with the growth of the cerebral hemispheres. The position of the transplant at P20 is typical of that found at all later ages.

organization and the connections manifested by the transplants varied substantially among each of the three types. In the remainder of this chapter, the histological organization and connections of each type of transplant are described, followed by a discussion of developmental rules and mechanisms that can be implied by comparing the results from the different transplant studies.

4. Retinal Transplants

Of the three types of tissue transplanted to the superior colliculus, the retinal transplants maintained the greatest similarity to the normal retina both in terms of their histology and in the connections they formed with the host brain. However, when retinae were transplanted in a similar fashion to the cortex of a newborn host, there were several significant differences in the transplant development. In this section, the histology and connections that retinae exhibit one or more months after transplantation to the superior colliculus are examined and then compared to retinae transplanted to the cortex.

4.1. Histological Characteristics

The histology of the retinal transplants was characteristic of the normal retina.[33] Nissl-stained sections of the transplants revealed cell and plexiform layers appropriate for the retina. The lamination was occasionally arranged in a round eyelike fashion with a photoreceptor cell layer around the outside and a ganglion cell layer along the inside (Fig. 2B). More typically, the photoreceptor cells were arranged in numerous rosettes surrounded by inner nuclear layer cells and a ganglion cell layer situated outside the inner nuclear layer (Fig. 2C). In some cases, the inner nuclear layer appeared doubled. Ganglion cell axons collected in numerous places throughout the transplant and exited as discrete bundles from the transplant. These bundles could often be discerned on the gross brain as they coursed rostrally and entered one of the superior colliculi. Each of the retinal cell types has been examined in some detail by various methods and compared to its counterpart in the normal retina.

The morphology of the photoreceptor cells has been studied with the electron

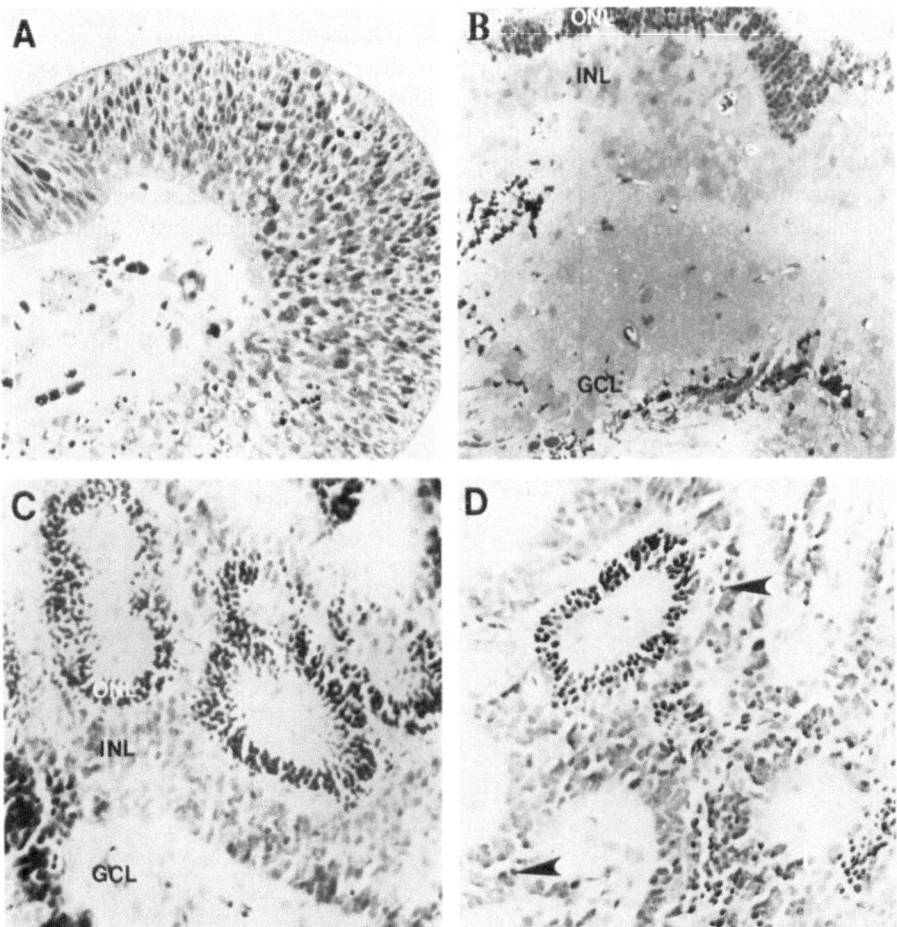

FIGURE 2. Photomicrographs of sections through retinal transplants stained with cresyl violet. (A) A retina on embryonic day 14, the time of transplantation. It is a typical pseudostratified neuroepithelium that is still relatively undifferentiated. (B) A retina 1 month after transplantation. This transplant is an eyelike structure with photoreceptors on the outside and ganglion cells and optic fibers on the inside [outer nuclear layer (ONL), inner nuclear layer (INL), and ganglion cell layer (GCL)]. (C) A more typical transplant with the photoreceptor cells arranged in rosettes surrounded by inner nuclear layer cells and ganglion layer cells. This retina was also dissociated and reaggregated prior to transplantation. (D) A retina that was transplanted to an adult host. It contrasts with the previous retinae that were transplanted to newborns in that the separation between the different cell layers is not complete. Notice the photoreceptor cells interspersed in the inner nuclear layer cells (arrowheads).

microscope.[23] The cell bodies and inner segments of these cells are relatively unremarkable (Fig. 3A). Both rod and cone cells have been encountered, although no attempt was made to compare the relative proportion of each type found in the transplants to the normal retina. Considerable variations have been observed in the morphology of the outer segments. In our earliest transplants, no pigment epithelium was included with the retinae during transplantation. In these transplants, the outer border of the photoreceptors was composed of cilia-like processes. More recently, however, some pigment epithelium was included so as to help locate the transplants for injection or to place a lesion in them for the studies on connectivity. In these transplants, irregular outer segments were present, and phagocytic cells were found in the lumen of the rosettes. Other studies have demonstrated a similar importance of pigment epithelium in the development and maintenance of outer segments.[44]

Amacrine cell morphology and distribution in the retinal transplants were studied with immunohistochemical techniques (McLoon and Karten, unpublished). Antibodies

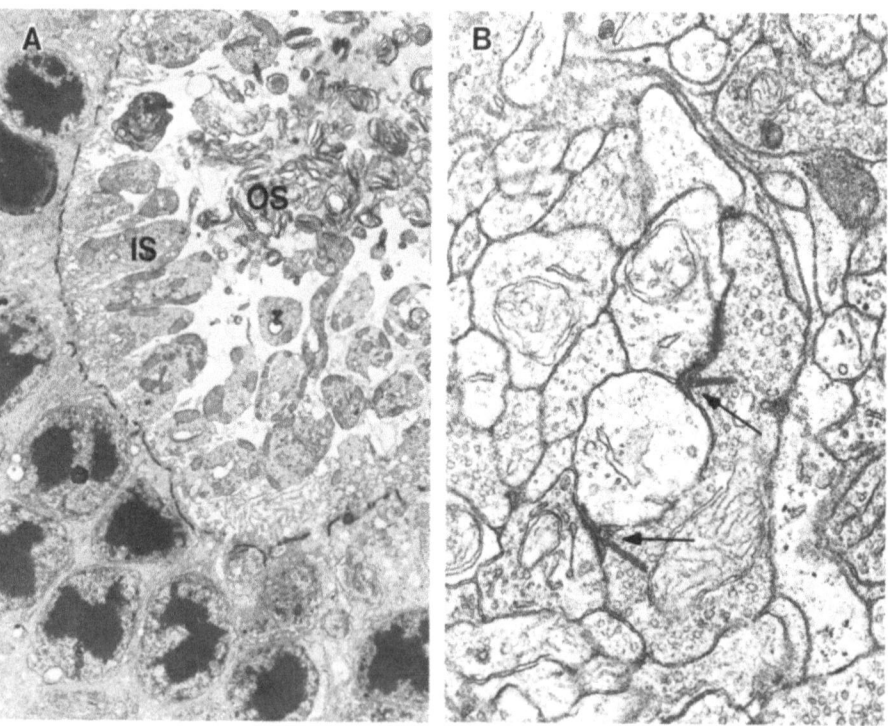

FIGURE 3. Electron micrographs of retinal transplants. (A) Photoreceptor cells. The cell bodies are on the left and bottom. An external limiting membrane separates the cell bodies from the inner segments (IS). Outer segments (OS) are present in the center of the rosette but are extremely disorganized. (B) A synaptic profile in the inner plexiform layer. The two ribbon synapses (arrows) are typical of synapses formed by bipolar cell processes onto amacrine and ganglion cells.

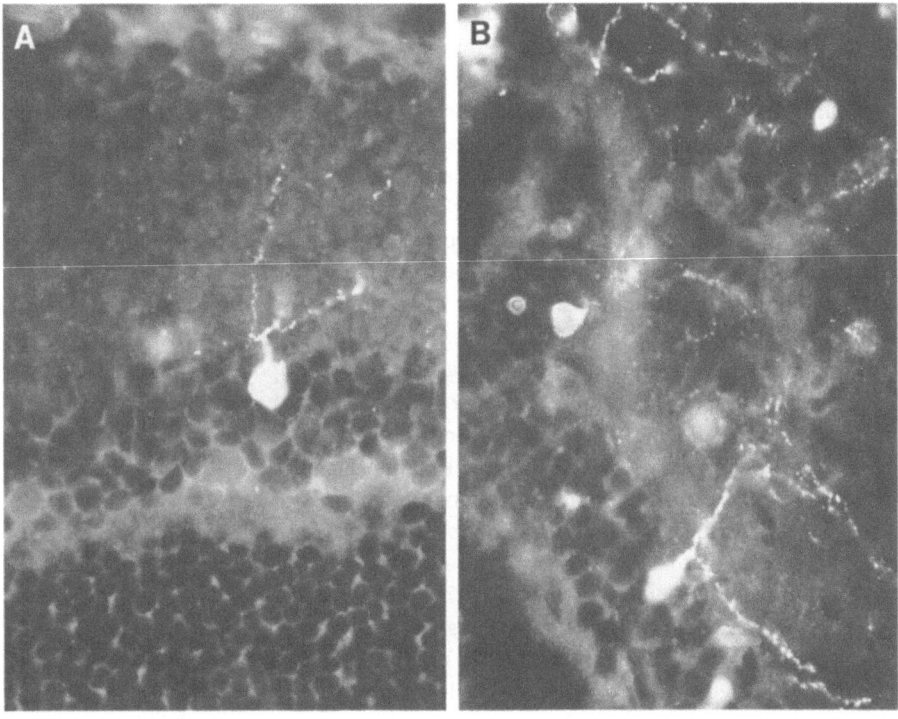

FIGURE 4. Photomicrographs of amacrine cells labeled immunohistochemically with an antibody against vasointestinal polypeptide in normal retina (A) and in a retinal transplant (B). The position, size, and distribution of the dendritic arbors for this cell type are comparable in normal retina and transplants.

against tyrosine hydroxylase, 5-hydroxytryptamine, somatostatin, vasoactive intestinal polypeptide, substance P, and glutamic acid decarboxylase were used to identify specific classes of amacrine cells and their processes. These were compared with the morphology of amacrine cells in normal retinae processed in the same manner. Based on the presence of specific immunoreactivity, all classes of amacrine cells identified in the normal retina were also found in the retinal transplants. All amacrine cell types were restricted to the inner portion of the inner nuclear layer or the ganglion cell layer. In instances where the transplant inner nuclear layer was doubled, the various amacrine cell populations were found in both layers. In these cases, the amacrine cells in the outer layer often had dendrites that arborized in the first and second inner plexiform layers. The general branching pattern of the dendritic arbors of each of the amacrine cell types matched very closely their counterparts in the normal retina (Fig. 4). There were minor differences mainly in the orientation and fine pattern of the dendrites, but this was probably the result of the gross distortion of the cell and plexiform layers in the transplants.

The morphology of the ganglion cells in retinal transplants was studied by back-filling cells with HRP to obtain a Golgi-like image.[34,42] HRP was injected into the superior colliculus of the host rats, and after a sufficient survival time the transplants were processed with standard histochemical techniques. Most of the retrogradely labeled cells were distributed in the ganglion cell layers of the transplants. Occasionally, labeled cells were found along the inner border of the inner nuclear layer, which probably represent displaced ganglion cells. Based on the size and shape of their dendritic arbors in the inner plexiform layer, filled ganglion cells were identified that corresponded to each of the four general subtypes distinguished by Perry in the normal rat retina.[40] Occasionally, filled ganglion cells were found that had an appropriate dendritic arbor in their adjacent inner plexiform layer and another dendritic branch that ran to a neighboring inner plexiform layer and formed another small arbor (Fig. 5A). This second arbor was of the same pattern as the main arbor. Even in these cases, all the dendrites from a given cell arose from the same side of that cell, attesting to the inherent polarity of ganglion cells.

Ultrastructural analysis of the synaptology within retinal transplants has revealed a full complement of synaptic types.[23] The outer nuclear layer contained conspicuous synaptic triad arrangements with associated synaptic ribbons typical of photoreceptor–bipolar–horizontal cell connections. Occasional anomalies appeared in these synapses. The one most often encountered was a doubled synaptic ribbon. The inner plexiform layer contained numerous dyad synapses also with a synaptic ribbon and conventional synapses typical of the normal retina (Fig. 3B). This suggests that appropriate amacrine–bipolar–ganglion cell interrelationships were maintained.

The immunoreactivity and distribution of glial cell processes in the retinal transplants were somewhat surprising.[32] The Müller cell processes of the normal retina run from outer to inner limiting membrane as can be demonstrated by the Müller cell-specific antibody G3 (developed by C. Barnstable),[2] while an antibody to glial fibrillary acidic protein (GFA) only stains processes in the optic fiber layer which appear to be astrocytes (Fig. 6A). The astrocytes migrate into the retina along the optic nerve during the first postnatal week. In the transplants, however, the pattern of glial staining was similar with both antibodies. Single glial fibers ran from the outer limiting membrane at the center of the photoreceptor cell rosettes through the ganglion cell layer which appeared to be both GFA and G3 positive (Fig. 6B). By injecting colchicine into the eye or by injuring the normal retina, Müller cells can develop a GFA antigenicity. It is not clear at this point if the glial cells in the retinal transplants are astrocytes that migrated in from the host brain or are Müller cells expressing an injurylike reactivity. Electron microscopic immunohistochemistry using the two antibodies should resolve this.

Retinae have been dissociated and reaggregated prior to transplantation.[28] Fetal retinae were removed from the eye cups and dissociated into single-cell suspensions. The suspensions were pelleted by centrifugation and fragments of the pellet were transplanted to newborn hosts as before. This procedure did not appear to seriously alter the development of normal retinal morphology. These transplants exhibited lamination and cellular morphology similar to those seen in retinae transplanted directly (Fig. 2C). Retinae have also been held in tissue culture for as long as 2 weeks prior to transplantation.[29] Again, these transplants developed relatively normal morphology. However, the longer-term cultures never developed as great a size as the other transplants presumably due to extensive cell death in the explants while in culture.

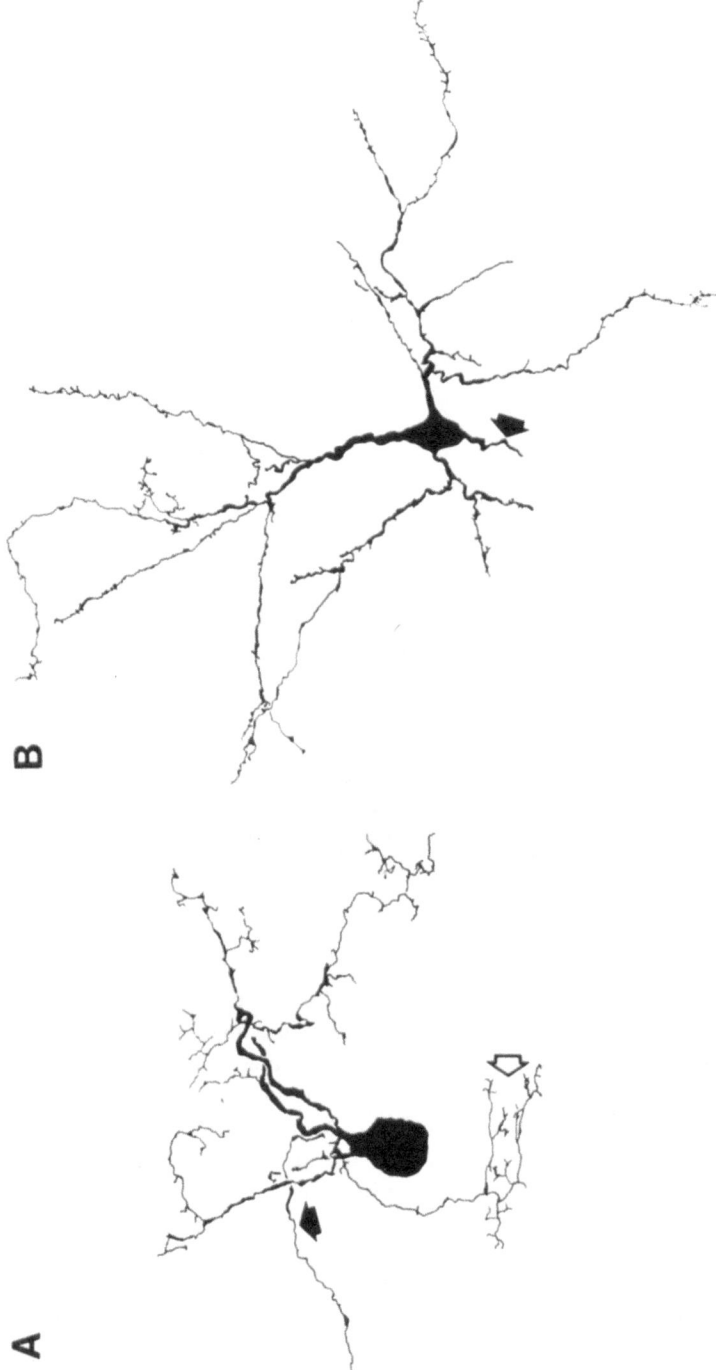

FIGURE 5. Camera lucida drawings of cells in the transplants filled by an injection of HRP. (A) A ganglion cell in a retinal transplant. The main dendritic arbor of the cell ramifies in the nearest inner plexiform layer while one branch runs across the optic fiber layer and arborizes in a distant inner plexiform layer (open arrow). (B) A pyramidal cell in a cortical transplant. The apical dendrite of the cell curves, and the basal dendrites are not restricted to one plane. The basal dendrites run in several directions and appear to be orienting to clumps of small cells. Axons of both cells are marked by closed arrows.

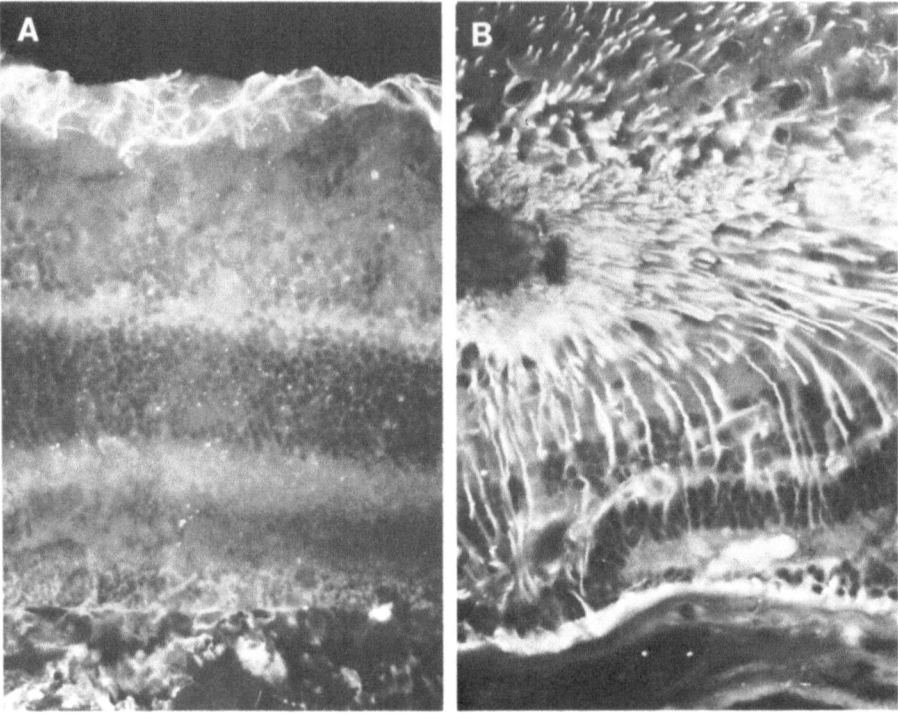

FIGURE 6. Photomicrographs of glial cells labeled immunohistochemically with an antibody against glial fibrillary acidic protein (GFA) in normal retina (A) and in a retinal transplant (B). In normal retinae, the GFA-positive cells are restricted to the optic fiber layer. In the transplants, the GFA-positive cells run between the inner and outer limiting membrane, a distribution typical of Müller cells.

4.2. Connections with the Host Brain

The connections of retinal transplants with the host brain have been investigated using a variety of techniques. Most of the studies on connections of retinal transplants have either involved placing a lesion in the transplant and then examining the host brain for degenerating axons using the Fink–Heimer silver stain or injecting the transplants with HRP and tracing the anterograde transport of the protein into the host brain.[33] In almost every case, a projection was identified to the superficial layers of the superior colliculus in the stratum (s.) zonale, s. griseum superficiale, and s. opticum. Often, the projection was traced rostrally along the optic tract and into the nucleus of the optic tract, posterior and olivary pretectal nuclei, dorsal and medial terminal nuclei, and dorsal lateral geniculate nucleus (Fig. 7). These are all nuclei that are normally retinorecipient. No nonretinal nuclei along the tract, such as lateral posterior nucleus, received a projection. For some unknown reason, the ventral lateral geniculate nucleus has never been observed to receive a retinal transplant projection even when the more rostrally located medial terminal nucleus was innervated. It is possible that the temporal aspects

of the development of the ventral lateral geniculate nucleus and the transplant are misaligned. Transplants to prenatal hosts will hopefully address this question. Retinae
transplanted to hosts with bilateral enucleations at the time of transplantation exhibited
the same ability to innervate the visual nuclei.

There is a clear suggestion of competition between the transplant and host retinal
projections in the superior colliculus. When the host eye contralateral to the colliculus
innervated by the transplant was removed at birth, the transplant terminals were found
throughout the s. griseum superficiale and s. opticum of the colliculus. However, when
the contralateral eye was present, the transplant efferents were restricted to a narrow
band along the border of the s. griseum superficiale and s. opticum. However, in contrast
to dually innervated superior colliculi in amphibians[4] or hamsters,[47] no significant segregation of fibers was seen between the host optic projections and the transplant projections to the superior colliculus. In this study, the retinal transplants received a lesion
and the host eyes were injected with [³H]proline.[33] Alternate sections of the host brain

FIGURE 7. Camera lucida drawings of a series of sections through a brain that received a projection fom a
retinal transplant. The transplant is caudal to the last sections. The transplant was injected with HRP and
the distribution of the label is indicated by the shading. The host eye contralateral to the left side was
enucleated at the time of transplantation. Notice the greater extent of the transplant projection to that side,
both within the superior colliculus (SC) and in its rostral extent. The other nuclei labeled are the nucleus
of the optic tract (NOT), dorsal terminal (DT), medial geniculate (MGN), posterior pretectal (PP), olivary
pretectal (PO), anterior pretectal (PA), lateral posterior (LP), medial terminal (MT), dorsal and ventral
lateral geniculates (dLGN and vLGN).

were processed with the Fink–Heimer degeneration technique and autoradiography. The distribution of the degeneration in the host colliculus overlapped areas with auto-radiographic label. The only exception to this was at sites where a fascicle of axons from the transplant entered the superior colliculus. In these areas, there was extensive evidence of degeneration and the autoradiographic label was nearly absent, suggesting there was local exclusion of the host retinal projection. It may be that the optic axons must surpass a certain minimum density before two populations will segregate. This was also suggested by studies on dually innervated hamster superior colliculus.[46]

After a lesion of the retinal transplants, electron microscopy was used to identify degenerating synaptic terminals in the host superior colliculus.[23] The transplants formed typical optic synapses in the colliculus. The degenerating terminals contained round vesicles, large pale mitochondria, and mainly formed axodendritic synapses with some serial synapses.

As described above with regard to ganglion cell morphology, an injection of HRP into the superior colliculus of the host rat retrogradely labeled the ganglion cells in the transplant.[34] Besides demonstrating that a projection exists from the transplant to the host superior colliculus, this technique has allowed us to address the question of topography in the projection. A focal injection in the superior colliculus in general labeled groups of cells throughout the transplants. Although this showed that there is no topography in the projection from the transplant as a whole, it does not rule out the possibility that some topographical order existed within the projection from localized regions of the transplant.

Unlike the tectal and cortical transplants, which are described below, the retinal transplants did not appear to receive a projection from the host. HRP injected into the transplants did not reveal any retrogradely labeled cells in the host brain, and [³H]proline injected into the host eyes did not result in any autoradiographic label in the transplants.

Both the retinae that were cultured and those that were dissociated and reaggregated prior to transplantation exhibited projections similar to those seen with retinae transplanted directly.[28,29] Again, these projections showed a remarkable specificity for central visual nuclei. The only significant difference with the connections of these transplants was that the retinae held in culture for 10 days or more had connections confined to the visual nuclei nearest to where the transplant axons entered the host brain. These projections were confined to the superior colliculus and pretectum.

4.3. Retina Transplanted to Other CNS Locations

Fetal retinae have also been transplanted to the cerebral cortex,[24,25,36] cerebellum,[27] and dorsal column nuclei[30] of newborn rats. The results are similar in each of these cases. Qualitatively, these transplants developed the same histological features as retinal grafts that developed adjacent to the superior colliculus. However, neither degeneration nor anterograde tract-tracing techniques revealed any projections from these transplants. Quantitative analysis of the cellular composition of the retina transplanted to the cortex showed several things. There was a reduced number of cells in the ganglion cell layer proportional to the number of cells in the inner and outer nuclear layers, and there was a loss of synapses in the inner plexiform layer involving ganglion cell processes.[25] Soma size analysis of the cells in the ganglion layer suggested that the only cells remaining in this layer were displaced amacrine cells or atrophied ganglion cells.[36] Several studies have shown that destruction of central visual nuclei during development leads to a loss of retinal ganglion cells in the eye.[10,31,41] The loss of ganglion cells in these transplants suggests that the ganglion cells were unable to form connections with appropriate nuclei. If, however, the tectum is transplanted along with the retina into the cortex, there appears to be survival of the ganglion cells.[24] Furthermore, the results show that the visual cortex, cerebellum, and dorsal column nuclei, another primary sensory nucleus like the superior colliculus, are not recognized as acceptable targets for the ganglion cell axons.

5. Tectal Transplants

Fetal tectal tissue also survived and differentiated when transplanted to the midbrain of newborn rats. Histologically, the tectal transplants were quite distinct from retinal transplants. Although the tectal transplants were not as well differentiated, features resembling the normal colliculus could be identified in the transplants. The tectal transplants also formed interconnections with the host brain that contrasted markedly with those formed by retinal transplants.

5.1. Histological Characteristics

Histological analysis of the tectal transplants revealed a complicated internal organization.[19,20] Neurofibrillar staining of the tectal transplants showed a rich plexus of fibers through much of the transplant (Fig. 8). There were also relatively fiber-sparse areas. The fiber-sparse areas appeared to have a high density of neurons with small somas when neurofibrillar preparations were correlated with Nissl-stained sections. These fiber-sparse patches were reminiscent of the s. griseum superficale in superior colliculi that has a reduced retinal projection such as that resulting from a contralateral enucleation at birth. The fiber-rich regions of the transplants had a lower density of cells and the cells were of a wider size spectrum. In some cases, the transplants remained as a sheet in the host brain. In these instances, a laminar organization was apparent that approached that seen in the normal superior colliculus. The fiber-sparse regions appeared on the surface of these transplanted sheets as does the s. griseum superficiale of the normal superior colliculus.

Electron microscopic examination of the tectal transplants revealed a neuropil appropriate for the superior colliculus.[19] Occasionally, anomalies were encountered such as synaptic densities with no presynaptic structure or presynpatic profiles lacking any obvious vesicles. These are features also seen in the partially deafferented superior colliculus.[21]

There did not appear to be any significant migration of neurons or glia between

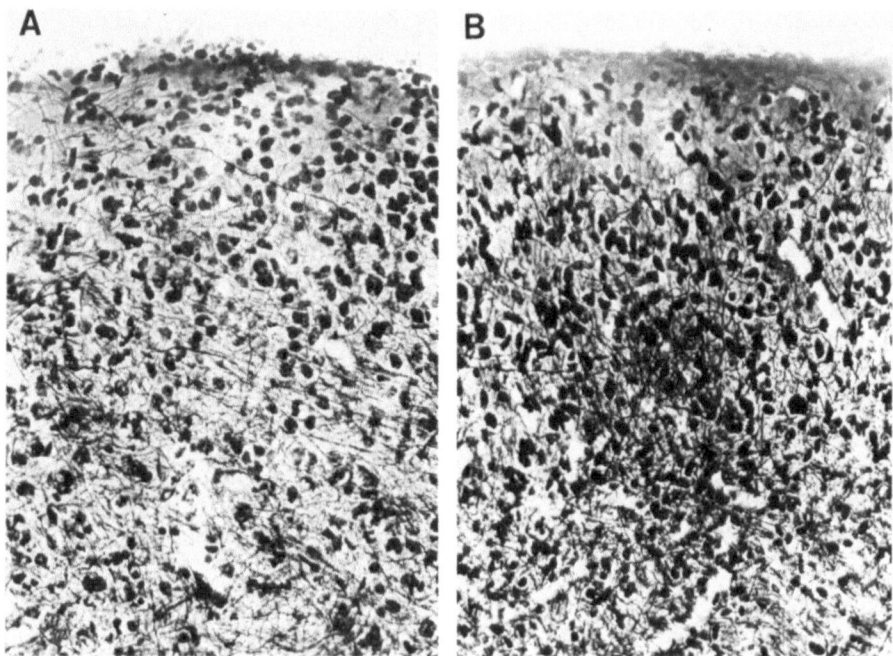

FIGURE 8. Photomicrographs of a section through a tectal transplant (A) and a normal superior colliculus (B) prepared with a Holmes stain to show neurofibrils. Notice the similar distribution of cells and fibrils in the two sections.

the host brain and transplants.[19] Exposing the donor tissue to [³H]thymidine prior to transplantion resulted in many of the neurons and glia within the transplant being labeled one month after transplantation. In most cases, no thymidine-labeled cells were found in the host brain. In a few cases, labeled cells were found separated from the transplant in host neuropil. However, these were only in cases where the transplant had been forced into the host colliculus, suggesting these cells may have been isolated during transplantation. The host brain has also been labeled with [³H]thymidine prior to receiving a transplant. Early thymidine administration was used to label neurons and later administration to label glia. In neither case were labeled neurons or glia found in the transplants.

5.2. Connections with the Host Brain

The projections of tectal transplants into the host brain followed a considerably different course than the efferents of retinal transplants.[22] The projection from tectal transplants identified by anterograde tract-tracing techniques entered via the host colliculus. Unlike the retinal transplants, the fibers did not immediately begin to ramify but continued into lower s. opticum and s. griseum intermediale (Fig. 9). Some fibers also continued deeper into the midbrain and ramified in the central gray and tegmentum. Projections have been identified to the pretectum and parabigeminal nucleus as well, although these were present variably. The usual longer projections of the superior colliculus such as those to the dorsal thalamus and spinal cord were not found with these transplants.

Unlike retinal transplants, the tectal transplants received projections from the host nervous system. Projections from the host eyes were consistently found using anterograde tract-tracing techniques. Injection of [3H]proline into the host eyes resulted in autoradiographic label in the tectal transplants.[19] Transplants located on the surface of the superior colliculus, inferior colliculus, or cerebellum received retinal projections in 93% of the cases studied, whereas the transplants completely embedded in the superior colliculus or in the ventricle did not receive retinal projections. When autoradiographic sections were correlated with alternate neurofibrillar-stained sections, it appeared that the host retinal projection was confined to the fiber-sparse patches in the transplants. Furthermore, the only fiber-sparse patches consistently innervated by the host eyes were those present on the surface of the transplants. Anterograde transport of [3H]proline was coupled with degeneration techniques to demonstrate the central projections of the individual host eyes. When both host eyes projected to the same fiber-sparse patch, there was little or no overlap of the two projections. After removal of the host eyes, degenerating retinal synapses were also identified in the tectal transplants with electron microscopy.[20]

Anterograde tract-tracing techniques were used to demonstrate host cortical projections to the transplants. These projections were distributed through most of the transplant except for the fiber-sparse regions.[19] When the host cortex was lesioned and the transplant was examined with the electron microscope, degenerating synapses were

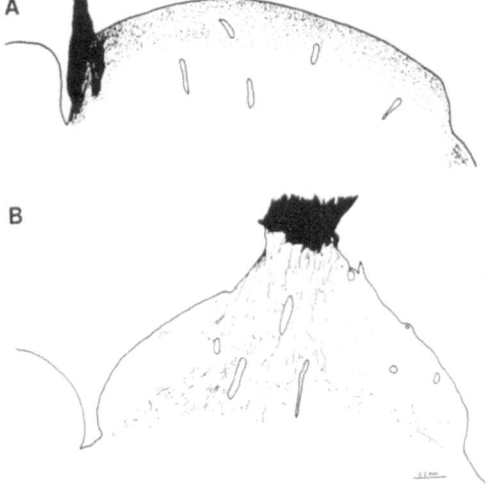

FIGURE 9. Camera lucida drawings through host superior colliculus receiving a projection from a retinal transplant (A) and from a tectal transplant (B). The transplants were injected with HRP, and the distribution of the label in the colliculus is drawn. Notice the retinal transplant projection distributes in the colliculus above the layers that receive the tectal transplant projection. (The section in "A" was processed with TMB resulting in a grainy label, and "B" was processed with DAB which shows the axons.)

found.[6] These were asymmetric synaptic contacts onto dendritic spines and small dendrites. There is also evidence that this projection is physiologically viable. Single units were recorded in the transplants that could be driven orthodromically by electrical stimulation of the cortex. The latency and percent of the units that could be activated by cortical stimulation were similar to those recorded in the normal superior colliculus.

HRP injections made into the tectal transplants not only confirmed the projection from the host cortex but also identified a number of other nuclei in the host brain with projections into the transplants.[7] These injections retrogradely labeled 2000 to 6000 cells per animal in 50 different host brain nuclei. With few exceptions, these were cells in nuclei that normally project to the superior colliculus. The few exceptions were projections appropriate to inferior colliculus, suggesting that in a few cases the tectal fragments that were transplanted included some presumptive inferior colliculus. The most consistently identified and largest projection to the transplants was from the occipital cortex. The cortical cells projecting to the transplants were layer V pyramidal cells, the cells that normally project to the superior colliculus. Labeled cells were also commonly identified in the pretectal nuclei and parabigeminal nucleus. Other areas that normally project heavily into the superior colliulus but that projected quite variably into tectal transplants included ventral lateral geniculate nucleus, substantia nigra, and locus coeruleus. The host superior colliculus also often contributed a projection to the tectal transplants.

6. Cortical Transplants

Embryonic cortex also survived and developed when transplanted to the superior colliculus of newborn rats. However, the interconnections this tissue developed with the host brain least resemble the projections of its site of origin of the three types of tissue transplanted.

6.1. Histological Characteristics

The cytology of cortical transplants was distinct from that of retinal and tectal transplants.[14] While a pattern of lamination comparable to that seen in the normal adult cortex was not present in the cortical transplants, they did have a cortexlike organization (Fig. 10). A stain for neurofibrils revealed a darkly stained central core reminiscent of cortical white matter and a thin fiber layer on the surface similar to layer I of the normal cortex. Cells in the transplants were organized into distinct groups that could be characterized by cell sizes, cell packing density, and cell staining characteristics. These groups were sometimes distributed as a recognizable layer but more often were found as clusters of cells. Cells labeled during development by a single injection of [^{3}H]thymidine were frequently found grouped together in the mature transplants.[13] This is a further suggestion that cells representing adult cortical layers have remained together, since in the normal cortex cells of a given layer are also generated around the same time.

Golgi analysis of the cortical transplants revealed the three main cell types of nor-

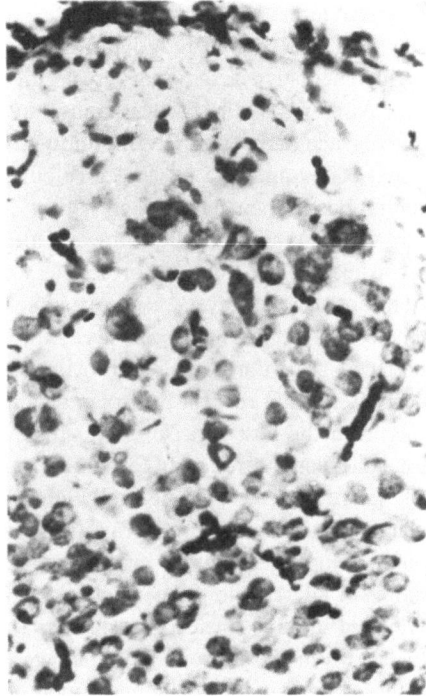

FIGURE 10. Photomicrograph of a cresyl violet-stained section through a cortical transplant. Although no correlation can be made with the layers of the normal cortex, differences in cell soma sizes and cell packing density suggest the presence of some laminar organization in these transplants. This is apparent in this section. The upper third of the section has a low cell density with mostly small cells. The middle third of the section has an intermediate cell density with large and medium-sized cells. The lower third of the section has a high cell density with mostly medium-sized cells.

mal cortex: pyramidal cells, spiny stellate cells, and nonspiny stellate cells.[15] Although the basic shape of each of these cell types was quite recognizable, the orientation of the dendrites was irregular and unpredictable. It is interesting that spiny stellate cells are present in these transplants. They are normally the thalamo-recipient cell in the normal cortex, but there is no thalamic input to these transplants (see below for details). It was also shown that the sequence of maturation of the dendritic arbors for each of the cortical cell types was the same as for the normal cortex.

6.2. Connections with the Host Brain

Efferents of the cortical transplants were studied by degeneration and autoradiographic tract-tracing techniques.[12] The most consistent projection was to the host superior colliculus. The axons entered from the transplant at the collicular surface as distinct bundles that then ramified in intermediate and occasionally deep layers of the host colliculus. The zone of termination for the cortical transplant efferents in the host colliculus was clearly below the area in which retinal transplant efferents would terminate. From the host colliculus, some transplant axons ran rostrally and caudally within two well-defined tracts. A dorsal group projected to the pretectum , nucleus of the optic tract, and lateral posterior nucleus. A ventral group ran rostrally and caudally adjacent to the central gray in the dorsal longitudinal fasciculus to innervate the midbrain tegmentum, reticular formation, and occasionally even the pontine gray. It is interesting that the transplant fibers tended to follow existing host tracts and yet ignored others. This seems highly suggestive of some sort of specificity.

Afferents of the cortical transplants from the host brain were demonstrated by anterograde and retrograde techniques.[14] A projection was frequently found from the visual cortex and to a lesser degree other sensory cortices. These projections were discontinuous in their termination within the transplants but no pattern to their distribution was discerned. While intracortical projections are prevalent in the normal brain, corticotectal projections are present as well. Since the projecion to the transplants arose from host cortical layer V, the usual source of subcortical projections, it suggests that the projection fom the host cortex to the cortical transplants is an expansion of the corticotectal projection rather than being an appropriate projection to the cortex. Subcortical afferents to cortical transplants were also identified. The most common projection was from the host superior colliculus and pretectum. Less common were projections from the pontine reticular formation, raphe nuclei, locus coeruleus, and central gray. These are all projections normally found to the superior colliculus and pretectum; reticular formation and central gray are structures that would not normally project to the cortex. No projection was ever found from the dorsal thalamus, the main subcortical source of afferents to the normal cortex. It is also interesting that the host retina never projected into the cortical transplants, a major source of afferents to tectal transplants. In fact, even those projections that were present to the cortical transplants were far reduced compared to their innervation of tectal transplants. A cortical transplant typically received projections from 100 cells in the host brain compared to the 2000–6000 cells that projected into tectal transplants.

7. Transplants to Adult Hosts

Fetal retina, tectum, and cortex have also been transplanted to the superior colliculus of adult rats.[35] The survival of these transplants was comparable to that found with newborn hosts. Each type of tissue differentiated histological characteristics appropriate for its site of origin, although the degree of differentiation was always much less than with transplants to newborns. For example, in some areas of retinal transplants, especially where the photoreceptor density was low, there was very poor if any segregation between photoreceptor and inner nuclear layer cells (Fig. 2D). This was never seen in retinal transplants to newborn hosts.

Transplants to adult hosts developed only local projections into the host superior colliculus. The maximum penetration of axons from these transplants was 2 mm compared to over 6 mm routinely seen with transplants to newborns. Many laboratories have suggested that glial scarring stands as a major barrier to axonal growth in the mature brain.[5] Staining sections of these transplants with an antibody against GFA protein failed to show any evidence of gliosis at the transplant–host brain interface. It seems more likely that axons do not grow well through the mature glial environment of the CNS regardless of any glial reaction to injury. Finally, no definitive projections from the host retina or brain were identified into any of the transplants in the adult host. This was particularly surprising since several studies have shown that certain adult CNS neurons will grow into fetal brain transplants.[1,17,39] If the primary barrier to CNS regeneration is the mature glial environment, then there should be no barrier to growth of axons from adult neurons into fetal CNS.

8. Discussion

Taken together, the different studies presented above offer insight into the rules governing the differentiation of a tissue and the cells within that tissue and the rules governing the formation of connections between neuronal populations. Let us consider each of these issues separately.

8.1. Differentiation of the Transplants

Fetal retina, tectum, or cortex transplanted to the superior colliculus of newborn or adult hosts will continue to develop and will differentiate appropriately for its site of origin. This unique differentiation of the three tissue types in anomalous locations supports numerous tissue culture studies. The tissue culture studies have shown that a neural tissue allowed to develop *in vitro* will acquire characteristics appropriate for its site of origin.[8] Particularly with regard to the retinal transplants, the transplanted tissue develops more completely than the tissue allowed to remain in culture. It also appears that the transplants placed into newborns differentiate better than those placed into adults. It is tempting to hypothesize the presence of some essential growth factor in the newborn animal that is not present *in vitro* or in the adult. On the other hand, it may be simply that vascularization is essential for adequate delivery of oxygen and nutrients. This would of course not be possible *in vitro* and might be somewhat slow to occur in the adult brain.

It appears that the basic shape of a given cell type is rigidly determined but that the final orientation and branching pattern of its processes are the result of interactions with its environment. Two of the best examples of this phenomenon were the ganglion cells of retinal transplants and the pyramidal cells of cortical transplants. The ganglion cells could always be classified as one of the four main ganglion cell types found in the normal retina. The dendrites always arose from one side of the cell, as is normal, but occasionally dendrites were found that ran through the ganglion cell layer and optic fiber layer to arborize in a distant inner plexiform layer. The pyramidal cells were a similar case. These cells always had a large apical dendrite and numerous smaller basal dendrites. However, unlike pyramidal cells in the normal cortex, the apical dendrites often curved and the basal dendrites fanned out in all directions from the cell to orient to islands of other cell types in the vicinity. It is still unknown exactly how the various presynaptic target cells attract appropriate dendrites. One thing that is quite clear is that the dendrites are not oriented by purely mechanical pressures exerted within the tissue.

One final aspect of transplant development that is most important to emphasize is that no significant movement of cells was seen between the transplant and host brain for any of the transplants described here. This was based on labeling the transplant or host brain with [³H]thymidine prior to transplantation and observing the distribution of labeled cells after transplantation. This is important in order to demonstrate that the interconnections that were described were indeed between the transplant and host cells and not simply between the transplant and transplant cells that migrated into the host brain. Also, the alternative possibility of connections between host and host cells that migrated into the transplant was shown not to be the case. However, in other transplant studies, this may be an issue (e.g., with cortex placed over the cerebellum where host granule cells migrated into the transplants and appeared to receive connections from the host brain[16]).

8.2. Specificity of Connections

The most significant information derived from the studies of visual system transplants is relative to the development of specific neuronal connections. In the normal animal, a given neuronal population sends its axons along certain pathways to connect with only certain other neuronal populations, and the axons terminate within their target population in a very precise pattern. This phenomenon is commonly referred to as *neuronal specificity*. The mechanisms responsible for the development of neuronal specificity have been the subject of intense investigation for several decades.[11] The transplant studies outlined here shed some light on those mechanisms that may be active in the developing visual system.

The development of specific neuronal connections can be broken down into three steps. First is guidance of the growing axons through the brain. Second is selection by the growing axons of the neuronal population in which to terminate. Third is determination of where to terminate within a target population. Each of these steps is considered here relative to the transplant studies. The majority of these studies pertain to retinal axons and how they find their proper targets within the brain.

8.2.1. Axon Guidance

It is quite remarkable that the efferents of the retinal transplants can course backwards along the optic tract and innervate the appropriate visual nuclei. The question is what guides these growing axons? Three mechanisms seem most possible. First, it is possible that the transplant axons follow the host retinal axons. It has been shown in other studies that retinal axons have an affinity for other retinal axons.[3] It does not appear, however, that this is the key mechanism that guides the transplant axons, for even retinae transplanted to hosts with bilateral enucleations at birth still connect correctly. Second, the transplant efferents might be following mechanical cues along the surface of the brain that lead right to the various visual nuclei. This, too, can probably be discounted. If simple mechanical cues were singularly responsible for guidance, any axons challenged with the same situation would be expected to follow the same course. However, the efferents of both the tectal and the cortical transplants follow a completely different path than the retinal axons. Third, retinal axons could recognize and follow the external limiting membrane of the brain. The course of the optic axons, whether from the transplants or normal eyes, is always along the external limiting membrane. This is in contrast to the output of the tectal and cortical transplants, which run deep in the brain. Also, the fiber-sparse patches in the tectal transplants were only innervated by host retinal axons if they were positioned on the surface of the transplants. If the tectal transplants were embedded in the brain so as not to be in contact with the brain surface, they were not innervated by the host retinal axons. It was thought that axons from retinal transplants in the cortex might contact the superior colliculus and lateral geniculate nucleus by growing along the major afferent and efferent tracts of the cortex. However, these transplants failed to connect, possibly because the retinal axons were only capable of following the external limiting membrane. If, as we suspect, retinal axons do have an affinity for the external limiting membrane, mechanical factors may still be important in defining the actual tract of the axons along this surface. Our current invetigations are aimed at identifying specific chemical markers on the outer portion of the radial glia in the developing brain for which retinal axons might have an affinity.

Other axonal types may recognize other guidance cues. For example, the efferents of cortical transplants reliably run in two rostrocaudal pathways of the host brain while ignoring other pathways physically available. This is highly suggestive of some marker being present in those pathways that is recognized by cortical axons.

8.2.2. Target Selection by Growing Axons

The transplant work gives some of the best evidence that retinal axons recognize some specific property of their appropriate postsynaptic target neurons. The efferents from retinae transplanted to the superior colliculus coursed considerable distances to innervate most of the primary visual nuclei. In no case were projections found to nuclei that are not normally retinorecipient. In retinae transplanted to the cortex where the retinal axons were unable to reach the primary visual nuclei, the ganglion cells degenerated rather than connect to inappropriate cells. With tecta transplanted to the superior colliculus, the host retinal axons innervated only those cell groups analogous to the retinorecipient portion of the normal superior colliculus. Yet cortical transplants, an inappropriate target for retinal axons, transplanted in the same manner received no host retinal projections. These results offer no clue as to the cellular mechanism for the recognition between retinal axons and their target. They leave little doubt that some form of active recognition is required and that strictly passive mechanisms, as have been recently popular,[9] could not suffice.

The high degree of specificity observed between retinal axons and their target neurons cannot be generalized to all systems. The tectal transplants both made and received appropriate connections. This was hardly a test of specificity, since the tectum was placed in a proper position to develop. The cortical transplants placed on the superior colliculus, however, received projections that were quite atypical for the cortex. All the afferents to the cortical transplants were normal projections to the superior colliculus. Several of these never project to the normal cerebral cortex. In the case of the cortical transplants, though, the degree of acceptance of these afferents appeared low. Where tectal transplants received projections from several thousand host neurons, the cortical transplants received projections from approximately 100 host neurons. Many of the major afferents to the cortex, such as the dorsal thalamus, did not project to the cortical transplants. Certainly, physical proximity of host axons to the transplant is important in that it limits those axons that have a chance of projecting into the transplant. A similar mechanism may be important in normal development. There could be neuron populations that can recognize each other but that normally would not have access to one another. This appears to be the case with the dopamine-containing axons from transplants of ventral mesencephalon to the hippocampus, which selectively innervated noradrenergic and serotoninergic hippocampal sites but avoided the cholinergic sites.[48] This mesencephalic dopamine projection to the hippocampus is not present in normal rats.

Clearly, many factors must be active in determining which axons will innervate the transplants. Tectal transplants placed in the superior colliculus might be expected to receive the majority of the different collicular afferents. Yet the only major and consistent projections were from the host retina and cortex. These are two of the last projections to grow into the superior colliculus. The transplantation was done when the host brain was in a relatively late stage of development. It is conceivable that an axon population must be actively growing in order to enter a transplant; thus, temporal factors may also influence transplant innervation.

8.2.3. Distribution of Axons within the Target

Many factors determine the pattern of termination of an axonal population within its target nuclei. These factors probably include mechanical constraints, competition, synaptic activity, and temporal patterns of ingrowth. The normal retina projects onto the central visual nuclei such as the superior colliculus in a retinotopic fashion. Thus, the entire two-dimensional map of the retina is reproduced in its pattern of termination on the superior colliculus. The retinal transplants, however, appeared to exhibit no topography in their projection onto the host superior colliculus. Current theories of how topography develops in a projection fall into two basic camps. First are theories requiring some sort of specific recognition molecules graded across the retina and colliculus. A matching of appropriate markers or strength of markers between retinal axons and tectal cells would result in synaptogenesis.[38] The second camp suggests that an orderly ingrowth of optic axons into the terminal field could generate the topographic projection.[9] The orderly ingrowth would be accomplished by an orderly growth of the axons from the eye to the central visual nuclei. This would require a certain degree of mechanical guidance of the growing axons. Neighbor-to-neighbor relationships are not seriously disturbed in the transplants, so that if specific position markers were present, they should be active in the transplants. Since no order was seen in the transplant projection, it seems unlikely that position markers are active. However, mechanical constraints on growing axons are completely disturbed in the transplants. Thus, if orderly ingrowth of axons mediated via mechanical factors is necessary for generating a topographical projection, it is not surprising that no topography was found.

Competition has a role in determining the distribution of axons in their terminal field. As was shown for the host retinal innervation of tectal transplants, when both host eyes innervated a single fiber-sparse patch in the transplants, they segregated into separate eye-specific terminal groups. In a similar paradigm in frogs and fish when two eyes innervated one tectum, the terminals separated into eye-specific bands (see Chapter 3). Furthermore, in these studies, if synaptic activity was silenced, no segregation occurred.[37] This correlates well with the retinal transplant innervation of the host superior colliculus. In this case, there was an overlap of transplant and host retinal innervation. Since the retinal transplants were enclosed by the skull and underlying skin, it is unlikely that they had any significant evoked activity. If activity is required for segregation of axons in the target, it is not surprising that no segregation was seen between retinal transplants and host retinal axons in the colliculus. In tectal transplants, however, where both host eyes innervated a single area, they segregated into eye-specific patches. In this case, both eyes are functioning. As described earlier, retinal transplant axons were relegated to deeper layers of the superficial portion of the superior colliculus (s. griseum superficiale–s. opticum border) when the host eyes were present. Removal of the host retinal projection allowed the transplant to terminate in the upper layers of the colliculus as well (throughout s. griseum superficiale and s. opticum). This has some interesting correlates with normal development. Work in the hamster has shown fine optic axons ramifying in upper s. griseum superficiale of the superior colliculus while large-diameter axons tend to ramify around the s. griseum superficiale–s. opticum boundary.[45] In development of the normal rat retina, it has been shown that the smaller ganglion cells reach an adultlike morphology before the larger ganglion cells.[43] It could be suggested that the axons of the large ganglion cells arrive later and fill in beneath the early ones. The transplant projection would thus be entering the colliculus relatively late and is relegated to the lower position except when the early arriving host projection is eliminated by enucleation.

9. The Future for Transplantation Studes

As the studies described above relative to the developing visual system have hopefully demonstrated, the transplantation paradigm is useful for defining various aspects of neuronal development. A number of additional transplant studies in the visual system are required to complete the present line of investigation. It would be useful to follow the sequence in development of interconnections between the host and the transplant. This could show such things as whether or not transplants form many connections initially, both appropriate and inappropriate, and subsequently retract inappropriate connections. Also, important clues about the substrates that different populations of growing axons track along could come from developmental studies. It would be useful to alter the age of donors and recipients in a more rigorous study. This may give more insight into the role of temporal factors in determining patterns of connections.

The most exciting potential use for the transplant paradigm in probing aspects of neuronal development is based on the studies showing transplants can be held in culture, dissociated and reaggregated without seriously affecting the subsequent development of the tissue. This opens the possibility of altering or removing specific cell types using drugs or antibodies prior to transplantation. It would be most interesting to alter specific cell surface properties or remove a specific cell class and influence the histotypic differentiation of that tissue or alter the pattern of connections it forms. This may offer a means for bridging the gap between those developmental rules that can only be defined by the current studies and the underlying cell biological mechanisms responsible for those phenomena.

Transplantation also offers a potential method for restoring lost function due to trauma, stroke, or a genetic disorder. It would be wonderful to replace a malfunctioning portion of the CNS with a transplant of fetal tissue and restore normal function. Various brain centers appear to have two functions that we are just beginning to understand. First, there is the serial processing of information. This is most important for systems such as the visual system. Any portion of the system receives an orderly array of information, processes the information, and then sends out information in another orderly array to the next link in the chain. The second function appears to be more neurohumoral and less involved in the precise serial processing of information. For example, the substantia nigra projects to the caudate nucleus and uses dopamine as the transmitter in this pathway. If this pathway is interrupted, it results in a very characteristic motor abnormality. However, with as little as 10% of the pathway intact, the system operates normally. Also, by simply administering dopamine in the complete absence of the pathway, the motor symptoms are alleviated. It appears that, in part, the function of the substantia nigra could be reduced to supplying dopamine to the caudate nucleus, and this does not necessarily require precise processing of any information. The possibility of using transplants in neurohumoral functions looks very promising, but problems remain to be solved before we can even consider the possibility of using transplants in systems involving precise serial processing of information. Two of the biggest hurdles that remain are to find a method that allows axons to grow significant distances in the mature CNS neuropil and to find a method that maintains the topography in the interconnections between transplants and host nervous system. Certainly, the task of restoring lost neuronal function is one of the great challenges of neuroscience, and transplantation appears to be an interesting and even promising approach to this problem.

ACKNOWLEDGMENTS. We would like to express our gratitude to Ray Lund, who has been a major collaborator in many of the studies reviewed here. Alan Harvey and Chris Jaeger were the driving forces behind the work on tectal and cortical transplants, respectively. The figures were prepared by Teresa Pickens and the manuscript was prepared by Marion Hinson. Preparation of this review and many of the studies described were supported by NIH Grants EY03713 and EY04627.

References

1. Azmitia, E. C., Perlow, M. J., Brennan, M. J., and Lauder, J. M., 1981, Fetal raphe and hippocampal transplants into adult and aged C57BL/6N mice: A preliminary immunocytochemical study, *Brain Res. Bull.* **7**:703.
2. Barnstable, C. J., 1980, Monoclonal antibodies which recognize different cell types in the rat retina, *Nature (London)* **286**:231.
3. Bodick, N., and Levinthal, C., 1980, Growing optic nerve fibers follow neighbors during embryogenesis, *Proc. Natl. Acad. Sci. USA* **77**:4374.

4. Constantine-Paton, M., and Law, M. I., 1978, Eye-specific termination bands in tecta of three-eyed frogs, *Science* **202**:639.

5. Guth, L., Reier, P. J., Barrett, C., and Donati, E., 1983, Repair of the mammalian spinal cord, *Trends Neurosci.* **6**:20.

6. Harvey, A. R., Golden, G. T., and Lund, R. D., 1982, Transplantation of tectal tissue in rats. III. Functional innervation of transplants by host afferents, *Exp. Brain Res.* **47**:437.

7. Harvey, A. R., and Lund, R. D., 1981, Transplantation of tectal tissue in rats. II. Distribution of host neurons which project to transplants, *J. Comp. Neurol.* **202**:505.

8. Hild, W., and Callas, G., 1967, The behavior of retinal tissue *in vitro*, light and electron microscopic observations, *Z. Zellforsch. Mikrosk. Anat.* **80**:1.

9. Horder, T. J., an Martin, K. A. C., 1979, Morphogenetics as an alternative to chemospecificity in the formation of nerve connections, *Proc. Soc. Exp. Biol. Med.* **32**:275.

10. Hughes, W. F., and McLoon, S. C., 1979, Ganglion cell death during normal retinal development in the chick: Comparisons with cell death induced by early target field destruction, *Exp. Neurol.* **66**:587.

11. Jacobson, M., 1978, *Developmental Neurobiology*, 2nd ed., Plenum Press, New York.

12. Jaeger, C. B., and Lund, R. D., 1979, Efferent fibers from transplanted cerebral cortex of rats, *Brain Res.* **165**: 338.

13. Jaeger, C. B., and Lund, R. D., 1980a, Transplantation of embryonic occipital cortex to the brain of newborn rats: An autoradiographic study of transplant histogenesis, *Exp. Brain Res.* **40**:265.

14. Jaeger, C. B., and Lund, R. D., 1980b, Transplantation of embryonic occipital cortex to the tectal region of newborn rats: A light microscopic study of organization and connectivity of the transplants, *J. Comp. Neurol.* **194**:571.

15. Jaeger, C. B., and Lund, R. D., 1981, Transplantation of embryonic occipital cortex to the brain of newborn rats: A Golgi study of mature and developing transplants, *J. Comp. Neurol.* **200**:213.

16. Jaeger, C. B., and Lund, R. D., 1982, Influence of grafted glia cells and host mossy fibers on anomalously migrated host granule cells surviving in cortical transplants, *Neuroscience* **7**:3069.

17 Kromer, L. F., Björklund, A., and Steveni, U., 1981, Regeneration of the septohippocampal pathways in adult rats is promoted by utilzing embryonic hippocampal implants as bridges, *Brain Res.* **210**:173.

18. Lund, R. D., 1981, Transplantation as a tool in the analysis of development of the rat's visual system in: *New Approaches in Developmental Neurobiology* (D. Gottlieb, ed.), Society of Neuroscience, Bethesda.

19. Lund, R. D., and Harvey, A. R., 1981, Transplantation of tectal tissue in rats. I. Organization of transplants and pattern of distribution of host afferents within them, *J. Comp. Neurol.* **201**:191.

20. Lund, R. D., and Hauschka, S. D., 1976, Transplanted neural tissue develops connections with host rat brain, *Science* **193**:582.

21. Lund, R. D., and Lund, J. S., 1971, Synaptic adjustment after deafferentation of the superior colliculus of the rat, *Science* **171**:804.

22. Lund, R. D., McLoon, L. K., McLoon, S. C., Harvey, A. R., and Jaeger, C. B., 1983, Transplantation of the developing visual system in rats, in: *Nerve, Organ and Tissue Regeneration: Research Perspectives* (F. J. Seil, ed.), Academic Press, New York, pp 303-324

23. Lund, R. D., and McLoon, S. C., 1980, Cytology and connections of retinal transplants in rats. *Neurosci. Soc. Abstr.* **6**:823.

24. Matthews, M. A., and West, L. C., 1982, Optic fiber development between dual transplants of retina and superior colliculus placed in the occipital cortex, *Anatomy and Embryology* **163**:417.

25. Matthews, M. A., West, L. C., and Riccio, R. V., 1982, An ultrastructural analysis of the development of foetal rat retina transplanted to the occipital cortex, a site lacking appropriate target neurons for optic fibres, *J. Neurocytol.* **11**:533.

26. McLoon, L. K., and Lund, R. D., 1981, Neuronal differentiation and cell division in cultured fetal retinae of rat, *Anat. Rec.* **199**:169A.

27. McLoon, L. K., and Lund, R. D., 1982, Embryonic retinae transplanted to the inferior colliculus of newborn rats, *Neurosci. Sco. Abstr.* **8**:452.

28. McLoon, L. K., Lund, R. D., and McLoon, S. C., 1982, Transplantation of reaggregates of embryonic neural retina to neonatal rat brain: Differentiation and formation of connections, *J. Comp. Neurol* **205**:179.

29. McLoon, L. K., McLoon, S. C., and Lund, R. D., 1981, Cultured embryonic retinae transplanted to rat brain: Differentiation and formation of projections to host superior colliculus, *Brain Res.* **226**:15.
30. McLoon, L. K., Sharkey, M. A., and Lund, R. D., 1983, Embryonic neural retina transplanted to spinal cord, *Neurosci. Soc. Abstr.* **9**: p. 373.
31. McLoon, S. C., and Hughes, W. F., 1978, Ganglion cell death during retinal development in chick eyes explanted to the chorioallantoic membrane, *Brain Res.* **150**:398.
32. McLoon, S. C., and Karten, H. J., 1983, Distribution of glial cell processes in retina transplanted to the rat brain, *Neurosci. Soc. Abstr.* **9**: p. 854.
33. McLoon, S. C., and Lund, R. D., 1980, Specific projections of retina transplanted to rat brain, *Exp. Brain Res.* **40**:273.
34. McLoon, S. C., and Lund, R. D., 1980, Identification of cells in retinal transplants which project to host visual centers: A horseradish peroxidase study in rats, *Brain Res.* **197**:491.
35. McLoon, S. C., and Lund, R. D., 1983, Development of fetal retina, tectum and cortex transplanted to the superior colliculus of adult rats, *J. Comp. Neurol.* **217**:376–389.
36. McLoon, S. C., and Lund, R. D., 1984, Loss of ganglion cells in fetal retinal transplants to cortex in rats *Devel. Brain Res.*, in press.
37. Meyer, R. L., 1983, Tetrodotoxin blocks the formation of ocular dominance columns in goldfish, *Science* **218**:589.
38. Meyer, R. L., and Sperry, R. W., 1976, Retinotectal specificity: Chemoaffinity theory, in: *Neural and Behavioral Specificity Studies on Development of Behavior and the Nervous System*, Vol. 3 (D. Gottlieb, ed.), pp.111–149, Academic Press, New York.
39. Oblinger, M. M., and Das, G. D., 1982, Connectivity of neural transplants in adult rats: Analysis of afferents and efferents of neocortical transplants in the cerebellar hemisphere, *Brain Res.* **249**:31.
40. Perry, V. H., 1981, Evidence for an amacrine cell system in the ganglion cell layer of the rat retina, *Neuroscience* **6**:931.
41. Perry, V. H., and Cowey, A., 1982, A sensitive period for ganglion cell degeneration and the formation of aberrant retino-fugal connections following tectal lesions in rats, *Neuroscience* **7**:583.
42. Perry, V. H., Lund, R. D., and McLoon, S. C., 1984, Characterization of the ganglion cells in retinal transplants, submitted for publication.
43. Perry, V. H., and Walker, M., 1980, Morphology of cells in the ganglion cell layer during development of the rat retina, *Proc. R. Soc. London Ser. B* **208**:433.
44. Robison, W. G., and Kuwabara, T., 1979, A new, albino–beige mouse: Giant granules in retinal pigment epithelium, *Invest. Ophthalmol. Vis. Sci.* **17**:365.
45. Sachs, G. M., and Schneider, G. E., 1980, The morphology of single axons innervating the hamster's superior colliculus, *Neurosci. Soc. Abstr.* **6**:749.
46. So, K.-F., 1979, Development of abnormal recrossing retinotectal projections after superior colliculus lesions in newborn Syrian hamsters, *J. Comp. Neurol.* **186**:241.
47. So., K.-F., and Schneider, G. E., 1978, Abnormal recrossing retinotectal projections after early lesions in Syrian hamsters: Age-related effects, *Brain Res.* **147**:277.
48. Stenevi, U., Björklund, A., and Svendgaard, N.-A., 1976, Transplantation of central and peripheral monoamine neurons to the adult rat brain: Techniques and conditions for survival, *Brain Res.* **114**:1.

5

Camera Bulbi Anterior

New Vistas on a Classical Locus for Neural Tissue Transplantation

LARS OLSON, HÅKAN BJÖRKLUND, AND BARRY J. HOFFER

1. Introduction

In experimental neurobiology, it is often desirable to decrease the level of complexity presented by the intact adult mammalian nervous system by various isolation procedures. In this chapter, we will summarize evidence that transplantation to the anterior chamber of the eye of rats and other rodents is an efficient means of obtaining such a decrease in complexity while preserving the functional integrity of the grafted tissues. A transplant can be defined for our purposes as a tissue piece in a living organism that has been completely physically isolated from its normal environment at some point during its life. When placed back into the original type of environment, it is referred to as a *homotopic transplant;* when grafted to other locations, it is called a *heterotopic transplant.* Thus, grafts of neuron-containing tissues to the anterior chamber of the eye are always considered heterotopic.

It was well over a century ago that the first written reports on intraocular transplantation experiments appeared. Transplantation of peripheral nervous tissues to the eye chamber was described by Faldino[14,15]; these studies were soon followed by reports on attempts to graft CNS tissues.[15,67] More systematic intracular grafting of CNS tissues in mice and rats was performed by May[34-38] and his collaborator, Chatagnon.[9] For references on recent uses of the intraocular approach by ourselves and others, the reader is referred to Olson et al.[59] The usefulness of the intraocular transplant method in stud-

---

LARS OLSON AND HÅKAN BJÖRKLUND ● Department of Histology, Karolinska Institute, Stockholm, Sweden.   BARRY J. HOFFER ● Department of Pharmacology, University of Colorado Medical Center, Denver, Colorado 80262.

ies of factors regulating growth of catecholamine-containing nerves and in differentiating between intrinsic and extrinsic determinants of development have recently been described.[41,52,55,85]

The specific advantages of intraocular grafting that makes it a highly versatile and useful experimental tool can be summarized as follows.

1. *Noninvasive visualization.* The tissue pieces to be grafted can be observed visually with the aid of a stereoscopic microscope during the grafting procedure and repeatedly during the postoperative course through the cornea of the host animal, without any physical intervention with the host other than a light inhalation anesthesia during observation periods. This is a unique feature of the intraocular grafts that permits observations of (1) growth, (2) vascularization, (3) functional responses if contractile tissues are included in the grafted material, and (4) early detection of any postoperative disturbances.

2. *Easy and rapid surgery.* With training, intraocular grafting takes only a few minutes per eye. One small cut is made through the cornea. Thus, no blood is shed; no suturing is necessary; the cornea heals rapidly and the intraocular pressure is reestablished within a few hours. Vision is usually preserved. Growing grafts expand in a liquid-filled chamber without being squeezed or exerting unnecessary pressure on host tissues. Most other forms of tissue grafting require more surgery and are liable to cause more stress to the host animal.

3. *Multiple grafts.* By repeated transplantation to the same eye, two or more tissue pieces can be grafted. Under visual inspection, grafts can be manipulated into contact with each other, or they can be placed at opposite ends of the eye chamber without physical contact. Sequential double-grafting has proven a useful method for producing intraocular isolated replicas of known CNS pathways and for studying whether neural connections can form when they are not known to occur *in situ.*

4. *Quantitation of two-dimensional fiber networks in host iris.* Nerve fibers in the host iris, whether intrinsic to the iris or derived from various types of neuronal tissue grafts, will form essentially two-dimensional networks. Many such nerve fiber types can be studied by specific histochemical methods in whole mounts of the host iris. This permits a precise quantitation of nerve fiber production that is not easily obtained in other transplantation locations. Recently, computerized image analysis of fiber outgrowth has been applied to various iris whole mount preparations (see below).

5. *Interactions with iris innervation.* After selective denervation procedures (see below), defined populations of nerve fibers will remain in the host iris that may interact with the grafts. Ingrowth of nerve fibers from the host iris into the graft itself may be readily evaluated. Alternatively, the interaction between nerve fibers entering the irides from the graft and the endogenous host iris innervation can be examined.

6. *Good survival.* Nonneuronal as well as neuronal tissues survive well in the eye chamber. Although rejection does occur in this site (e.g., of heterografts), it is possible that the eye chamber is endowed with a certain degree of immunological privilege. Thus, allografts are usually not rejected. As far as brain tissue is concerned, other factors may also contribute to the successful taking. There are marked similarities between the intraocular and the intracranial fluid environments. Moreover, oxygenation during the initial nonvascular phase of grafting may be partly effected directly through the cornea.

In the following sections, we shall describe the technical details of the grafting and provide examples of how the specific advantages listed above can be used to study factors that influence neuronal growth, development, and function.

7. *Possibilities for electrophysiological recordings.* As will be detailed below, transplants located in the anterior eye chamber are readily accessible for electrophysiological experiments. After unfolding the cornea, electrodes can be visually guided into the grafts. Using standard recording techniques, data can be obtained from intraocular brain tissue grafts for many hours.

2. Transplantation Procedure

Most of our own experience is derived from transplantations to the anterior chamber of the eye of adult rats. Adult mice or young rats may also be used as well as hamsters and guinea pigs. Whenever iris whole mounts are planned to study nerve fiber production by the grafts, albino animals will have to be used. The grafting technique has been described in detail by Olson and Malmfors[46] and Olson et al.[59] The technique (Fig. 1) can be summarized as follows: Recipient eyes are topically treated with one drop of a 1% atropine solution before grafting to ensure dilation of the pupil. This minimizes the risk of iris injury or prolapse during transplantation. Tissues to be grafted are dissected using sterile instruments and kept in sterile saline.

The opening of the cornea is a crucial step. A thin, sharp, and pointed object, such as a piece of a razor blade broken at an acute angle, is necessary. The head is held firmly, eyelids retracted, and the eyeball supported with the index finger while the point of the blade is inserted horizontally through the cornea near its vertex. A cut is then made in a medial direction with an angle of about 30° between blade and cornea. The length of the cut through the cornea is varied such that the particular pipette to be used for grafting fits snugly into the opening. The tissue piece to be grafted is drawn up into the tip of a Pasteur pipette having a smooth oblique opening and a maximal outer diameter of 1.5 mm. The pipette is inserted through the opening in the cornea, with the ovoid pipette opening facing downwards, and the graft injected into the chamber by slight pressure on the rubber bulb of the pipette. After removing the pipette, the position of

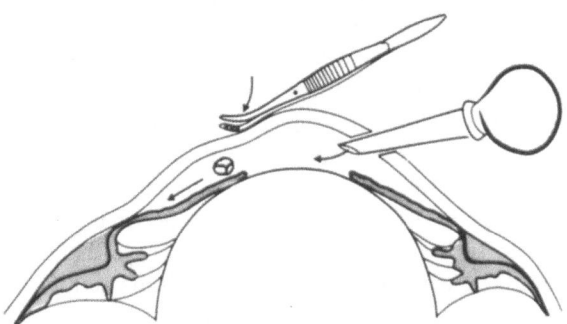

FIGURE 1. Schematic illustration of the intraocular transplantation technique. (For details, see Ref. 59.)

the graft may be altered by pushing on the outside of the temporarily collapsed eye. Bilateral grafting usually takes only 2–3 min per animal. There is no need for any special postoperative care; grafted animals can be caged together.

Several variations of the above procedure are possible. Multiple grafts may be made in the eye chamber by repeating the above sequence at several-week intervals during which time healing of the cornea occurs. It is also possible to introduce two or more pieces together through the same opening in the cornea. This latter experimental protocol increases the risk of postoperative disturbances. There may also be large variations in the location of multiple grafts in relation to each other, if they are introduced simultaneously.

It is also possible to place grafts in the posterior chamber. In this case, the atropine pretreatment should be omitted and the pipette advanced through the pupil to the posterior side of the iris before releasing the graft. It is our experience that survival and growth of grafts are poorer in this position than in the anterior chamber.

Finally, when very fine tissue fragments, and, in particular, cell suspensions, are to be grafted, the material can be injected into the eye chamber. Injections should be made starting from behind the limbus, and then proceeding through the sclera, the lens capsule, the pupil, and finally, into the anterior chamber. Adult rat eyes can accept a maximum of approximately 5 μl of injected material. The retrolimbal approach minimizes leakage of injected material. For a more detailed description of the transplantation technique, the reader is referred to Olson *et al.*[59]

3. Observations through the Cornea

Grafts become rapidly attached to and vascularized from the anterior surface of the host iris. The cut in the cornea heals, usually without any permanent changes in corneal transparency. The development of graft vascularization can be followed in some detail. When fetal brain tissue is grafted, the capillarization can be followed and the density of the capillary network estimated directly through the cornea (Fig. 2). The first signs of circulation in the grafts may be observed within 24–48 hr, depending on graft types. Gray and white brain matter can be distinguished, and the shape of the grafts observed. A relatively accurate estimation of the growth is usually obtained by measuring the long and short diameter of the ovoid or rounded grafts. Since the depth of the anterior eye chamber is fixed and can be approximated to 1 mm, the product of the long and short diameter of a brain tissue graft can be used as a simple and fast estimation of graft volume, which correlates very well with actual measurements of graft weights performed at sacrifice. Adult peripheral tissues and ganglia usually decrease their size slightly during the first week after grafting and then remain constant in size. Fetal brain tissue, on the other hand, will continue to grow *in oculo* for about the same time as it would have growth *in situ*. However, such grafts seldom outgrow the eye chamber; instead, they accommodate to the available space in the lateral angle. Grafts performed between animals of the same strain (e.g., albino Sprague–Dawley) will remain viable and appear healthy as long as the host itself lives. Thus, intraocular grafts can be kept for several years. During such long periods, aging, e.g., of brain tissue grafts, may be observed also through the cornea in the form of opaque white calcifications.

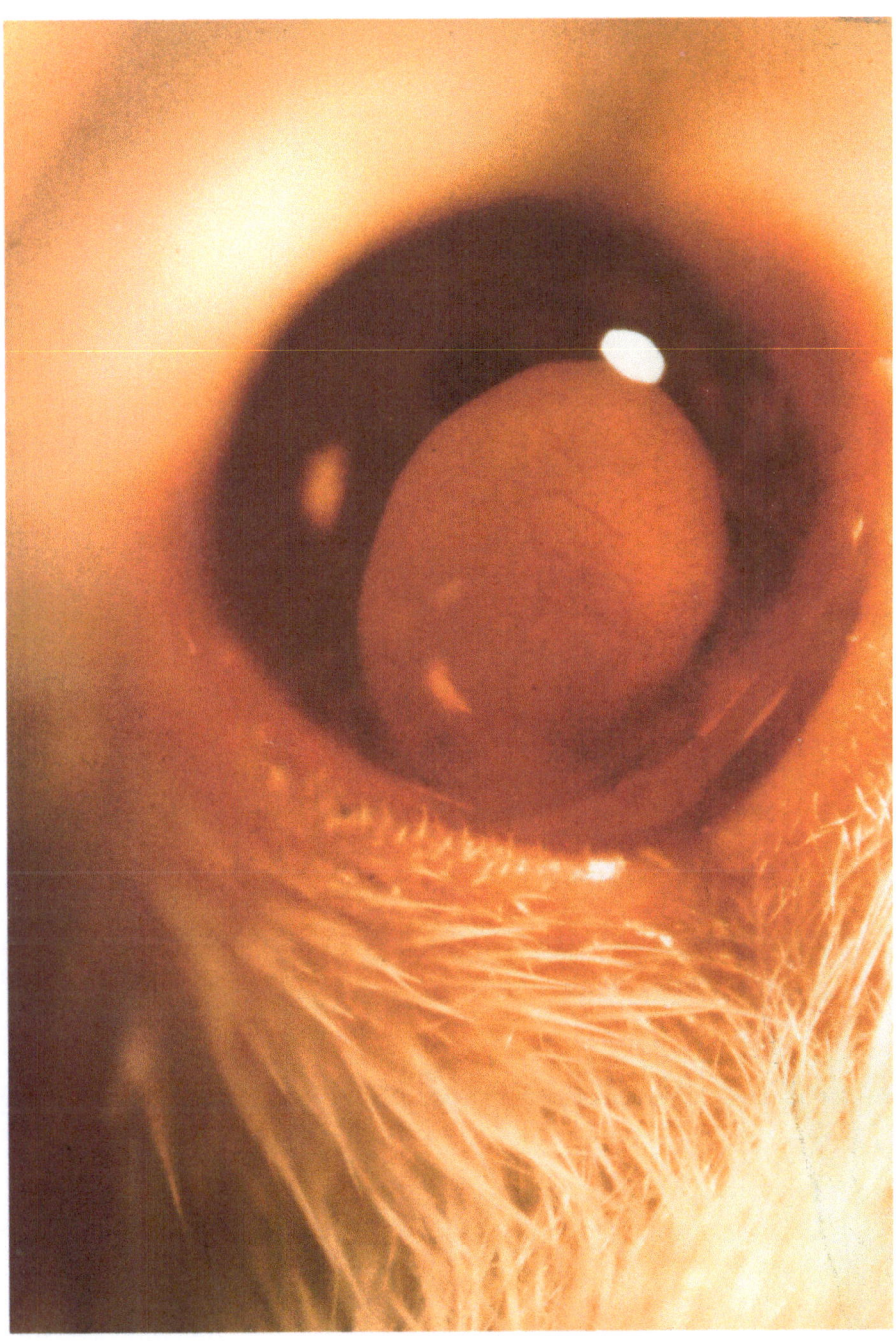

FIGURE 2. Example of an intraocular graft, as seen through the cornea of the living host animal. This dentate gyrus graft was photographed 2 months after the grafting. Note the dense vascularization of the graft.

Operative complications and postoperative disturbances should be infrequent and minor. Presently, we discard less than one out of 20–30 brain tissue grafts. A review of possible disturbances and their effects on grafts is found in Olson et al.[59]

Finally, when contractile tissues are grafted to the eye chamber, it is often possible to observe their contractions through the cornea. Thus, functional innervation of intraocular heart grafts by autonomic nerves from the host iris will result in a "heart rate" of the grafts that is dependent upon the illumination to the host eye, for the same set of autonomic nerves regulate the pupillary diameter.

4. Nerve Fibers in the Rat Iris

At least seven different defined types of nerves are known to be present in the rat iris and can be studied in whole mounts using various histological and histochemical techniques. The sources for these various nerve types and their putative transmitters, when known, as well as the techniques used by us to visualize them, are summarized in Table I.

The adrenergic innervation of the iris can be selectively removed by extirpation of the ipsilateral superior cervical ganglion. The ganglion is reached through a ventral midline incision. This will cause permanent denervation. Adrenergic denervation can also be accomplished by systemic 6-OH-DA injections. In this case, nonterminal axon bundles may be spared and give rise to new terminal networks. A cholinergc denervation is obtained by removal of the ciliary ganglion located on a branch of the ophthalmic nerve deep in the orbit. We reach the ciliary ganglion through a cut below the lower eyelid, via the orbit after removing parts of the Harderian gland. The trigeminal gan-

TABLE I
Nerve Fibers in the Rat Iris

Source	Fiber type	Putative transmitter	Histochemical techniques for visualization in whole mounts
Superior cervical ganglion	Postganglionic sympathetic, nonmyelinated	Noradrenaline	Falck–Hillarp histofluorescence Tyrosine-hydroxylase immunofluorescence
Ciliary ganglion	Postganglionic parasympathetic, nonmyelinated	Acetylcholine	Acetylcholinesterase histochemistry Neurofilament immunofluorescence?
Trigeminal ganglion	Sensory myelinated	?	Neurofilament immunofluorescence Linder silver stain, myelin stains
	Sensory nonmyelinated	Substance P	Substance P immunofluorescence Linder stain?
	Sensory nonmyelinated	?	Linder stain?
Trigeminal ganglion or central to the ganglion	Catecholamine-containing, nonadrenergic	?	Falck–Hillarp histochemistry
?	Nonmyelinated?	Enkephalin	Enkephalin immunohistochemistry
?	Nonmyelinated?	VIP	VIP immunohistochemistry

glion is difficult to extirpate, but the sensory innervation of the iris can be removed by a stereotactic electrothermal lesion. A complete sensory ablation will also denervate the cornea, lending to a loss of the blinking reflex. The eyes of such animals will have to be treated with extreme care using artificial tears and antibiotics several times daily to avoid destruction of the cornea. Lesioning the trigeminal input to the eye will also remove all substance P-positive nerve fibers, and the catecholamine-containing nonadrenergic nerve fibers.[51] In addition, there will be a partial adrenergic denervation because parts of the adrenergic axons reach the eye via the same route as the sensory innervation. Apparently, a more specific lesion of the substance P fibers can be obtained by capsaicin treatment.

5. Electrophysiological Recordings from Intraocular Grafts

For electrophysiological experiments, host animals are anesthetized (urethane, 1.25 g/kg i.p.) and intubated. The head is fixed in a head-holder. The cornea is cut open and reflected, and the anterior chamber superfused with Earle's balanced salt solution (Fig. 3A). A ring-shaped pressure foot is placed over the eye and sealed to the sclera with agar, as shown in Fig. 3A. The superfusion fluid is heated to 35–37°C and allowed to fill the superfusion bath.

5.1. Possibilities to Add Drugs

Drugs can be added to the system in several different ways. If drugs are injected into the host animal (i.p., i.v., or s.c.), they will reach the intraocular brain graft to approximately the same extent as the host brain, since graft and host brain share circulation. The question of a blood–brain barrier in the grafted brain tissue has not been fully resolved, but available evidence suggests that there is a blood–brain barrier, at least for catecholamines, in the intraocular grafts, although the barrier may be less efficient than in normal brain tissue. A second way to introduce drugs is to add them to the superfusion fluid. Precisely defined concentrations of various drugs will influence the graft directly. This technique has the additional advantage of causing effects only in the graft. Thus, drugs in concentrations that might affect the host animal unfavorably can still be used when added to the superfusion fluid. Third, drugs may be administered locally within the graft through microiontophoresis or pressure ejection through multibarreled glass electrodes inserted into graft tissue (Fig. 3B).

5.2. Possibilities to Stimulate Nerves in Intraocular Grafts

The surface of the transplant can be stimulated electrically using fine bipolar electrodes (tip separation 50 µm). In cases where autonomic nerves from the host iris have invaded the grafted tissue, these inputs can be activated. Sympathetic activation can be achieved through stimulation of the cervical sympathetic trunk using a bipolar elec-

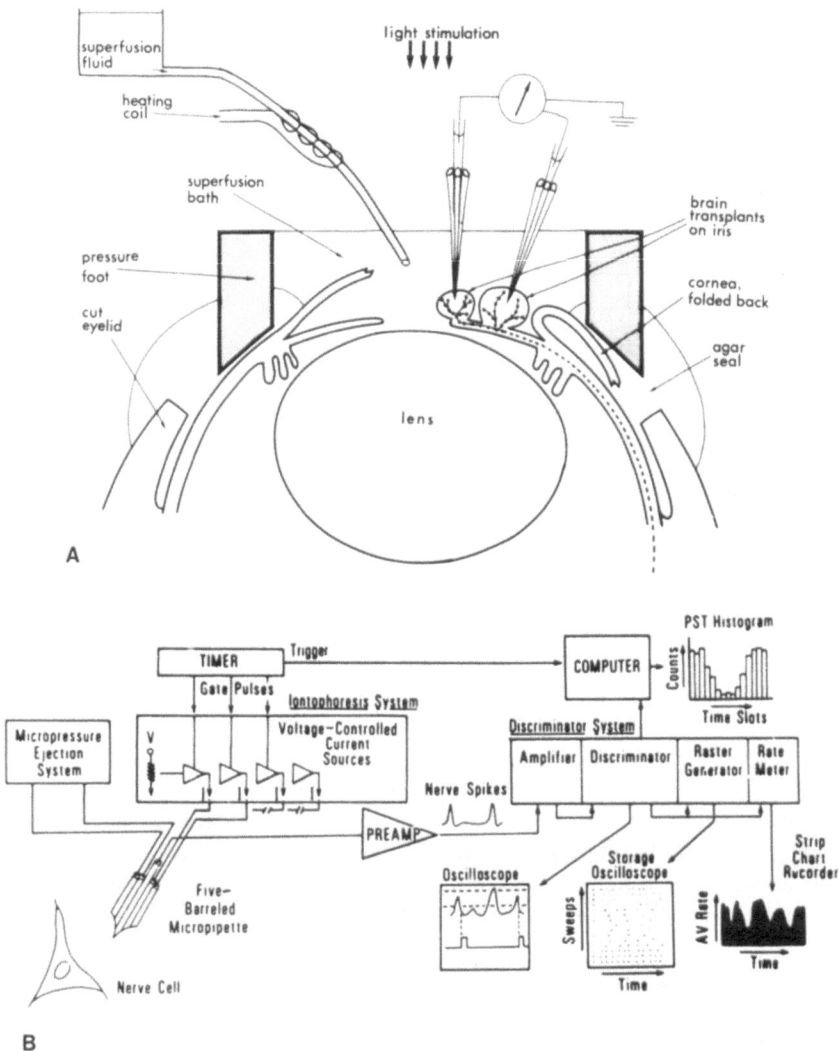

FIGURE 3. (A) Schematic diagram of the recording setup. From Ref. 81. (B) Block diagram of typical instrumentation used for electrophysiological recording and for local drug application.

trode.[24] Sympathetic and parasympathetic activity can also be controlled noninvasively by changing the light influx to the eye carrying the graft (Fig. 4).[18,83]

5.3. Recording from Intraocular Brain Tissue Grafts

Extracellular recording of nerve cell activities can be made by single- or multibarrel micropipettes with the recording barrels filled with 3–5 M NaCl, having resistances of 1–4 MΩ. Under optimal conditions, the recording system is stable enough to allow

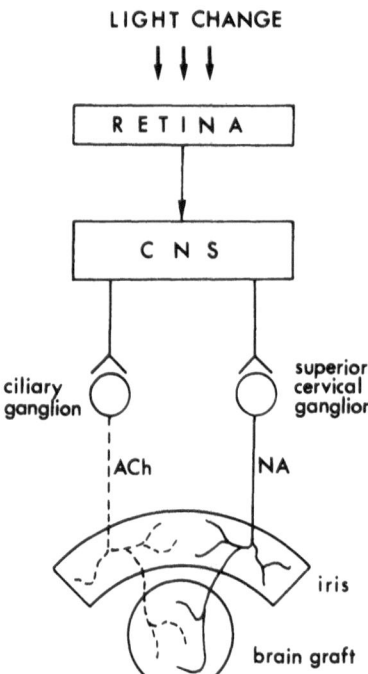

FIGURE 4. Schematic illustration of the use of the light reflex to activate specifically adrenergic and cholinergic pathways to intraocular grafts. From Ref. 52.

recordings from the same cell during several hours. In addition to single-neuron electrical activity, slow waves (the equivalent of electroencephalograms and field potentials) may be recorded from intraocular grafts. When desirable, simultaneous recording from nerve cells in the host brain may be performed, e.g., in order to compare drug effects on an isolated system versus the intact brain. Techniques for electrophysiological experimentation have been described in detail elsewhere (Fig. 3B).[18,24-26,44,58,62,63,81,82]

6. Intraocular Grafts of Peripheral Nervous Tissues

6.1. Adrenergic Ganglia

Grafts of the superior cervical ganglion and other adrenergic ganglia to the anterior chamber have been used extensively to characterize factors and circumstances controlling peripheral adrenergic nerve growth.[46,56] When parts of the adult superior cervical ganglion are grafted to a sympathetically denervated eye, the graft will rapidly reinnervate the iris until, after a month, an approximately normal nerve density is reached. This adrenergic nerve fiber proliferation is inhibited when the ganglion is grafted to a sympathetically innervated eye. Thus, endogenous adrenergic nerves inhibit outgrowth of adenergic nerves from the graft. When immature superior cervical ganglia are grafted to a sympathetically denervated eye, they not only reinnervate the host iris, but also down-regulate the number of nerve cell bodies, thus adjusting to the limited peripheral field available. Adrenergic transmitter mechanisms of the newly formed nerves are similar to those of mature nerves. Adrenergic ganglia other than the superior cervical ganglia are also able to reinnervate a sympathetically denervated host iris. This shows a lack of target tissue specificity for peripheral adrenergic neurons. Indeed, regardless of the neuronal source, the pattern of innervation is always determined by the target tissue.

The superior cervical ganglion has also been combined with fetal brain tissue in the eye chamber. Such experiments show a striking innervation of an adjacent graft of cortex cerebri from a superior cervical ganglion graft. Similarly, grafts of CNS cortical areas may be innervated by peripheral adrenergic nerves directly from the host iris (see below).

6.2. Sensory Ganglia

Survival and cellular immune responses to transplantation of sensory ganglia to the eye chamber have been described by Zalewski and collaborators.[88-96] Sensory ganglia were shown to induce the formation of taste buds in simultaneously grafted tongue tissue. Recently, survival and sprouting of substance P-containing nerve fibers into the host iris from grafts of sensory ganglia have also been described. The trigeminal ganglion also survives grafting to the eye chamber where its interactions with central and peripheral monoamine neurons have been described by Seiger.[69,71]

6.3. Auerbach's Plexus

When pieces of the intestinal muscular wall are grafted to the eye chamber, the smooth muscle layers and the intervening plexus of Auerbach survive.[46] The neurons of Auerbach's plexus receive a new adrenergic innervation from the host iris. Four neuropeptides, substance P, vasoactive intestinal polypeptide (VIP), enkephalin, and somatostatin, have been identified in the nerve layers of intestinal grafts.[68] Available evidence suggests outgrowth of peptide-containing nerve fibers from the intestinal graft into the host iris. Thus, a reciprocal innervation between Auerbach's plexus and the host iris is established.

6.4. Adrenal Medulla

Transplantation of chromaffin tissue from the adrenal medulla to the anterior chamber of the eye has demonstrated remarkable plasticity of medullary tissue. When grafted to a sympathetically denervated host iris, the chromaffin graft will form adrenergic nerve fibers that reinnervate the host iris (Fig. 5).[40,41,45] Recently, more detailed fluorescence histochemical studies of the site of origin of these fibers using either multiple miniature grafts or dissociated cells from the adrenal medulla injected into the eye chamber suggest the presence of at least three cell types. The most interesting of these is perhaps transformed chromaffin cells that are often elongated or polygonal, and show processes that are thick and strongly fluorescent close to their cellular origin, and thin,

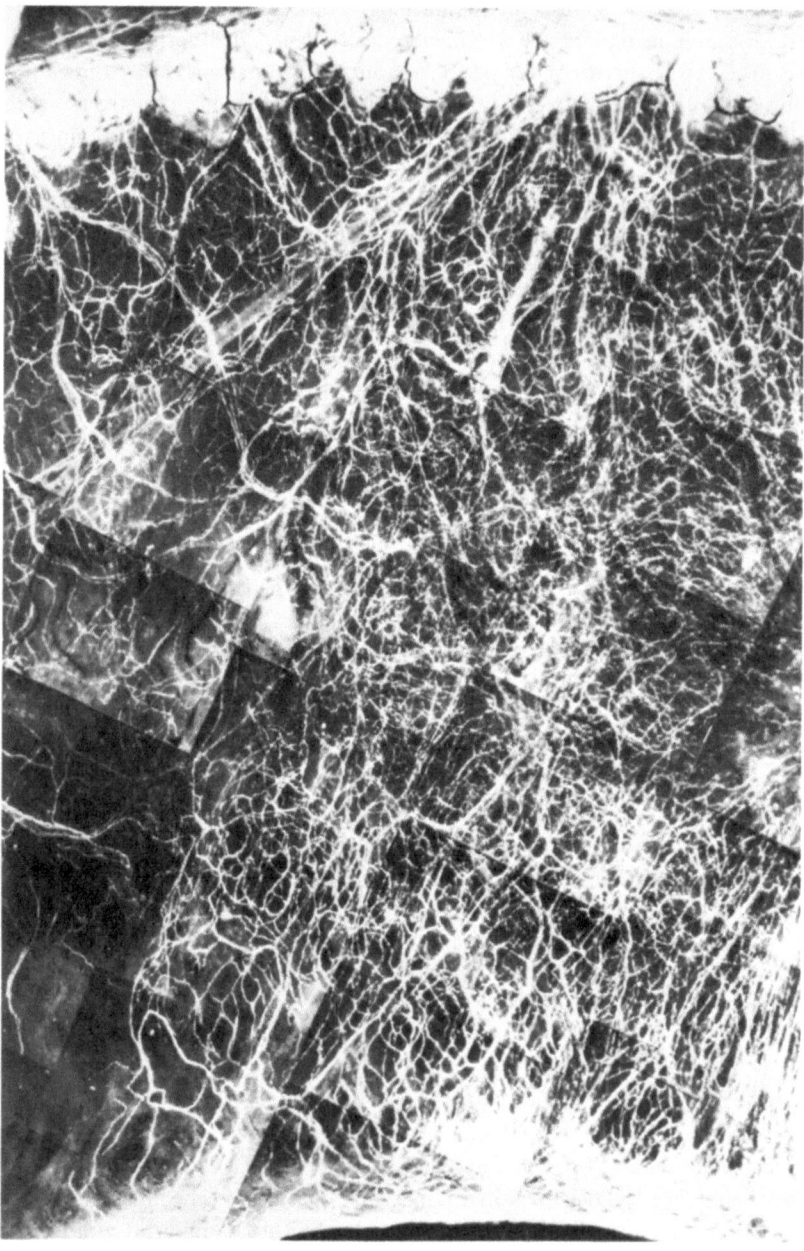

FIGURE 5. Outgrowth of adrenergic nerve fibers from an adrenal medulla graft into a sympathetically denervated host iris. From the area of attachment of the graft (right margin), a plexus of nerve terminals has
formed that is very similar to the normally present sympathetic ground plexus. Corpus ciliare at top, sphincter margin at bottom. Fluorescence microphotomontage, $\times$ 10. From Ref. 52.

varicose, and indistinguishable from normal adrenergic nerve terminals in the periphery. A second cell type consists of unchanged chromaffin cells attached to the iris, intensively fluorescent, but without processes. Third, nerve cell bodies are found. They are considerably larger than the chromaffin cells, and elaborate numerous fluorescent nerve fibers. It is unclear at the moment what proportion of the nerve fibers formed after grafting are derived from cells that still retain a partially chromaffin appearance of their cell bodies as opposed to cells with an entirely neuronal phenotype. In all probability, the cells with a neuronal phenotype seen after grafting of small fragments or dissociated cells were already present in the donor material, although it cannot be excluded that they represent fully transformed chromaffin cells. Although there seem to be some differences between species and between developmental stages (see Ref. 84), the general consensus seems to be that chromaffin cells are channeled toward a more neuronal phenotype (noradrenaline, nerve fibers) by the presence of nerve growth factor, and the absence of corticosteroids, while the absence of nerve growth factor and the presence of high local concentrations of corticosteroids, as in the normal adrenal medulla, helps maintain a chromaffin phenotype (noradrenaline and adrenaline, no nerve fibers).

Sequential double-grafting has shown that intraocular chromaffin tissue is able to innervate added peripheral targets such as an iris graft.[56,57] Chromaffin tissue *in oculo* can also effectively innervate an adjacent graft of cerebral cortex or the hippocampal formation. Interestingly, two types of nerve fibers formed in the grafts. One type, a rich plexus of varicose fibers, was similar to the normally present locus coeruleus-derived fibers in the cortices. The other type consisted of thicker, strongly fluorescent fibers that were usually adjacent to blood vessels. The degree of ingrowth of catecholamine-containing nerve fibers to the host iris as well as to the adjacent brain tissue graft was negatively correlated with the presence of adrenal cortical tissue in contact with the medullary tissue.[56,57]

7. Intraocular Grafts of CNS Tissues

In the following sections, we shall describe properties of a series of different CNS areas grafted alone or in various combinations to the anterior chamber of the eye. Although some areas can be grafted with relatively good results from postnatal donor material, optimal survival is always obtained from prenatal stages of development. When staged animals are not easily available from local animal dealers, evaluation of pregnant females by palpation under narcosis (see Table II),[59] usually permits staging to within 1 day of pregnancy and gives an approximate estimate of the number of fetuses carried by each pregnant rat. After sacrifice, fetuses can be staged by measuring the crown–rump length, which relates to gestation day, as shown in Table II. Each area has an optimal stage for grafting; usually larger and better developed grafts will be obtained from more immature donor material. A guide to useful stages for grafting various CNS areas and their expected growth *in oculo* is given in Table III.

Fetuses are kept *in utero* at room temperature until used. Pieces are dissected from

TABLE II
Relationship of Crown–Rump Length to Gestational Age

Gestation day	Crown–rump length (mm)	Signs at palpation in narcoses
4–5	1–2	Uterine horns are difficult to find. Have variable thickness.
8–9	4–6	Uterine horns have small, often closely spaced, distinct swellings.
12	8	
13	9	
14	11–12	Small distinct firm spheres with an increasing diameter that approximates the corresponding CRL stage.
15	13–14	
16	16–17	
17	18–19	Elastic, somewhat soft ovoid enlargements. Width less than CRL.
18	22	Fetal structures begin to become palpable. Head becomes identifiable. Still distinct borders between adjacent fetuses. Softer than at 17 days.
19	24–25	Fetal indurations appear.
20		Uterine horns are thick, soft, continuous tubes if litter is large.
21		
22		Birth

TABLE III
Staging for CNS Grafts

Region	Optimal stage Day			Optimal stage CRL			Expected growth % volume increase	Remarks
Parietal cerebral cortex	(15–)	16–18	(–22)	(15–)	18–24	(–45)	200–500	Transient blood sinusoids
Cerebellum		13–14	(–15)		11–13	(–15)	400–800	Myelinated parts conspicuous after maturation. By far most vulnerable to disturbances during grafting procedure
Hippocampus	(17–)	18–20	(–22)	(20–)	26–34	(–40)	200–600	
Dentate gyrus	(18–)	19–20	(–22)	(26–)	30–36	(–45)	300–600	
Entorhinal cortex		15–18			14–25		200–400	
Olfactory bulb		17–18	(–20)		19–25	(–35)	0–100	Easily fragmented at grafting
Caudate nucleus	(15–)	16	(–17)	(14–)	15–17	(–20)	200–300	Flat appearance on iris after maturation
Septal nuclei	(15–)	16–18	(–19)	(14–)	16–24	(–30)	0–100	
Tectum		18			23–26		0–100	
Substantia nigra region	(15–)	16	(–17)	(14–)	16–18	(–25)	100–200	
Dorsal raphe nucleus region	(16–)	17–18	(–20)	(16–)	18–25	(–35)	100–200	
Locus coeruleus region	(15–)	17–19	(–22)	(14–)	20–30	(–45)	50–200	
Spinal cord	(14–)	15–17	(–20)	(12–)	14–20	(–35)	100–400	

the fetal CNS under sterile conditions using dissection microscopes and standard micro-dissection instruments. Care should be taken to avoid unnecessary pinching, squeezing, or pulling the pieces to be grafted. Grafts are kept in Ringer's solution, and are drawn into Pasteur pipettes for transfer. Good survival of dissected material will be obtained when grafting is performed within 1–2 hr after sacrificing the pregnant female.

7.1. Cerebellar Grafts Maintain Organization and Function

To obtain good survival of cerebellar tissue, the graft must be taken from a very early developmental stage.[26,28] The bud will then develop into typical cerebellar tissue including trilaminar organotypic cortex (Fig. 6). In optimal cases, foliation of the cortex will also occur *in oculo*.[61] While the gross organization of the cerebellar grafts is rela-tively normal, Golgi studies reveal impairment of dendritic development of the Purkinje neurons (Fig. 7), showing a critical dependence upon extrinsic factors such as the pres-ence of mossy and climbing fiber afferents, for normal dendritic development.[85] Inter-neurons in the grafts had a more normal appearance in Golgi-stained material.[85]

Despite the morphological disturbances, the electrophysiological development of cerebellar grafts was surprisingly normal. After intraocular maturation, extracellular recordings demonstrate Purkinje cells with a spontaneous discharge pattern closely mim-icking the well-known *in situ* activity, regarding both time course and pattern of firing.[26] Furthermore, antidromic spike responses, parallel fiber excitation, and basket-stellate

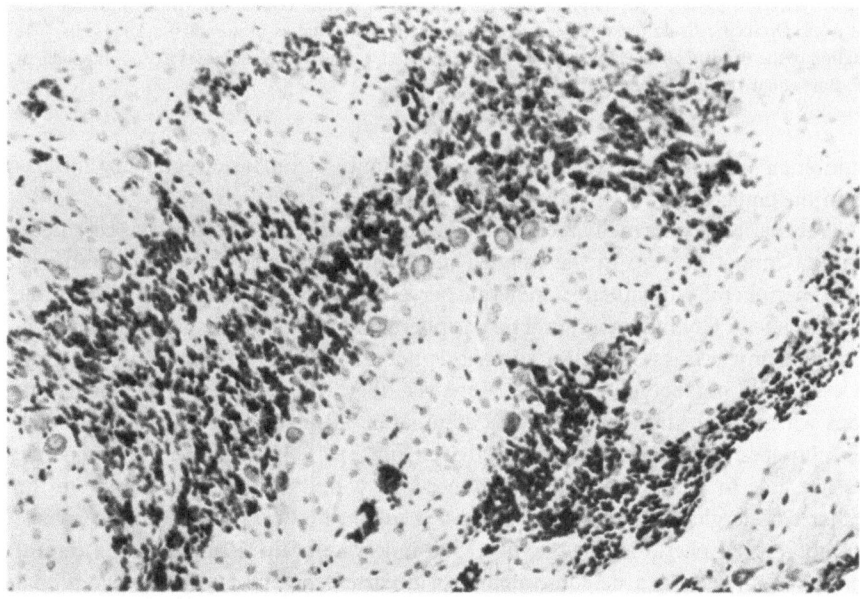

FIGURE 6. Cresyl violet-stained section of an intraocular cerebellar graft. The normal trilaminar organiza-tion is essentially preserved. Note rows of large Purkinje neurons.

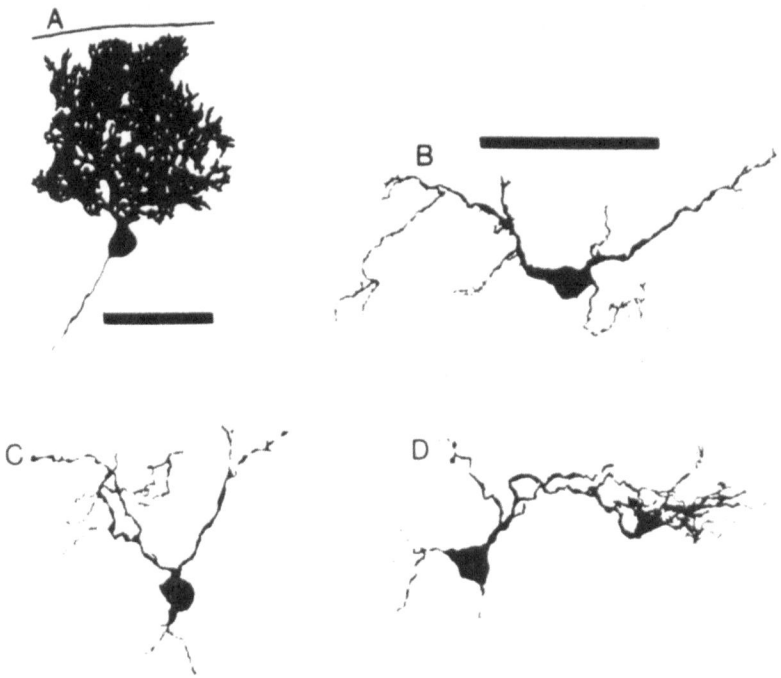

FIGURE 7. (A) Drawing of a fully impregnated Purkinje cell in normal cerebellum. (B–D) Purkinje cells stained *in oculo*. Such cells do not possess extensive dendritic branching as in normal Purkinje cells. The termination of the dendritic branch extending to the right in (D) shows an elaborate system of filopodia at its end. Bars equal 100 μm. From Ref. 85.

cell inhibition were found, underlining the importance of intrinsic regulatory mechanisms in the ontogeny of the function of this brain area.[26]

   Cerebellar cortex normally receives an adrenergic innervation from locus coeruleus. This input cannot develop in the isolated intraocular cerebellar grafts. Interestingly, however, peripheral sympathetic adrenergic nerve fibers present in the host iris are able to invade the fetal CNS tissue graft. The peripheral adrenergic nerves will take on morphological characteristics of central adrenergic nerve terminals and innervate the graft in an essentially organotypic manner. Moreover, they will form fully functional connections with the Purkinje cells such that stimulation of the cervical sympathetic trunk will inhibit spontaneous Purkinje cell activity (Fig. 8).[24] Thus, peripheral adrenergic nerves are able to establish functional connections with Purkinje cells similar to the cerebellar adrenergic synapses normally formed *in situ* by fibers from locus coeruleus. This ability of peripheral adrenergic nerves to innervate brain tissue lacking a central adrenergic input has been demonstrated also in cortex cerebri, and the hippocampal formation.[57] The mechanism underlying ingrowth of peripheral adrenergic nerve fibers in brain tissue is not known. The fact that the innervation becomes organotypical suggests that the ingrowing peripheral fibers are able to locate vacant adrenergic synaptic sites, but other possibilities remain. More recently, morphological evidence for

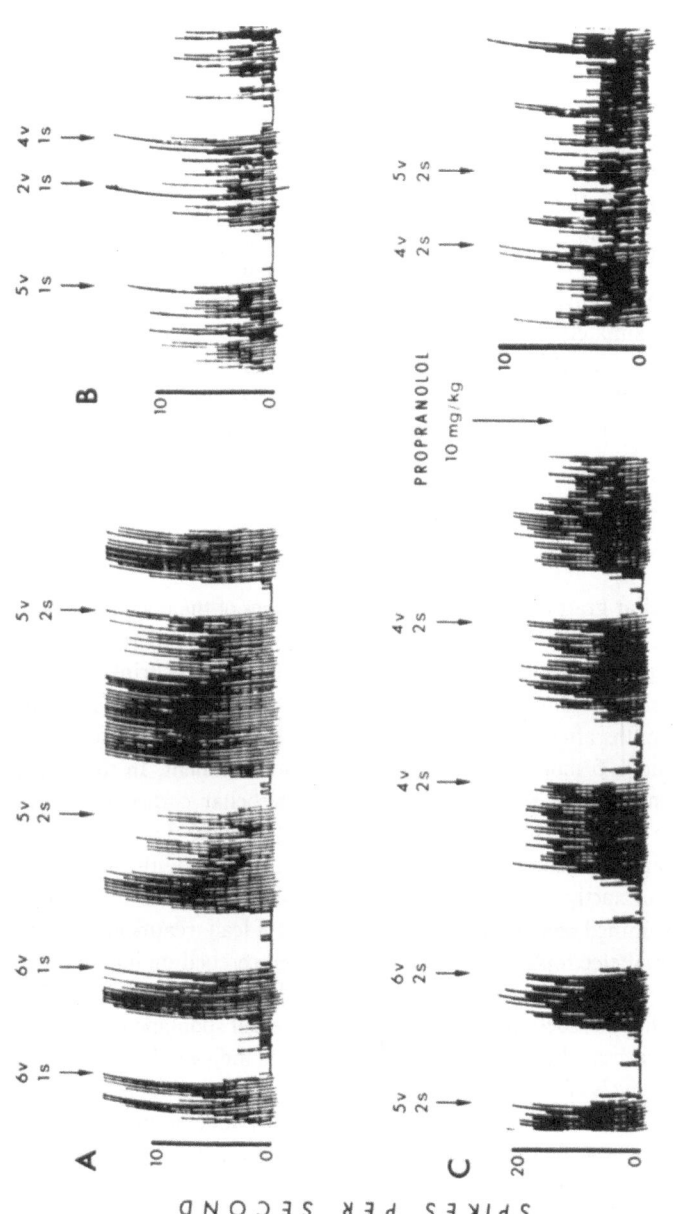

FIGURE 8. Effects of propranolol on Purkinje cell inhibition produced by pulse train stimulation of the superior cervical trunk. (A–C) Three different cells from the same transplant. Arrows above record show time of onset of stimulation at 10/sec; V, stimulation strength (voltage); and s, stimulation time (sec). Note prompt inhibition, usually within 2 or 3 seconds after start of stimulation. For cell (C), propranolol was given i.p. and the right-hand record was made 10 min after injection. Note that, despite a decrease in discharge rate, previously supramaximal stimuli are now ineffective. From Ref. 24.

an ingrowth of sympathetic fibers into the CNS has been found *in situ* in the hippocampal formation following cholinergic denervation.[10–12,29,30,78,79]

7.1.1. Ethanol Effects

The intraocular cerebellar graft has also been used to delineate the locus of genetic difference between two lines of mice that differ in ethanol sensitivity.[63] Cerebellar anlagen from fetal long-sleep and short-sleep donors survive and mature when grafted within and across lines. Mature cerebellar transplants from all four groups contain Purkinje cells that show sustained spontaneous discharge; excitation and inhibition are readily evoked by electrical stimulation of the graft surface. Superfusion of ethanol into the anterior chamber depresses Purkinje cell discharge in all grafts, but neurons from long-sleep donors are approximately one order of magnitude more sensitive than neurons from short-sleep donors. The differential sensitivity is unaltered by the host recipient line used. The data therefore strongly suggest that differential sensitivity of Purkinje cells to ethanol in long-sleep and short-sleep mouse lines is an intrinsic property of the cerebellum.

7.1.2. Lead Effects

Finally, the cerebellar graft model has been used in studies of the potentially harmful effects of chronic lead exposure on brain development. In these experiments, adult recipient rats were exposed to 1% lead acetate or sodium acetate in the drinking water.[5] Cerebellar anlagen were grafted to the anterior chamber. Animals were exposed to lead for approximately 2 months after grafting. Histological and electrophysiological studies of the grafts were made 4–5 months after cessation of lead treatment. In this experimental design, each animal thus contains two types of cerebellar cortex, the normal cerebellar cortex of the host, which has been exposed to lead during 2 months of adult life, and the intraocular cerebellar cortex, which shares circulation with the host and thus has been exposed to exactly the same amount of lead during 2 months of intraocular development. When examined several months after cessation of lead treatment, host cerebellum was found normal electrophysiologically. The grafted cerebellum had developed organotypically into a foliated trilaminar structure typical of cerebellar cortex.[6,61] In marked contrast to the host cerebellum, virtually all Purkinje cell spontaneous discharge was absent from the lead-treated cerebellar grafts.[6,61] The data indicate that lead administration, eliciting blood levels of 450–550 μg/liter, produces a long-lasting selective electrophysiological deficit in the developing cerebellum and, thus, underscore the potential hazards of even small amounts of environmental lead for the immature organism.

The findings *in oculo* have led to a study in which newborn rats were injected daily with lead during the first 20 days of life. Cerebellar function was then studied several months after cessation of lead treatment. We could confirm an impairment of spontaneous activity in the Purkinje cells, although in the whole animal studies, the effect was much less dramatic.[4] Thus, the intraocular model has demonstrated a direct effect of lead on the developing cerebellum, which might have escaped detection in whole animal studies.

7.2. Hippocampal Grafts: Structure and Function, Inputs from Host Iris

The immature hippocampal formation survives transplantation to the eye chamber well from a wide range of donor stages (crown–rump length 11–42 mm). This is in contrast to cerebellum, which survives grafting well only from a relatively restricted early developmental period. Hippocampal grafts become vascularized from the host iris, proliferate extensively, and develop an organotypical cytoarchitecture. Pyramidal cells form a typical layer surrounded by stratum oriens- and stratum radiatum-like areas in most parts of the grafts (Fig. 9).[44] Golgi impregnation studies show a longitudinal orientation of the apical and basal dendritic systems of the pyramidal cells similar to that described in situ.[85] In a few transplants that also contained the dentate gyrus, a dentate granule cell layer separate from the hippocampal formation was observed. Interestingly, however, the presence or absence of the dentate gyrus seemed to make little difference in growth or development of the hippocampal formation. Likewise, the presence or absence of an adrenergic input from the host iris did not influence the histological differentiation of the hippocampal graft. Responses of the hippocampal transplant to electrical stimulation of its surface correlate well with the histological organization. Large negative and positive field potentials were found at various depths in the transplant, and could be associated with excitation and inhibition, respectively, of pyramidal neurons. Much of the excitation was shown to be orthodromic in nature. In contrast to in situ findings, pyramidal cells in oculo had negligible spontaneous electrical activity. With repeated electrical stimulation, however, a slow background discharge gradually developed in many cells. The data thus suggest that hippocampal pyramidal neuron "spontaneous activity" is critically dependent on some external input.[44] Further histochemical and electrophysiological studies revealed that cholinoceptivity and a cholinergic afferent input from the iris was present in the hippocampal transplants. The pyramidal neurons possess an excitatory muscarinic receptor, and they are also activated by application of cGMP. Pyramidal neurons show an elevated discharge rate with superfusion of phosphodiesterase inhibitors, and the excitatory responses to acetylcholine and cGMP are potentiated to a greater extent than the elevation in resting discharge. After removal of the ciliary ganglion, acetylcholinesterase-positive fibers are not seen in the host iris or in the transplant. Excitatory responses to acetylcholine and cGMP remain, however, after both ciliary and superior cervical ganglionectomy. These studies[25] give further credence to a role of cGMP in cholinergic transmission in the mammalian CNS.

7.2.1. Epileptiform Activities

Because of its low seizure threshold,[1] the hippocampus has been extensively used in research on epileptic mechanisms. In view of the difficulties in distinguishing endogenous from transmitted epileptiform activity in situ, many investigators have attempted

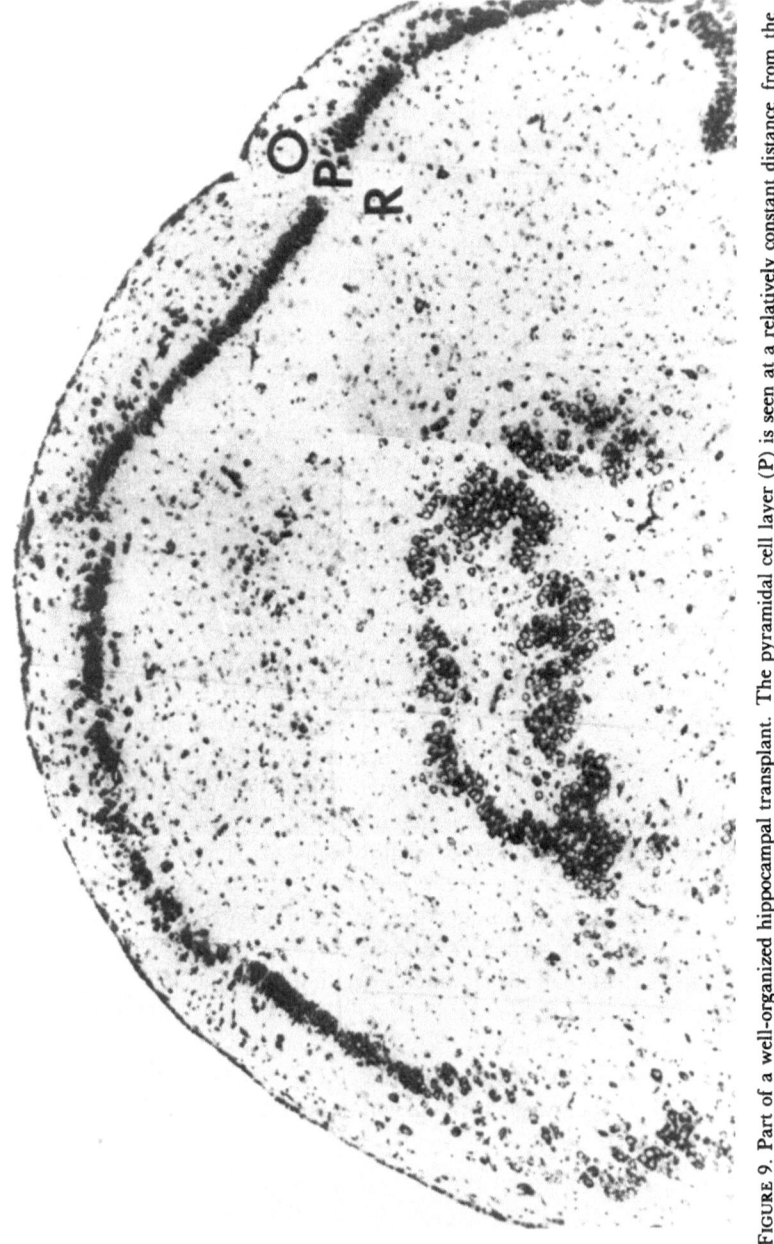

FIGURE 9. Part of a well-organized hippocampal transplant. The pyramidal cell layer (P) is seen at a relatively constant distance from the surface (110–150 μm), and with all neurons arranged perpendicular to the surface curvature. Superficial to this cell layer is a rather cell-poor layer suggestive of stratum oriens (O), and beneath it is a very cell-poor layer suggestive of stratum radiatum (R). Postoperative time is 33 days. CRL of donor fetus is 22–24 mm. Montage of microphotographs of a 6-μm toluidine blue-stained section, × 45. From Ref. 44.

to use isolated hippocampal systems to define the sites of abnormal discharge. While isolated from the rest of the CNS, the intraocular hippocampal transplant receives cholinergic and adrenergic fibers from the autonomic ground plexus of the iris. This preparation has greatly facilitated delineation of factors regulating various forms of epileptic activity. It is ideally suited to answer questions such as what is the minimum local neuronal substrate necessary to produce normal and abnormal activity, and how do adrenergic and cholinergic inputs influence epileptic activities?

In a healthy, unstimulated hippocampal transplant, the "EEG" is usually of a low voltage, giving a flat appearance. With the appropriate electrical or chemical stimulation, various high-voltage phenomena can be elicited resembling seizures, interictal spikes, and hypersynchronous slow-wave activity. Electrical stimulation, penicillin superfusion, and cobalt iontophoresis were all effective in eliciting these high-voltage phenomena. Immediately after a seizure, the transplant was generally completely refractory to repeated stimulation for up to 30 min. Parenteral administration of barbiturates could block seizures and interictal spikes, as could superfusion of diazepam. Thus, the epileptic activity could be counteracted by classical anticonvulsant drugs. The ability to generate seizurelike activity is not a general phenomenon of intraocular brain tissue grafts. Cerebellar transplants do not generate such activity following electrical stimulation or penicillin superfusion, for example. The hypersynchronous activity seen in hippocampus but not the cerebellum *in oculo* corresponds to the normal situation *in situ,* where hippocampus is seizure-prone, while cerebellum generally does not exhibit seizure activity. The various forms of epileptiform activities in the hippocampal transplants, such as seizures, paroxysmal depolarizing shifts, interictal spikes, and hypersynchrony all resemble phenomena seen *in situ,* and indicate that these phenomena can be generated by the isolated hippocampus without connections to other brain areas.[23]

A study was also made of the influence of cholinergic and adrenergic inputs on hippocampal seizure activity *in oculo.*[18] Cholinomimetics were shown to initiate seizures and hypersynchronous neuronal activity in the hippocampal transplants. cGMP derivatives and isobutylmethylxanthine elicited similar changes. Reflex activation of the cholinergic parasympathetic input to the iris and transplant by illumination of the retina induced seizures (Fig. 10), or increased the rate of penicillin-induced interictal spike discharge. β-Adrenergic agonists inhibited interictal spikes and paroxysmal depolarizing shifts induced by penicillin. Fluorescence histochemical studies showed that host sympathetic adrenergic fibers from the ground plexus of the iris invade the transplant to form fine varicose nerve terminals. Activation of these adrenergic afferents to the transplant diminished both the amplitude and the frequency of penicillin-induced epileptiform activity. Thus, it was concluded that epileptiform activity in hippocampal transplants *in oculo* is strongly modulated by cholinergic and adrenergic inputs, the former exerting a facilatory influence and the latter an inhibitory influence.[18,83]

It has also been shown that β-endorphin or methionine enkephalin increases the firing rate of pyramidal neurons in hippocampal grafts, increases the EEG amplitude, and ultimately elicits epileptiform activity. Enkephalin-induced excitations can be counteracted by naloxone. These results suggest that excitatory responses to opiate peptides in the hippocampal formation are due to changes in intrinsic neuronal circuitry.[82]

Moreover, the hippocampal graft model has also been used to study ethanol with-

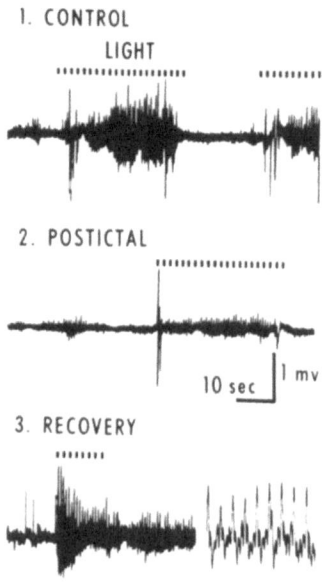

PCN 1600u/ml

1. CONTROL

LIGHT

2. POSTICTAL

10 sec | 1 mv

3. RECOVERY

FIGURE 10. Light-induced hypersynchronous electrographic activity in a hippocampal transplant. (1) High-voltage 4- to 7-Hz slow waves induced by ipsilateral illumination of the retina. (2) After retina bleaching by 20 min of bright light, illumination-induced hypersynchrony is markedly attenuated. (3) Recovery after 30 min of darkness.

drawal seizures.[27] Chronic administration of alcohol to the host animal, followed by sudden alcohol withdrawal, induced clear-cut epileptiform activity in hippocampal grafts. Thus, the conclusion could be drawn that ethanol withdrawal seizures are an intrinsic property of this brain region that can be triggered by falling ethanol levels. Excitability of the transplants returns to normal several hours after withdrawal, when blood ethanol levels have stabilized at a low value.

7.2.2. Adrenergic Inputs

The intraocular hippocampal graft has been used to compare growth of central and peripheral adrenergic neurons. Recent evidence suggests that there are important differences in growth regulatory mechanisms governing central and peripheral adrenergic neurons.[56,58,72,74-76] Using sequential double brain tissue grafts, it is possible to innervate an intraocular hippocampal graft with central adrenergic fibers from a locus coeruleus transplant.[58] Thus, both the locus coeruleus and the superior cervical ganglion (via the sympathetic plexus of the host iris, or directly from a ganglionic cograft) can innervate an intraocular hippocampal graft. Various types of axonal sprouting in response to partial deafferentation have been described in several different brain areas, particularly in the hippocampal formation.[33,65,66,80] Using monoamine fluorescence histochemistry, it is now well established that peripheral sympathetic axons can invade CNS areas. Ingrowth occurs into intraocular grafts of cortex cerebri,[57,73] cortex cerebelli,[24] and hippocampus.[18,44,83] Likewise, sympathetic adrenergic fibers from grafts of the superior cervical ganglion[2,3,78] or from the local sympathetic innervation of blood vessels[30,79] can invade the hippocampal formation *in situ*. In the case of intraocular cerebellar and hippocampal grafts, the ingrowing peripheral sympathetic fibers have been shown to establish functional synaptic contacts with the CNS tissue.[18,24,83] Sympathetic ingrowth reached a steady "paranormal" level after approximately 1 month in the hippocampus, regardless of whether the innervation came from sympathetic fibers in the host iris, or from a ganglionic graft added to the hippocampal transplant after intraocular maturation of the latter. In marked contrast to this innervation by peripheral sympathetic fibers, the ingrowth from locus coeruleus continues *in oculo* until the hippocampal graft has received a marked degree of adrenergic hyperinnervation. Calculations of cell numbers indicate that grafted locus coeruleus neurons increase their terminal fields to at least three times normal levels. The intraocular hippocampal graft can be assumed to have a much greater paucity of afferent connections than the hippocampal formation in the *in situ* experiments of Björklund *et al.*[3] In view of the known effects of various deafferentation procedures on fiber sprouting in the hippocampus, it is possible that the extreme degree of adrenergic hyperinnervation seen *in oculo*[60] represents a response by locus coeruleus to the lack of afferent input to the hippocampal target. Such data demonstrate that the regulation of growth in central and peripheral adrenergic fibers, exposed to the same target tissue and the same surrounding milieu, is quite different. Similar conclusions have been reached from studies on the influence of nerve growth factor.[13,43]

It is uncommon in studies of lesion-induced neuronal plasticity *in situ* to correlate the abundant anatomical data with studies of the functional consequences of a changed neuronal connectivity (see Ref. 80). The intraocular approach is one system that permits studies of the functional consequences of changing synaptology. Recordings from locus coeruleus neurons in double grafts of the locus coeruleus and hippocampus demonstrate an atropine-sensitive excitatory response to illumination, suggesting innervation of the locus coeruleus graft by cholinergic nerve fibers from the host iris. Exposure of the double grafts to an epileptogenic agent such as penicillin causes marked excitation of the locus coeruleus neurons without, however, inducing epileptiform activity in the adjacent hippocampus graft. This is in marked contrast to the results with single hippocampal grafts, which seize readily after penicillin, as described above. Local application of the inhibitory agent GABA into the locus coeruleus part of the double graft allows penicillin-induced epileptiform activity to be expressed in the hippocampus part, suggesting that functional inhibitory innervation develops between adrenergic fibers from the locus coeruleus and pyramidal neurons in hippocampus. In support of this, subsequent excitation of the locus coeruleus neurons by local iontophoresis of glutamate terminates the hippocampal seizure. Thus, sequential transplantations of locus coeruleus and hippocampus to the anterior chamber of the eye create a functional, yet isolated, neuronal pathway that can be utilized to study the development of neuronal connections. The adrenergic hyperinnervation demonstrated histochemically[60] is paralleled by an adrenergic inhibitory hyperfunction.[81]

Immunohistochemistry of the astrocytic population of hippocampal grafts using antibodies raised against glial fibrillary acidic protein (GFA) demonstrates a higher than normal amount of GFA-like immunoreactivity in hippocampal grafts. The pattern of GFA-positive astrocytes has been evaluated in detail only in grafts of cortex cerebri, as described below, but it seems fair to conclude that many different brain areas, including the hippocampal formation, develop a relative gliosis when isolated *in oculo*. This higher than normal amount of GFA does not, however, prevent near-normal development of several other morphological features of the grafts, and of their electrophysiological properties.

7.3. Area Dentata Grafts: Intrinsic Organization

The dentate gyrus, like the hippocampus, has a characteristic and distinct organization of its nerve cell bodies, facilitating comparisons between grafts and the *in situ* organization. The developmental potential of a region termed *area dentata* including the anlage of the dentate gyrus, the hilus, and the medial regio inferior, was studied *in oculo*. Using a variety of histological and histochemical techniques, we conclude[19] that the cytoarchitectonics of the area dentata graft was similar to the area dentata *in situ*. There was a tightly packed C-shaped layer of granule cells with mossy fibers contacting proximal dendrites of hilar and pyramidal cells. Axons of pyramidal and hilar cells in turn gave rise to a rich innervation of the granule cell dendrites. Granule cell dendritic spine density was nearly normal, and synapses were common in the granule cell neuropil. Sympathetic fibers from the host iris innervated the area dentata grafts in an organotypic fashion. Thus, pieces of isolated area dentata displayed a marked intrinsic developmental capacity in terms of cellular and afferent organization. There were, however, some differences between area dentata grafts and their *in situ* counterparts. They included lesser arborization of granule cell dendrites, an increased density of fibrous astrocytes, pyramidal cells with reduced kainic acid sensitivity, and a hypervascularized neuropil.

7.4. Cortex Cerebri Grafts: Tropic Influences of Other Brain Areas during Development

A wealth of experimental evidence suggests that neocortex is an area of the brain in which development is most critically dependent upon normal contacts with other brain areas, and through them, upon influences from the external milieu. It should therefore be particularly interesting to study the degree of structural and funtional development of this area *in oculo,* and the extent to which the development of the cortex cerebri may be influenced by other brain areas cografted to the eye chamber. The developing cerebral cortex will survive grafting from gestational days 15 to 22. Optimal stages for grafting are gestational days 16–18. Transient blood sinusoids will form within a few days. In another few days, a capillary network is established in the growing gray matter. The grafted piece will increase its volume *in oculo* two- to fivefold. The gross organization of cortical grafts *in oculo* is similar to their *in situ* counterparts.[73] The layering of rat parietal cerebral cortex is not distinct *in situ,* nor is it distinct in the grafts. However, an outer cell-poor molecular layer can be clearly distinguished in the grafts followed by some neuron-rich layers underneath. Several different neuron types are found in the grafts, including typical pyramidal neurons. Many pyramidal neurons had their long axis oriented perpendicular to the surface, and it was sometimes possible to observe a columnar arrangement of neurons within the grafts.

7.4.1. Catecholamine Inputs to Cortical Grafts

Fibers from the sympathetic adrenergic plexus of the host iris were able to innervate the developing parietal cortex cerebri transplants in an organotypic manner. Sympathetic fibers in the iris have large varicosities, and clearly visible intervaricose parts; several run together in each strand of the plexus. When entering the transplants, the fibers changed their morphology to a CNS type including small round varicosities with almost invisible intervaricose parts; the fibers run singly rather than in bundles. In this way, the heterotopic sympathetic innervation of the graft resembled the distribution, morphology, and density of innervation normally present in cortex cerebri and derived from locus coeruleus.

Individual sympathetic fibers could be seen to enter the graft tissue together with blood vessels from the host iris. As long as the sympathetic fibers followed blood vessels, they had a peripheral-type morphology. At various points along the blood vessels, they would branch off into CNS neuropil and rapidly assume a CNS-type morphology. Adrenergic innervation of cortex cerebri grafts was also obtained from a cografted superior cervical ganglion. In these experiments, the normal sympathetic innervation of the host eye was removed by superior cervical ganglionectomy. Adult as well as immature superior cervical ganglia in contact with the brain grafts gave rise to a relatively organotypic innervation of the cortex tissue. When immature ganglia were used, migration of sympathetic neuroblasts into brain tissue was sometimes observed to the extent that single, completely integrated sympathetic nerve cell bodies were seen in the cortex neuropil.

The cerebral cortex has also been combined with central noradrenergic neurons of locus coeruleus and with central DA neurons from substantia nigra using sequential grafting. Locus coeruleus gave rise to a rich adrenergic innervation of the cerebral cortex that was of organotypic density within a month. Cortex grafts also became abundantly innervated by DA nerve terminals from adjacent substantia nigra grafts. The terminals were fine, curved, and generally confined to deeper layers of the cortical grafts and, thus, again strikingly similar in morphology and distribution to the normally present DA innervation. Coeruleo-cortical and nigro-cortical pathways were formed both when cortex was added to grafts already containing the mature central monoamine neurons, and when cortex was grafted first and the central monoamine neurons added later.

7.4.2. GFA Immunohistochemistry

Although several aspects of the morphology of single cortical grafts in oculo are reminiscent of normal cortex cerebri in situ, there are also important differences. The cortex grafts contain an excessive amount of astrocyte processes filled with GFA (Fig. 11). In addition to the abnormally high amount of GFA, the single grafts showed several more normal features such as a GFA-positive glia limitans on the free surfaces of the

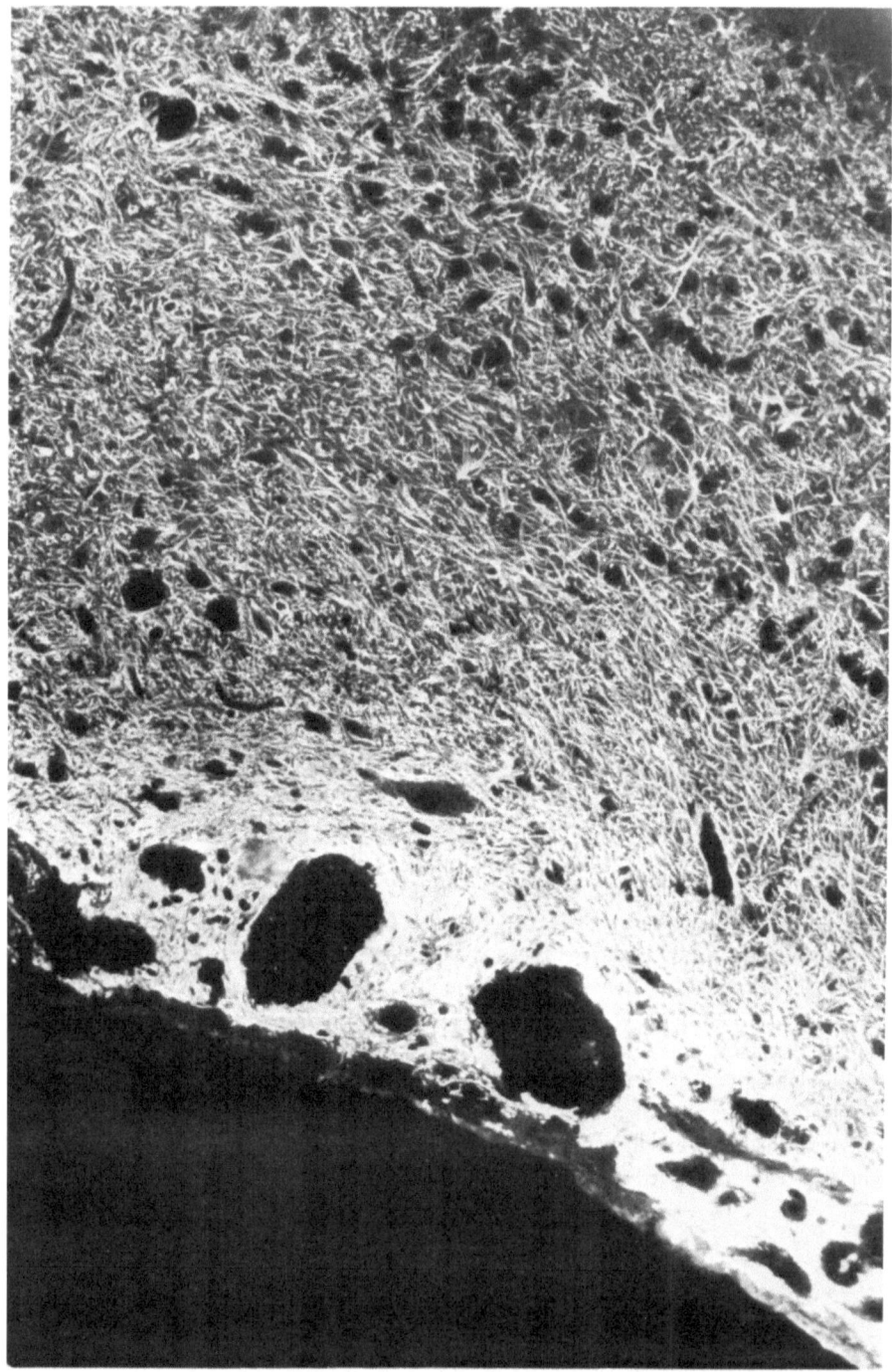

FIGURE 11. Close-up of a single cortex graft with an excessive amount of astrocytic filaments, especially toward the iris. The border between the graft and the host iris cannot be detected due to the strong GFA-like immunofluorescence in the latter. However, the posterior epithelium of the iris does not contain any immunofluorescence. Fluorescence microphotograph, × 260.

grafts, and perivascular membranes around blood vessels. Astrocyte processes terminating in glia limitans and on the perivascular membranes were frequently noted.

7.4.3. Trophic Influences

A most striking stimulatory effect was seen when fetal parietal cerebral cortex was grafted to eyes carrying one of several different types of previously grafted CNS areas such as locus coeruleus, tectum, or cerebral cortex (Fig. 12). When the cortex graft was in direct physical contact with one of these latter areas, the cortex graft had a final size that was 50–100% larger (Fig. 13) than corresponding grafts placed in sham-operated control eyes. The increased growth was observed already by 1 week after transplanting the test cortex graft, and continued to be manifest during the rest of the intraocular growth phase, after which graft sizes, and hence, the difference between control and stimulated test grafts, remained constant. DNA measurements and histological examinations suggested that the increased volume of stimulated cortex grafts was due both to hyperplasia and to hypertrophy. Contact between a stimulating and the stimulated graft was found to be critical in eliciting a trophic response. Growth-stimulated cortical grafts had a better organized cytoarchitecture with larger neurons, including typical pyramidal cells, more neuropil, a lower cell density, and a more organotypic distribution of cell bodies than nonstimulated controls. The distribution of GFA-like immunoreactivity was also more normal in stimulated cortex grafts (Fig. 14). A search for other areas of the CNS that might be similarly stimulated, such as locus coeruleus, hippocampus, and spinal cord, has so far yielded mainly negative results, suggesting that the potential for growth stimulation might be a unique property of cortex cerebri. The in oculo model has thus demonstrated a profound effect of adjacent neural tissue on development of neocortex.[8] Stimulated and nonstimulated cortex grafts were also compared electro-physiologically.[62] A number of differences were found between single grafts and locus coeruleus-stimulated cortical grafts. Neurons in growth-stimulated grafts manifested a slow sustained spontaneous discharge similar to that found in rat cortex in situ. Local administration of two putative excitatory transmitters, glutamate and acetylcholine, markedly augmented this discharge. In contrast, neurons from nonstimulated grafts fired in high-frequency bursts separated by long pauses, and this discharge was comparatively insensitive to glutamate. Poststimulus inhibition after local stimulation of the transplant surface was readily observed in the growth-stimulated grafts, but absent in all non-stimulated grafts tested. Moreover, superfusion of picrotoxin, which antagonizes GABA-mediated inhibitory pathways, reversibly converted the growth-stimulated graft discharge pattern into one more characteristic of nonstimulated grafts. These data demonstrate the importance of extrinsic inputs for functional development of neuronal circuits within the neocortex, and suggest that, in the absence of such inputs, local inhibitory mechanisms may not develop normally.

7.4.4. Thyroid Hormone Dependency

The importance of thyroid hormones for brain development is well known. Intra-ocular transplantation of fetal brain areas to thyroidectomized recipients may help in

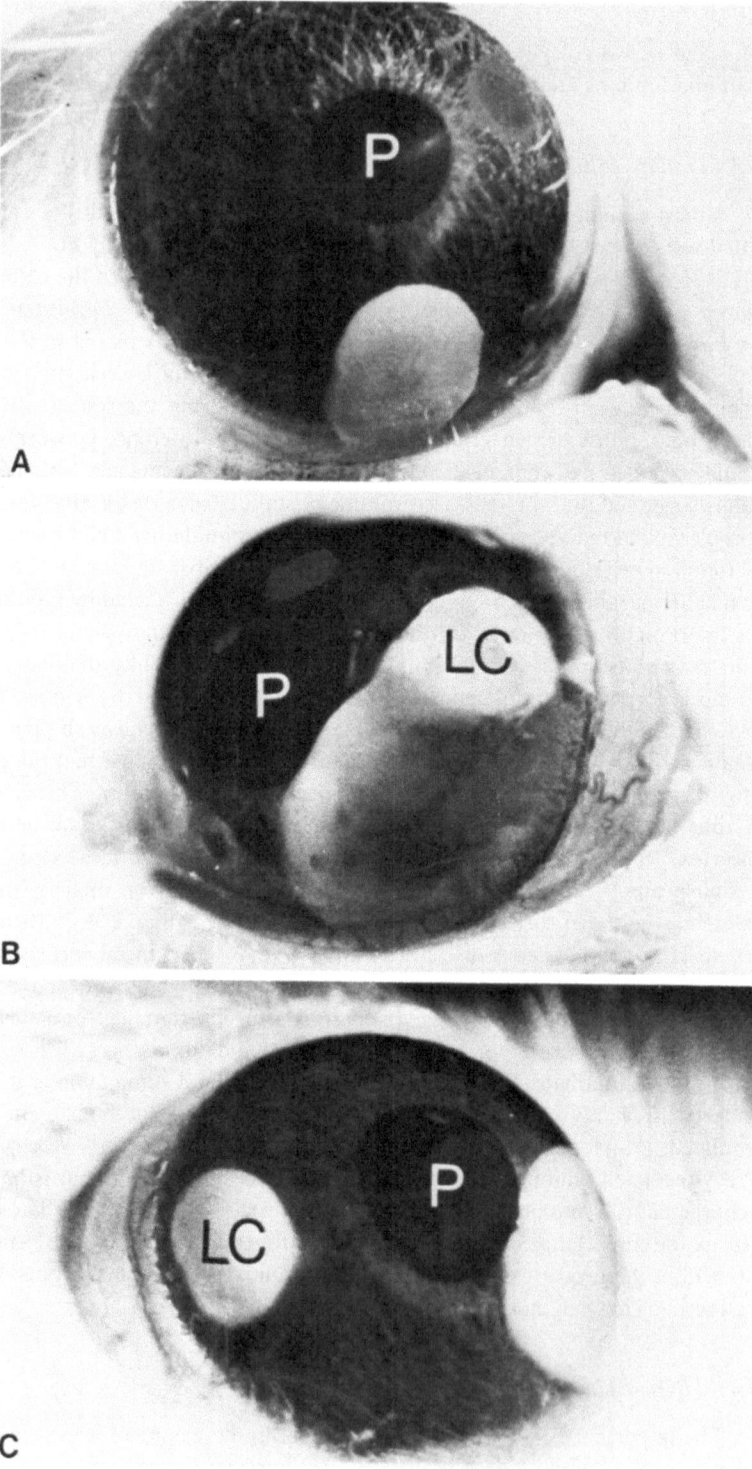

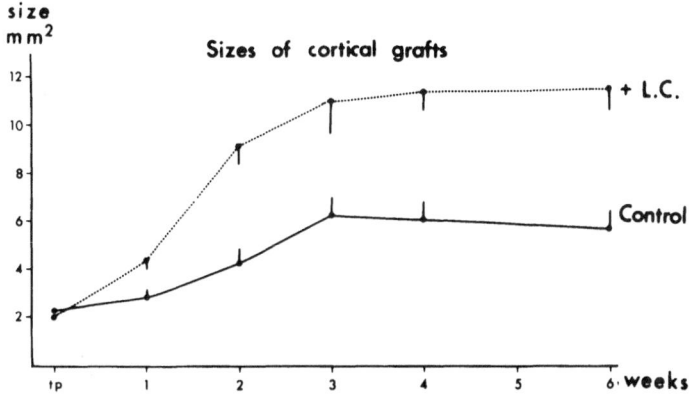

FIGURE 13. Diagram illustrating the stimulatory effect elicited by the presence in the eye of a mature locus coeruleus graft on the growth of cortex cerebri grafts. Size measurements were made *in vivo* through the cornea of the host animals.

resolving the spatial and temporal effects of thyroid hormone deficiency. Thus, it was recently shown that cortex cerebri has a donor stage-dependent thyroid hormone interaction such that 18-day donor fetuses give rise to larger cortex grafts, whereas full-term donor fetuses give rise to smaller cortex grafts in thyroidectomized hosts.[20] Interestingly, the density of sympathetic adrenergic innervation of the cortex cerebri grafts from the host iris was always reduced by 50% in thyroidectomized hosts. This histochemically detectable reduction was paralleled by a reduction in noradrenaline concentration in cortex cerebri grafts in the experimental groups.

7.4.5. Lead Effects

Cortex cerebri grafts have also been used in our studies of the effects of chronic perinatal lead exposure on brain development. Lead levels of intraocular grafts were considerably higher than lead levels of host cortex cerebri following exposure of the hosts to lead acetate in the drinking water.[5] The higher lead levels found in grafts might be explained by a more immature state of a possible blood–brain barrier for lead in the grafts. Maturation of cortex cerebri grafts was sensitive to lead treatment. In general, high lead levels (2% lead acetate in drinking water) caused a decreased growth of cortex cerebri grafts, while 1% lead acetate caused relatively selective changes in the developing CNS tissue, leading to a net increase in tissue volume by hyperproliferation of certain neuronal or glial components. This dose may, however, be close to a higher level where less specific, generally deleterious lead effects on the host animal and/or the graft

←——

FIGURE 12. Photographs showing the *in vivo* appearance of brain tissue grafts after intraocular maturation. Single cortical graft (A) is much smaller than cortical graft grown in contact with locus coeruleus (LC) graft (B). When the cortex graft is placed opposite the LC graft (C), the final size of the cortex graft is similar to that of a single cortical graft. From Ref. 8.

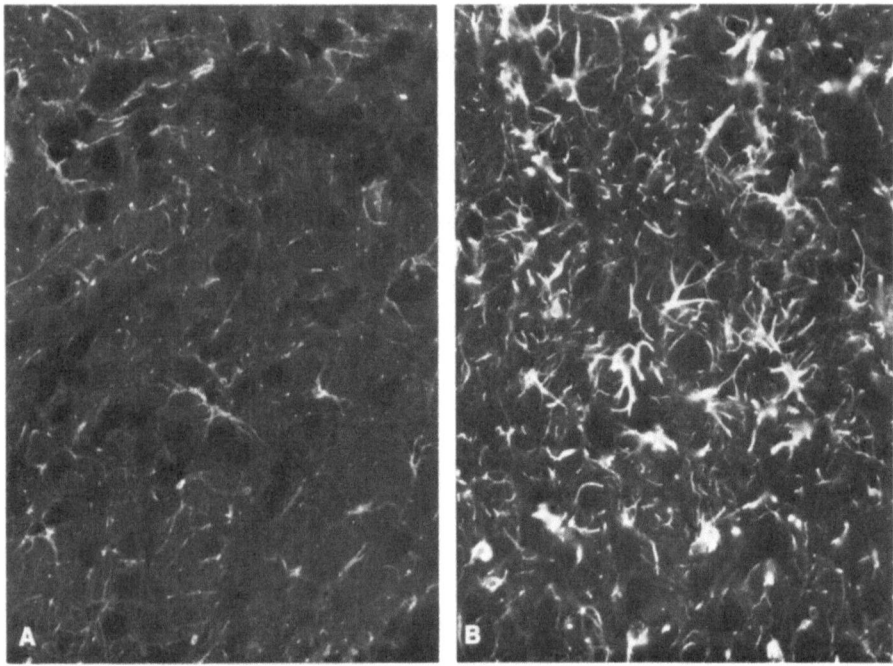

FIGURE 14. Comparison of the amount of GFA-positive structures in a locus coeruleus-stimulated (A), and a single(B) intraocular cortex cerebri graft. Whereas the density of GFA-like immunoreactivity is extremely high in the single grafts, the locus coeruleus-stimulated cortex grafts show a lower, more normal amount of GFA-like immunoeactivity, although the density is still increased compared to normal cortex cerebri *in situ.* Fluorescence microphotographs, $\times$ 330.

appear, manifested by the prominent decrease in graft growth in host animals exposed to 2% lead acetate.[6] Preliminary results indicate that the density of adrenergic innervation of cortical grafts is also abnormal following chronic lead treatment.

7.5. Spinal Cord Grafts: Initial Observations

When defined segments of fetal rat spinal cord were grafted to the anterior chamber of the eye, such grafts became vascularized from the host iris, grew and developed neuron subtypes, myelinated fiber bundles, astroglial populations, and electrical activity reminiscent of normal spinal cord tissue (Figs. 15 and 16).[42,87] Preliminary extracellular recordings from spinal cord grafts have revealed large spikes possibly generated by α-motoneurons (Fig. 17). The intraocular technique permits studies of intrinsic circuitry as well as conditions for formation of afferent and efferent connections with the host iris and other central or peripheral tissues, which can be grafted into contact with the spinal cord transplants. An example of an intrinsic system preserved in the grafts is a rich network of nerve fibers with enkephalinlike immunoreactivity (Fig. 18).[7] When the spinal cord is combined with the cerebral cortex *in oculo,* enkephalin-positive neurons of the spinal cord graft are able to innervate the adjacent cortex graft.[7] Special emphasis has been given the possible formation of adrenergic afferents to spinal cord grafts. In contrast to many other brain areas, no appreciable ingrowth of peripheral sympathetic nerves occurred into spinal cord grafts. Locus coeruleus grafts seemed to be able to innervate adjacent spinal cord grafts provided that the sensory innervation of the host iris was removed. Thus, spinal cord grafts differ from grafts of brain cortices because a prominent adrenergic innervation from a locus coeruleus cograft can only be obtained under special conditions. Experiments of this kind may provide clues to enigmas of spinal cord regeneration. A speculation from these experiments is that the inhibition of ingrowth of adrenergic nerves into spinal cord tissue is related to the unique relationship between spinal ganglia and spinal cord, and that this in turn may relate to the fact that recovery of function across a complete mechanical lesion of the spinal cord *in situ* does not occur.

7.6. Caudate Grafts

In contrast to most other brain regions, caudate grafts do not easily form thick, rounded, well-delineated structures *in oculo.* Instead, they tend to flatten out and spread

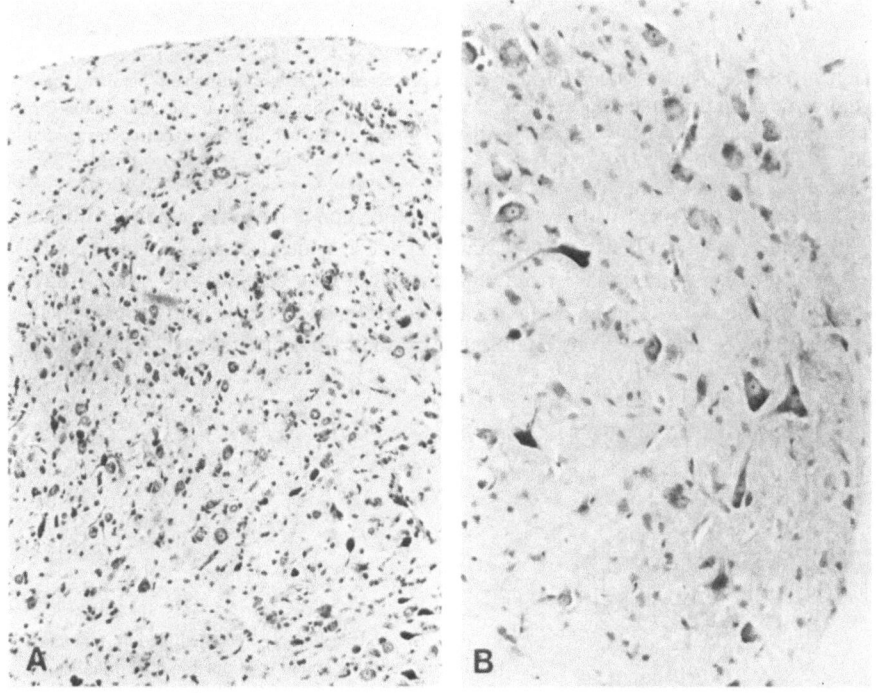

FIGURE 15. Photographs of toluidine blue-stained sections of an intraocular spinal cord graft. (A) Low-power photo showing a cell-poor outer layer reminiscent of white matter, (B) Close-up; note the abundance of large motor neuron-like cell bodies. (A) × 135; (B) × 330.

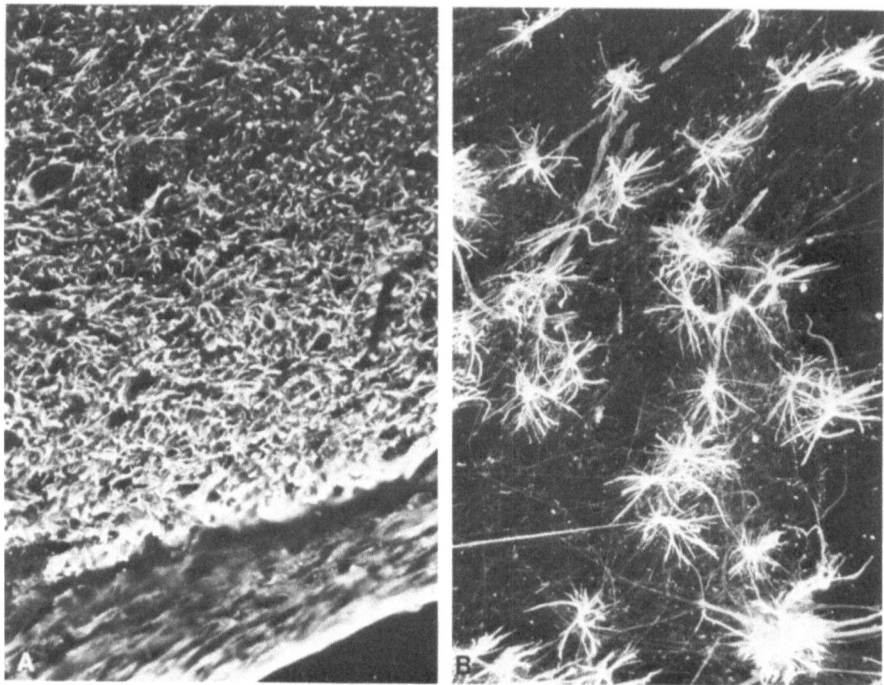

FIGURE 16. (A) Section of an intraocular spinal cord graft processed for GFA-immunohistochemistry showing an astrocytic gliosis. Iris in lower part of the picture; × 330.(B) Smear of an intraocular spinal cord graft processed for GFA-immunohistochemistry. A large number of reactive fluorescent astrocytes is seen; × 135.

over a large portion of the host iris. More immature donor material produces thicker grafts than older donors. The youngest gestational age at which it is possible to unequivocally separate the caudate anlage from adjacent developing cerebral cortex is 14 days (crown–rump length 11–12 mm). Thicker caudate grafts had a dense homogeneous neuropil, with scattered small neurons often found in small rows or clusters.[58] The caudate

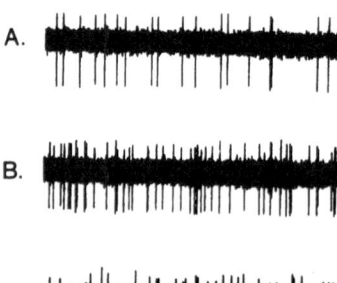

FIGURE 17. Extracellularly recorded action potentials from neurons in the spinal cord portion of a spinal cord-neocortex double graft. (A–C) Three neurons from two different transplants. These firing activities were driven by a small glutamate-leak from the recording pipette. From Ref. 42.

grafts became well fused with the stroma of the anterior surface of the iris. Interestingly, almost no sympathetic fibers were seen to invade caudate grafts (Fig. 19). When studied electrophysiologically, the initial spontaneous discharge rate ranged from 0 to 7 spikes/sec (mean 2.3 $\pm$ 0.4). With repeated electrical stimulation or iontophoresis of appropriate drugs, sustained elevations in "spontaneous" activity were obtained. Local electrical stimulation of the surface of the transplant usually elicited several driven action potentials associated with a small field potential. Iontophoresis of DA slowed most neurons in the caudate grafts, as did iontophoresis of noradrenaline. Glutamate excited all cells tested, while, somewhat surprisingly, acetylcholine did not alter the discharge rate in most cells. Both DA and noradrenaline at suitable iontophoretic doses could also suppress driven activity as well as background activity induced by electrical stimulation. These electrophysiological observations of the caudate grafts are similar in most respects to what has been demonstrated in the normal caudate except perhaps for the lack of effects of acetylcholine in the grafts. Perhaps the excitatory responses to this transmitter seen *in situ* may reflect a presynaptic or indirect action. When caudate grafts were combined with locus coeruleus grafts, the caudate grafts received a moderate to high number of fine varicose catecholamine-containing nerve terminals from the locus coeruleus graft. Innervation occurred both when the caudate was grafted first, and when it was added to an already present locus coeruleus graft.[58] The density of locus coeruleus-derived nerve terminals in the caudate seemed to be much higher than that of locus coeruleus-derived fibers in the caudate *in situ*. When caudate grafts were combined with substantia nigra grafts, the caudate portion had a variable dopaminergic innervation. In some instances, a dense and diffuse pattern of innervation, similar to that normally produced

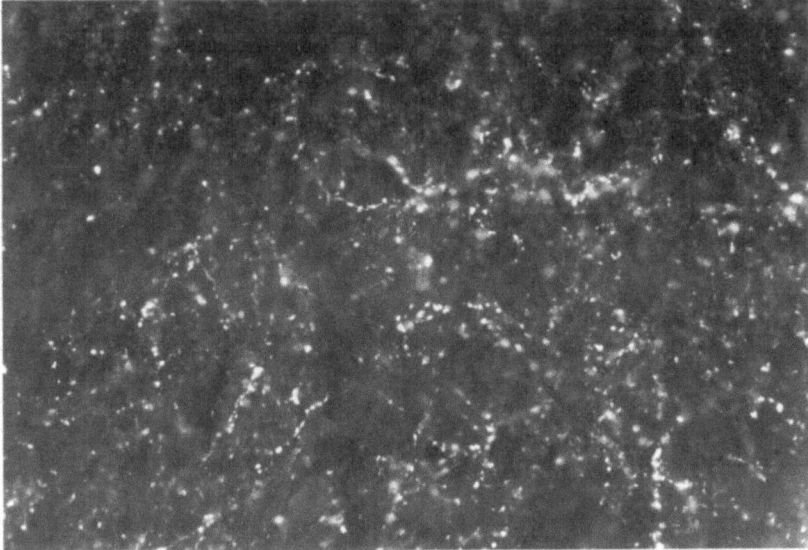

FIGURE 18. Presence of enkephalinlike immunoreactive fibers in an intraocular spinal cord graft. Fluorescence microphotograph, $\times$ 330.

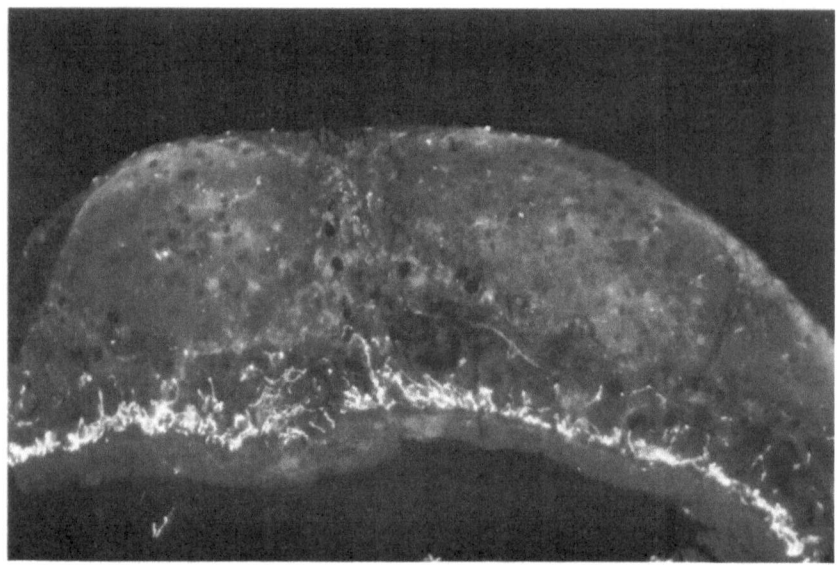

FIGURE 19. Caudate transplant on an intact iris, as seen in a 6-μm section. Fluorescent fibers of the sympathetic adrenergic ground plexus are clearly visible in the dilator plate (below), whereas the caudate transplant is totally devoid of fluorescent fibers. Specimen preincubated in α-m-NE, 10^{-5}M. Fluorescence microphotograph, × 135. (From Ref. 58).

by nigral axons in the caudate, was present. Such areas were generally found close to the nigral grafts. Other areas, more distant from the nigral grafts, contained a sparse innervation of fine varicose terminals.

7.7. Central Noradrenaline, Dopamine, and 5-Hydroxytryptamine Neurons Grafted to the Eye Chamber

Relatively extensive studies of factors that influence fiber production by grafted central monoamine neurons have been performed using the intraocular model.[21,22,41,47-50,52-54,56,57,69,72,74-77] All three monoamine neuron types (mainly dissected from locus coeruleus, substantia nigra, and raphe nuclei, respectively) survive grafting to the anterior eye chamber. The monoamine neurons innervate the neuropil of the grafts and extend a "halo" of nerve fibers into the host iris, where they can be studied by Falck–Hillarp histochemistry of whole mounts following sympathetic denervation of the host eye. In this way, two-dimensional plexuses of nerve fibers are formed in the host iris, which can be precisely quantified. In addition to the host iris, other peripheral tissues cografted to the eye chamber may be innervated by the central monoamine neurons. As has been partly described in several sections above, cografts of potential CNS target areas to the eye chamber will also become innervated by the grafted central monoamine neurons (Fig. 20). When central monoamine neurons invade peripheral tissues, they tend to change their morphology to mimic that of autonomic nerve fibers normally present in the peripheral tissue.

When comparing central and peripheral adrenergic neurons in the *in oculo* model (Fig. 21) it becomes evident that although the superior cervical ganglion and locus coeruleus are able to substitute for each other morphologically and functionally in different target tissues, there are profound differences between these two types of adrenergic neurons. Thus, locus coeruleus is not able to innervate heart tissue, which is readily innervated by the superior cervical ganglion. Further, when locus coeruleus innervates the host iris, it is not influenced by the presence or absence of a peripheral sympathetic innervation of the iris. This is in marked contrast to a transplanted superior cervical

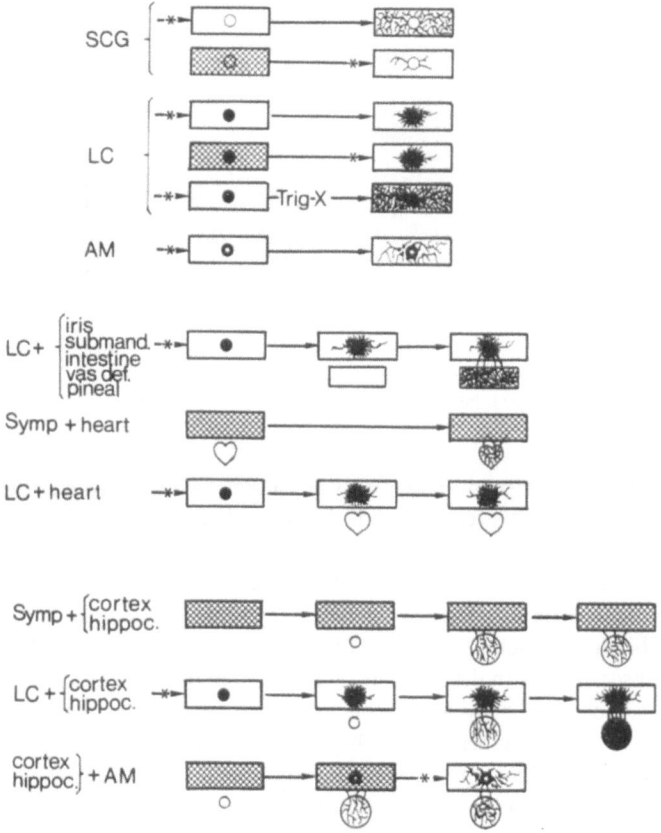

FIGURE 20. Summary of the main results. In all diagrams, host irides are depicted as large rectangles, and the presence of a normal sympathetic nerve plexus on the host irides is indicated by a raster. Time flows from left to right as indicated by arrows. Removal of the sympathetic nerve plexus in the host iris is performed at various stages of the experiments and indicated by an asterisk. Open circles on host irides: grafts of the superior cervical ganglion. Filled circles: grafts of the developing locus coeruleus. Filled circles with a star: grafts of the adrenal medulla. All drawings demonstrate the amount and distribution of central adrenergic nerve fibers as seen in whole mounts of irides and freeze-dried sections of other types of grafts with Falck–Hillarp fluorescence histochemistry. For further descriptions, see the text. From Ref. 56.

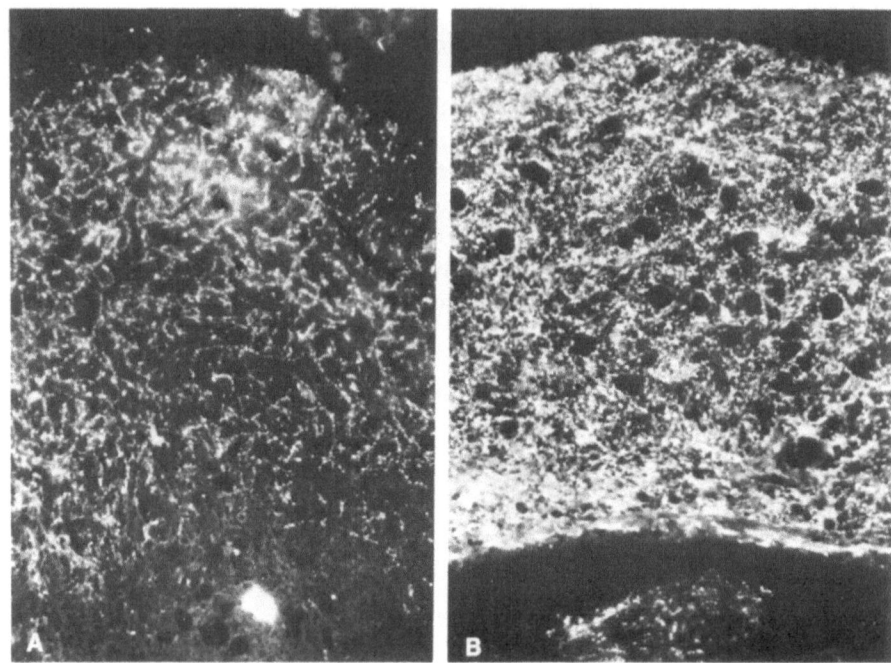

FIGURE 21. Dense innervation by NE fibers from locus coeruleus transplants of a cerebral cortex transplant (A), and of a caudate transplant (B), Both transplants were preincubated in α-m-NE 10^{-5}M. Fluorescence microphotoraph $\times$ 135. From Ref. 58.

ganglion, which will only produce new nerve terminals in a sympathetically denervated host iris. Interestingly, locus coeruleus will form a halo of nerve fibers on a sympathetically denervated host iris covering approximately one-third of the host iris. Nerve fiber production will then cease. If, however, the sensory innervation of the host iris is removed by lesioning the trigeminal ganglion, the grafted locus coeruleus neurons will immediately begin a sprouting response and innervate the entire iris (Fig. 22). Thus, presence of sensory nerves in the iris, derived from or via the trigeminal ganglion, inhibits nerve fiber proliferation by locus coeruleus. This interesting relationship between locus coeruleus and sensory nerves can also be demonstrated by adding an iris graft to a locus coeruleus transplant-carrying eye. In this case, a vigorous fiber outgrowth will be initiated by the iris graft leading to a complete adrenergic innervation of the iris graft from the locus coeruleus graft, while the host iris will continue to be only partially innervated. Obviously, this innervation of the iris graft is induced because it becomes totally denervated during transplantation. The interaction between locus coeruleus and nerve fibers derived from the trigeminal ganglion has been further elucidated by grafting the trigeminal ganglion and locus coeruleus together into the eye chamber.[69]

One principal difference between peripheral and central adrenergic neurons that may explain some of the differences in the nerve fiber growth responses *in oculo* is the insensitivity to nerve growth factor of locus coeruleus. Thus, the locus coeruleus is not stimulated by NGF, is not inhibited by anti-NGF, does not contain NGF, and does not seem to synthesize NGF.[43]

A further difference betweeen the superior cervical ganglion and locus coeruleus is seen when the two adrenergic neuron-containing areas are allowed to innervate the hippocampal formation *in oculo*. As described above, peripheral adrenergic neurons will give rise to an adrenergic innervation with a relatively normal density in a hippocampal graft while locus coeruleus, if given enough time, will cause a marked adrenergic hyperinnervation of the same target. These results suggest that the ingrowing central adrenergic nerve fibers, in contrast to the ingrowing peripheral adrenergic nerve fibers, are able to recognize many more vacant synaptic sites in the isolated hippocampal formation.

The influence of thyroxin on developing central monoamine neurons has been demonstrated in a series of experiments where locus coeruleus and substantia nigra were grafted to eyes of thyroidectomized hosts. The immature locus coeruleus has been shown to be unable to form fiber bundles in the host iris in the absence of thyroxin.[70] Similarly, the immature substantia nigra is inhibited in its ability to form axon bundles and nerve terminals by thyroxin deficiency.[22] After intraocular maturation, however, locus coeruleus can be stimulated to produce nerve fibers by the addition of an iris transplant. In this case, bundle formation is possible even in the absence of thyroxin. Paralleling the

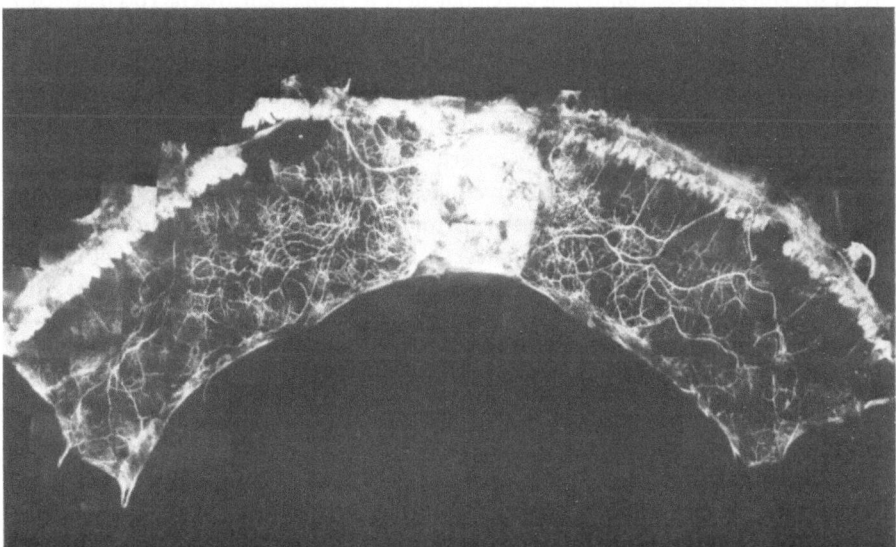

FIGURE 22. Survey of an iris whole mount. Brightly fluorescent area (center) represents the locus coeruleus transplant with heavily packed fluorescent NA-containing nerve fibers. Three weeks after trigeminotomy, the fluorescent monoamine nerve fibers are here seen to spread out over the entire surface of the host iris. This contrasts with the distribution on the control specimens, where a restricted halo of fluorescent fibers is found centered around the CNS tissue attachment. Montage of approximately 200 fluorescence microphotograhs, × 15. From Ref. 53.

impaired axon bundle formation demonstrated by locus coeruleus grafted to thyroidec-tomized hosts, there was an increased size of the locus coeruleus graft *in oculo*.[21] Finally, the impaired axon bundle formation demonstrated by locus coeruleus during thyroxin deficiency could be counteracted by a sensory denervation of the host iris.[71]

8. General Conclusions

Transplantation of small defined areas of the developing central and peripheral nervous system to the anterior chamber of the eye in rodents is a simple and efficient way of isolating tissues from their normal surroundings while maintaining their structural and functional integrity.

We have described how several different areas of the developing nervous system are able to survive grafting and continue development *in oculo* in a strikingly organo-typic manner. Such experiments stress the importance of intrinsic genetic determinants of development in various brain areas. Intraocular grafting thus becomes a unique way of dissecting the complex question of brain development into smaller, somewhat more accessible problems of regional development. The technique also permits a differentia-tion between intrinsic and extrinsic determinants of development. In cases where a dis-turbed structural and/or functional development of a given brain area has been detected, as was the case with single cortex cerebri grafts, multiple transplantations may be per-formed to see if connections with other brain areas will allow a given region to develop more normally. In the case of the cortex cerebri, a strong trophic, stimulatory effect was exerted by several, but not all areas of the CNS. Grafting of monoamine neurons to the eye chamber has revealed several new aspects of adrenergic fiber growth regulation.

The intraocular approach has been applied to several problems of disturbed devel-opment of the CNS. Thus, the effects of thyroid hormone deficiency on brain develop-ment have been elucidated. Moreover, severe, hitherto unrecognized effects of chronic perinatal lead exposure on brain development have been disclosed. The method has also proven useful in studies of the effects of ethanol on brain function. The possibility of transplantation between various genetically defined lines of mice or rats, such as the long-sleep and short-sleep mice which differ in ethanol sensitivity, or between various mutants, should prove helpful in determining the precise locus of a given disturbance of CNS structure or function.

Grafting of CNS tissues to an ectopic locus such as the anterior chamber also serves as a "benchmark" for grafting of the same tissues to the CNS. Knowledge about optimal stages for survival and maturation of brain areas in complete isolation from the rest of the CNS constitutes critical background information when studying the development of the corresponding tissues grafted to various intracranial sites and spinal cord[39] sites. Thus, we have used data from the substantia nigra and adrenal medulla transplanted to the eye chamber in order to graft the same areas to the brains of animals with uni-lateral 6-OH-DA-induced Parkinson's disease, and demonstrated permanent reduction of the motor abnormalities by such grafts.[16,17,64,86] For these experiments, by us and others, the reader is referred to Chapter 16 of this volume.

ACKNOWLEDGMENTS. Supported by the Swedish Medical Research Council (14X-03185, 14P-5867), the Swedish Council for Planning and Coordination of Research, Magnus Bergvalls Stiftelse, Karolinska Institutets fonder, the "Expressen" Prenatal Research Foundation, and USPHS Grants ES-02011, DA-02702, and MH-00289.

References

1. Ajmone-Marsan, C., 1972, Focal electrical stimulation, in: *Experimental Models of Epilepsy* (D. Purpura, J Penry, D. Tower, D. Woodbury, and R. Walter, eds.), pp. 147–172, Raven Press, New York

2. Björklund, A., and Stenevi, U., 1977, Experimental reinnervation of the rat hippocampus by grafted sympathetic ganglia. I. Axonal regeneration along the hippocampal fimbria, *Brain Res.* **135**:259.

3. Björklund, A., Stenevi, U., and Svendgaard, N. A , 1976, Growth of transplanted monoaminergic neurones into the adult hippocampus along the perforant path, *Nature (London)* **262**:787

4. Björklund, H., Palmer, M. R., Lind, B., Hoffer, B., and Olson, L., 1983, Postnatal lead exposure alters spontaneous cerebellar Purkinje neuron discharge, *Environ. Res.* **31**:448.

5. Björklund, H., Lind,B., Piscator, M., Hoffer, B., and Olson, L., 1981, Lead, Zinc, and copper levels in intraocular brain tissue grafts, brain, and blood of lead-exposed rats, *Toxicol. Appl. Pharmacol.* **60**:424.

6. Björklund, H., Olson, L , Seiger, Å., and Hoffer, B , 1980, Chronic lead and brain development: Intraocular brain grafts as a method to reveal regional and temporal effects in the central nervous system, *Environ. Res.* **22**:224.

7. Björklund, H., Palmer, M. R., Seiger, Å., Hoffer, B. J., and Olson, L., 1982, Survival and growth of neurons with enkephalin-like immunoreactivity in fetal brain areas grafted to the anterior chamber of the eye, *Neuroscience.* **10**:1387.

8. Björklund, H., Seiger, Å., Hoffer, B., and Olson, L., 1983, Trophic effects of brain areas on the developing cerebral cortex. I. Growth and histological organization of intraocular grafts, *Dev. Brain. Res.* **6**:131.

9. Chatagnon, P. A., 1952, Recherches sur la differenciation du neurone dans la greffe brephoplastique endocularie chez le rat blanc, *Arch. Biol.* **63**:199.

10. Crutcher, K., 1982, Neonatal septal lesions result in sympathohippocampal innervation in the adult rat, *Exp. Neurol.* **76**:1.

11. Crutcher, K. ., and Davis, J. N., 1981, Sympathohippocampal sprouting is directed by a target trophic factor, *Brain Res.* **204**: 410.

12. Crutcher, K . A., and Davis, J. N., 1982, Target regulation of sympathetic sprouting in the rat hippocampal formation, *Exp. Neurol.* **75**:347.

13. Ebendal, T., Olson, L., Seiger, Å., and Hedlund, K.-O., 1980, Nerve growth factors in the rat iris, *Nature (London)* **286**:25.

14. Faldino, G., 1923, Ulteriore contributo allo studio dello sviluppodelle articolazioni, *Chir. Organi Mov.* **7**:1.

15. Faldino, G., 1924, Sullo sviluppo dei tessuti embrionali omoplastici innestati nella camera anteriore dell occhio del coniglio, *Arch. Sci. Biol. (Bologna)* **5**:328.

16. Freed, W., Morihisa, J., Spoor, E., Hoffer, B., Olson, L., Seiger, Å., and Wyatt, R., 1981, Transplanted adrenal chromaffin cells in rat brain reduce lesion-induced rotational behavior, *Nature (London)* **292**:351.

17. Freed, W., Perlow, M., Karoum, F., Seiger, Å., Olson, L., Hoffer, B., and Wyatt, R., 1980, Restoration of dopaminergic function by grafting of fetal rat substantia nigra to the caudate nucleus: Long-term behavioral, biochemical, and histochemical studies, *Ann. Neurol.* **8**:510.

18. Freedman, R., Taylor, D., Seiger, Å.,Olson, L., and Hoffer, B., 1979, Seizures and related epileptiform activity in hippocampus transplanted to the anterior chamber of the eye: Modulation by cholinergic and adrenergic input, *Ann. Neurol.* **6**:281.

19 Goldowitz, D., Seiger, Å., and Olson, L., 1982, Anatomy of the isolated area dentata grown in the rat anterior eye chamber, *J. Comp. Neurol.* **208**:382.

20. Granholm, A.-C., Olson, L., and Seiger, Å., 1982, Intraocular development and adrenergic innervation of cortical and subcortical brain areas: Influence of thyroid hormone deficiency, *Dev. Neurosci.* **5**:436.

21. Granholm, A.-C., and Seiger, Å., 1981, Thyroid hormone dependency in immature but not mature grafted locus coeruleus neurons: Evidence from intraocular innervation of iris transplants, *Med. Biol.* **59**:51.

22. Granholm, A.-C., and Seiger, Å., 1981, Thyroxine dependency of the developing substantia nigra: Evidence from intraocular grafting experiments in rats, *Neurosci. Lett.* **22**:279.

23. Hoffer, B., Olson, L., Freedman, R., and Seiger, Å., 1977, Seizures and related epileptiform activity in hippocampus transplanted to the eye. I. Characterization of seizures, interictal spikes, and synchronous activity, *Exp. Neurol.* **54**:233.

24. Hoffer, B., Olson, L., Seiger, Å., and Bloom, F., 1975, Formation of a functionl adrenergic input to intraocular cerebellar grafts: Ingrowth of inhibitory sympathetic fibers, *J. Neurobiol.* **6**:565.

25. Hoffer, B., Seiger, Å., Freedman, R., Olson, L., and Taylor, D. 1977, Electrophysiology and cytology of hippocampal formation transplants in the anterior chamber of the eye. II. Cholinergic mechanisms, *Brain Res.* **119**:107.

26. Hoffer, B., Seiger Å., Ljungberg, T., and Olson, L., 1974, Electrophysiological and cytological studies of brain homografts in the anterior chamber of the eye: Maturation of cerebellar cortex *in oculo*, *Brain Res.* **79**:165.

27. Hoffer, B., Taylor, D., Baker, R., Deitrich, R., Seiger Å., and Olson, L., 1980, Ethanol withdrawal seizures in hippocampus transplanted to the anterior chamber of the eye, *Life Sci.* **26**:239.

28. Ljungdahl, Å., Seiger, Å., Hökfelt, T., and Olson, L., 1973, ^{3}H-GABA uptake in growing cerebellar tissue: Autoradiography of intraocular transplants, *Brain Res.* **61**:379.

29. Loy, R., Milner, T. A., and Moore, R. Y., 1980, Sprouting of sympathetic axons in the hippocampal formation: Conditions necessary to elicit ingrowth, *Exp. Neurol.* **67**:399.

30. Loy, R., and Moore, R. Y., 1977, Anomalous innervation of the hippocampal formation by peripheral sympathetic axons following mechanical injury, *Exp. Neurol.* **57**:645.

31. Lund, R. D., and Lund, J. S., 1971, Synaptic adjustment after deafferentation of the superior colliculus of the rat, *Science* **171**:804.

32. Lynch, G., and Cotman, C. W., 1975, The hippocampus as a model for studying anatomical plasticity in the adult brain, in: *The Hippocampus: A Comprehensive Treatise*, Vol. I (R. L. Issacson and K. H. Pribam, eds.) pp. 123–154, Plenum Press, New York.

33. Lynch G., Rose, G., Gall, C., and Cotman, C. W., 1975, The response of the dentate gyrus to partial deafferentation, in: *Golgi Centennial Symposium, Proceedings* (M. Santini, ed.), pp. 305–317, Raven Press, New York.

34. May, R. M, 1930, La greffe dans l'oeil de rat blanc adulte du tissu cerebral de rat nouveau-ńe, *Arch. Anat. Microsc. Morphol. Exp.* **26**:433.

35. May, R. M., 1945, Régeneration cérebrale provoquiée par la greffe intraoculaire simultannée de tissu cérebral de nouveau-ńe et de nerf sciatique chez la souris, *Bull. Biol. Fr. Belg.* **79**:151.

36. May, R. M., 1949, Connexions entre des cellules cérebrales et des muscles de la cuive dans leut greffe bréphoplastique intraoculaire, simultanée chez la souris, *Arch. Anat. Microsc. Morphol. Exp.* **38**:145.

37. May, R. M., 1952, La greffe bréphoplastique intra-oculaire simultanée de tissu cérebral et de thymus vivant ou mort chez la souris *Arch. Anat. Microsc. Morphol. Exp.* **41**:237.

38. May, R. M., 1954, La greffe bréphoplastique intraoculaire du cervelet chez la souris, *Arch. Anat. Microsc. Morphol. Exp.* **43**: 42.

39. Nygren, L. G., Olson, L., and Seiger, Å., 1977, Monoaminergic reinnervation of the transected spinal cord by homologous fetal brain grafts, *Brain Res.* **129**:227.

40. Olson, L., 1970, Fluorescence histochemical evidence for axonal growth and secretion from transplanted adrenal medullary tissue, *Histochemie* **22**: 1.

41. Olson, L., Björklund, H., Ebendal, T., Hedlund, K–O., and Hoffer, B., 1981, Factors regulating growth of catecholamine-containing nerves, as revealed by transplantation and explantation studies, in: *Development of the Autonomic Nervous System* (K. Elliott and G. Lawrenson, eds.) pp. 213–226, Pitman Medical, London.

42. Olson, L., Björklund, H., Hoffer, B. J., Palmer, M. R., and Seiger, Å., 1982, Spinal cord grafts: An intraocular approach to enigmas of nerve growth regulation, *Brain Res. Bull.* **9**:519.

43. Olson, L., Ebendal, T., and Seiger, Å., 1979, NGF and anti-NGF: Evidence against effects on fiber growth in locus coeruleus from cultures of perinatal CNS tissues, *Dev. Neurosci.* **2**:160.

44. Olson, L., Freedman, R., Seiger, Å., and Hoffer, B., 1977, Electrophysiology and cytology of hippocampal formation transplants in the anterior chamber of the eye. I. Intrinsic organization, *Brain Res.* **119**:87.

45. Olson, L., Hamberger, B., Hoffer, B., Miller, R., and Seiger, Å., 1981, Nerve fiber formation by grafted adult adrenal medullary cells, in: *Chemical Neurotransmission, 75 Years. Second Nobel Conference* (L. Stjärne, P. Hedqvist, H. Lagercrantz, and Å. Wennmalm, eds.), pp. 35–48, Academic Press, New York.

46. Olson, L., and Malmfors, T., 1970, Growth characteristics of adrenergic nerves in the adult rat: Fluorescence histochemical and ³H-noradrenaline uptake studies using tissue transplantations to the anterior chamber of the eye, *Acta Physiol. Scand. Suppl.* **348**:1.

47. Olson, L., and Seiger, Å., 1972, Brain tissue transplanted to the anterior chamber of the eye. I. Fluorescence histochemistry of immature catecholamine and 5-hydroxytryptamine neurons reinnervating the rat iris, *Z. Zellforsch. Mikrosk. Anat.* **135**:175.

48. Olson, L., and Seiger, Å., 1974, Nerve growth specificity and regulation as revealed by intraocular brain tissue transplants, in: *Dynamics of Degeneration and Growth in Neurons,* (K. Fuxe, L. Olson, and Y. Zotterman, eds.), pp. 499–507, Pergamon Press, Elmsford, N.Y.

49. Olson, L., and Seiger, Å., 1975, Brain tissue transplanted to the anterior chamber of the eye. II. Fluorescence histochemistry of immature catecholamine and 5-hydroxytryptamine neurons innervating the rat vas deferens, *Cell Tissue Res.* **158**:141.

50. Olson, L., and Seiger, Å., 1976, Locus coeruleus. Fiber growth regulation *in oculo, Med. Biol.* **54**:142.

51. Olson, L., and Seiger, Å., 1980, A system of atypical catecholamine-containing nerve fibers in the rat iris present after total superior cervical ganglionectomy, *Med. Biol.* **58**:94.

52. Olson, L., and Seiger, Å., 1983, Nerve fiber formation by the superior cervical ganglion, the adrenal medulla, and locus coeruleus: Similarities and differences as revealed by grafting, in: *Autnomic Ganglia* (L. G. Elfvin, ed.), pp 507–522 Wiley, New York.

53. Olson, L., Seiger, Å., and Ålund, M., 1978, Locus coeruleus fiber growth *in oculo* induced by trigeminotomy, *Med. Biol.* **56**: 23.

54. Olson, L., Seiger, Å., and Ålund, M., 1978, How nerve fiber outgrowth from intraocular locus coeuleus grafts is controlled by the state of innervation of the host iris, in: *Formshaping Movements in Neurogenesis* (C.-O. Jacobson and T. Ebendal, eds.), pp. 245–256, Almqvist and Wiksell, Stockholm.

55. Olson, L., Seiger, Å., Ålund, M., Freedman, R., Hoffer, B., Taylor, D., Woodward, D., 1979, Intraocular brain grafts: A method to differentiate between intrinsic and extrinsic determinants of structural and functional development in the central nervous system, in: *Neural Growth and Differentiation* (E. Meisami and M. Brazier, eds.), pp. 223–235, Raven Press, New York.

56. Olson, L., Seiger, Å., Ebendal, T., and Hoffer, B., 1980, Comparisons of nerve fiber growth from three major catecholamine-producing cell systems: Adrenal medulla, superior cervical ganglion and locus coeruleus, in: *Histochemistry and Cell Biology of Autonomic Neurons, SIF Cells and Paraneurons* (O. Eränkö, S. Soinila, and H. Päivärinta, eds.), pp. 27–34, Raven Press, New York

57. Olson, L., Seiger, Å., Freedman, R., and Hoffer, B., 1980. Chromaffin cells can innervate brain tissue: Evidence from intraocular double grafts, *Exp. Neurol.* **70**:414.

58. Olson, L., Seiger, Å., Hoffer, B., and Taylor, D., 1979, Isolated catecholaminergic projections from subtantia nigra and locus coeruleus to caudate, hippocampus and cerebral cortex formed by intraocular sequential double brain grafts, *Exp. Brain. Res.* **35**:47.

59. Olson, L., Seiger, Å., and Strömberg, I. 1983, Intraocular transplantation in rodents: A detailed account of the procedure and examples of its use in neurobiology with special reference to brain tissue grafting, in: *Advances in Cellular Neurobiology,* Vol. 4 (S. Fedoroff, and L. Hertz, eds.), pp. 407–442, Academic Press, New York.

60. Olson, L., Seiger, Å., Taylor, D., and Hoffer, B., 1980, Conditions for adrenergic hyperinnervation in hippocampus. I. Histochemical evidence from intraocular double grafts, *Exp. Brain Res.* **39**: 277.

61. Palmer, M. R., Björklund, H., Freedman, R., Taylor, D. A., Marwaha, J., Olson, L., Seiger, Å., and Hoffer, B. J., 1981, Permanent impairment of spontaneous Purkinje cell discharge in cerebellar grafts caused by chronic lead exposure, *Toxicol. Appl. Pharmacol.* **60**:431.

62. Palmer, M., Björklund, H., Olson, L., and Hoffer, B., 1983, Trophic effects of brain areas on the developing cerebral cortex. II. Electrophysiology of intraocular grafts, *Dev. Brain Res.* **6**:141.

63. Palmer, M. R., Sorensen, S. M., Freedman, R., Olson, L., Hoffer, B., and Seiger, Å., 1982, Differential ethanol sensitivity of intraocular cerebellar grafts in long-sleep and short-sleep mice, *J. Pharmacol. Exp. Ther.* **222**:480.

64. Perlow, M., Freed, W., Hoffer, B., Seiger, Å., Olson, L., and Wyatt, R., 1979, Brain grafts reduce motor abnormalities produced by destruction of nigrostriatal dopamine system, *Science* **204**:643.

65. Raisman, G., 1969, Neuronal plasticity in the septal nuclei of the adult rat, *Brain Res.* **14**:25.

66. Raisman, G., and Field, P. M., 1973, A quantitative investigation of the development of collateral reinnervation after partial deafferentation of the septal nuclei, *Brain Res.* **50**:241.

67. Sartori, C., 1926, Sugli innesti di tessuti embrionali, *Arch. Ital. Chir.* **15**:3.

68. Schultzberg, M., Hökfelt, T., Olson, L., Ålund, M., Nilsson, G., Terenius, L., Elde, R., Goldstein, M., and Said, S., 1980, Substance P, enkephalin and somatostatin immunoreactive neurons in intestinal tissue transplanted to the anterior eye chamber, *J. Auton. Nerv. Syst.* **1**:291.

69. Seiger, Å., 1980, Growth interaction between locus coeruleus and trigeminal ganglion after intraocular double grafting, *Med. Biol.* **58**:149.

70. Seiger, Å., and Granholm, A.-C., 1981, Thyroxin dependency of the developing locus coeruleus: Evidence from intraocular grafting experiments, *Cell Tissue Res.* **220**:1.

71. Seiger, Å., and Granholm, A.-C., 1982, Intraocular fiber growth of grafted locus coeruleus neurons: Sensory denervation counteracts alterations induced by thyroid hormone deficiency, *Med. Biol.* **60**:159.

72. Seiger, Å., and Olson, L., 1977, Quantitation of fiber growth in transplanted central monoamine neurons, *Cell Tissue Res.* **179**: 285.

73. Seiger, Å., and Olson, L., 1975, Brain tissue transplanted to the anterior chamber of the eye. III. Substitution of lacking central noradrenaline input by host iris sympathetic fibers in the isolated cerebral cortex developed *in oculo*, *Cell Tissue Res.* **159**: 325.

74. Seiger, Å., and Olson, L., 1977, Reinitiation of directed nerve fiber growth in central monoamine neurons after intraocular maturation, *Exp. Brain Res.* **29**:14.

75. Seiger, Å., and Olson, L., 1977, Growth of locus coeruleus neurons *in oculo* independent of simultaneously present adrenergic and cholinergic nerves in the iris, *Med. Biol.* **55**:209.

76. Seiger, Å., and Olson, L., 1978, Innervation of peripheral tissue grafts by locus coeruleus neurons *in oculo:* Only partial correspondence with degree of sympathetic innervation, *Brain Res.* **139**:233.

77. Seiger, Å., Olson, L., and Farnebo, L.-O., 1976, Brain tissue transplanted to the anterior chamber of the eye. IV. Drug-modulated transmitter release in central monoamine nerve terminals lacking normal presynaptic receptors, *Cell Tissue Res.* **165**:157.

78. Stenevi, U., Björklund, A., and Svendgaard, N.-A., 1976, Transplantation of central and peripheral monoamine neurons to the adult rat brain: Techniques and conditions for survival, *Brain Res.* **224**:1.

79. Stenevi, U., and Björklund, A., 1978, Growth of vascular sympathetic axons into the hippocampus after lesions of the septo-hippocampal pathway: A pitfall in brain lesion studies, *Neurosci. Lett.* **7**:219.

80. Steward, O., 1982, Assessing the functional significance of lesion-induced neuronal plasticity, *Int. Rev. Neurobiol.* **23**:197.

81. Taylor, D., Freedman, R., Seiger, Å., Olson, L., and Hoffer, B. J., 1980, Conditions for adrenergic hyperinnervation in hippocampus. II. Electrophysiological evidence from intraocular double grafts, *Exp. Brain Res.* **39**:289.

82. Taylor, D., Hoffer, B., Zieglgänsberger, W., Siggins, G., Ling, N., Seiger, Å., and Olson, L., 1979, Opioid peptides excite pyramidal neurons and evoke epileptiform activity in hippocampal transplants *in oculo*, *Brain Res.* **176**:135.

83. Taylor, D., Seiger, Å., Freedman, R., Olson, L., and Hoffer, B., 1978, Functional reinnervation of transplants in the anterior chamber of the eye by the autonomic ground plexus of the iris, *Proc. Natl. Acad. Sci. USA* **75**:1009.

84. Tischler, A., Perlman, R., Nunnemacher, G., Morse G., DeLellis, R., Wolfe, M., and Sheard, B., 1982 Long-term effects of dexamethasone and nerve growth factor on adrenal medullary cells cultured from young adult rats, *Cell Tissue Res.* **225**:525.

85. Woodward, O., Hoffer, B., Olson, L., and Seiger, Å., 1977, Intrinsic and extrinsic determinants of

dendritic development as revealed by Golgi studies of cerebellar and hippocampal transplants *in oculo*, *Exp. Neurol.* **57**:984.

86. Wuerthele, S. M., Freed, W. J., Olson, L., Morihisa, J., Spoor, L., Wyatt, R. J., and Hoffer, B. J., 1981, Effect of dopamine agonists and antagonists on the electrical activity of substantia nigra neurons transplanted into the lateral ventricle of the rat, *Exp. Brain Res.* **44**:1.

87. Yellin, H., 1976, Survival and possible trophic function of neonatal spinal cord grafts in the anterior chamber of the eye, *Exp. Neurol.* **51**:579.

88. Zalewski, A. A., 1971, The cellular immune reaction to transplanted sensory ganglia, *Exp. Neurol.* **32**:218.

89. Zalewski, A. A., 1971, The effect of Ag–B locus compatibility and incompatibility on neuron survival in transplanted sensory ganglia in rats, *Exp. Neurol.* **33**:576.

90. Zalewski, A. A., 1972, Regeneration of taste buds after transplantation of tongue and ganglia grafts to the anterior chamber of the eye, *Exp. Neurol.* **35**:519.

91. Zalewski, A. A., 1972, Trophic function of homografted neurons of Ag–B-histocompatible rats, *Transplantation* **14**:618.

92. Zalewski, A. A., 1974, Neuronal and tissue specifications involved in taste bud formation, *Ann. N.Y. Acad. Sci.* **228**:344.

93. Zalewski, A. A., 1980, Survival, regeneration and trophic function of neurons in 1-year transplants of sensory ganglia, *Exp. Neurol.* **68**:390.

94. Zalewski, A. A., and Silvers, W. K., 1973, Trophic function of neurons in homografts of ganglia in immunologically tolerant rats, *Exp. Neurol.* **41**:777.

95. Zalewski, A. A., and Silvers, W. K., 1974, Survival of neurons in homografts of ganglia in adult rats neonatally treated with bone marrow or lymph node cells, *Anat. Rec.* **178**:243.

96. Zalewski, A. A., and Silvers, W. K., 1977, The long-term fate of neurons in allografts of ganglia in Ag–B compatible normal and immunologically tolerant rats, *J. Neurobiol.* **8**:207.

6

The Olfactory Organ
Neural Transplantation

G. A. Monti Graziadei and P. P. C. Graziadei

1. Introduction

The vertebrate CNS acquires the morphology observed in the adult animal by selective differentiation of groups of neurons, occurring at specific times during embryogenesis, and by subsequent establishment of specific neuronal connections.[1-4]

The nervous system, composed of several billion units, deals with information transfer and processing (storage and retrieval) at the organism level. At its maturity, it is extremely complex; indeed, it represents the anatomical basis of consciousness. Because knowledge of the development of the functional relationships between groups of neurons is essential to our understanding of the nervous system as a whole, several analytical methods of study are implemented by the neurobiologist in an attempt to provide clues to its working mechanisms. Of these methods, the technique of transplantation, widely used by experimental embryologists, was applied to the study of the nervous system from the beginning of this century.[5] While in amphibians, transplantation of neural tissue has been successful resulting in several contributions from many laboratories, the mammalian pioneering work of Saltykow,[6] Marinesco,[7] Altobelli,[8] Dunn,[9] and others met with limited success. Only recently has the use of mammals for neural transplantation been reintroduced, primarily by investigators such as Das,[10] Björklund,[11] and Olson.[12] These workers demonstrated that a high rate of success can be obtained from meeting relatively simple technical requirements. Consequently, it is not surprising that the establishment of a reliable and successful transplantation technique has resulted in its application to a variety of experimental designs addressing problems such as development and repair of the nervous system (See Chapter 1).

G. A. Monti Graziadei and P. P. C. Graziadei ● Department of Biology, Florida State University, Tallahassee, Florida 32306.

Due to the lack of neuron regeneration in the nervous system of adult vertebrates, a major goal of neural transplantation has been to substitute for damaged neurons. However, transplantation provides the opportunity to address additional problems complementary to cell replacement such as: (1) the ability of neuronal elements, transplanted at selected ages, to connect with a number of targets (specificity and plasticity of connections), and (2) the influence of an abnormal environment on the transplanted neuronal population.

As outlined below, transplantation techniques can be successfully applied to the study of the olfactory system. In fact, these studies can address not only problems specifically related to the olfactory organ, but also problems of neuronal development and replacement, as well as problems of specificity of connection pertinent to our understanding of the nervous system as a whole.

In the last decade, it has been shown with morphological, autoradiographic, and experimental methods that the olfactory and vomeronasal sensory neurons are replaced in adult vertebrates. Replacement occurs because of the persistence in the neuroepithelium of stem cells that divide and differentiate into new neurons. The axons of the new neurons subsequently reach the olfactory and accessory olfactory bulb, where new synaptic contacts are established with the appropriate target, assuring continuity of function.[13-25]. Generation and replacement of neurons in adult animals, consequently, allows in this system new experiments of autoplastic as well as homoplastic transplantation between adult animals.

2. Transplantation in Amphibians

Transplantation studies of the olfactory organ were among the first to be reported at the beginning of this century and the amphibians were the animals of choice. The major goals of these experiments were to establish: (1) the role of the brain on the induction and development of the placode, and (2) the degree of influence exercised by the olfactory placode on the development of the brain.

Bell[26] was the first to transplant the olfactory placode of frog embryos into ectopic regions of the head. He observed that the nasal anlage can develop independent of its connections with the prosencephalon and that the sensory nerve originating from it may ectopically terminate into the lateral wall of the diencephalon. Similar results were shown by Burr[27-30] and May.[31] Luna[32] reported that development of the olfactory organ occurs when the placode is transplanted with or without brain fragments and that connections of the sensory nerves with the mesencephalon results in formation of glomeruli. He also observed independent development of the organ transplanted on the abdominal wall and formation of connections of the olfactory sensory axons with the striated muscle fibers of the abdomen. Carpenter,[33] working with *Ambystoma*, utilized local vital staining with transplantation, which provided data of particular significance with regard to the influence of the brain on the development of the nasal placode. He determined that the capacity of self-differentiation is present in the nasal anlage even before the appearance of the neural folds. Recently, transplantation experiments in our laboratory have shown that the olfactory placode can be successfully transplanted in *Xenopus* larvae at specific developmental stages and that it can survive and develop even when separated from its own target.[34]

Transplantation experiments in amphibians also have played a major role in determining the degree of influence of the olfactory placode on the development of the forebrain. There is general agreement that the olfactory organ is essential in the development of the entire telenephalon.[35] Conversely, transplanting supernumerary olfactory placodes in the olfactory region of the head results in hyperplasia of the telencephalon and often in the appearance of additional lobes (Fig. 1). Moreover, placodes transplanted outside the nasal area of the head can form interesting connections with several brain regions. Hyperplasia of the neuronal population and formation of glomerular structures (formed by the sensory terminals and the dendritic branches of the local neurons) can be observed when the diencephalon and mesencephalon are invaded by the olfactory axons of the ectopically located placodes (Fig. 2). Hyperplasia also was observed in the sensory ganglia of cranial nerves when olfactory fibers, from transplanted placodes, were experimentally directed to penetrate these peripheral centers.[31]

Sensory inputs determine, to a large extent, the development and organization of their own target areas; however, the experiments of transplantation in amphibians indi-

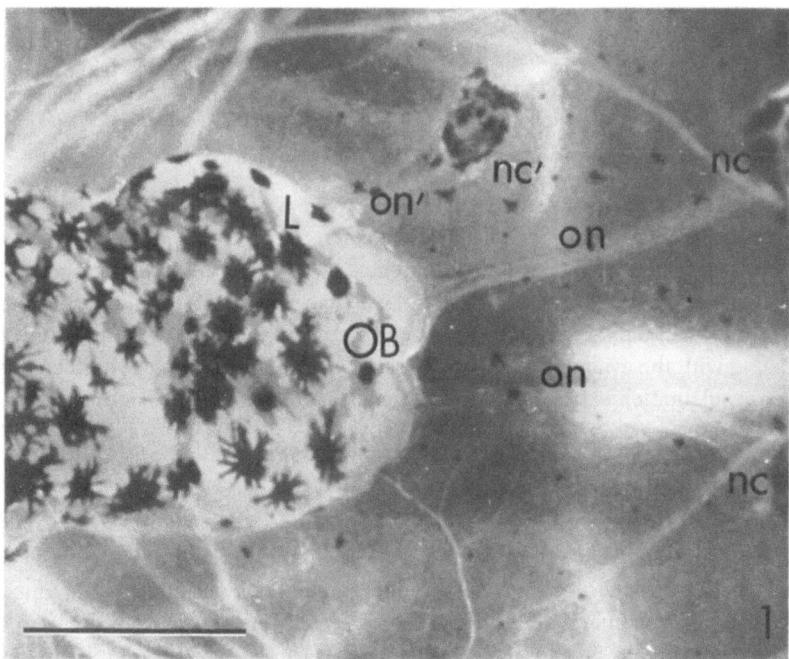

FIGURE 1. The dorsal view of the brain of *Xenopus laevis* tadpole, sacrificed at stage 50, shows the normal nasal capsules (nc) and the olfactory nerves (on) penetrating the olfactory bulbs (OB). From a supernumerary nasal capsule (nc'), transplanted as a placode when both donor and host were at stage 29/30, the sensory nerve (on') has projected toward the brain, inducing the formation of an extra lobe (L). Bar = 1 mm.

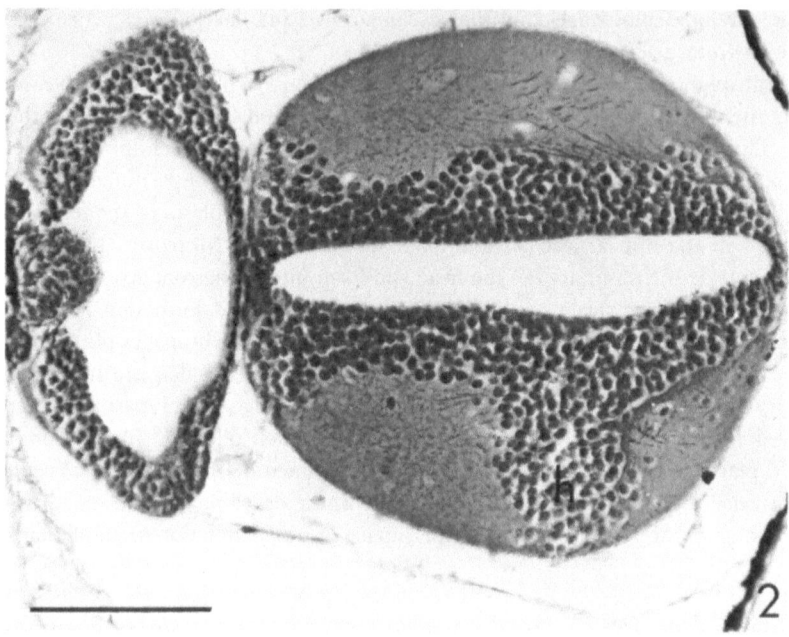

FIGURE 2. Coronal section through the dorsal thalamus of a stage 48 *Xenopus laevis*. A marked unilateral neuronal hyperplasia (h) has been induced by a supernumerary olfactory nerve (the nerve enters cranially to the plane of section). At the time of transplantation both donor and host were at stage 29/30. Silver stain. Bar = 100 μm.

cate that the olfactory neurons are able to affect the organization of areas in the CNS not necessarily connected to the olfactory pathway.

The experiments conducted on amphibians by means of transplantation techniques have provided interesting observations on the olfactory organ. Recently, the finding that the olfactory neuroepithelium is a neurogenetic matrix that persists in the adult mammal, along with the availability of new and improved techniques, has stimulated a variety of transplantation experiments on mammals using the olfactory organ.

3. Transplantation in Mammals

The transplantation technique offers the possibility to investigate the connections that can be established: (1) between neuronal populations not directly associated, and (2) between neuronal populations normally associated yet confronted at different developmental times. In the olfactory system, these goals can be achieved simply by partially or totally removing the olfactory bulb (which induces degeneration of the mature sensory neurons), and by allowing the axons of the newly differentiating sensory neurons to regrow intracranially and to reestablish connections with the experimentally modified target. For this reason, some observations not technically belonging in the transplantation category, yet achieving similar results, will also be briefly summarized here.

3.1. Partial and Total Removal of the Olfactory and Accessory Olfactory Bulb

Partial removal of the olfactory and accessory olfactory bulb in neonatal as well as in mature animals damages the sensory axons and results in the degeneration of their perikarya, which are located in the nasal and vomeronasal epithelia. In both peripheral neuroepithelia, reconstitution of the neuronal population is assured by the persistence of a neurogenetic matrix. The newly differentiated neurons send their axons past the lamina cribrosa into the spared bulbar formations.[36] In this experimental condition, we can study the pattern of connections between the population of reconstituted olfactory neurons and the already differentiated targets, namely, the olfactory bulb proper and the accessory olfactory bulb. When the sensory axons reinnervate the damaged targets, two main phenomena can be observed: (1) the sensory fibers from the olfactory and vomeronasal neurons form glomerular structures in a disorderly fashion in all layers of the spared olfactory bulb, and (2) the large neurons of the olfactory bulb (mitral and tufted cells) reorient their dendrites toward the ectopic glomeruli, where synaptic contacts with the sensory axon terminals can be demonstrated (Figs. 3-6).

Total removal of the main and accessory olfactory bulbs in mice and rats results in the exposure of a large forebrain area. The soft brain of the neonatal rodents extends anteriorly toward the lamina cribrosa and becomes easily accessible to the invasion of

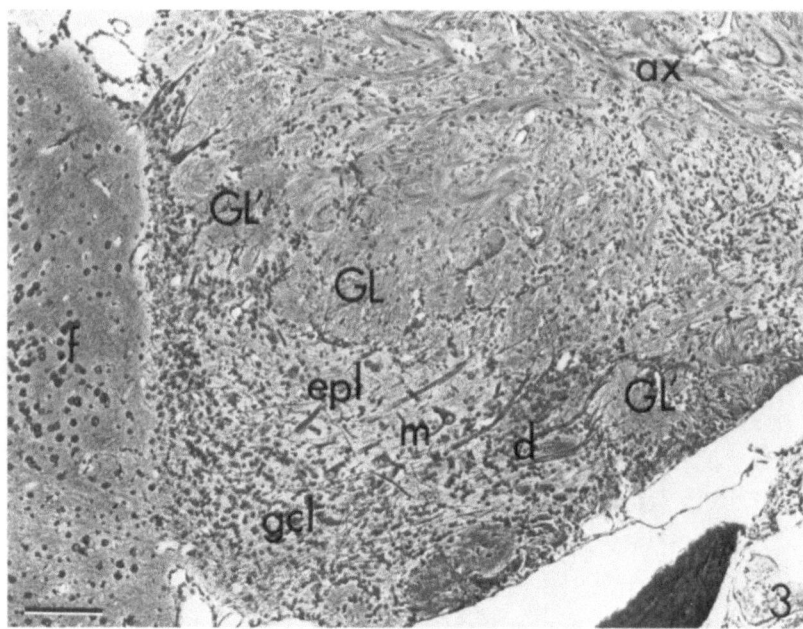

FIGURE 3. Horizontal section through the brain of a mouse partially bulbectomized 1 day after birth and sacrificed after 60 days. A small portion of bulb is recognizable. Glomeruli (GL) have reformed around the bulbar fragment, invading also the granule cell layer (GL′). Some mitral cells (m) have directed their dendrites (d) toward the ectopic glomeruli. gcl, granule cell layer; ax, olfactory axons; epl, external plexiform layer; f, forebrain. Silver stain. Bar = 100 μm.

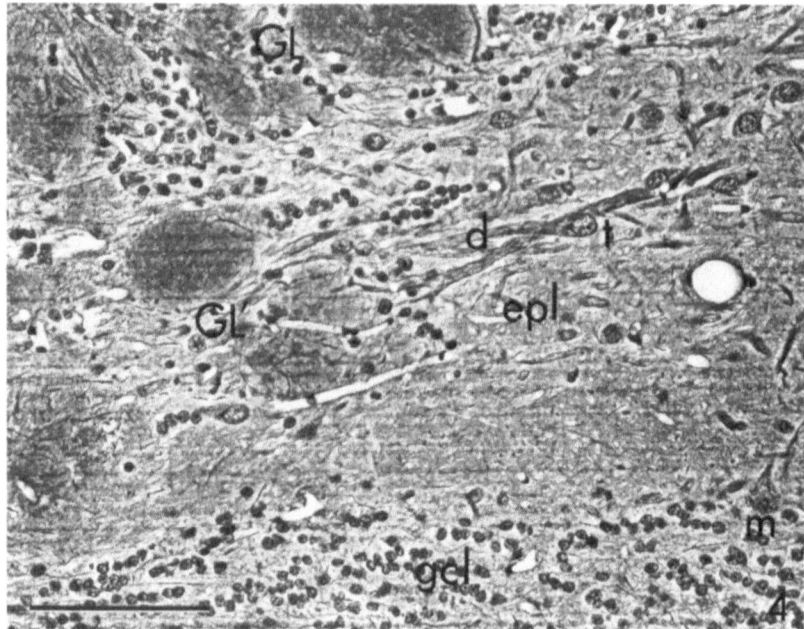

FIGURE 4. Horizontal section through the head of a mouse partially bulbectomized 7 days after birth and sacrificed after 220 days. Ectopic glomeruli (GL') are in the external plexiform layer (epl); tufted cells (t) have directed their dendrites toward them. d, dendrite; m, mitral cell; gcl, granule cell layer; GL, normal glomeruli. Silver stain. Bar = 100 μm.

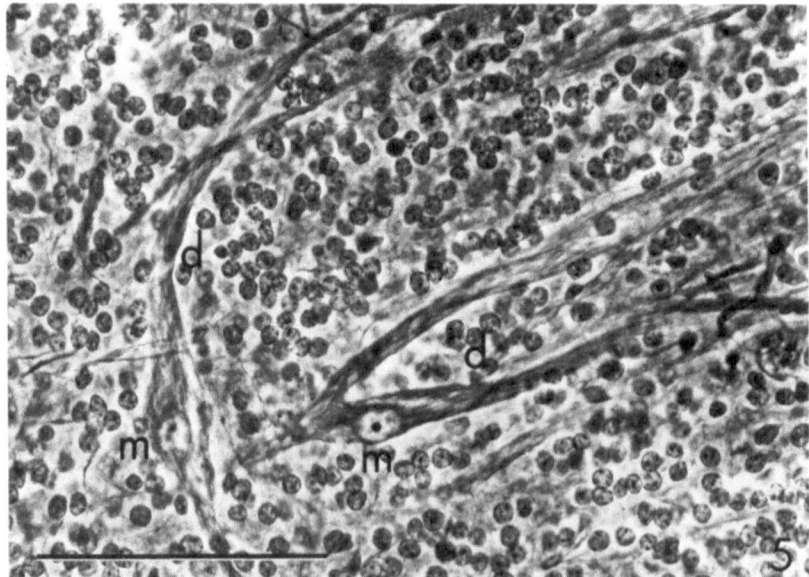

FIGURE 5. Horizontal section, tangential to the dorsal border of the granule cell layer, from a mouse partially bulbectomized 2 days after birth and sacrificed after 75 days. The mitral cell dendrites (d) have an unusual orientation, also tangential, and they branch into ectopic glomeruli (not shown) that have grown into the granule cell layer. m, mitral cells. Silver stain. Bar = 100 μm.

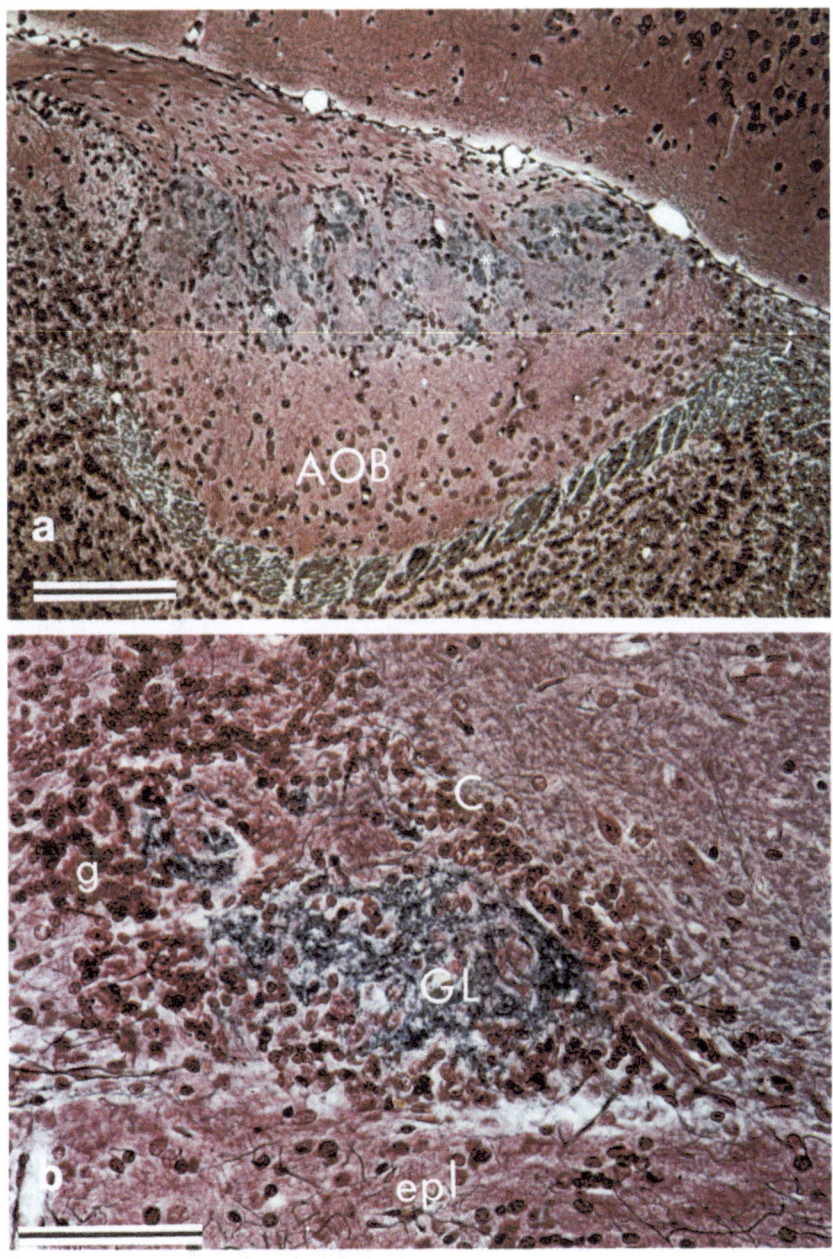

FIGURE 6. (a) Horizontal section through the accessory olfactory bulb (AOB) of an adult mouse, stained with the silver method of Loots. This staining method has characteristically colored in blue the axon terminals of the vomeronasal neurons. Asterisks indicate the specifically stained glomeruli. (b) Parasagittal section through the head of a rat. The cerebellar cortex from a 19-day embryo was transplanted into the olfactory bulb of a 9-week-old host that was sacrificed after 90 days. The developed cerebellar tissue (C) is continuous with the external plexiform layer (epl) of the host's olfactory bulb. Vomeronasal axons have formed a glomerulus (GL), characteristically stained in blue, among the granules (g) of the transplanted cerebellum. Bars = 100 μm.

the newly developed olfactory axons.[37,38] Accessibility of an adult brain to the sensory axons is often impaired by the rapid formation of scar tissue that develops after bulbectomy. Seldom in adult (in the absence of scar tissue) and routinely in neonatal animals, formation of glomerular structures occurs along the exposed forebrain surface (Figs. 7 and 8). In neonatal animals, where our observations have been more comprehensive, we have observed some large argyrophilic neurons whose dendrites penetrate the glomeruli; at ultrastructural levels, synaptic apparatus can be seen between the olfactory axon terminals and the dendrites.

Partial and total removal of the olfactory bulb have shown the ability of the olfactory sensory axons to establish glomerular structures in unconventional areas of the forebrain and to induce morphological changes of neurons in the invaded regions. The data so far collected in partial bulbectomy experiments seem to indicate that after damage to the target, the regrowing olfactory axons do not follow the orderly pathways established during development (lack of mechanical guidance?) and that they are unable to recognize a specific locus in their target as it occurs during ontogenesis (lack of chemoaffinity?). These axons, however, maintain the capacity to modify the neurons of the bulb in their spatial displacement and in their synaptic arrangement. The possibility that this characteristic can be expressed postnatally because of the persistence of embryonic

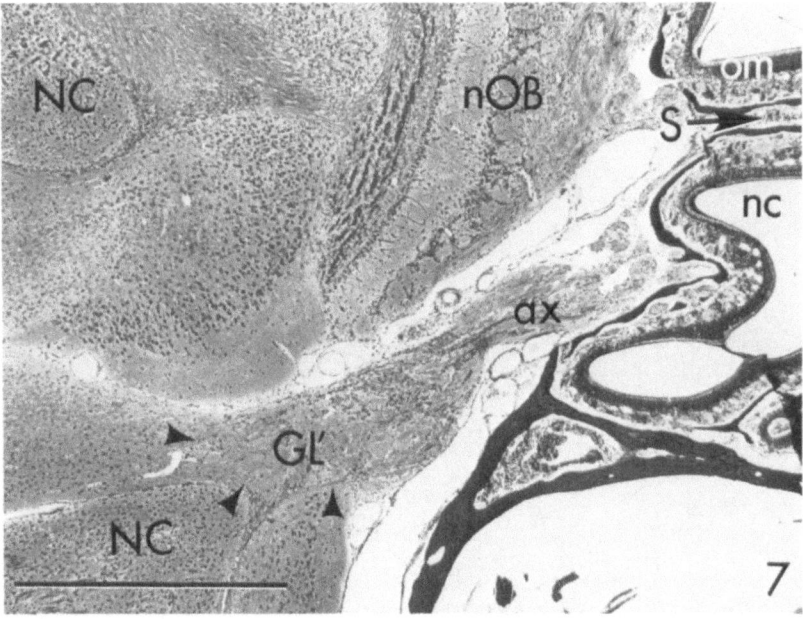

FIGURE 7. Same animal of Figure 6b. Showing the extent of the bulbectomy. The regrown olfactory axons (ax) have extensively formed glomerular structures (GL') along the exposed surface of the spared forebrain (arrowheads). Due to the large cavity left by the removal of the right olfactory bulb, the left cerebral hemisphere has extended past the midline, indicated by "S" on the right of the micrograph. om, olfactory mucosa; nc, nasal cavity; NC, nucleus caudatus; nOB, normal olfactory bulb. Bar = 1 mm.

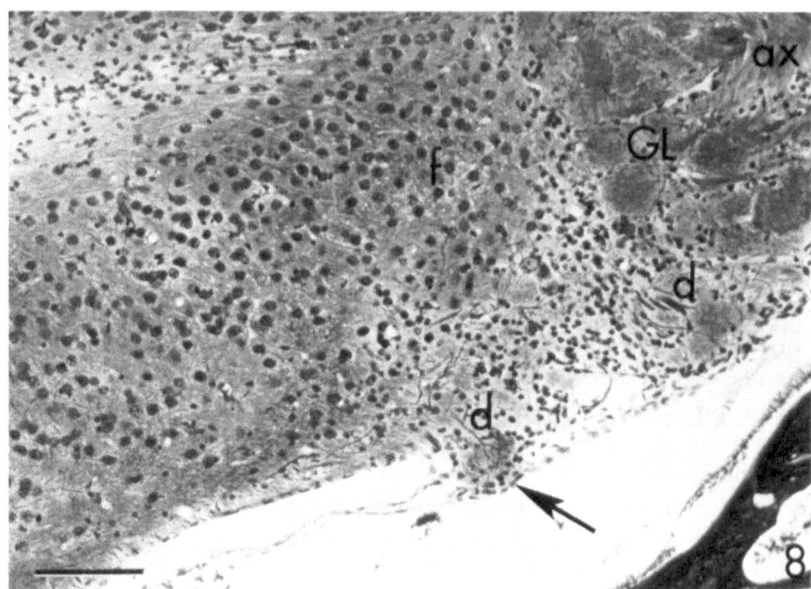

FIGURE 8. Horizontal section through the forebrain of a mouse totally bulbectomized 14 days after birth and sacrificed after 85 days. The olfactory axons (ax) have penetrated the spared forebrain (f) and formed glomerular structures (GL). One glomerulus (arrow) is directly localized under the pia. Large dendrites (d) have branched in some of the glomeruli. Bar = 100 μm.

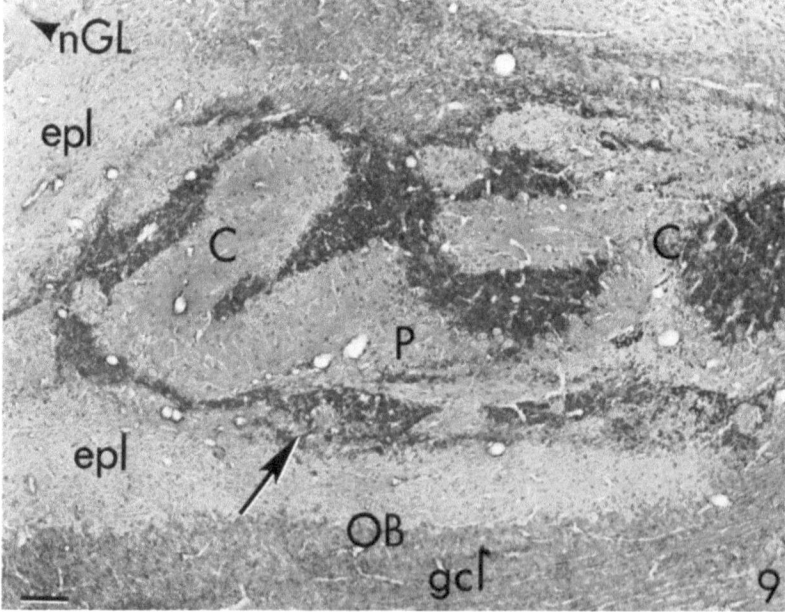

FIGURE 9. Same animal of Fig. 6b in a different plane of section. The illustration shows the integration of the cerebellar transplant (C) with the host's olfactory bulb (OB). The arrow indicates the region where the vomeronasal axons have formed glomeruli (see Fig. 6b). epl, external plexiform layer; gcl, granule cell layer; P, Purkinje cells; nGL, normal glomeruli. Bar = 100 μm.

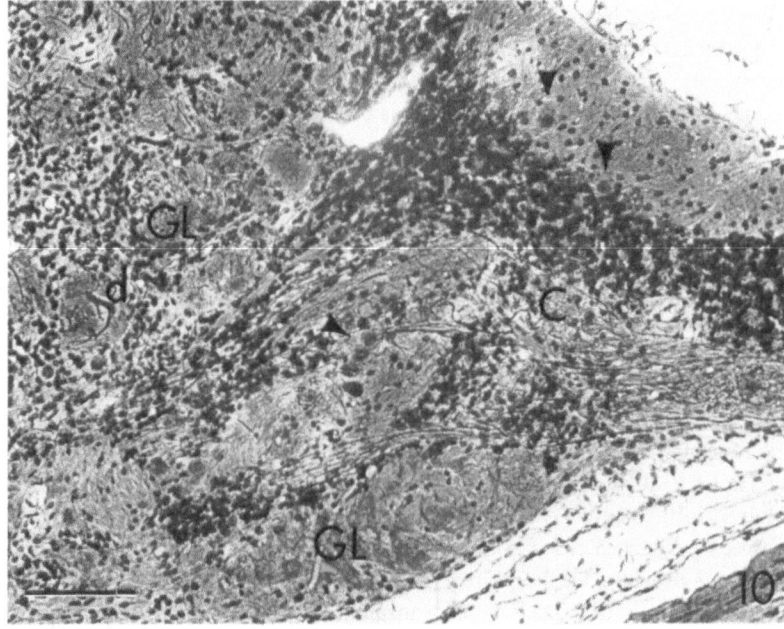

FIGURE 10. Fragment of cerebellum (C) transplanted into a mouse that underwent extensive partial bulbectomy. The cerebellar tissue, from a 16-day embryo, was transplanted into a neonatal host brain. The animal was sacrificed 50 days after surgery. Glomerular structures (GL) have reformed in proximity of and into the transplanted cerebellum. In some glomeruli, dendritic branches (d) are present. Purkinje cells are indicated by arrowheads. Bar = 100μm.

potentials needs further investigation. Our observations demonstrate that all regions of the forebrain, when experimentally exposed by the removal of the olfactory bulb, can be invaded by both olfactory and vomeronasal axons which form glomeruli without apparent regional discrimination. The presence of large argyrophilic neurons close to the glomeruli and the fact that they establish synaptic contacts with the sensory axons suggest a functional interaction.

3.2. Embryonic Brain and Olfactory Mucosa Grafts

The remarkable presence of an apparently inexhaustible neurogenetic matrix in the olfactory neuroepithelium has allowed the series of experiments presently implemented in our laboratory:

a. Embryonic brain fragments, derived from several specific CNS regions (cerebellum and occipital cortex), have been transplanted close to the lamina cribrosa in a space made available by the partial or total removal of the olfactory and accessory olfactory bulb.

b. Olfactory mucosa, from neonatal to adult donors, has been transplanted alone or with fragments of selected regions of embryonic CNS tissue in the anterior chamber of the eye.

c. Olfactory mucosa, from neonatal rats, has been transplanted from the nasal cavity into the lateral or fourth ventricles or directly into several cortical areas of the telencephalon.

These experiments allow us to explore how and to what extent the olfactory neurons can interact with different CNS portions in two essentially different experimental conditions, namely, when they are left *in situ* (a) and when they are removed from their natural environment in the nasal cavity (b, c).

3.2.1. Transplantation of Occipital Cortex and Cerebellum

The transplantation techniques used in our laboratory follow the steps outlined by Das and Hallas.[39] Successful transplantation of the occipital cortex and cerebellum has been obtained in better than 90% of transplants when partial removal of the olfactory bulb has been performed. In partially bulbectomized hosts, the vascularization to the transplants was provided by the spared portions of bulb. In totally bulbectomized animals, only a small number (ca. 10%) of transplants survive. The large cavity produced by the removal of the olfactory bulb often impairs the vascularization of the brain frag-

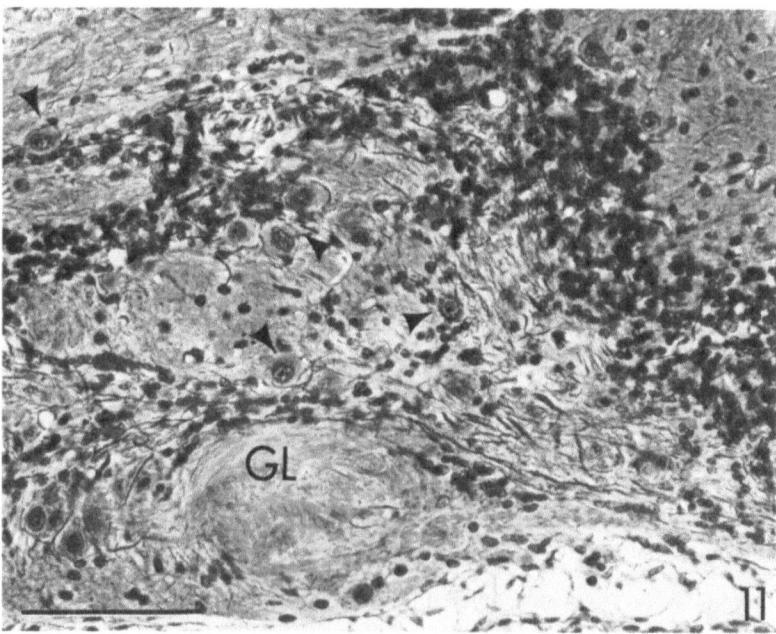

FIGURE 11. Other section from the preparation illustrated in Fig. 10 showing numerous Purkinje cells (arrowheads) and a large glomerular structure (GL). Bar = 100 μm.

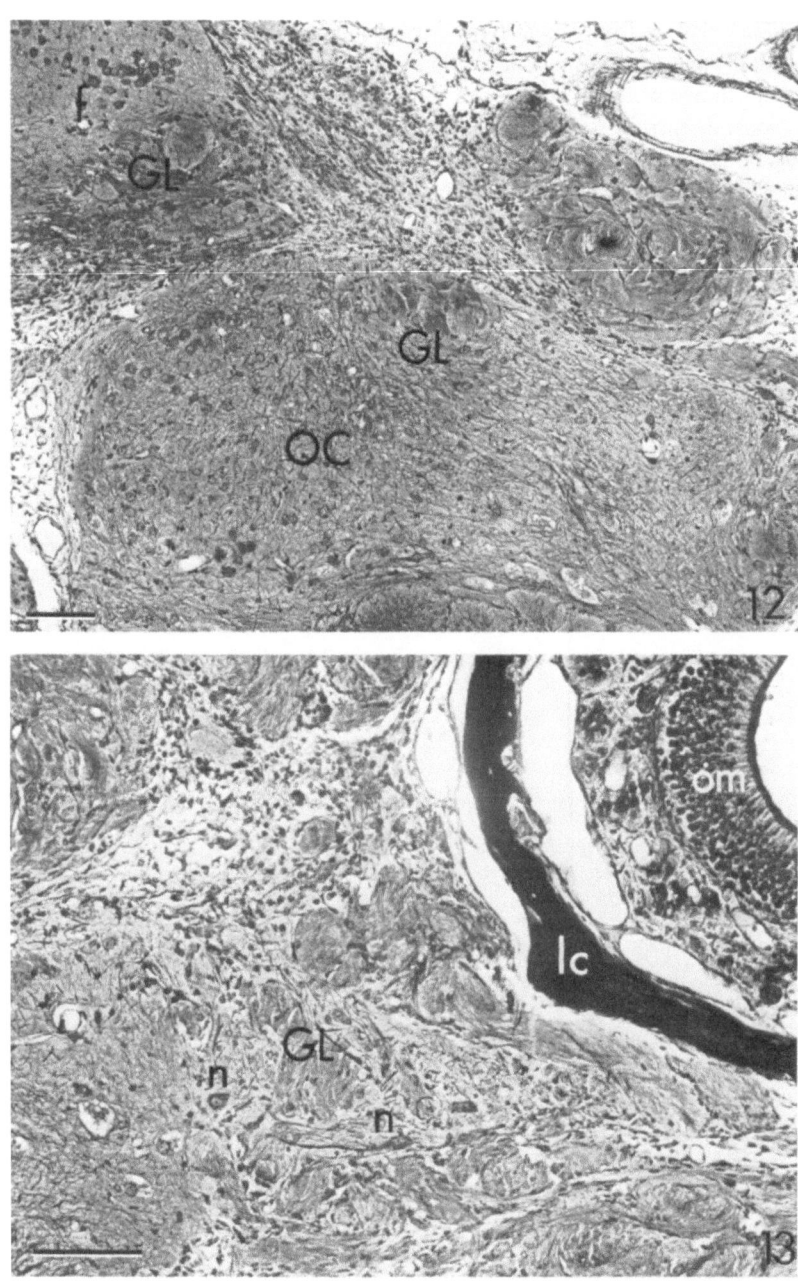

FIGURES 12 AND 13. Occipital cortex (OC) transplanted from a 15-day embryo to a 10-day-old mouse after extensive bulbectomy. In Fig. 12, glomerular formations (GL) are present in the transplanted tissue and in the host forebrain (f). In Fig. 13, some large neurons (n) of the transplant are surrounded by glomerular structures (GL). om, olfactory mucosa; lc, lamina cribrosa. Bar = 100 μm.

ments which derives from the meningeal covering of the lamina cribrosa. Both mouse (E15 to E18) and rat (E18 to E22) embryos have been used as donors for homoplastic transplants in neonatal and adult hosts. The morphological features of the transplanted brain tissue, at sacrifice, are indistinguishable from the comparable tissues of control animals. This is particularly noticeable for the cerebellar transplants. No obvious separation is noticeable between the transplanted fragment and the surrounding area of the host brain; rather, fusion and exchange of nerve bundles seems to occur. In occipital cortex and cerebellum transplants (Figs. 6, 9–13), the olfactory axons penetrate, forming glomerular structures that disrupt the normal arrangement of the transplanted brain.[40,41]

The blue staining of the vomeronasal terminals has been found to be very specific, and in Fig. 6b it shows how these terminals are arranged in a discrete glomerular fashion among the granules of the transplanted cerebellum.

3.2.2. Intraocular Transplant of Olfactory Mucosa

Transplants of olfactory mucosa into the anterior chamber of the eye are performed using tissues from neonatal and adult animals. For the transplant study, we have followed the technique of Olson et al.[42] (see also Chapter 5). The olfactory mucosa survives

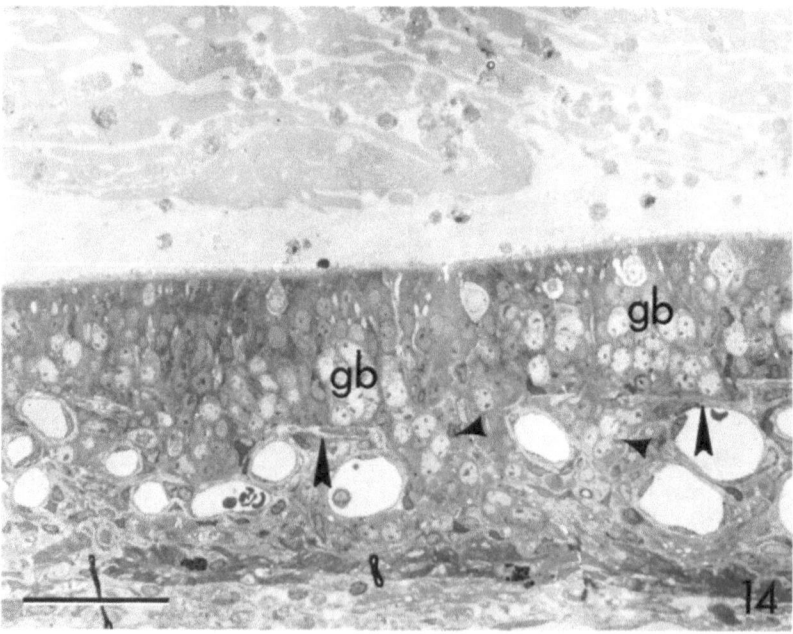

FIGURE 14. Olfactory mucosa transplanted in the anterior chamber of the eye of an adult rat (40 days' survival). Several globular basal cells (gb) can be identified both inside the neuroepithelium and in the lamina propria (small arrowheads). Basal lamina is indicated by large arrowheads. Araldite-embedded material stained with toluidine blue. Bar = 50 μm.

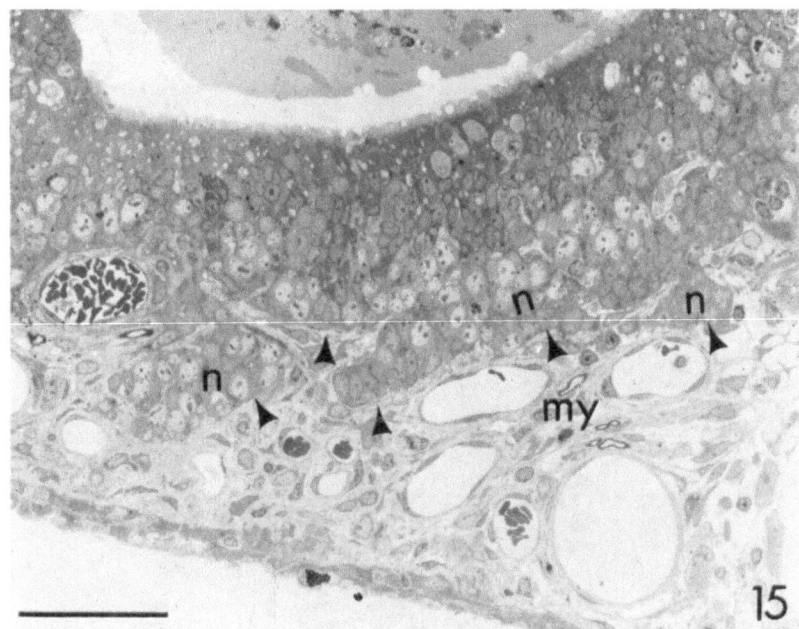

FIGURE 15. Same preparation as in Fig. 14. The base of the neuroepithlium has an irregular outline (arrowheads) due to the migration of neuronal elements (n) into the lamina propria. Myelinated fibers (my) are in the lamina propria. Bar = 50 μm.

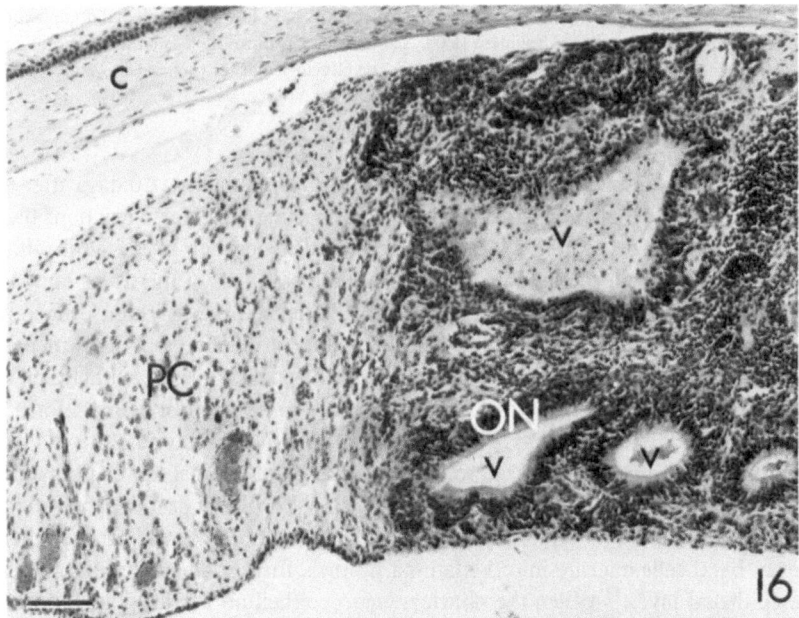

FIGURE 16. Olfactory mucosa from a neonatal donor and parietal cortex from a 14-day embryo have been transplanted into the anterior chamber of the eye of an adult rat. The olfactory neuroepihelium (ON) is arranged in vesicles (v) and is attached to the brain transplant (PC). c, cornea. Bar = 100 μm.

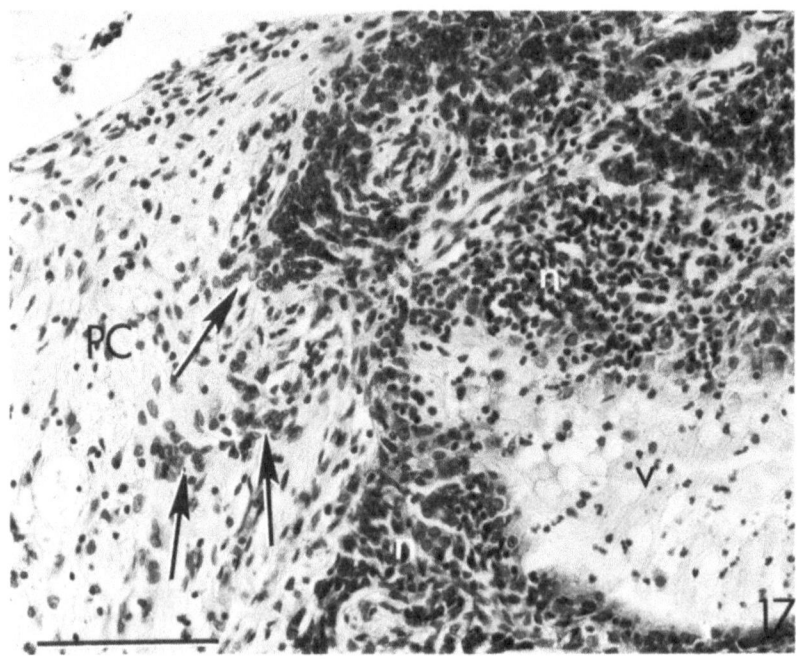

Figure 17. Detail from Fig. 16. Olfactory neuronal elements (n) have lost their epithelial organization and have penetrated (arrows) the brain fragment (PC). The olfactory neurons have been identified in adjacent sections by means of immunohistochemical staining for the olfactory marker protein. Bar = 100 μm.

in the anterior chamber of the eye for up to 1 year. During the first 10 days after transplantation, the epithelial layer arranges in interconnecting vesicular formations lined by respiratory and sensory epithelium. Following the degeneration of the adult olfactory neurons, resulting from the lesion of the axons, the basal cells of the transplanted neuroepithelium rapidly divide and a new population of neurons is reconstituted. With immunohistochemical methods, it has been possible to observe that mature neurons stained for olfactory marker protein are present in the olfactory neuroepithelium. When the olfactory mucosa is transplanted alone, the axons of the mature neurons form large neuromas in the lamina propria, or they penetrate into the iris musculature of the host. Within the lamina propria, adjacent to the epithelial basal lamina, there are myelinated and large unmyelinated axons (Figs. 14 and 15). Capillary loops deeply penetrating into the epithelium disrupt its typical organization. From the base of the neuroepithelium, groups of basal cells migrate into the lamina propria, further altering the arrangement of the epithelial layer.[43] When the olfactory neuroepithelium is transplanted in association with fragments of brain tissue (Figs. 16 and 17), the system of vesicles becomes more complex. The olfactory axons penetrate into the fragments of brain tissue where, however, they fail, even at long postoperative survivals, to form glomerular structures.

3.2.3. Intracerebral Transplants of Neuroepithelium

Transplantation of olfactory neuroepithelium from neonatal and adult donors into the cerebral ventricles or directly into the cerebral cortex has a rate of survival of 90%. When the neuroepithelium is transplanted in the ventricles, it attaches at random to the walls of the cavity and receives its vascular supply through the ependymal lining, and/or directly from the choroid plexuses. Portions of the neuroepithelium are organized in vesicles (Figs. 18 and 19); others are layered over the ventricular walls with the free surface facing the ventricular cavity. Following degeneration of the mature neurons, the basal cells of the neuroepithelium actively divide and new neurons are formed. The axons of the reconstituted neurons penetrate the host brain, and are olfactory marker protein positive; however, they do not form glomerular structures in the host brain even at long survival times. Olfactory neurons, as previously observed in the intraocular transplants, migrate into the host brain often along the sensory fiber bundles. Only a few of these morphologically identifiable neurons are positively stained for the olfactory marker protein.

The neuroepithelium transplanted directly into the brain parenchyma is also

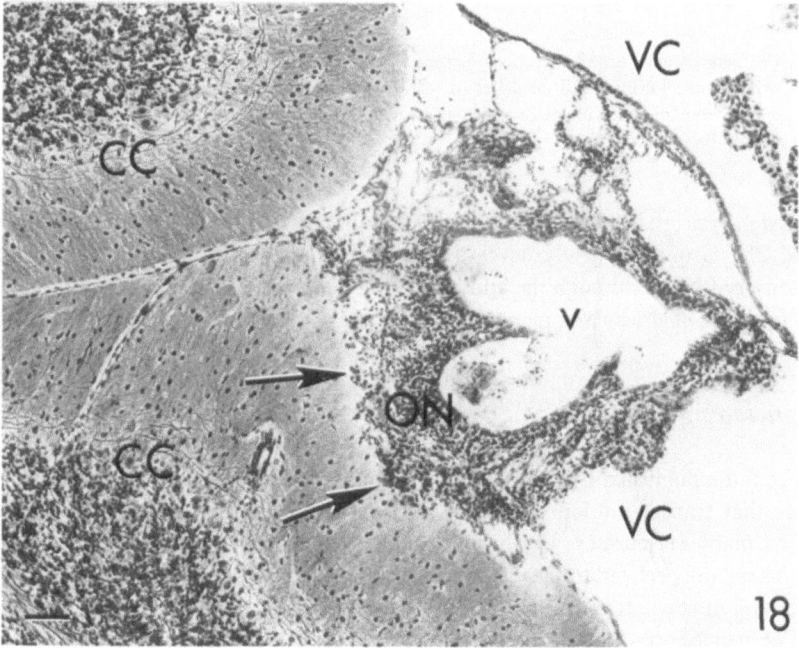

FIGURE 18. Intraventricular transplant of olfactory mucosa from a neonatal rat to an adult host. The neuroepithelium (ON), arranged in vesicles (v), has attached to the cerebellar cortex (CC). Clusters of neurons (arrows) have streamed toward the cerebellar cortex, losing their epithelial organization. VC, ventricular cavity. Bar = 100 μm.

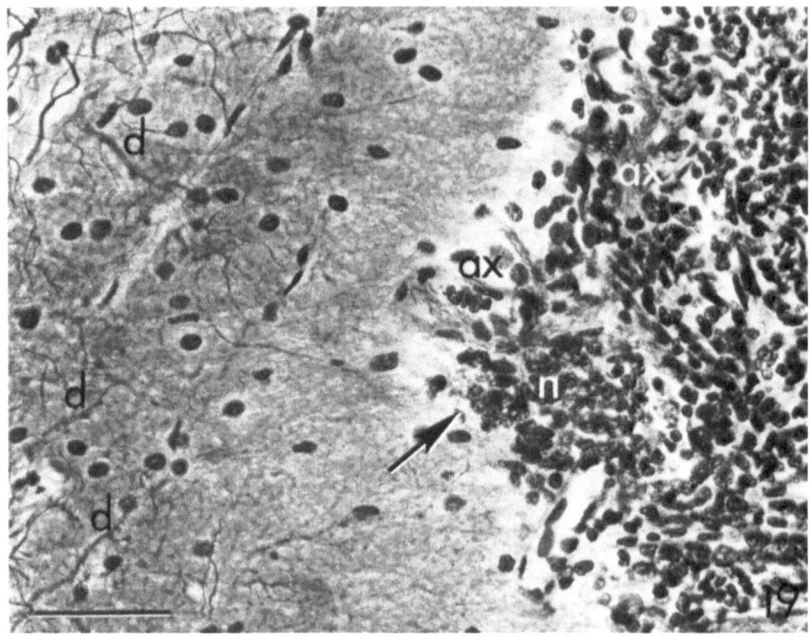

FIGURE 19. Detail from Fig. 18. Olfactory elements have invaginated into the cortex of the cerebellum (arrow), where some Purkinje cell dendrites (d) can be recognized. The olfactory neurons (n) and their axons (ax) have been identified in adjacent sections by means of immunohistochemical stain for the olfactory marker protein. Bar = 50 μm.

arranged in vesicles; however, portions of it are directly invaginated into the tissue (Figs. 20 and 21). Groups of newly developed neurons migrate into the parenchyma along their sensory fibers. Although the olfactory fibers penetrate into the host brain, they also fail to form the characteristic glomerular structures.

4. Concluding Remarks

From the published literature and the recent work carried out in our laboratory, it appears that transplantation techniques can be applied to the study of the olfactory organ in many vertebrates. Furthermore, these studies can be utilized to investigate problems not only related to the olfactory organ, but also to problems of neuronal development and interaction concerning the nervous system as a whole.

The persistence of a neurogenetic matrix in adult vertebrates producing neurons that show the ability to profoundly modify the development and morphology of neurons in several CNS areas (not necessarily belonging to the olfactory pathway) proposes the question of the exceptionality of the olfactory sensory neurons and their role in the normal development of the nervous system. We are now trying to understand the mechanisms controlling the migration of the olfactory cells (intraocular and brain transplant

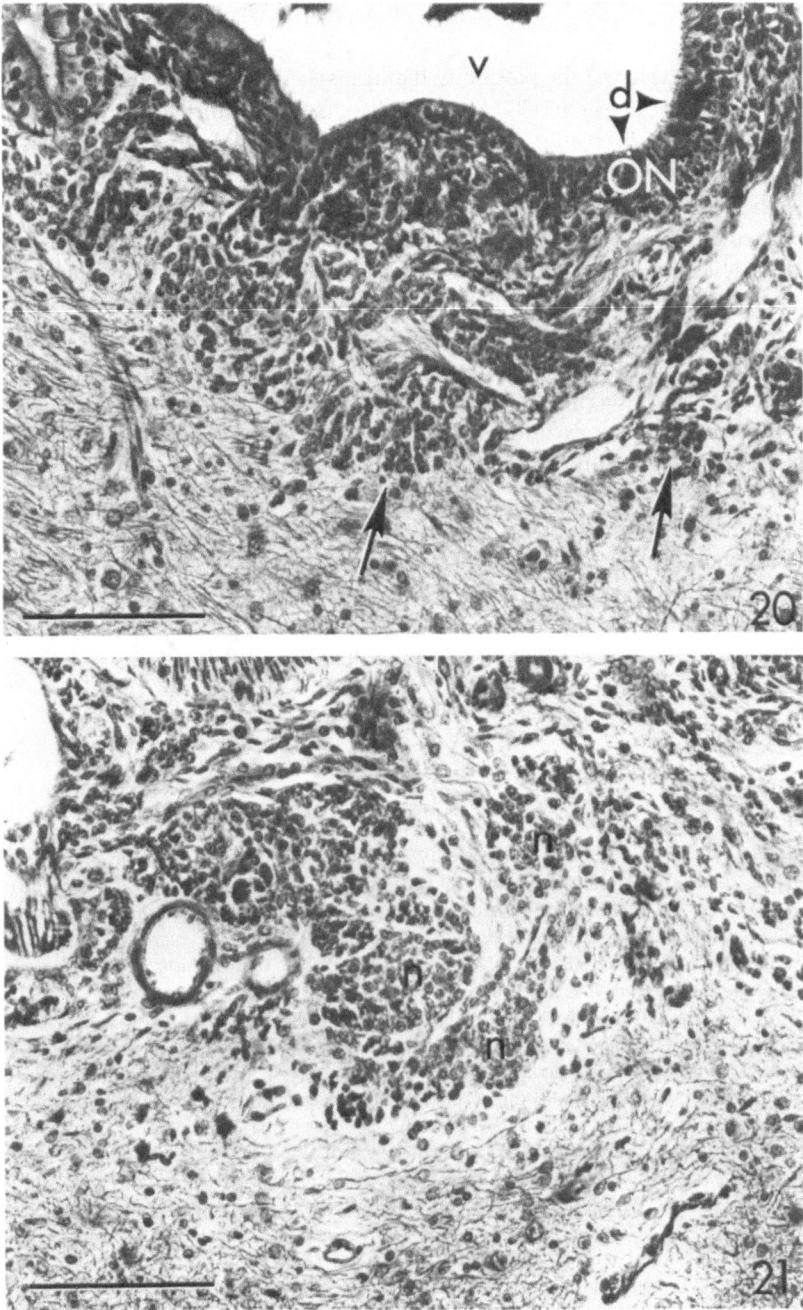

FIGURES 20 AND 21. In Fig. 20, a vesicle (v) is lined with olfactory mucosa that has attached to the floor of the fourth ventricle. The olfactory neuroepithelium (ON) is partially organized with dendrites (d) emerging into the lumen of the vesicle. Other olfactory elements have migrated into the floor of the fourth ventricle (arrows),where they intermingle with the host CNS. In Fig. 21, clusters of olfactory neurons (n) are embedded in the host CNS. Bars = 100 μm.

experiments) by exploring the possibility that transplantation may influence the differentiation of the olfactory stem cells. Although there is still insufficient evidence to sustain this hypothesis, the observation that transplanted olfactory neurons survive and grow their axons into the CNS without forming glomerular structures (contrary to what occurs when the olfactory neurons reside in the nasal cavity and penetrate a variety of transplanted CNS areas) indicates that the transplantation influences in some ways the characteristic of these neurons to form glomeruli. While the capacity of the olfactory neurons for turnover and the viability of their matrix remain unchanged, the ability to influence the experimentally proposed targets seems to depend on their environment.

It has been nearly a century since Bell[26] performed the first transplants utilizing the olfactory system. Recently, renewed interest by neurobiologists in transplantation techniques has offered an exciting avenue of neuronal research. Further developments will depend on the ingenuity and imagination of the researchers.

ACKNOWLEDGMENTS. Intracerebral transplants of olfactory neuroepithelium are carried out in our laboratory by Dr. E. Morrison. Intraocular transplants of olfactory mucosa are carried out by J. A. Heckroth in partial fulfillment of the Ph.D. degree. Their collaboration is gratefully acknowledged.

This work was supported by a grant from the NIH (NS 16421).

References

1. Sidman, R. L., Miale, I. L., and Feder, N., 1959, Cell proliferation and migration in the primitive ependymal zone: An autoradiographic study of histogenesis in the nervous system. *Exp. Neurol.* **1:** 322.
2. Angevine, J. B., Jr., And Sidman, R. L., 1961, Autoradiographic study of cell migration during histogenesis of cerebral cortex in the mouse, *Nature (London)* **192:**766.
3. Angevine, J. B., Jr., and Sidman, R. L., 1962, Autoradiographic study of histogenesis in the cerebral cortex of the mouse, *Anat. Rec.* **142:**210.
4. Sidman, R. L., 1970, Autoradiographic methods and principles for study of the nervous system with thymidine-H^3, in: *Contemporary Research Techniques of Neuroanatomy* (S.O.E. Ebbesson and W. J. Nauta, eds.), Springer-Verlag, Berlin.
5. Harrison, R. G., 1908, Embryonic transplantation and development of the nervous system, *Anat. Rec.* **2:**385.
6. Saltykow, S., 1905, Versuche über Gehirnplantation, zugleich ein Beitrag zur Kenntniss der Vorgange an den Zelligen Gehirnelementen, *Arch. Psychiatr.* **40:**329.
7. Marinesco, G., 1907, Quelques recherches sur la transplantation des ganglions nerveus, *Rev. Neurol.* **15:**241.
8. Altobelli, R., 1914, Innesti cerebrali, *Gazz. Int. Med. Chir.* **17:**25.
9. Dunn, E., 1917, Primary and secondary findings in a series of attempts to transplant cerebral cortex in the albino rat, *J. Comp. Neurol.* **27:**565.
10. Das, G. D., and Altman, J., 1972, Studies on the transplantation of developing neural tissue in the mammalian brain. 1. Transplantation of cerebellar slabs into the cerebellum of neonate rats, *Brain Res.* **38:**233.
11. Kromer, L. F., Björklund, A., and Stenevi, U., 1979, Intracephalic implants: A technique for studying neuronal interactions, *Science* **204:**1117.
12. Olson, L., and Seiger, Å., 1972, Brain tissue transplanted to the anterior chamber of the eye. I. Fluorescence histochemistry of immature catecholamine and 5-hydroxytryptamine neurons reinnervating the rat iris, *Z. Zellforsch. Mikrosk. Anat.* **135:**175.

13. Graziadei, P. P. C., and Metcalf, J. F., 1971, Autoradiographic and ultrastructural observations on the frog's olfactory mucosa, *Z. Zellforsch. Mikrosk, Anat.* **116**:305.
14. Graziadei, P. P. C., 1973, Cell dynamics in the olfactory mucosa, *Tissue Cell* **5**:113
15. Graziadei, P. P. C., and DeHan, R. S., 1973, Neuronal regeneration in frog olfactory system, *J. Cell Biol.* **59**:525.
16. Moulton, D. G., 1974, Dynamics of cell populations in the olfactory epithelium, *Ann. N. Y. Acad. Sci.* **237**:52.
17. Graziadei, P. P. C., and Monti Graziadei, G. A., 1978, Continuous nerve cell renewal in the olfactory system, in. *Handbook of Sensory Physiology,* Vol. IX (M. Jacobson, ed.), pp. 55–83, Springer-Verlag, Berlin.
18. Graziadei, P. P. C., and Monti Graziadie, G. A., 1979, Neurogenesis and neuron regeneration in the olfactory system of mammals. 1. Morphological aspects of differentiation and structural organization of the olfactory sensory neurons, *J. Neurocytol.* **8**:1.
19. Graziadei, P. P. C., and Monti Graziadei, G. A., 1980, Plasticity of connections in the olfactory pathway: Transplantation studies, in: *Olfaction and Taste VII,* (H. van der Sterre, ed.,) pp. 155–158, IRL Press, London.
20. Barber, P. C., and Raisman, G., 1978, Replacement of receptor neurones after section of the vomeronasal nerves in the adult mouse, *Brain Res.* **147**:297
21. Barber, P. C., and Raisman, G., 1978, Cell division in the vomeronasal organ of the adult mouse, *Brain Res.* **141**:57.
22. Monti Graziadei, G. A., and Graziadei, P. P. C., 1979, Neurogenesis and neuron regeneration in the olfactory system of mammals. II. Degeneration and reconstitution of the olfactory sensory neurons after axotomy, *J. Neurocytol.* **8**:197.
23. Wilson, K. C. P., and Raisman, G., 1980, Age-related changes in the neurosensory epithelium of the mouse vomeronasal organ: Extended period of post-natal growth in size and evidence for rapid cell turnover in the adult, *Brain Res.* **185**:103.
24. Wang, R. T., and Halpern, M., 1982, Neurogenesis in the vomeronasal epithelium of adult garter snakes. I. Degeneration of bipolar neurons and proliferation of undifferentiated cells following experimental vomeronasal axotomy, *Brain Res.* **237**:23.
25. Wang, R. T., and Halpern, M., 1982, Neurogenesis in the vomeronasal epithelium of adult garter snakes. 2. Reconstitution of the bipolar neuron layer following experimental vomeronasal axotomy, *Brain Res.* **237**:41.
26. Bell, E. T., 1907, Some experiments on the development and regeneration of the eye and nasal organ in frog embryos, *Arch. Entwicklungsmech. Org.* **23**:457.
27. Burr, H. S., 1926, Some experiments on the transplantation of the olfactory placode in *Amblystoma, J. Comp. Neurol.* **37**:455.
28. Burr, H. S., 1923, An experimentally produced aberrant olfactory nerve in *Amblystoma, Anat. Rec.* **25**:121.
29. Burr, H. S., 1920, The transplantation of the cerebral hemispheres of *Amblystoma, J. Exp. Zool.* **30**:159.
30. Burr, H. S., 1916, The effects of the removal of the nasal pits in *Amblystoma* embryos, *J. Exp. Zool.* **20**:27.
31. May, R. M., 1927, Modifications des centres nerveux dues a la transplantation de l'oeil et de l'organe olfactif chez les embryons d'anoures, *Arch. Biol. (Liege)* **37**:336.
32. Luna, E., 1915, Ricerche sperimentaly sulla morfologia dell'organo dell'olfatto negli anfibi, *Arch. Ital. Anat. Embriol.* **14**:609.
33. Carpenter, E., 1937, The head pattern in *Ambystoma* studied by vital staining and transplantation methods, *J. Exp. Zool.* **75**: 103.
34. Stout, R. P., and Graziadei, P. P. C., 1980, Influence of the olfactory placode on the development of the brain in *Xenopus laevis* (Daudin). 1. Axonal growth and connections of the transplanted olfactory placode, *Neuroscience* **5**:2175.
35. Clairambault, P., 1976, Development of the prosencephlon, in: *Frog Neurobiology* (R. Llinas and W. Precht, eds.), pp. 924–944, Springer-Verlag, Berlin.
36. Graziadei, P. P. C., and Samanen, W., 1980, Ectopic glomerular structures in the olfactory bulb of neonatal and adult mice, *Brain Res.* **187**:467.

37. Graziadei, P. P. C., Levine, R. R., and Monti Graziadei, G. A., 1978, Regeneration of olfactory axons and synapse formation in the forebrain after bulbectomy in neonatal mice, *Proc. Natl. Acad. Sci. USA* **75**:5230.

38. Graziadei, P. P. C., Levine, R. R., and Monti Graziadei, G. A., 1979, Plasticity of connections of the olfactory sensory neuron: Regeneration into the forebrain following bulbectomy in the neonatal mouse, *Neuroscience* **4**:713.

39. Das, G. D., and Hallas, B. H., 1978, Transplantation of brain tissue in the brain of adult rats, *Experientia* **34**:1304.

40. Graziadei, P. P. C., and Monti Graziadei, G. A., 1980, Neurogenesis and neuron regeneration in the olfactory system of mammals. III. Deafferentiation and reinnervation of the olfactory bulb following section of the *fila olfactoria* in rat, *J. Neurocytol.* **9**:145.

41. Graziadei, P. P. C., and Kaplan, M. S., 1980, Regrowth of olfactory sensory axons into transplanted neural tissue. 1. Development of connections with the occipital cortex, *Brain Res.* **201**:39.

42. Olson, L., Seiger, Å., and Strumberg, I., 1981, Intraocular transplantation in rodents: A detailed account of the procedure and examples of its use in neurobiology with special reference to brain tissue grafting, Neural Transplantation and Explantation: Techniques and Applications, Workshop 5th Neuroscience Meeting, Liège, Belgium, pp. 51–84.

43. Heckroth, J. A., Monti Graziadei, G. A., and Graziadei, P. P. C., 1983, Intraocular transplants of olfactory neuroepithelium in rat, in: *Int. J. Dev. Neuroscience*, **1**:273–287.

7

Correction of Genetic Gonadotropic Hormone-Releasing Hormone Deficiency by Preoptic Area Transplants

DOROTHY T. KRIEGER AND MARIE J. GIBSON

The hypogonadal (hpg) mutant mouse appeared in a breeding colony at Harwell, England. The mutant mice of both sexes fail to show any sign of postnatal gonadal or accessory sexual tissue development. However, sexual differentiation proceeds normally up to birth. Hormone assays and immunoperoxidase staining indicate that the primary cause of the hypogonadism is due to a severe deficiency in the hypothalamic releasing hormone [gonadotropin-releasing hormone (GnRH)] that governs pituitary luteinizing hormone (LH) and follicle-stimulating hormone (FSH) release. In view of previous studies demonstrating that transplants of embryonic mammalian CNS tissue survive, differentiate, and become anatomically integrated with the CNS of neonatal and adult mammalian recipients,[1-16] we were interested to see if transplants of normal fetal tissue could correct this genetically induced CNS defect. For the transplant material, we used fetal preoptic area (POA), a primary site of GnRH cell bodies. To more fully understand the nature of the results obtained, it is necessary to first briefly review the normal regulation of the hypothalamic–pituitary–gonadal axis and the details of the nature of the gonadal deficiency of the hpg animal.

DOROTHY T. KRIEGER AND MARIE J. GIBSON ● Division of Endocrinology, Mount Sinai School of Medicine, New York, New York 10029.

1. Factors Involved in the Regulation of the Hypothalamic–Hypophyseal–Gonadal Axis

Figures 1 and 2 present schematic diagrams of the gonadotropin control system in males and females. [In this diagram, LHRH (luteinizing hormone-releasing hormone) is used interchangeably for GnRH.] In the rodent, cell bodies that contain GnRH are principally located within the septal and medial preoptic nuclei.[17] There are also recent reports on the presence of GnRH neurons and fibers in the olfactory bulb of hamster[18] and rat.[19] The main GnRH pathway concerned with gonadotropin secretion originates from the septal and medial POA, coursing through axons in the mediobasal hypothalamus to terminate in the external layer of the median eminence; GnRH is then released into the capillaries of the hypothalamo–hypophyseal portal system to reach the anterior pituitary (Fig. 3). Such release of GnRH, which occurs in a pulsatile fashion, concomitant with LH pulses,[20] is essential for normal gonadotropin secretion. Additional evidence for the role of the POA has been obtained with (1) electrochemical stimulation of this site—which results in proportional increases of GnRH in hypophyseal portal blood, and of pituitary LH release[21,22]—as well as (2) from studies in female animals that demonstrate correlation of POA GnRH concentrations with those in hypophyseal portal blood, coincident with the LH surge[23,24] that occurs on the afternoon of proestrus.

GnRH neurons, as indicated in Figs. 1 and 2, receive afferent inputs from other brain areas. Among these afferents are various neurotransmitter systems. In the present consideration of neurotransmitter regulation of gonadotropin release, major emphasis will be placed on the adrenergic system. A wealth of evidence has accumulated supporting a stimulatory role of catecholamines in GnRH release.[25–31] These studies have been performed mostly in female animals. There is relatively little information with regard to effects of the adrenergic system on control of male reproductive function. Orchidectomy is followed by an increase in hypothalamic content of norepinephrine,[32] which precedes the rise in gonadotropins. Phenoxybenzamine, an α-blocker, suppresses the postcastration increase in gonadotropin release in male rats.[33] The role of dopamine

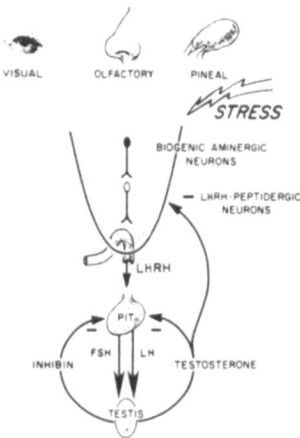

FIGURE 1. Schematic diagram of gonadotropin control systems in males, showing interactions of neuronal and hormonal feedback controls. Minus indicates negative feedback link. LH stimulates testicular secretion of testosterone; FSH stimulates maturation and growth of spermatogenic tubule cells. The exact role of inhibin remains to be elucidated. LHRH secretion is regulated by a biogenic amine neural system that links gonadotropin regulation to the remainder of the CNS. Also indicated are several types of stimuli that have been demonstrated to affect gonadotropic function. Visual and olfactory influences presumably act via LHRH peptidergic neurons; the locus of action of pineal effects is either via the hypothalamus or on the pituitary and is an inhibitory one. Reproduced from Ref. 69 with permission.

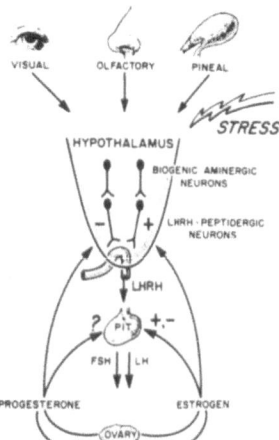

FIGURE 2. Schematic diagram of gonadotropin control systems in females, showing the interactions of neuronal and hormonal feedback controls. Minus indicates negative feedback; plus indicates positive feedback. FSH controls the development of the ovarian follicle; secretion of estrogen by the developing follicle is both FSH- and LH-dependent. LH secretion brings about ovulation and stimulates progesterone secretion. Reproduced from Ref. 69 with permission.

in both sexes is controversial, with reports of both a stimulatory and an inhibitory role; such discrepancies may be secondary to variations in the hormonal milieu at the time of study or in doses employed, which may differentially affect subclasses of dopamine receptors.[29,34–38] Although serotonin, histamine, and acetylcholine have also been implicated in GnRH regulation, additional studies are required to clarify the nature of such control.

Noradrenergic innervation of the hypothalamus, including the supraoptic, preoptic, and arcuate nuclei, as well as the internal layer of the median eminence, is exerted mainly via the ventral tegmental pathway, which originates from two major groups of norepinephrine perikarya located in medulla and pons. Colocalization of tyrosine hydroxylase (the catecholamine-synthesizing enzyme) with GnRH using a dual immunoperoxidase technique showed a juxtaposition of catecholamine fibers on GnRH cells and their dendrites.[39] The ventral tegmental pathway also contributes catecholamine axons, which terminate in close proximity to GnRH axon terminals in the median eminence, although these two kinds of terminals do not coexist in the same neuroanatomical loci[40]; catecholamine terminals are restricted to the outer layer of the median eminence, while GnRH terminals are seen in the internal layer, in close contact with the portal blood vessels. These findings have supported the suggestion that, in addition to directly stimulating GnRH neurons, catecholamines control or modulate the release of GnRH into portal blood.

The foregoing anatomical observations, together with physiological studies of catecholamine turnover,[23] responses to drugs affecting catecholamine synthesis and turnover or adrenergic receptors,[41–43] and lesion experiments, have led to the hypothesis that catecholamines control or modulate tonic, pulsatile, and cyclic release of GnRH into portal blood, as well as mediating those aspects of the negative feedback effect of gonadal steroids which are effected via a hypothalamic locus of action. Many of these studies favoring catecholaminergic regulation of GnRH, however, have been pharmacological ones, employing drugs that later have been shown to act in multiple sites and neutralize the activity of other substances, while lesion experiments not only interrupt catechol-

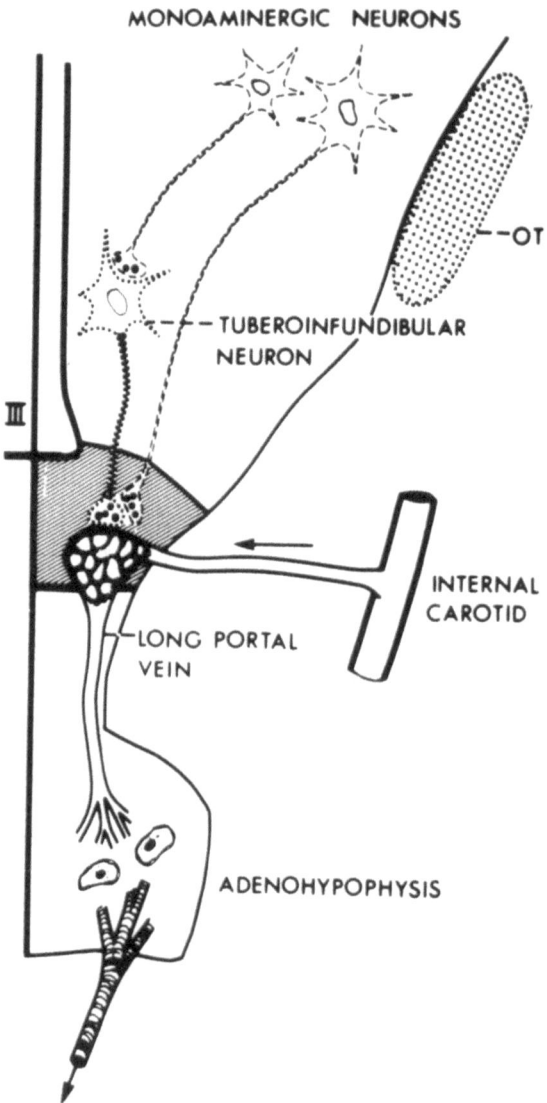

FIGURE 3. Diagram of the hypothalamic–pituitary axis in coronal section. The tuberoinfundibular neuron is representative of a neuron that is the source of a hypothalamic regulatory hormone. Such neurons, which receive axosomatic and axodendritic projections from monoaminergic neurons, terminate on the capillary plexus in the median eminence. The long portal veins drain the capillary plexus into the sinusoids of the anterior pituitary (adenohypophysis). OT, optic tract. Reproduced from Ref. 69 with permission.

amine pathways but also interfere with GnRH pathways. Such considerations indicate that studies in POA-transplanted hpg animals may offer a novel approach in helping to shed further light on this question.

In addition to neural influences on gonadal hormone secretion, neural regulation of sexual behavior has been described in the male and female rodent. An intrahypothalamic role for GnRH in facilitating steroid effects on such behavior has been suggested.[44-46] This GnRH effect occurs at an extrapituitary site, as it persists after removal of the hypophysis,[44] and is evidenced after local intrahypothalamic GnRH infusion.[45] It is abolished by administration of GnRH antibodies within the mesencephalic central gray,[46] a site containing GnRH axons. A similar role for GnRH with regard to mounting behavior and intromission has been demonstrated for the male,[47,48] though other reports suggest a lesser role for the decapeptide.[49,50] This behavioral effect may be mediated by the GnRH fiber tracts which project from the preoptic and septal areas to the limbic forebrain and midbrain regions.[51]

2. Nature of the hpg Animal and Its Endocrine Status

It is not known whether the GnRH deficiency present in these animals is a result of gene deletion, failure of gene expression, or due to synthesis of an abnormal GnRH. Secondary to such GnRH deficiency, pituitary LH and FSH content is markedly reduced in both sexes, compared to normal littermates, as is prolactin content in female hpg mice, in animals studied up to 1 year of age. In females, ovarian follicles rarely reach the antral stage, and in males, spermatogenesis rarely advances beyond the diplotene stage. Both male and female mutants appear to possess normal olfactory function.

The pituitary gland of hpg mice can respond to GnRH injections by releasing LH,[52] and the gonads can respond with androgen production and spermatogenesis to purified preparations of LH and/or FSH, but only when the gonadotropins are administered as multiple doses over the 24-hr period.[53] Such injections, however, do not bring about restoration of normal weights of accessory reproductive tissue. The gonads of both sexes of hpg mice develop normally when transplanted to normal animals.[54] LH and FSH receptors are present in the gonads of hpg mice.[55] Pituitary GnRH receptor levels in hpg mice are approximately 30% less than in normal mice, but are rapidly stimulated by GnRH injections.[56] There are no available data on the content of other neuropeptides or neurotransmitters in or projecting to the POA in such hpg animals, or even in their normal counterparts. (Most of the available data noted above regarding neurotransmitters have been obtained in other rodent species.) Such data will be an essential prerequisite in analyzing the full nature of the defect present and the results of any experimental manipulation.

Preliminary studies of feedback responsiveness in hpg animals indicate that the hypothalamus is a major site of testosterone negative feedback, while that of estradiol seems to be (in contrast to normal animals) at the pituitary level.

Although testosterone implants alone have stimulated full spermatogenesis in hpg males after 60 days, none of these animals were able to mate successfully with normal females. However, in mutants given a single subcutaneous injection of testosterone on day 1 after birth, followed by implants when adult, 95% of males succeeded in mating and siring hpg offspring. The normal pattern of female mating behavior was displayed in hpg females given estradiol benzoate and progesterone injections, and the quality of lordosis was further enhanced after concurrent GnRH injections.[57]

3. Effect of Fetal Preoptic Area Transplants in Male hpg Animals

We have studied the effect of implantation of fetal (16- to 18-day-old fetuses) POA (a site of GnRH production) from unaffected animals of the hpg strain on hypothalamic–pituitary–gonadal function in adult (5- to 6-month-old male) hpg mice.[58] For control tissue transplants to hpg and normal animals, the cortex from prepubertal mice was utilized. Two POA tissue segments were used per implant; cortical fragments were of the same dimensions as the preoptic fragments.

For POA grafts, an anterior cut 1.0 mm wide was made at the midline of the ventral side of the fetal brain just caudal to the bifurcation of the anterior cerebral artery; a second cut was made ~ 0.7 mm posterior to the first, just rostral to the hypothalamus. Lateral cuts were made 0.5 mm from the midline and the tissue block was finally cut 0.5 mm deep. The tissue was then placed in a drop of sterile saline in a Petri dish on a bed of ice and bisected with iris scissors. Two fetal preoptic tissue segments (comprising four pieces of tissue) were used per implant; cortical fragments of the same dimensions as the preoptic fragments were dissected from the frontal cortex of prepubertal mice. Graft recipients were anesthetized with chloral hydrate and placed in a Kopf stereotaxic instrument. Grafts, in 2–4 μl saline, were injected into the anterior third ventricle using a 22-gauge needle. Stereotaxic coordinates of the injection site were 5.5–6.0 mm down from the dura at the midline of the bregma. There were five experimental groups: normal animals, hpg animals, hpg animals with POA transplants, hpg animals with transplants of cortex, and normal animals with transplants of cortex.

The growth rates of grafted animals were similar to those of unaffected mutants or normal animals. Two months postimplantation, blood was obtained via retroocular puncture for plasma gonadotropin and testosterone determinations, and the animals then killed by decapitation. On inspection at the time of sacrifice, it was apparent that of the hpg animal groups, only in those with POA implants were testes descended into the scrotum. In a given experimental group, half of the brains were processed for immunocytochemical studies, and the other half for GnRH immunoassay.[59] Immunoassays

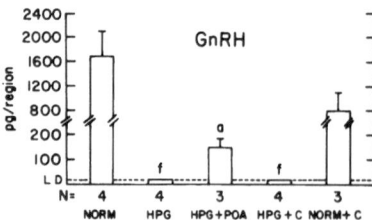

FIGURE 4. Effect of implants of preoptic area (POA) or cortex (C) on hypothalamic GnRH concentrations. Data analyzed with nonparametric methods (Kruskal–Wallis one-way analysis of variance by ranks, followed by Ryan's procedure, with $\alpha = 0.05$; these methods were utilized due to unequal variances). GnRH values for HPG and HPG + C groups are combined, as they are the same. HPG + POA compared with HPG and HPG + C: a = $p < 0.05$. NORM compared with HPG: f = $p < 0.05$. Reproduced from Ref. 58 with permission.

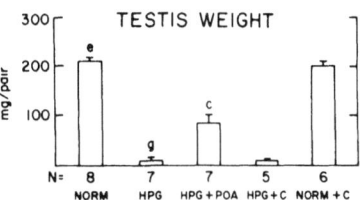

FIGURE 5. Effects of implants of preoptic area or cortex on testicular weight. Data analyzed using planned orthogonal comparisons. HPG + POA compared with HPG and HPG + C: c = $p < 0.01$. NORM compared with HPG + POA: e = $p < 0.001$. NORM compared with HPG: g = $p < 0.001$. Reproduced from Ref. 58 with permission.

were performed on sections that consisted of hypothalamus alone, and on another pool that represented remaining brain sections, excluding the brain stem from the midbrain down. For immunocytochemical studies, brains were divided into three coronal blocks, the first coronal cut being made 2 mm in front of the optic chiasm and the second behind the mammillary bodies. The anti-GnRH serum employed has been previously reported.[60]

For assays for GnRH, after removal of the brain from recipients, the hypothalamus was dissected by coronal cuts 1 mm in front of the optic chiasm and behind the median eminence; laterally, to 1 mm lateral to the hypothalamic sulcus; and dorsally to the top of the third ventricle. It was placed in 0.5 ml of 0.1 M HCl. The remaining brain sections (excluding the brain stem from the midbrain down) were placed in 1 ml of 0.1 M HCl. Speciments were frozen until GnRH assay, at which time they were homogenized and neutralized.

Pituitary glands were bisected *in situ,* one-half of each being frozen until time of assay for gonadotropin concentrations, and the other processed for further immunocytochemical studies.

Figures 4 through 8 indicate the effect of POA or control cortical implants on hypothalamic GnRH concentration, testicular weight, and pituitary and serum gonadotropin concentrations. Evidence of functional grafts was found in seven of the eight POA-implanted hpg animals, while none was present in hpg animals implanted with cortex. It is apparent from these figures that such grafts do not totally correct GnRH or LH and FSH (pituitary and serum) levels to those seen in normal animals. Whether this can be achieved by implantation of larger amounts of POA tissue remains to be seen. Another possibility is that lack of complete normalization is secondary to the absence of appropriate afferent connections to transplanted GnRH cells. Further immunocytochemical and electron microscopic studies are necessary to determine the nature of any such afferent connections to the grafted GnRH cells.

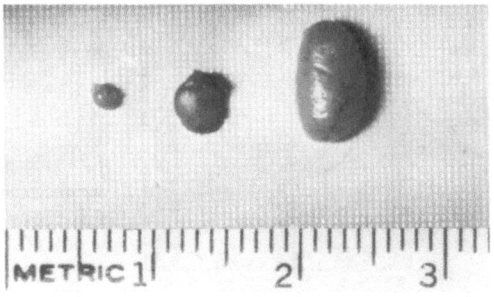

FIGURE 6. Effect of implants of preoptic area on testicular size. Testis depicted on left is that from an adult hpg mouse. Testis depicted on right is that from a normal littermate. Testis depicted in middle is from an hpg animal that received a preoptic area implant.

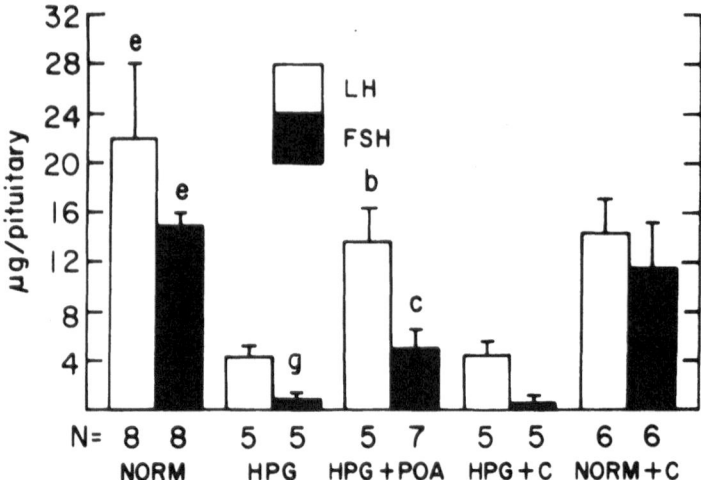

FIGURE 7. Effect of implants of preoptic area or cortex on pituitary gonadotropin concentrations. Data analyzed as in Fig. 5. Symbols as in previous figures HPG + POA compared with HPG and HPG + C: b = $p < 0.02$. Reproduced from Ref. 58 with permission.

Serum testosterone concentrations were also incompletely corrected (normal 1.0 ng/ml, hpg and hpg + cortex both undetectable, and hpg + POA 0.3 ng/ml). These testosterone levels correlated with the observed differences in seminal vesicle weight, used as an index of biological testosterone efficacy. Such weights (mg/pair) were: normal, 258.3 ± 25.6; normal + cortex, 218.0 ± 24.6; hpg, 17.2 ± 1.6; hpg + cortex, 14.0 ± 2,9; and hpg + POA, 66.1 ± 16.2.

Histological examination of the testes revealed full spermatogenesis in the hpg POA recipients, while hpg animals with cortical implants were indistinguishable from those of untreated hpg animals (Fig. 9). The extent of spermatogenesis in the grafted recipients is also illustrated in the scanning electron micrograph (Fig. 10).

Studies of brain sections correlated well with individual observations of grafted animals. In the single animal without hormonal evidence of a functional transplant, a cannula scar, but no evidence of transplanted tissue, was seen. In all four remaining POA-grafted animals that were examined immunocytochemically, evidence of viable grafts

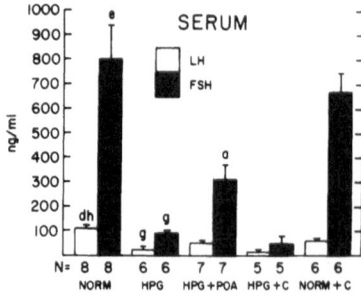

FIGURE 8. Effect of implants of preoptic area or cortex on serum gonadotropin concentrations. Data analyzed as in Fig. 5. Symbols as in previous figures. NORM compared with HPG + POA d = $p < 0.01$. NORM compared with NORM + C: h = $p < 0.05$. Reproduced from Ref. 58 with permission.

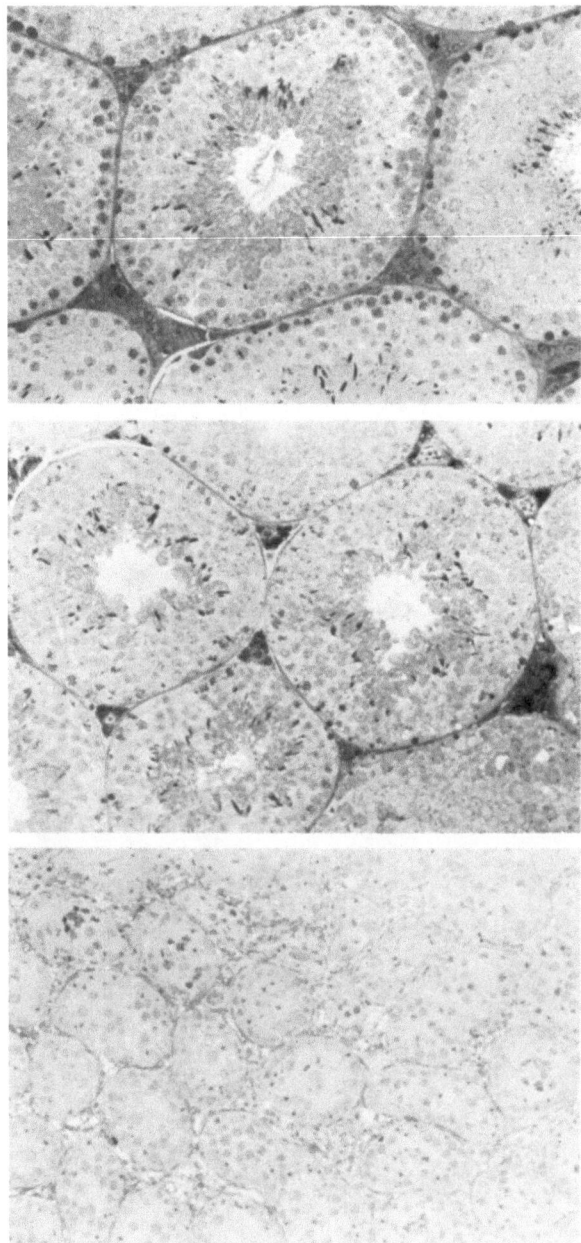

FIGURE 9. Testicular sections from normal animal (top); hpg animal with POA implant (middle); and hpg animal with cortex implant (bottom). Full spermatogenesis and evidence of interstitial cell development are present in POA recipients, while sections from animals with cortical implants are identical to those from untreated hpg animals. × 154. Reproduced from Ref. 58 with permission.

FIGURE 10. SEM photomicrograph of testis from hpg animal with POA implants, showing full spermatogenesis.

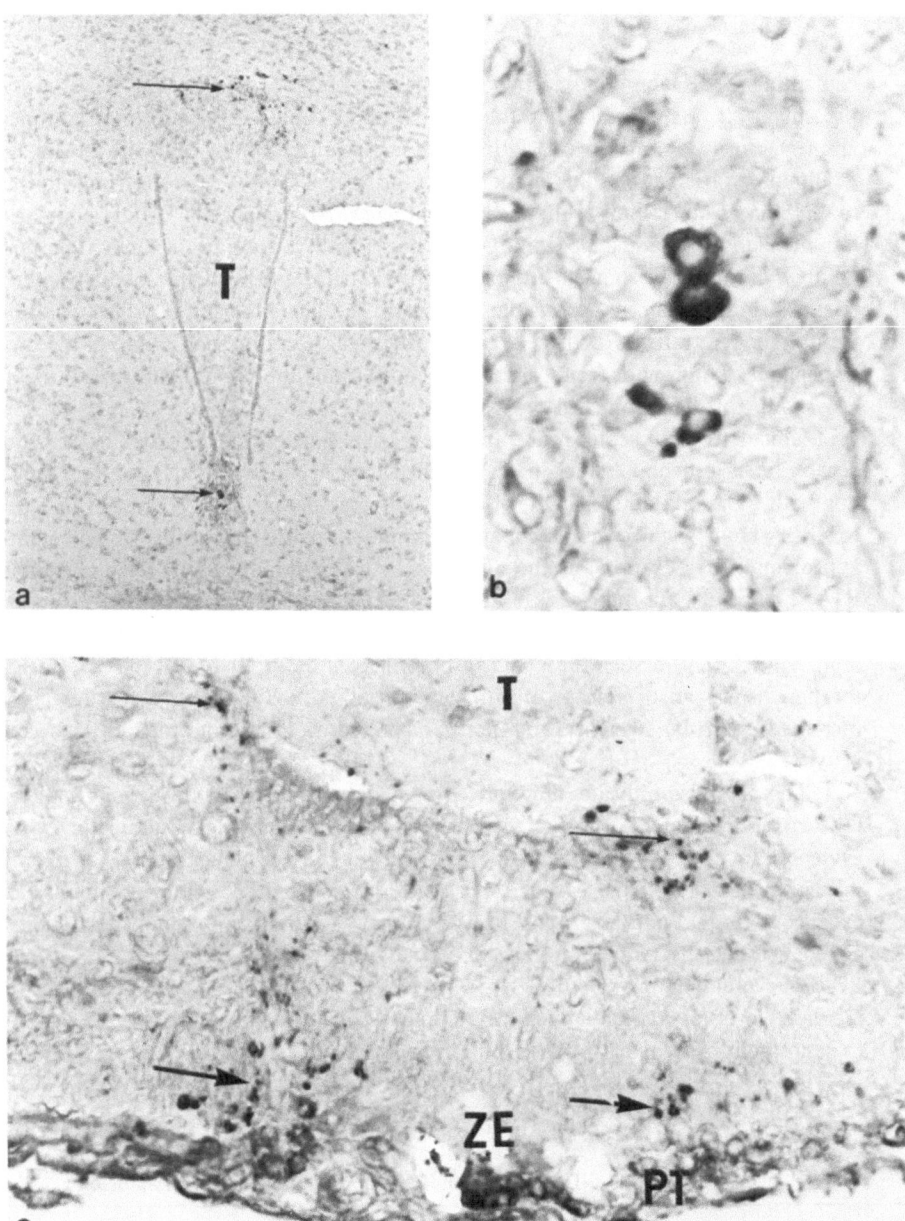

FIGURE 11. Localization of GnRH in an hpg mouse hypothalamus containing a third ventricular transplant (T) of normal fetal mouse preoptic area. (a) Coronal section at the level of the midhypothalamus just rostral to the median eminence. The transplant fills the third ventricle and extends into the brain tissue just dorsal and ventral of it (arrows), where reactive neuronal perikarya were found. (b) Higher magnification of three reactive neurons indicated by the lower arrow in (a). (c) Section more caudal to (a), again showing the transplant within the third ventricle at the level of the mid-median eminence. Reactive fibers enter the dorsal median eminence at the ventrolateral corners of the graft (upper, thinner arrows), and some can be traced to the zona externa (ZE) (lower, thicker arrows). Some background staining can be seen in the pars tuberalis (PT) just external to the ZE. a, × 112; b, × 764; c, × 360. Reproduced from Ref. 58 with permission.

containing GnRH-immunoreactive neurons was present (see Fig. 11). Perikarya were present in all of the grafts; immunoreactive fibers could be seen entering the median eminence in its lateral aspects, with others innervating capillaries in the zona externa. Preliminary observations did not indicate outgrowth of GnRH neurons from transplants to CNS areas other than to the median eminence. No evidence of GnRH staining was found in untransplanted hpg animals or those with cortical transplants. The normal adult mice similarly showed immunoreactive GnRH fibers in the median eminence, particularly in its lateral aspects, i.e., a location similar to the site of projection of fibers from the POA grafts. No GnRH-reactive perikarya were present in any of the cortical transplants. Outgrowths from cortical grafts infiltrated the surrounding area but did not grow to the median eminence.

Gonadotropin receptor concentrations were also determined in 20,000g fractions of decapsulated testicular homogenates obtained from normal animals, hpg mutants, and hpg with POA implants, using [^{125}I]hCG and receptor-purified [^{125}I]hFSH in Dulbecco's PBS. Homogenates were incubated overnight at room temperature under conditions approximating saturation and the supernatants assessed individually for testosterone by RIA. The results are indicated in Table I.[61]

These data indicated that partial restoration of hypothalamic GnRH levels in hpg animals (deficient in such GnRH) can be produced by grafted GnRH cells. Such restoration brings about release of sufficient pituitary LH and FSH to cause greatly increased testicular weights and normal spermatogenesis and interstitial secretory activity.

The data also indicate, therefore, the presence of intact and functioning pituitary GnRH receptors and also receptivity of the previously undeveloped testicular tissue, as evidenced by development of gonadotropin receptors therein. Such findings are compatible with the above-noted earlier reports of LH responsiveness (albeit diminished) to exogenous GnRH administration in hpg animals[52] and of the normal function of hpg gonads when transplanted into normal animals.[54]

In evaluating our results, we considered the possibility that the presence of increased levels of hypothalamic GnRH in the POA-grafted mutant could be secondary to restoration of a factor tropic to GnRH production that was lacking in the mutant.

TABLE I
Effects of POA Transplants on Testicular Gonadotropin Receptors

Group (n)	Testes weight (mg)	Seminiferous tubule area (mm² × 10⁻³)	Receptor HCG (pmole)	Receptor FSH (pmole)	Testosterone T (ng)	Testosterone Serum[a] (ng/ml)
Normal (8)	104 ± 11	62.0 ± 2.1	0.76 ± 0.2	1.53 ± 0.2	14.3 ± 4.0	1.0
hpg (7)	3.8 ± 1.0	6.1 ± 0.3	—[b]	0.003 ± 0.0005	0.35 ± 0.1[a]	—[c]
hpg + POA (7)	41.8 ± 14.0	40.2 ± 2.4	0.22 ± 0.1	0.50 ± 0.05	1.91 ± 1.3	0.3

[a]Pooled single samples.
[b]Insufficient.
[c]Undetermined.

This, however, seems unlikely, as the only GnRH cells evident were present within the grafted tissue, not in the host tissue. We also discounted the possibility that any non-specific neural tissue could stimulate GnRH production, as there were no hormonal or testicular changes in the hypogonadal mutants with cortical implants.

4. Effect of Fetal Preoptic Area Transplants in Female hpg Animals

These studies were performed in a fashion similar to that described for male animals, utilizing transplants of fetal POA and cortex. Our preliminary observations indicate that while normal animals showed evidence of estrous cyclicity, as determined by vaginal smears, hpg animals or hpg animals with cortical transplants showed evidence only of diestrus smears. The hpg animals with POA transplants either remained in diestrus (n = 5) or entered a state of constant vaginal estrus (n = 9). The mice with estrous smears were later found to have increased ovarian and uterine weights, pituitary LH and FSH concentrations similar to those present in normal animals, and immunocytochemical evidence of viable grafts containing GnRH-immunoreactive neurons.

5. Possible Role of Donor Sex in Functioning of CNS Transplants

The question arises as to whether the sex of the donor, as compared to that of the recipient animal, is an important variable. In the first experiments, this was not considered. However, a group of female hpg mice in the experiment discussed above received POA implants from known male or female fetuses. There were no significant differences in the success of the transplants between the two groups, so for the parameters studied it appears that male or female fetal tissue is equally effective. This should be further investigated, in view of the large number of reports of sexual dimorphism in several aspects of CNS organization. These include the initial report of Raisman and Field indicating sexual dimorphism in the ratio of synapses on dendritic shafts and the number of dendritic spines in the POA,[62] and the large number of subsequent reports of sexual dimorphism in many other aspects of CNS organization, i.e., POA nuclear volume, neuronal dendritic branching, and synaptic organization (see Ref. 63 for review). It is now established that the mammalian brain is inherently female or bipotential, with sexual differentiation of its reproductive functions largely determined by testicular hormones secreted during perinatal development in the male. It is felt that aromatization of testosterone to estrogen through an interaction with estrogen receptors in the medial POA is critical in this regard.[64,65] Such "masculinization" of the brain results in suppression both of the capacity to support cyclic feminine patterns of gonadotropin secretion and of female behavioral patterns, with enhancement of patterns characteristic of the male. The noted sex differences in CNS structure may represent the expression of the growth-promoting effects of gonadal steroids that have been observed in cultures of fetal brain areas.[66]

The major period of cytoplasmic differentiation and synaptic formation in the rat POA begins by the time of birth and continues throughout the first days of postnatal life.[63] If this were so, the implants (fetal day 15–18) used in the experiments described in this chapter should represent grafting of "female" CNS areas. It is now realized, however, that the occurrence of "masculinization" of the CNS exists over a continuum, when considered as times critical for determination of gonadotropin release patterns (which are cyclic in the female, acyclic in the male), for patterns of sexual behavior, or for morphometric criteria. It might then be expected that the grafts of "masculinized" "acyclic" brains into female recipients should be more disruptive of development of normal hypothalamic–pituitary–ovarian function (which requires cyclic gonadotropin release) than grafts of "female" "cyclic" brains into male recipients. There is recent evidence that suggests sex-specific development in neuroregulatory mechanisms. The sphinx moth *(Manduca sexta)* is characterized by specialized olfactory receptor cells found only on male antennae, which project into a macroglomerular complex characteristic of male but not female antennal lobes, such afferents presumably synapsing with male-specific antennal lobe neurons.[67] When antennal discs of the opposite sex were implanted into larvae, while the transplanted antennae exhibited structures characteristic of the donor sex, neurons resembling the male-specific antennal lobe neurons appear in female antennal lobes, innervated by sensory axons from the grafted male antennae. Another study[68] (see also Chapter 9, this volume) has indicated that transplantation of preoptic tissue from male rat neonates into the POA of female neonates increased masculine and feminine sexual behavior in the recipients during adulthood. It was suggested that the day-old male POA implants already have the potential to behaviorally masculinize the neonatal female brain. No control studies, however, were done transplanting female POA into female littermates, and there is at present no valid explanation for the increased feminine behavior observed.

6. Conclusion

The prolonged maintenance of relatively normal hormone parameters and the high success rate achieved in the (male) transplanted animals in this study indicate that the hpg grafted animal is a valuable model in which to study neural factors involved in the neuroendocrinology of gonadal function. The model can also further delineate the role of sexually dimorphic structures. This can be achieved with the use of grafts obtained from donors before and after the period of POA differentiation transplanted to both neonatal and adult recipients of the same or opposite sex. The hpg mouse therefore serves as another model, in addition to the others cited in this volume, for the study of factors involved in the transplantation of neural tissue to correct surgically or genetically induced CNS defects.

References

1. Azmitia, E. C., Perlow, M. J., Prennan, M. J., and Lauder, J. M., 1981, Fetal raphe and hippocampal transplants into adult and aged C57BL/6n mice: A preliminary immunocytochemical study, *Brain Res. Bull.* **7**:703.

2. Björklund, A., Kromer, L. F., and Stenevi, U., 1979, Cholinergic reinnervation of the rat hippocampus by septal implants is stimulated by perforant path lesion, *Brain Res.* **173**:57.

3. Björklund, A., and Stenevi, U., 1979, Regeneration of monoaminergic and cholinergic neurons in the mammalian central nervous system, *Physiol. Rev.* **59**:62.

4. Das, G. D., 1974, Transplantation of embryonic neural tissue in the mammalian brain. I. Growth and differentiation of neuroblasts from various regions of the embryonic brain in the cerebellum of neonate rats, *Life Sci.* **4**:93.

5. Hallas, B. H., Das, G. D., and Das, K. G., 1980, Transplantation of brain tissue in the brain of rat. II. Growth characteristics of neocortical transplants in hosts of different ages, *Am. J. Anat.* **158**:147.

6. Jaeger, C. B., and Lund, R. D., 1980, Transplantation of embryonic occipital cortex to the tectal region of newborn rats: A light microscopic study of organization and connectivity of the transplants, *J. Comp. Neurol.* **194**:571.

7. Jaeger, C. B., and Lund, R. D., 1980, Transplantation of embryonic occipital cortex to the brain of newborn rats: An autoradiographic study of transplant histogenesis, *Exp. Brain Res.* **40**:265.

8. Kromer, L. F., Björklund, A., and Stenevi, U., 1979, Intracephalic implants: A technique for studying neuronal interactions, *Science* **204**:1117.

9. Kromer, L. F., Björklund, A., and Stenevi, U., 1980, Innervation of embryonic hippocampal implants by regenerating axons of cholinergic septal neurons in the adult rat, *Brain Res.* **210**:153.

10. Low, W. C., Dunnett, S. B., Bunch, S. T., Thomas, S. R., Lewis, P. R., Iversen, S. D., Björklund, A., and Stenevi, U., 1981, Restoration of synaptic and behavioral function with embryonic transplants of cholinergic neurons, *Neurosci. Abstr.* **7**:259.

11. Lund, R. D., and Hauschka, S. D., 1976, Transplanted neural tissue develops connections with host rat brain, *Science* **193**:582.

12. Lund, R. D., and Harvey, A. R., 1981, Transplantation of tectal tissue in rats. I. Organization of transplants and pattern of distribution of host afferents within them, *J. Comp. Neurol.* **201**:191.

13. McLoon, L. K., McLoon, S. C., and Lund, R. D., 1981, Cultured embryonic retinae transplanted to rat brain: Differentiation and formation of projections to host superior colliculus, *Brain Res.* **226**:15.

14. Oblinger, M., Hallas, B. H., and Das, G. D., 1980, Neocortical transplants in the cerebellum of the rat: Their afferents and efferents, *Brain Res.* **189**:228.

15. Stenevi, U., Björklund, A., Kromer, L. F., Paden, C. M., Gerlach, J. L., McEwen, B. S., and Silverman, A. J., 1980, Differentiation of embryonic hypothalamic transplants cultured on the choroidal pia in brains of adult rats, *Cell Tissue Res.* **205**:217.

16. Wells, J., and McAllister, J. P., II, 1982, The development of cerebellar primordia transplanted to the neocortex of the rat, *Dev. Brain Res.* **4**:167.

17. Merchenthaler, I., Kovacs, G., Lovasz, G., and Setalo, G., 1980 The preoptico-infundibular LH-RH tract of the rat, *Brain Res.* **198**:63.

18. Jennes, L., and Stumpf, W. E., 1980, LHRH neuronal projections to the inner and outer surface of the brain, *Neuroendocrinol. Lett.* **2**:241.

19. Witkin, J. W., Paden, C. M., and Silverman, A. J., 1982, the LHRH systems in the rat brain, *Neuroendocrinology* **35**:429–438.

20. Carmel, P. W., Araki, S., and Ferin, M., 1976, Pituitary stalk portal blood collection in rhesus monkeys: Evidence for pulsatile release of gonadotropin-releasing hormone (GnRH), *Endocrinology* **99**:243.

21. Eskay, R. L., Mical, R. S., and Porter, J. C., 1977, Relationship between luteinizing hormone releasing hormone concentration in hypophysial portal blood and luteinizing hormone release in intact, castrated, and electrochemically-stimulated rats, *Endocrinology* **100**:263.

22. Turgeon, J., and Barraclough, C. A., 1973, Temporal patterns of LH release following graded preoptic electrochemical stimulation in proestrous rats, *Endocrinology* **92**:755.

23. Rance, N., Wise, P. M., Selmanoff, M. K., and Barraclough, C. A., 1981, Catecholamine turnover rates in discrete hypothalamic areas and associated changes in median eminence luteinizing hormone-releasing hormone and serum gonadotropins on proestrus and diestrous day 1, *Endocrinology* **108**:1795.

24. Araki, S., Ferin, M., Zimmerman, E. A., and Vande Wiele, R. L., 1975, Ovarian modulation of immunoreactive gonadotropin-releasing hormone (GnRH) in the rat brain: Evidence for a differential effect on the anterior and mid-hypothalamus, *Endocrinology* **96**:644.

25. Kalra, S. P., 1977, Suppression of serum LHRH and LH in rats by an inhibitor of norepinephrine synthesis, *J. Reprod. Fertil.* **49**:371.

26. Leung, P. C. K., Arendash, G. W., Whitmoyer, D. I., Gorski, R. A., and Sawyer, C. H., 1982, Differential effects of central adrenoreceptor agonists on luteinizing hormone release, *Neuroendocrinology* **34**:207.

27. Lofstrom, A., Eneroth, P., Gustafsson, J. A., and Skett, P., 1977, Effects of estradiol benzoate on catecholamine levels and turnover in discrete areas of the median eminence and the limbic forebrain, and on serum luteinizing hormone, follicle stimulating hormone, and prolactin concentrations in ovariectomized female rats, *Endocrinology* **101**:1559.

28. Ojeda, S. R., Negro-Vilar, A., and McCann, S. M., 1979, Release of prostaglandin Es by hypothalamic tissue: Evidence for their involvement in catecholamine induced luteinizing hormone releasing hormone release, *Endocrinology* **104**:617.

29. Sarkar, D. S., and Fink, G., 1981, Gonadotrophin-releasing hormone surge: Possible modulation through postsynaptic α-adrenoreceptors and two pharmacologically distinct dopamine receptors, *Endocrinology* **108**:862.

30. Kordon, C., and Glowinski, J., 1972, Role of hypothalamic monoaminergic neurons in the gonadotropin release-regulatory mechanisms, *Neuropharmacology* **11**:153–162.

31. Vijayan, W., and McCann, S. M., 1978, Re-evaluation of the role of catecholamines in control of gonadotropin and prolactin release, *Neuroendocrinology* **25**:150.

32. Chiocchio, S. R., Negro-Vilar, A., and Tramezzani, J. H., 1976, Acute changes in norepinephrine content in the median eminence induced by orchidectomy or testosterone replacement, *Endocrinology* **99**:629.

33. McCann, S. M., and Ojeda, S. R., 1976, Synaptic transmitters involved in the release of hypothalamic releasing and inhibiting hormones, in: *Review of Neuroscience,* Vol. 2 (S. Earenpreis and I. J. Kopin, eds.), pp. 91–110, Raven Press, New York.

34. Beck, W., Hancke, J. L., and Wuttke, W., 1978, Increased sensitivity of dopaminergic inhibition of luteinizing hormone release in immature and castrated female rats, *Endocrinology* **102**:837.

35. Wilson, C., 1974, Hypothalamic amines and their release of gonadotrophins and other anterior pituitary hormones, *Adv. Drug Res.* **8**:110.

36. Gudelsky, G. A., Simpkins, J., Muller, G. P., Meites, J., and Moore, K. E., 1976, Selective actions of prolactin on catecholamine turnover in the hypothalamus and on serum LH and FSH, *Neuroendocrinology* **22**:206.

37. Krulich, L., 1979, Central neurotransmitters and the secretion of prolactin, GH, LH and TSH, *Annu. Rev. Physiol.* **41**:603.

38. Vijayan, E., and McCann, S. M., 1978, The effect of systemic administration of dopamine and apomorphine on plasma LH and prolactin concentration in conscious rats, *Neuroendocrinology* **25**:221.

39. Hoffman, G. E., Wary, S., and Goldstein, M., 1982, Relationship of catecholamines and LHRH: Light microscopic study, *Brain Res. Bull.* **9**:417.

40. Palkovits, M., 1981, Catecholamines in the hypothalamus: An anatomical review, *Neuroendocrinology* **33**:123.

41. Sawyer, C. H., Markee, J. E., and Hollinshead, W. H., 1947, Inhibition of ovulation in the rabbit by the adrenergic blocking agent dibenamine, *Endocrinology* **41**:395.

42. Ratner, A., and Solomon, S., 1971, Effect of reserpine on plasma LH levels in ovariectomized and cycling proestrus rats, *Proc. Soc. Exp. Biol. Med.* **138**:995.

43. Kalra, S. P., Simpkins, J. W., and Kalra, P. S., 1980, Effects of pentobarbital on hypothalamic catecholamines and LRH activities, *Acta Endocrinol. (Copenhagen)* **95**:1.

44. Pfaff, D. W., 1973, Luteinizing hormone-releasing factor potentiates lordosis behavior in hypophysectomized ovariectomized female rats, *Science* **182**:1148.

45. Foremann, M. M., and Moss, R. L., 1977, Effects of subcutaneous injection and intrahypothalamic infusion of releasing hormones upon lordotic response to repetitive coital stimulation, *Horm. Behav.* **8**:219.

46. Sakuma, Y., and Pfaff, D. W., 1980, LH-RH in the mesencephalic central grey can potentiate lordosis reflex of female rats, *Nature (London)* **283**:566.

47. Dorsa, D. M., and Smith, E. R., 1980, Facilitation of mounting behavior in male rats by intracranial injections of luteinizing hormone-releasing hormone, *Regul. Pept.* **1**:147.

48. Moore, F. L., Miller, L. J., Spielvogel, S. P., Kubiak, T., and Folkers, K., 1982, Luteinizing hormone-

releasing hormone involvement in the reproductive behavior of a male amphibian, *Neuroendocrinology* **35**:212.

49. Dudley, C. A., Vale, W., Rivier, J., and Moss, R. L., 1981, The effect of LHRH antagonist analogs and an antibody to LHRH on mating behavior in female rats, *Peptides* **2**:393.

50. Davies, T. F., Mountjoy, C. W., Gomez-Pan, A., Watson, M. J., Hanker, J. P., Besser, G. J., and Hall, R., 1976, A double blind cross over trial of gonadotropin releasing hormone (LHRH) in sexually impotent men, *Clin. Endocrinol.* **5**:601.

51. Silverman, A. J., and Krey, L. C., 1978, The luteinizing hormone-releasing hormone (LH-RH) neuronal networks of the guinea pig brain. I. Intra- and extrahypothalamic projections, *Brain Res.* **157**:233.

52. Iddon, C. A., Charlton, H. M., and Fink, G., 1980, Gonadotrophin release in hypogonadal and normal mice after electrical stimulation of the median eminence of injection of luteinizing hormone releasing hormone, *J. Endocrinol.* **85**:105.

53. Charlton, H. M., Fink, G., and Halpin, G. M. G., 1981, The effects of multiple injections of gonadotrophin-releasing hormone (GnRH) in the hypogonadal mouse, *J. Physiol. (London)* **320**:106P.

54. Bamber, S., Iddon, C. A., Charlton, H. M., and Ward, B. J., 1980, Transplantation of the gonads of hypogonadal *(hpg)* mice, *J. Reprod. Fertil.* **58**:249.

55. Charlton, H. M., Parry, D., Halpin, D. M. G., and Webb, R., 1982, Distribution of I^{125}-labelled follicle stimulating hormone and human chorionic gonadotrophin in the gonads of hypogonadal *(hpg)* mice, *J. Endocrinol.* **93**:247.

56. Young, L. S., Charlton, H. M., and Clayton, R. N., 1982, GnRH receptor regulation in the hypogonadotrophic hypogonadal *(hpg)* and normal mouse, *Proceedings of the 1st Joint Meeting of British Endocrine Societies*, p. 58.

57. Ward, B. J., and Charlton, H. M., 1981, Female sexual behaviour in the GnRH deficient hypogonadal *(hpg)* mouse, *Physiol. Behav.* **27**:1107.

58. Krieger, D. T., Perlow, M. J., Gibson, M. J., Davies, T. F., Zimmerman, E. A., Ferin, M., and Charlton, H. M., 1982, Brain grafts reverse hypogonadism of gonadotropin-releasing hormone deficiency, *Nature (London)* **298**:468.

59. Nett, T. M., Akbar, A. M., Niswender, G. D., Hedlund, M. T., and White, W. F., 1973, A radioimmunoassay for gonadotropin-releasing hormone (Gn-RH) in serum, *J. Clin. Endocrinol. Metab.* **36**:880.

60. Schwanzel-Fukuda, M., and Silverman, A. J., 1980, The nervus terminalis of the guinea pig: A new luteinizing hormone-releasing hormone (LHRH) neuronal system, *J. Comp. Neurol.* **191**:213.

61. Davies, T. F., Platzer, M., Perlow, M. J., Gibson, M., Zimmerman, E. A., Ferin, M., and Charlton, H. M., 1982, Gonadotropin receptor development following hypothalamic brain transplantation in hypogonadal (hpg) mutant mice with GnRH deficiency, *Proceedings of the Endocrine Society 64th Meeting*, No. 963.

62. Raisman, G., and Field, P. M., 1971, Sexual dimorphism in the preoptic area of the rat, *Science* **173**:731.

63. MacLusky, N. J., and Naftolin, F., 1981, Sexual differentiation of the nervous system, *Science* **211**:1294.

64. Plapinger, L., and McEwen, B. D., 1978, Gonadal steroid–brain interactions in sexual differentiation, in: *Biological Determinants of Sexual Behavior* (J. Hutchison, ed.), pp. 153–218, Wiley, New York.

65. Naftolin, F., Ryan, D. J., Davis, I. J., Reddy, V. V., Flores, F., Petro, Z., Kuh, M., White, R. J., Takaoka, Y., and Wolin, L., 1975, The formation of estrogens by central neuroendocrine tissues, *Rec. Prog. Horm. Res.* **31**:295.

66. Toran-Allerand, C. D., 1976, Sex steroids and the development of the newborn mouse hypothalamus and preoptic area *in vitro:* Implications for sexual differentiation, *Brain Res.* **106**:407.

67. Schneiderman, A. M., Matsumoto, S. G., and Hildebrand, J. G., 1982, Trans-sexually grafted antennae influence development of sexually dimorphic neurones in moth brain, *Nature (London)* **198**:844.

68. Arendash, G. W., and Gorski, R. A., 1982, Enhancement of sexual behavior in female rats by neonatal transplantation of brain tissue from males, *Science* **217**:1276.

69. Martin, J. B., Reichlin, S., and Brown, G. M. (eds.), 1977, *Clinical Neuroendocrinology*, Davis, Philadelphia.

8

Hypothalamic Grafts and Neuroendocrine Cascade Theories of Aging

Immunocytochemical Viability of Preoptic Hypothalamic Transplants from Fetal to Reproductively Senescent Female Rats

JOSEPH ROGERS, GLORIA E. HOFFMAN, STEVEN F. ZORNETZER, AND WYLIE W. VALE

1. Introduction

Transplantation of fetal brain tissue has been reported for several structures including the cortex,[1] supraoptic and paraventricular hypothalamic nuclei,[2] and substantia nigra.[3] We report here transplantation of another brain structure, the preoptic area of the hypothalamus (POA), from 17- to 19-day prenatal rats to reproductively senescent female rats, aged 15-17 months. On the basis of luteinizing hormone-releasing hormone (LHRH) immunocytochemistry initiated 9 weeks after transplantation, grafted POA appear to have survived and to exhibit LHRH-positive cell bodies and fibers both afferent to and efferent from the host.

Research conducted over the last decade has begun to suggest a primary role of

JOSEPH ROGERS ● Department of Neurology, University of Massachusetts Medical School, Worcester, Massachusetts 01605.   GLORIA E. HOFFMAN ● Department of Anatomy, University of Rochester School of Medicine and Dentistry, Rochester, New York 14642.   STEVEN F. ZORNETZER ● Department of Pharmacology, University of California, Irvine, California 92668.   WYLIE W. VALE ● Peptide Biology Laboratory, The Salk Institute, La Jolla, California 92138.

CNS deterioration in mammalian senescence. In particular, the discovery and continuing elaboration of neural mechanisms for control of the endocrine system has spawned a "neuroendocrine cascade" hypothesis of aging, wherein age-dependent neurologic changes prompt a vicious cycle of endocrine imbalance, derangement of target organ responses, and, ultimately, further pathologic change in the nervous system.[4,5] For example, the primary basis of reproductive senescence in female rodents may lie in a failure of neural[6,7] rather than reproductive tract,[8] ovarian,[9] or pituitary[10] mechanisms. The latter systems do, of course, exhibit age-specific pathology; but there is growing evidence that these changes are secondary to senescent neural alterations.[4,5] Thus, it has been found that intraventricular injections of catecholamine precursors and agonists can produce LH surges and vaginal cycling in aged female rats.[11,12] Such cycling, however, is only temporary, and the mechanism by which it is achieved remains to be firmly established.

We believe brain transplant techniques offer several unique advantages for testing neuroendocrine cascade and other hypotheses of aging. First, brain transplants can serve to bridge gaps in our knowledge about neural mechanisms. Even a cursory review of the literature will reveal that most aging studies flounder not so much from an inability to detect altered function in old animals as from an inability to define what constitutes normal function in young animals. Consider, for example, research into the loss of estrous cycling in old females. There is probably no single neurotransmitter (including dopamine,[13] norepinephrine, [14] serotonin,[15] Met-enkephalin,[16] and a host of others[17,18]) that has not, at one time or another, been proposed to play a major role in regulating ovulatory cycles in young animals. Moreover, even groups who agree on the same transmitter may differ over whether its action on the system is excitatory or inhibitory.[13,19] Transplant techniques neatly sidestep these difficulties, making it possible to examine senescent deficits in complex systems such as the neuroendocrine axis without necessarily knowing all the details of how those systems work. Thus, to test a neuroendocrine cascade hypothesis of reproductive aging using brain transplants, it may be enough to know that some step in hypothalamic LHRH regulation becomes deranged in anestrous rats; whether that step involves dopamine, norepinephrine, serotonin, or some other neurotransmitter need not initially be specified. On the contrary, the functional effect of the transplant may provide leads as to which neurotransmitters, which anatomic loci, and which physiologic processes are critical to senescent reproductive dysfunction.

A second potential transplant use in aging research is as a technique complementary to brain lesion strategies. Previously, lesion experiments have typically sought to relate a particular senescent dysfunction with deterioration of a particular brain structure by showing that removal of the structure in young animals mimics the senescent deficit observed in old animals. Unfortunately, questions always remain as to whether the induced deficit is due to destruction of the avowed target area, destruction of tissues near the target area, or destruction of fibers passing through the target area. By demonstrating that a brain transplant subsequently restores function to a lesioned area, many of these difficulties vanish. For example, a grafted POA will probably not contain viable fibers of passage; hence, transplant restoration of function after a lesion is an elegant proof that the lesion deficit was not due to interruption of fiber tracts coursing through the damaged structure.

A third transplant strategy for aging research is more heuristic than conceptual.

Consider the finding that acute administration of certain drugs prompts acute restoration of estrous cycling.[11,12] How might we improve on such a result? One obvious step would be to make the drug administration chronic, in the hope of providing for the complete restoration of function. As such, brain transplants can in many ways be considered something of an ultimate implantable minipump. Transplants are self-maintaining. They release only the most "natural" of substances, and they are limited neither to a single substance nor to a single channel for delivery. Moreover, there is now some reason to believe that transplants may be able to deliver the right neurotransmitters at the right times to the right neural tragets,[2,3] thereby enhancing efficacy and minimizing unwanted side effects.

The overall goal of the present research is to use brain transplant strategies to test a neuroendocrine cascade hypothesis of aging: namely, that senescent disruption of estrous cycling is due primarily to senescent deterioration of brain, rather than target organ or pituitary, mechanisms. The POA is the major source for hypothalamic LHRH, with immunocytochemically positive cell bodies projecting to such structures as the olfactory bulbs, organum vasculosum of the lamina terminalis (OVLT), arcuate nucleus, and median eminence.[20] There is evidence that estradiol exerts a feedback action on rodent LHRH neurons.[21] The subsequent action of LHRH is then to stimulate release of LH, surges of which are essential in the normal estrous cycle.[22] Thus, LHRH neurons of the POA are a critical link in the neuroendocrine axis between peripheral reproductive organs and the hypophysis. As such, they are a possible contributor in the disruptions in estrous cycling found in senescence. LHRH neurons of the POA are also implicated in senescent reproductive dysfunction by a process of elimination: although the reproductive organs and pituitary do show age-specific pathology, they appear capable of normal reproductive function if appropriately stimulated.[4-10] By default, attention is directed to the remaining link in the chain, the hypothalamus. Finally, there is evidence that POA lesions in young and adult female rodents induce anestrous (pseudopregnant) conditions similar in several dimensions to those observed in reproductively senescent, 24-month-old female rodents.[7]

The present chapter represents a first step in determining whether senescent deterioration of the POA is a critical link in senescent disruption of estrous cycling. We report here that fetal rat POA survive and contain LHRH-producing neurons when grafted to the third ventricle of reproductively senescent female rat hosts. Although data on restoration of estrous cycling in transplanted hosts have so far been only slightly encouraging, tests of such function have been cursory and limited in scope compared to our immunocytochemical assays. Preliminary studies on castrated adult male rats, however, do show a significant elevation of immunoreactive plasma LH after POA transplantation.

2. Methods

2.1. Subjects and Experimental Design

Female Long-Evans rats were obtained as retired breeders from the Zivic-Miller Company, and were kept until they reached 13–15 months of age. At that time, we began monitoring their estrous cycles using vaginal smears.[8] Measurements were taken over two separate 2-week periods spaced 2 weeks apart. Three abnormal estrous conditions were observed: irregular cycles, constant estrus, and pseudopregnancy. Rats from each condition were randomly assigned to one of three transplant treatments. Group 1 received POA grafts. Group 2 received only intraventricular injection of the carrier vehicle (Dulbecco's medium). Group 3 was untreated. Estrous cycles were again monitored 15–30 and 45–60 days postsurgery; then the rats were perfused for LHRH immunocytochemistry as described below.

Preliminary biochemical studies on plasma levels of LH were also carried out using adult male Sprague–Dawley rats castrated by the supplier (Zivic–Miller) 2 weeks prior to transplant surgery. One group of these rats received POA transplants; the other group received either a cerebellar graft or carrier vehicle (Eagle's minimum essential medium) alone. A second set of intact (noncastrate) male Sprague–Dawley rats was also included in the study. These animals received either POA or cerebellar grafts. Tail vein blood samples were taken from each subject 21, 23, 25, 50, and 54 days postsurgery. Nine weeks after transplantation, five castrate, POA recipients and three intact, cerebellar recipients were perfused for LHRH immunocytochemistry (see below).

2.2. Transplant Surgery

The transplant protocol we employed is based on the method of Gash and Sladek.[2] Briefly, female rats, 17 days pregnant on the day of surgery, were cesarean delivered, with the pups removed together *in utero* to a beaker on ice. Long–Evans pups were used for Long–Evans recipients, and Sprague–Dawley pups for Sprague–Dawley recipients. One at a time, each fetus was cut free from the uterus and embryonic sac, its brain rapidly removed, and the POA dissected under a microscope based on visual landmarks. Each POA was then suspended in a drop of either Dulbecco's medium or Eagle's minimum essential medium, cut into four to six smaller pieces, aspirated into a 20-gauge spinal needle (along with 15–20 μl of vehicle), and stereotaxically injected into the midline third ventricle of a 4% chloral hydrate-anesthetized recipient. All dissection procedures were conducted on a 1-cm-thick gelatin bed mounted on a cold plate. For any given animal, the entire procedure, from dissection of the donor pup to stereotaxic implantation of the graft, was accomplished in a maximum of 9 min. Fetuses not used within 1 hr of cesarean section were discarded.

2.3. LHRH Immunocytochemistry

Nine weeks after transplant surgery, recipient rats were transcardially perfused, and their brains removed and processed for visualization of LHRH somata and neurites as previously described.[20] Briefly, serial 75μm sections were cut in the frontal plane using a vibrating microtome, and were treated for immunocytochemistry of LHRH by the procedure of Grzanna *et al.*[24] Antiserum L1, kindly supplied by R. Benoit and R.

Guillemin, was used at a concentration of 1:40,000. Incubation of sections in the antisera was for 24 hr. Goat anti-rabbit immunoglobulin (Antibodies, Inc.) was then employed at a dilution of 1:20. For final staining, peroxidase–antiperoxidase complex was used at a concentration of 1:50, followed by a 33 mg% solution of 3,3′-diaminobenzidine tetrahydrochloride (Sigma Chemical Co.) with 0.01% H_2O_2 in Tris-HCl buffer (pH 7.2). After staining, sections were mounted on subbed slides, air-dried overnight, cleared in xylene, and mounted with Permount. Specificity of the serum for LHRH was determined by testing the ability of 2.5 μg synthetic LHRH added to 1 ml diluted serum (as indicated above) to block completely all staining.

2.4. LH Radioimmunoassay

Tail vein blood samples were radioimmunoassayed for plasma LH using NIADDK kits provided by the National Hormone and Pituitary Program.

3. Results and Discussion

3.1. LHRH Immunocytochemical Viability

Of 11 POA recipients examined, 10 showed clearly viable, nonnecrotic hypothalamic grafts. Of three cerebellar recipients examined, two showed clearly viable, nonnecrotic cerebellar grafts. Successful implantation was observed in castrate males, intact males, and reproductively senescent females, the three experimental conditions surveyed immunocytochemically. Ocasionally, as many as three pieces of hypothalamus could be found grafted to different places in the same host. In all but two cases, where the graft had lodged and attached to the midline lateral ventricle (Fig. 1), transplants were attached to the roof, floor, and/or sides of the third ventricle (Fig. 2). There did not appear to be any consistent rostral–caudal preference for implantation. Transplants were well vascularized, with large and small blood vessels evident in many sections (Fig. 3). Damage to surrounding host tissue as a result of transplant surgery was seldom observed.

Two forms of transplant–host attachment were evident. At some point, all grafts firmly fused to the host ventricle such that a substantial portion of the wall of the ventricle was provided by the transplant. In these cases, the transplant merged almost completely with the host, leaving only a faint outline of the previous ventricular border (Fig. 4). Interestingly, in these cases a new ependymal layer apparently developed. This suggests several possibilities. It could be that ependymal cells from the host are induced to grow around the ventricle border of the transplant to provide a new ependymal lining. Alternatively, the grafted tissue may develop ependymal cells wherever it is in contact with the ventricle.

In many sections rostral or caudal to firm implantation, the transplant appeared to float in the ventricle, attached only by thin tissue bridges (Fig. 5). Occasionally, LHRH-positive fibers could be observed traversing into and out of the transplant via one of these

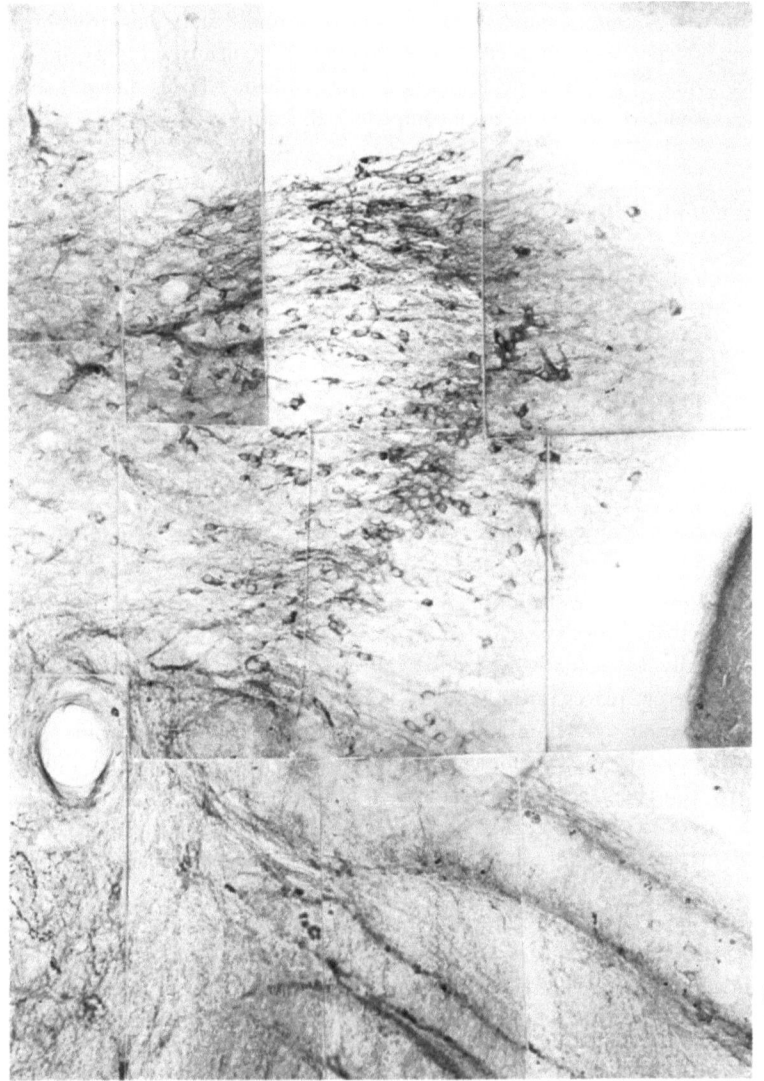

FIGURE 1. Photomontage of a fetal hypothalamic transplant attached to the roof of the lateral ventricle of a 16-month-old female host rat. A second hypothalamic graft (not illustrated) was found in the third ventricle of this subject, considerably caudal to the lateral ventricle implant. This suggests that transplants may be carried some distance, perhaps by flow of CSF within the ventricles, before they lodge and implant. LHRH stain, frontal section.

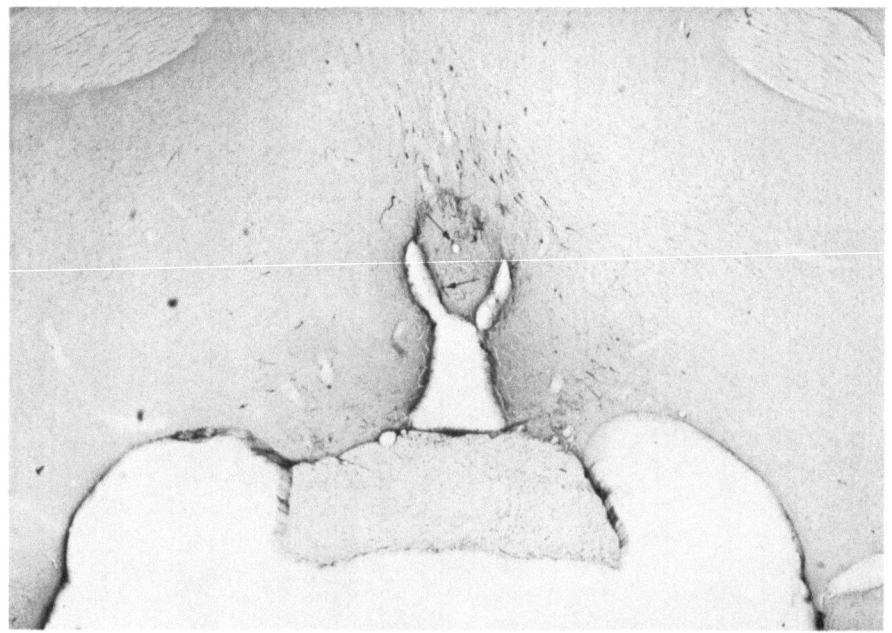

tissue bridges (Fig. 6). Whether the bridges are provided by the host or the transplant cannot now be determined.

All hypothalamic grafts exhibited LHRH-positive cell bodies (Fig. 7) with the bipolar, fusiform shape typical of LHRH neurons of the POA.[23] Varicose, LHRH-positive fibers were also present in all hypothalamic grafts as well as in the parenchyma of the host brain in the majority of sections containing grafts. In some sections, the fibers appeared to whorl within the transplant as if confined there (Fig. 8). In many other sections, however, numerous fibers could be seen crossing the host–transplant border (Fig. 9). Indeed, in a few favorable sections, LHRH-positive axons could be traced from their LHRH-positive cells of origin within the transplant out across the host –transplant border and well into the surrounding host tissue. Figures 10 and 11 show such a cell body and fiber, whereas in Fig. 6 an LHRH-positive cell in the host tissue is seen sending its axon into the transplant. Thus, it is clear that POA grafts are to some extent becoming structurally integrated with the host brain in the sense that they send afferents to the host and receive efferents.

In addition to developing an ependymal lining at appropriate places, hypothalamic transplants also appear to develop another feature appropriate to their new location. One of the major trajectories for LHRH fibers from the POA to the median eminence is along the wall of the third ventricle, just under the layer of ependymal cells.[20] Like

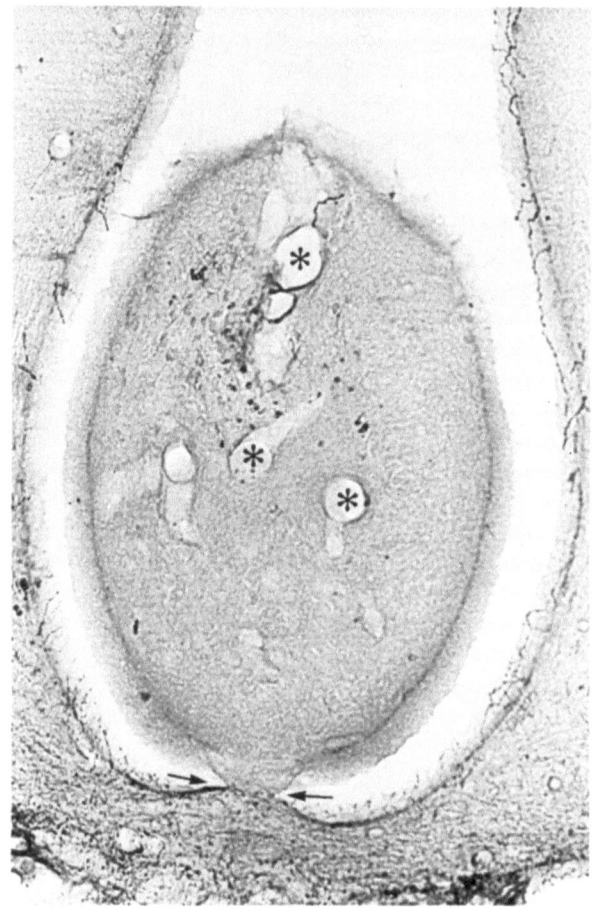

FIGURE 3. Typical vascularization of a fetal hypothalamic transplant (15-month-old female host). At this level, the graft appears barely attached (arrows) to the floor of the third ventricle. However, more caudally, it fused with the host. The presence of large blood vessels (*) in minimally attached areas of the transplant suggests that blood vessels may actively grow throughout the transplant to vascularize regions that are poorly fused with the host. LHRH stain, frontal section.

fibers from the host POA, fibers from POA grafts also tend to take this subependymal pathway (cf. Figs. 2, 7, 8, and 9).

3.2. Functional Effects of POA Grafts

3.2.1. Estrous Cycling

As female rodents age, they go through three distinct stages of reproductive impairment.[8] In the first stage, estrous cycles become irregular in onset and duration. After a

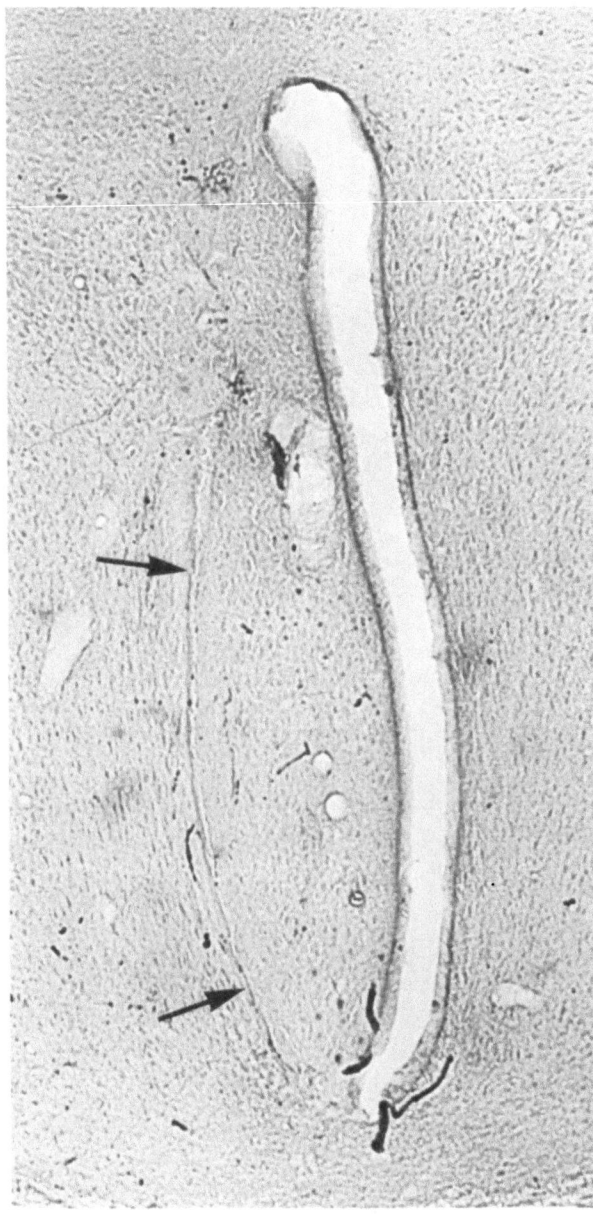

FIGURE 4. Fetal hypothalamic transplant fully fused to the lateral wall of the third ventricle (15-month-old female host). Note how the ependymal lining of the ventricle apparently remains continuous from host to transplant rather than following the former line of host ependymal cells (arrows). LHRH stain, frontal section.

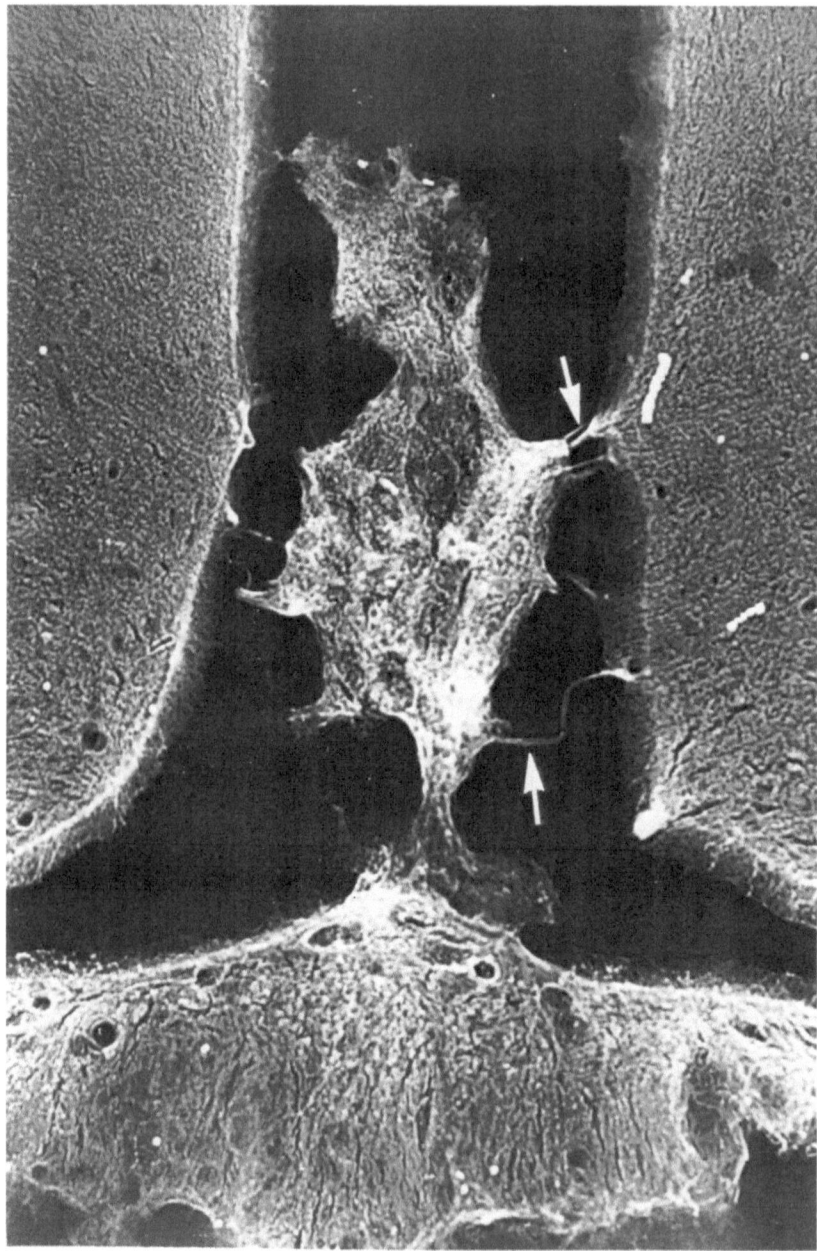

FIGURE 5. Section of a fetal hypothalamic transplant loosely attached to the third ventricle in an adult male castrate host. Thin tissue bridges (arrows) appear to stabilize transplants in such cases. LHRH stain, dark-field image, frontal section.

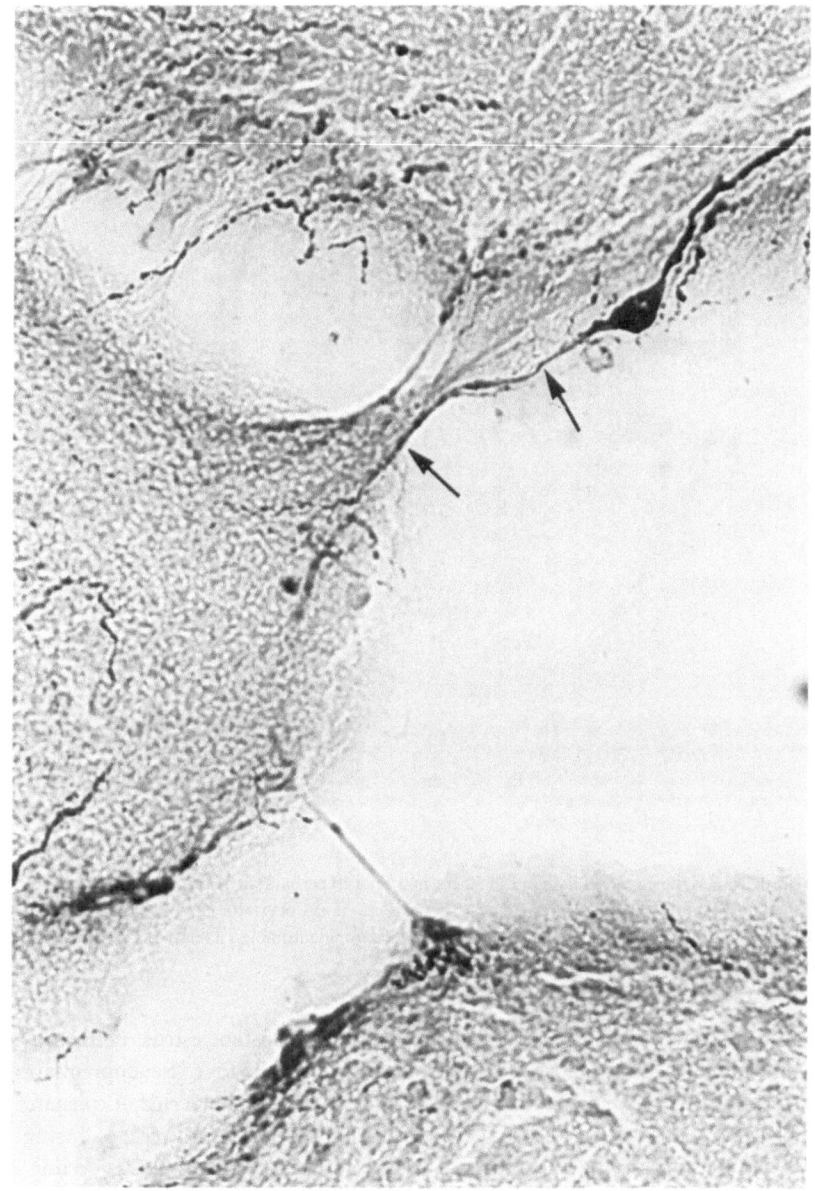

FIGURE 6. Thin tissue bridges between transplant and host may not only stabilize the graft, but also provide a potential route of entry and egress for vasculature and neurites. Here, an LHRH-positive cell body from a 16-month-old female host is seen sending a neurite (arrows) into the transplant via a tissue bridge. LHRH stain, frontal section.

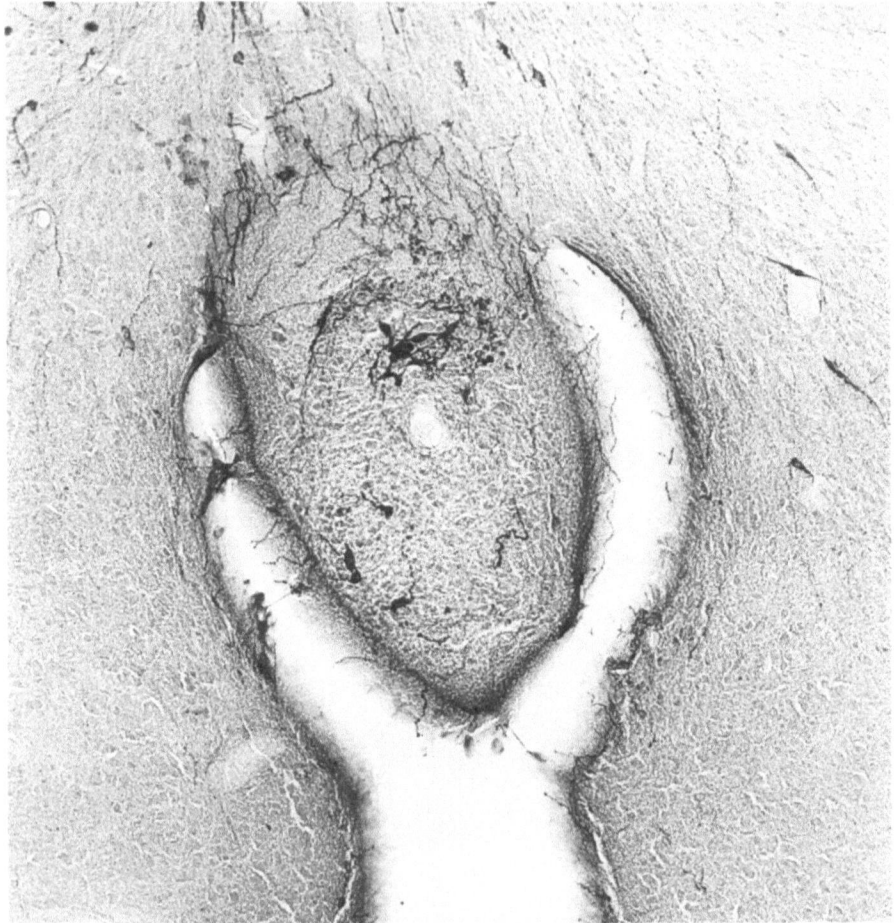

FIGURE 7. Typical LHRH-positive cell bodies and fibers in a fetal hypothalamic transplant. The host is a 16-month-old female rat. As is true for normal host tissue, some areas of grafts exhibit many LHRH-positive cell bodies (cf. Fig. 1) whereas other areas have few, if any, identifiable LHRH cell bodies in the plane of section. LHRH stain, Nomerski optics, frontal section.

few weeks to a few months of irregular cycling, a state of constant estrus, defined as vaginal cornification for 15 or more consecutive days, is reached. A pseudopregnant state, characterized by persistent diestrous vaginal smears, follows the period of constant estrus. As expected, reproductive function of control rats worsened, on average, during the 3 months between initial baseline and final postsurgery measures of estrous cycling. POA transplant recipients on the other hand, improved by an average of one stage (i.e., from pseudopregnancy to constant estrus, constant estrus to irregular cycling, or irregular cycling to regular cycling) over the same time period. A one-tailed Mann–Whitney U test of these differences proved significant at the $p < 0.05$ level. It should be cautioned, however, that only 2 of 14 POA recipients actually maintained regular estrous cycles for 4 weeks or more postsurgery. Other factors clearly need to be considered in order to achieve more complete and long-lasting restoration of reproductive function. For example, catecholamine reinitiation of estrous cycling[11,12] might be considerably more powerful given a POA-transplanted subject. In fact, there is no reason why catecholamine application could not itself be accomplished by transplant, with nuclei of catecholamine-containing neurons transplanted at the same time as POA grafts.

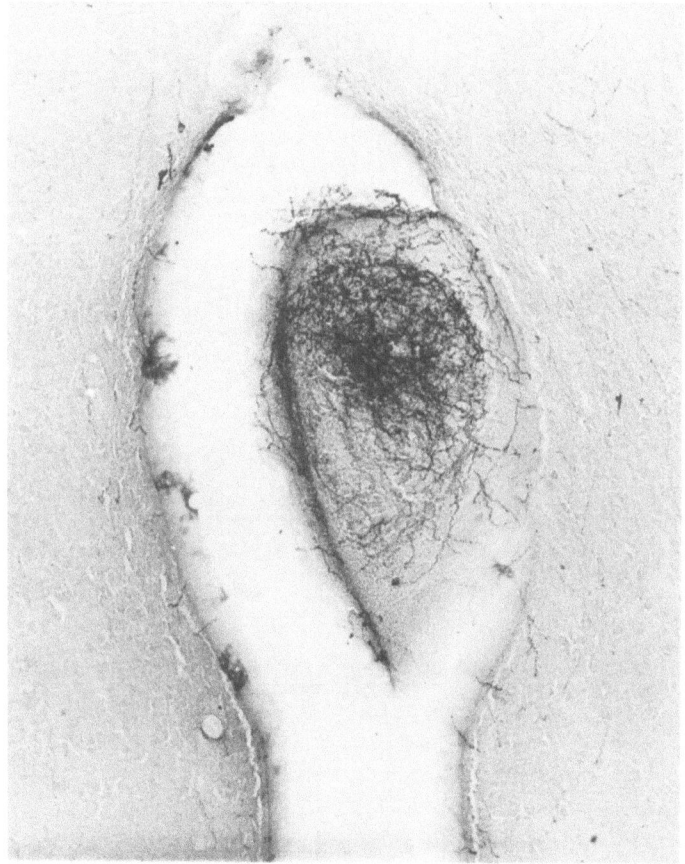

FIGURE 8. In some areas of transplants, LHRH-positive fibers appear to form whorls within the graft as if confined there. Inability to penetrate host tissue may be due to scarring of portions of the ventricular wall. However, this situation appears to be the exception rather than the rule. For example, Fig. 2, a more rostral section from the same transplant, clearly shows positive fibers crossing the host–transplant border. Numerous other instances of neurites freely entering and leaving grafts were observed (cf. Figs. 1, 2, 6, 9, and 10), and were more common than the whorling pattern evident in the present section. LHRH stain, frontal section.

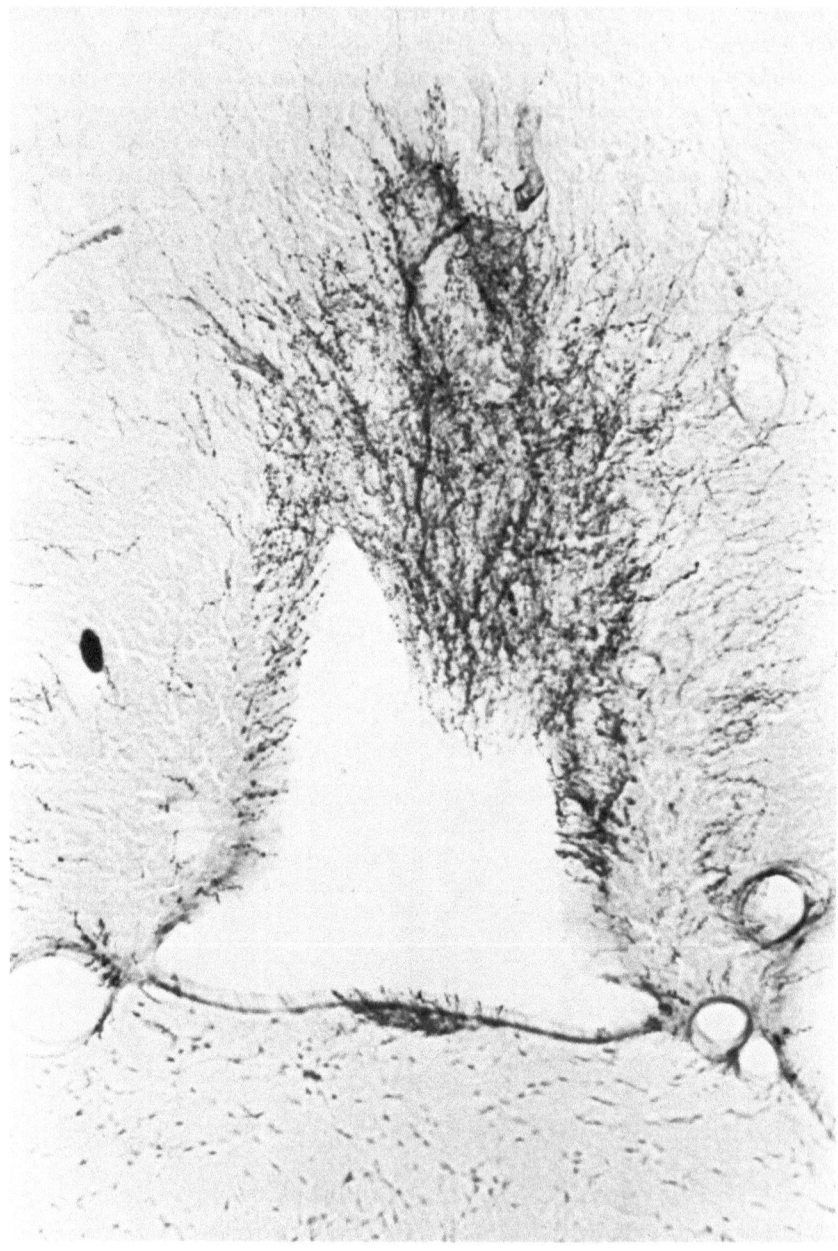

FIGURE 9. Dense plexus of LHRH-positive fibers entering and/or leaving a fetal hypothalamic graft lodged in the roof of the third ventricle. The host is a 16-month-old female rat. LHRH stain, frontal section.

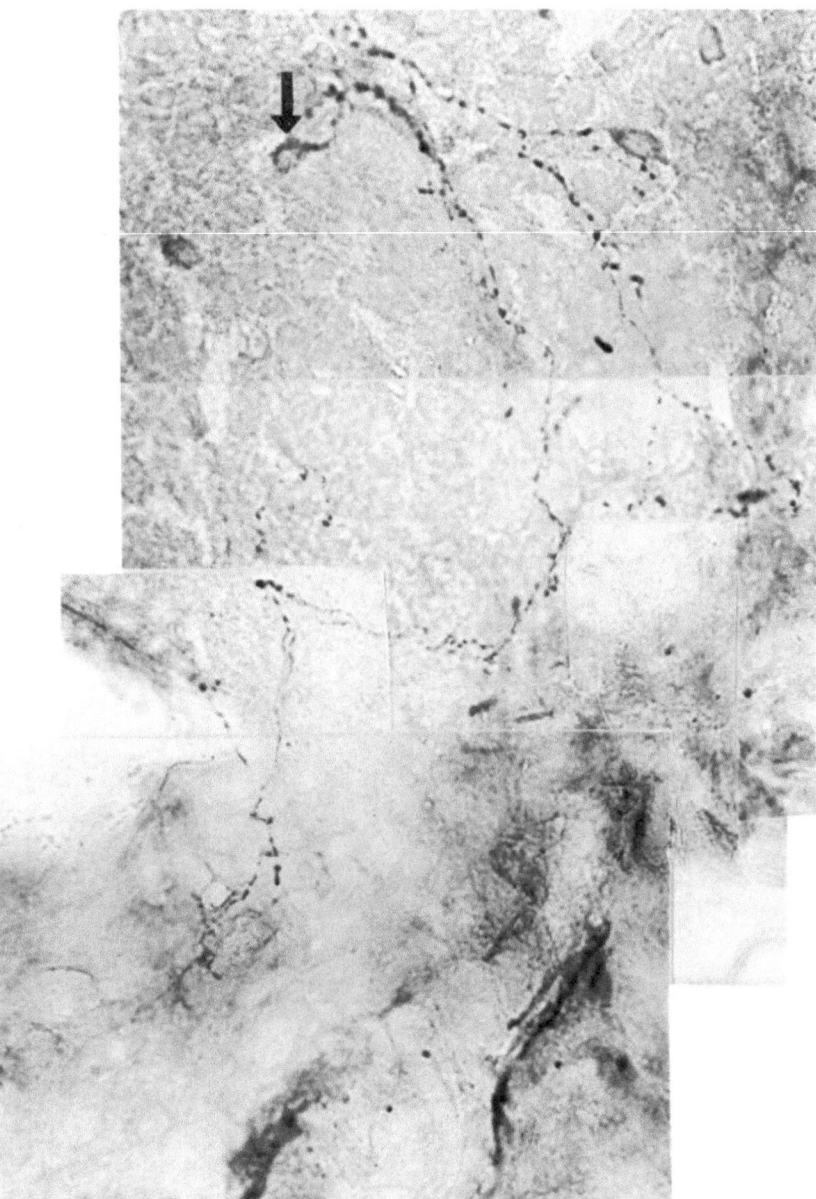

FIGURE 10. Photomontage of micrographs tracing the trajectory of an LHRH-positive fiber from a grafted cell body in the transplant out into host tissue. LHRH stain, frontal sections.

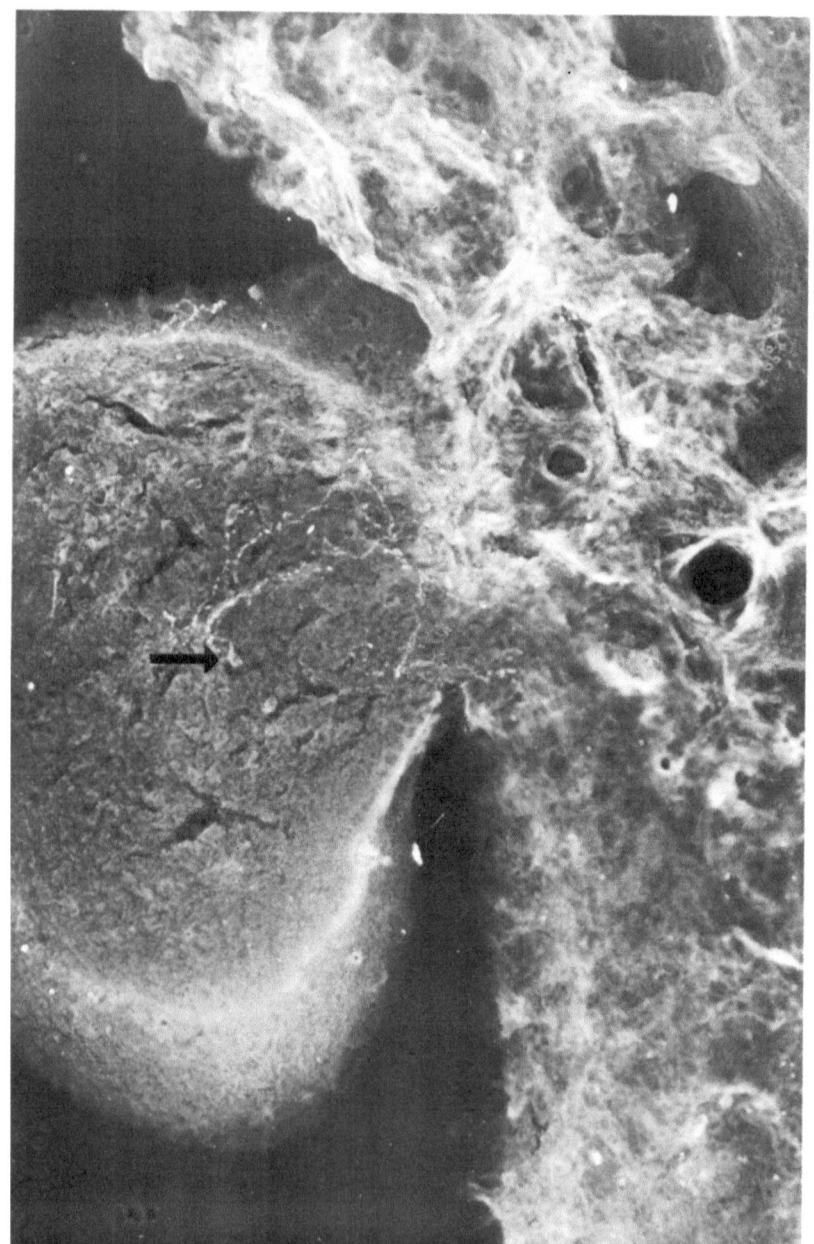

FIGURE 11. Lower-magnification, dark-field image of a portion of the field illustrated in Fig. 10. The same neuron seen in Fig. 10 is indicated (arrow) for orientation. The position of the transplant in relation to the floor of the third ventricle is evident. The host is an adult male castrate rat. LHRH stain, frontal section.

3.2.2. Plasma Immunoreactive LH Levels

Plasma LH levels of castrate POA rats averaged 749 $\pm$ 77 ng/ml in tail vein samples taken 14 and 31 days after transplant surgery. Castrate cerebellar and vehicle-injected controls averaged 537 $\pm$ 49 ng/ml during this time, significantly less ($t = 2.42$, $p < 0.02$) than POA recipients. These data replicated an earlier pilot study in which castrate POA rats averaged 1218 $\pm$ 120 ng/ml LH compared to 701 $\pm$ 148 ng/ml LH for castrate control rats. However, in the present study, plasma LH levels of castrate POA rats fell to values comparable to controls from 4 to 8 weeks postsurgery, so that the combined data for all five measures of LH did not differ significantly between the two groups for the full 60 days after transplantation. Whether this decline in LH of POA rats during the second month after surgery reflects deterioration of the transplant or compensatory changes in other parts of the neuroendocrine axis remains unclear. Given the immunocytochemical identification of LHRH neurons in the POA grafts, the former possibility, deterioration of transplants, seems less likely than a compensatory response.

As with studies of estrous cycling, we believe that challenging the transplants (e.g., with catecholamine agonists) may reveal sharper contrasts in plasma LH levels. Having taken the necessary first step, demonstrating viability of LHRH neurons in POE transplants, we now look forward to pursuing such functional correlates as restored estrous cycling and restored cyclic fluctuations of reproductive hormones.

ACKNOWLEDGMENTS. We thank Dr. Don Gash for his patience and generosity in sharing with us details of his transplant methods. Reproductively senescent female rats were kindly provided by Dr. Mahlon Wilkes, then at the University of California, San Diego. Dr. Michael Silver assisted with transplant surgery in a preliminary experiment.

References

1. Jaeger, C. B., and Lund, R. D., 1980, Transplantation of embryonic occipital cortex to the brain of newborn rats, *Exp. Brain Res.* **40**:265.
2. Gash, D., and Sladek, J. R., 1980, Vasopressin neurons grafted into Brattleboro rats: Viability and activity, *Peptides* **1**:11.
3. Björklund, A., Stenevi, U., Dunnet, S. B., and Iversen, S. D., 1981, Functional reactivation of the deafferented neostriatum by nigral transplants, *Nature (London)* **289**:497.
4. Finch, C. E., 1976, Endocrine and neural factors of reproductive aging—A speculation, in: *Aging,* Vol. 3 (R. D. Terry and S. Gershon, eds.), pp. 335–338, Raven Press, New York.
5. Finch, C. E., 1976, The regulation of physiological changes during mammalian aging, *Q. Rev. Biol.* **51**:49.
6. Finch, C. E., Felicio, L. S., Flurkey, K., Gee, D. M., Mobbs, C., Nelson, J. F., and Osterburg, H. H., 1980, Studies on ovarian–hypothalamic–pituitary interactions during reproductive aging in C57B1/6J mice, *Peptides* **1**(Suppl.):163.
7. Clemens, J. A., and Bennet, D. R., 1977, Do aging changes in the preoptic area contribute to loss of cyclic endocrine function?, *J. Gerontol.* **32**:19.
8. Ingram, D. L., 1959, The vaginal smear of senile laboratory rats, *J. Endocrinol.* **19**:182.
9. Talbert, G., and Krohn, P., 1966, Effect of maternal age on viability of ova and uterine support of pregnancy in mice, *J. Reprod. Fertil.* **11**:399.

10. Peng, M. T., and Huang, H., 1972 Aging of hypothalamic–pituitary–ovarian function in the rat, *Fertil. Steril.* **23**: 535.
11. Linnoila, M., and Cooper, R. L., 1976, Reinstatement of vaginal cycles in aged non-cycling female rats, *J. Pharmacol. Exp. Ther.* **199**:477.
12. Huang, H., and Meites, J., 1975, Reproductive capacity of aging female rats, *Neuroendocrinology* **17**:289.
13. Fuxe, K., Hokfelt, T., and Nilsson, O., 1967, Activity changes in the tubero-infundibular dopamine neurons of the rat during various states of the reproductive cycle, *Life Sci.* **6**:2057.
14. Sawyer, C. H., Hilliard, J., Kanematsu, S., Scaramuzzi, R., and Blake, C. A., 1974, Effects of intraventricular infusions of norepinephrine and dopamine on LH release and ovulation in the rabbit, *Neuroendocrinology* **15**:328.
15. Walker, R. F., 1981, Reproductive senescence and the dynamics of hypothalamic serotonin metabolism in the female rat, in: *Aging,* Vol. 17 (S. J. Enna, T. Samorajski, and B. Beer, eds.), pp. 95–106, Raven Press, New York.
16. Steger, R. W., Sonntag, W. E., Van Vugt, D. A., Forman, L. J., and Meites, J., 1980, Reduced ability of naloxone to stimulate LH and testosterone release in aging male rats: possible relation to increase in hypothalamic Met^5-enkephalin, *Life Sci.* **27**:747.
17. Harms, P. G., Ojeda, S. R., and McCann, S. M., 1973, Prostaglandin involvement in hypothalamic control of gonadotropin and prolactin release, *Science* **181**:760.
18. Krulich, L., 1979, Central neurotransmitters and the secretion of prolactin, GH, LH, and TSH, *Annu. Rev. Physiol.* **41**:603.
19. Vijayan, W., and McCann, S. M., 1978, Re-evaluation of the role of catecholamines in control of gonadotropin and prolactin release, *Neuroendocrinology* **25**:150.
20. Hoffman, G. E., and Gibbs, F. P., 1982, LHRH pathways in rat brain: 'Deafferentation' spares a subchiasmatic LHRH projection to the median eminence, *Neuroscience* **7**:1979.
21. Yen, S. C., 1977, Regulation of the hypothalamic–pituitary–ovarian axis in women, *J. Reprod. Fertil.* **51**:181.
22. Shirley, B. Wolinsky, J., and Schwartz, N. B., 1968, Effects of a single injection of an estrogen antagonist on the estrous cycle of the rat, *Endocrinology* **82**:959.
23. Hoffman, G. E., Wray, S., and Goldstein, M., 1982, Relationship of catecholamines and LHRH: Light microscopic study, *Brain Res. Bull.* **9**:417.
24. Grzanna, R., Molliver, M. E., and Coyle, J. T., 1978, Visualization of central noradrenergic neurons in thick section by the unlabeled antibody method: A transmitter-specific Golgi image, *Proc. Natl. Acad. Sci. USA* **75**:2502.

9

Brain Tissue Transplants and Reproductive Function

Implications for the Sexual Differentiation of the Brain

GARY W. ARENDASH AND ROGER A. GORSKI

1. Introduction

In many species, there are marked sex differences in the neural control of a variety of endocrine and behavioral processes. These sex differences in CNS function in the adult are due, in large part, to the hormonally induced sexual differentiation of the brain during development; in fact, the mammalian brain is inherently female or at least bipotential. Specifically, sexual differentiation of the neural mechanisms controlling reproductive function in mammals, with respect to both sexual behavior and gonadotropin secretion, results from exposure of the brain of the male to his testicular hormones during a restricted period of brain development.[1,2] In the rat, this "critical period" for sex steroid action apparently begins several days before birth and extends into the early postnatal period.[3,4] In contrast, it is the absence in the female of testicular secretions during the critical period that permits the development by adulthood of a neural substrate that subserves feminine behavior and the cyclical pattern of luteinizing hormone (LH) secretion necessary for ovulation. If, however, the inherently female brain is exposed to exogenous testicular hormones (i.e., testosterone) during the critical period, sexual differentiation (masculinization) of the brain will occur, resulting in the development of masculine sexual behavioral potential and a noncyclic or tonic pattern of LH

GARY W. ARENDASH ● Department of Biological Sciences, University of South Florida, Tampa, Florida 33620. ROGER A. GORSKI ● Department of Anatomy and Laboratory of Neuroendocrinology of the Brain Research Institute, University of California School of Medicine, Los Angeles, California 90024.

secretion during adulthood.[1,2] It must be emphasized that considerable evidence suggests that, in the rat, testosterone does not act directly to masculinize the brain, but rather that it is aromatized within steroid-sensitive regions to estrogen intraneuronally, which then acts to masculinize brain function.[4]

In the rat, it appears likely that estrogen (produced by the aromatization of testosterone) is acting at least in part on the medial preoptic area (MPOA) of the brain to bring about sexual differentiation as: (1) this brain area concentrates label after the administration of [³H]testosterone to the newborn rat[5]; (2) testosterone implants into the MPOA of neonatal females selectively increase masculine sexual behavior[6]; and (3) the MPOA is of critical importance for the expression of masculine sexual behavior in the adult male rat[7,8] and for normal ovarian cyclicity in the adult female rat.[9,10] The relatively recent demonstration of a marked structural sex difference in the MPOA—the sexually dimorphic nucleus of the preoptic area (SDN-POA)—dramatically reinforces this view.[11] The SDN-POA of the male rat is some three-to fivefold larger in volume and is comprised of more neurons than is the corresponding nucleus of the female (Fig. 1). Moreover, the adult SDN-POA is influenced by the hormone environment perinatally[12]; in fact, prolonged hormone treatment during this period can completely sex-reverse (i.e., masculinize) the SDN-POA of the female.[13] In intact rats, the sex difference in the volume of the SDN-POA develops gradually during the first 10 days of postnatal life,[14] essentially the same period during which steroids influence the functional sexual differentiation of the brain. Thus, the SDN-POA represents a distinct morphological signature of gonadal steroid action on the developing brain.

Although the developmental importance of testosterone (and/or estrogen) on sexual differentiation of the brain is well documented, the exact mechanism whereby gonadal steroids "organize" or program the development of the brain is currently unknown. In this regard, we believe that the SDN-POA can provide an important neural model for

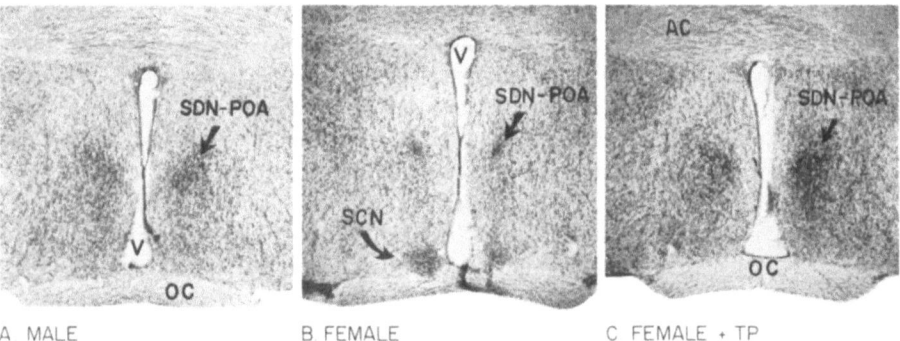

A MALE   B FEMALE   C FEMALE + TP

FIGURE 1. The sexually dimorphic nucleus of the preoptic area (SDN-POA) of the (A) male and (B) female rat, and (C) from a genetic female who was exposed to testosterone propionate (TP) daily for a prolonged period perinatally (See Ref. 13). AC, anterior commissure: OC, optic chiasm; SCN, suprachiasmatic nucleus; V, third ventricle. Thionin stain; A, B, and C at same magnification. Reprinted from Ref. 2 with permission.

study. Although the specific function(s) of the SDN-POA is currently unknown, the nucleus is easily identified histologically, which facilitates neuroanatomical, cytochemical, and neurochemical studies. The dependence of its development on steroid hormones which can be administered exogenously provides a unique opportunity to study developmental processes irrespective of the eventual function of these neurons. Moreover, the location of the SDN-POA in constant relationship to other landmarks, e.g., the optic chiasm, anterior commissure, and third ventricle, makes it possible to remove the SDN-POA by the punch technique for biochemical studies. It is this last fact, coupled with the survival of neural transplants in recipient animals, that led us to attempt to transplant the SDN-POA of the neonatal male into the MPOA of littermate females. This approach if successful, would permit us to investigate the influence of gonadal steroids on such transplanted tissue including its possible differentiation. Moreover, the clear differences in reproductive behavioral and hormonal secretion between adult male and female rats make these systems excellent ones in which to test the ability of transplanted tissue to establish functional connections with the host brain. By utilizing cross-sex transplants, transplant-induced changes in reproductive processes should be apparent, if present. The results of our initial studies are very promising and do support the conclusion that transplanted brain tissue can indeed make behaviorally effective connections with the host's brain.

2. Transplantation of Male Brain Tissue into Female Recipients

2.1. Intraparenchymal Transplants

Although there have been recent attempts to transplant cell suspensions of neural tissue,[15] most transplant studies have placed donor tissue into the ventricular system of the host, presumably in an attempt to promote graft viability. We argued, however, that functional connections might be more readily formed if the transplant was placed intraparenchymally, directly in contact with host brain tissue. In the case of the SDN-POA, for example, we considered it important to transplant male tissue as close as possible to the developing SDN-POA of the female. Thus, we developed an intraparenchymal transplantation technique that involves the use of a small metal cannula to punch out the appropriate brain tissue from a fresh slice of donor brain, and the extrusion of this transplant tissue from the cannula after it has been stereotaxically positioned into the appropriate region of the recipient rat's brain. The technique is highly versatile and has the advantage that: (1) discrete brain areas or subareas can be accurately punched out and (2) donor tissue can be transplanted stereotaxically into any brain location of a recipient animal. This latter fact permits the investigator to place the transplant into what he assumes to be the most natural environment for the development of functional connectivity.

Nevertheless, the present studies, at least, face one caveat. The SDN-POA is relatively small, and to ensure that the transplants included this nucleus, a larger region of the MPOA was punched and transplanted. Although evidence will be presented

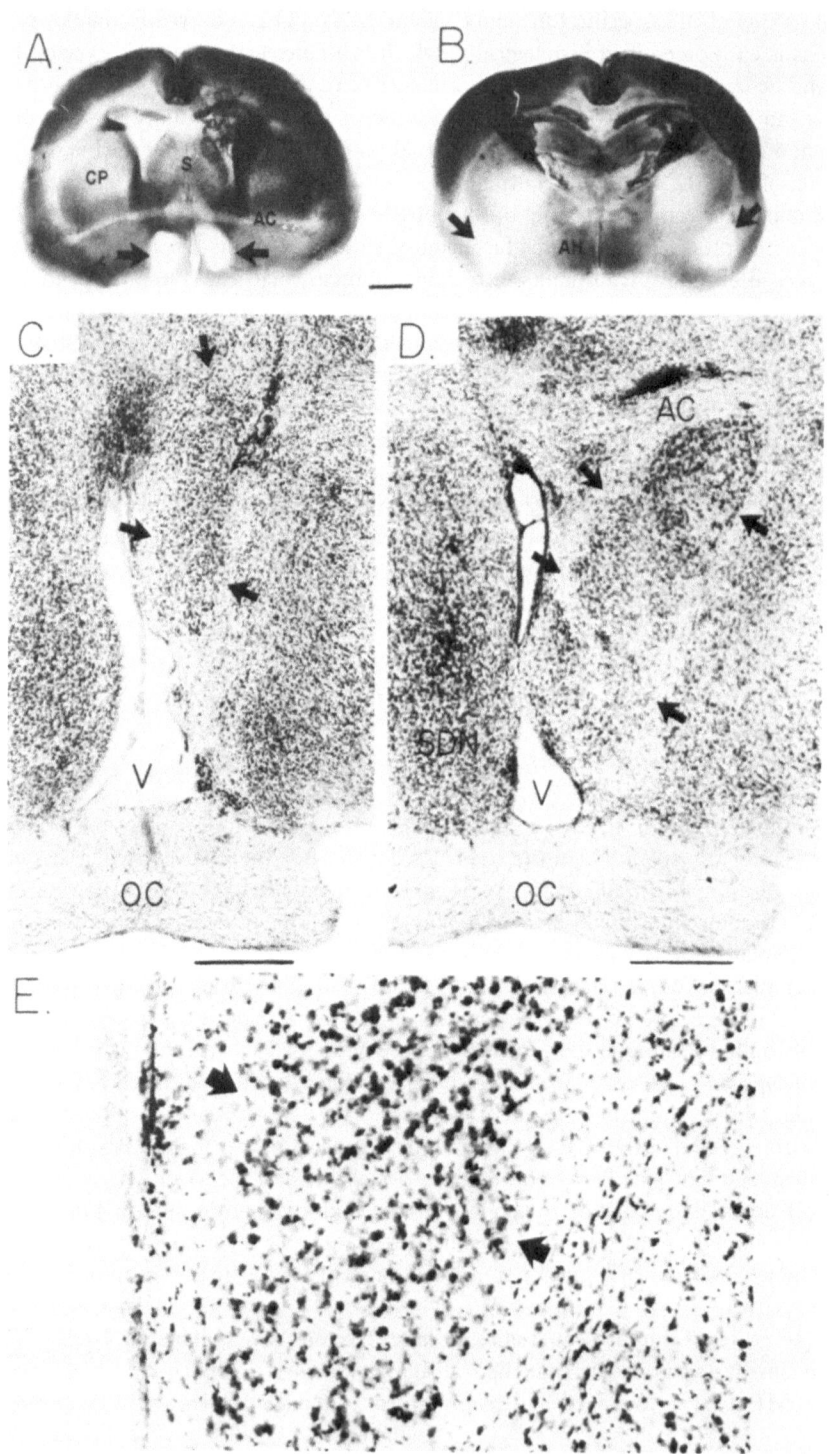

below (Section 6) that the SDN-POA did survive, at the present time we cannot determine that the functional effects of such transplants can be attributed to the neurons of the SDN-POA specifically vis-a-vis other MPOA neurons included in the transplant. In recognition of this uncertainty, we will refer only to the effects of transplanting MPOA tissue rather than to specific effects of surviving neurons of the donor's SDN-POA. Future modifications of techniques may permit the transplantation of only SDN-POA tissue and/or the identification of those neurons that establish functional connections.

2.2. Transplantation Protocol and Graft Viability

Since the MPOA, which contains the SDN-POA, is critical for the expression of male sexual behavior and is a prime target for gonadal steroid actions resulting in CNS sexual differentiation, we felt that new knowledge could be obtained by following the development of immature male MPOA tissue after transplantation into the MPOA of female recipients and by challenging the ability of such male MPOA transplants to alter the recipient female's reproductive functions. During the first five postnatal days, MPOA neurons are relatively undifferentiated and make few synaptic connections.[16,17] Indeed, the major period of cytoplasmic differentiation and synaptic formation in the rat MPOA appears to occur after the critical period for sexual differentiation. Therefore, the MPOA of 1-day-old male Sprague/Dawley rats, containing such undifferentiated neurons, was bilaterally punched out of the appropriate brain slice using a 0.7-mm-diameter metal cannula (Fig. 1A) and stereotaxically implanted bilaterally into the MPOA of 1-day-old female rats (See Ref. 18 for further details). Additional females received transplants of amygdala or caudate nucleus tissue from male donors or were sham-operated. Immediately following surgery, all recipients were injected subcutaneously with either testosterone propionate (TP; 8 μg) or oil vehicle (0.5 ml).* After ovariectomy in adulthood, recipients were tested for female sexual behavior, then implanted subcutaneously with a testosterone-filled capsule for subsequent male behavioral testing (see Ref. 18 for additional details). Following all testing, histological analysis of recipient brains indicated excellent transplant survival rates; with 89% of MPOA (N = 54), 100% of amygdala (N = 32), and 88% of caudate nucleus (N = 8) surviving the 6-month study, as indicated by the normal appearance of thioninstained neurons within the grafts (Fig. 2).

*In planning the studies of Sections 4 and 5, it was initially thought that a small dose of TP might be necessary to provide a more suitable hormonal milieu for normal masculine differentiation of neurons in the male tissue transplants. The results indicate, however, that this neonatal TP treatment was not necessary for transplant survival or the observation of transplant-induced reproductive modifications.

←——————————————————————————————

FIGURE 2. (A, B) Representative coronal brain slices from neonatal donors of (A) medial preoptic area (MPOA) and (B) amygdala tissue. Arrows indicate the tissues punched out and transplanted into neonatal female recipients. Bar = 1 mm. (C, D) Photomicrographs of representative sections through the MPOA of two female recipients implanted neonatally with male MPOA (C) or amygdala (D) tissues. Arrows indicate the transplants. Bars = 0.5 mm. Reprinted from Ref 18 with permission

2.3. Effects on Masculine Sexual Behavior

Analysis of experimental data indicated substantial transplant-induced modifications in masculine behavior. Those females that received bilateral male MPOA transplants neonatally showed a dramatic enhancement in adulthood levels of masculine sexual behavior (Fig. 3). A two-way analysis of variance for repeated measures revealed a highly significant group effect ($p < 0.0005$) as well as a significant measures effect ($p < 0.0005$) in a four-test sequence for copulatory activity. Female recipients with bilat-

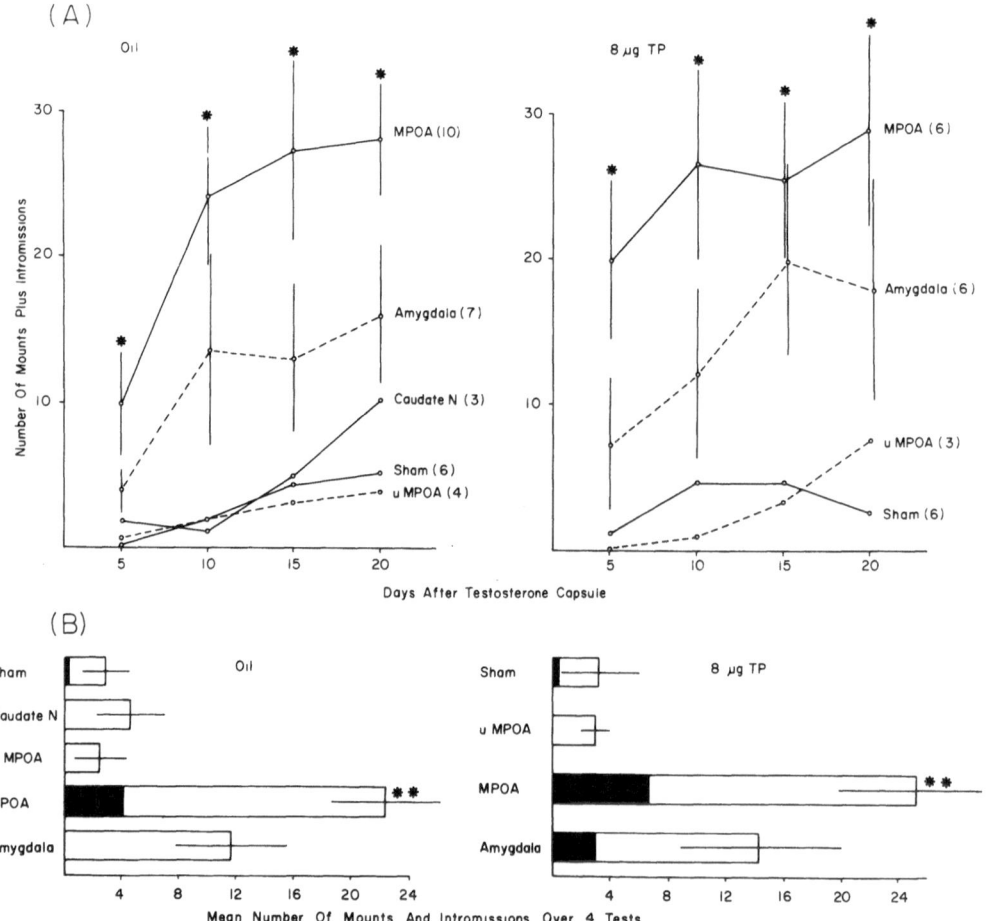

FIGURE 3. (A) Masculine sexual behavior shown during four tests by adult female rats given male brain tissue transplants and oil or 8 μg testosterone propionate (TP) as neonates. The number of animals in each group is shown in parenthesis; vertical lines are S.E. Asterisk indicates a significant difference ($p < 0.05$) from the control groups. (B) Mean number of mounts (open bers) and intromissions (closed bars) shown by these females during the tests for masculine behavior. Double asterisks indicate a significant difference from the group that received amygdala and oil ($p < 0.05$) and from all other groups ($p < 0.01$) except the one that received amygdala and TP. Modified from Ref. 18 with permission.

eral MPOA transplants showed substantially more mount and intromission responses in any single test than control females receiving caudate nucleus, sham, or unilateral MPOA transplants (Fig. 3A). Importantly, this marked enhancement of sexual behavior specifically required the use of MPOA tissue for transplantation, for only those females implanted with this brain area showed consistently elevated levels of masculine sexual behavior. Male sexual behavior in amygdala-implanted females, although elevated in a few recipients, did not differ significantly from control groups during any of the tests (Fig. 3A) or overall (Fig. 3B). Furthermore, *bilateral* transplants of MPOA directly into the female recipient's MPOA are apparently required to obtain an enhancement of masculine behavior, since unilateral MPOA transplants (i.e., only one MPOA transplant surviving into adulthood or only one transplant located within the recipient's MPOA) were ineffective. Indeed, the fact that two MPOA transplants are required to be placed "intraparenchymally" within the female's MPOA suggests not only that an "extraparenchymal" transplantation approach would have been unsuccessful in enhancing masculine sexual behavior, but also that the stereotaxic technique we used to position the transplanted MPOA tissue was crucial to the behavioral enhancement observed.

From the above behavioral results, we may conclude that male MPOA tissue can apparently develop functional connections with the recipient female brain, resulting in a dramatic enhancement in her display of male sexual behavior during adulthood. Furthermore, this behavioral enhancement was observed even if the female recipients were not given testosterone concurrently with the male MPOA transplants. Therefore, these data suggest that the 1-day-old male MPOA already has the potential to behaviorally masculinize the neonatal female brain, presumably because of the donor male's prenatal exposure to testosterone. In this regard, Weisz and Ward[19] have found that testosterone levels in males are consistently higher than those of females only on the fifth and fourth days prior to birth (i.e., at the beginning of the critical period in rats), and have suggested that this brief prenatal exposure to high testosterone levels may sensitize the developing male CNS to the masculinizing or organizing actions of testosterone circulating in relatively lower amounts (not consistently higher than females) at later stages of development. Such a previous testosterone-induced sensitization of male MPOA transplants could readily explain the enhanced male behavior observed in recipient females. Thus, the immature male MPOA neurons used for transplantation may have been programmed prenatally via testosterone exposure to express a "male" circuitry pattern that became functional when transplanted bilaterally into the appropriate neural and hormonal environment. Further support of this view requires that the transplantation of the neonatal *female* MPOA into a recipient female has no behavioral effect. Although this additional control procedure is important for any final interpretation of the mechanisms of sexual differentiation, the fact remains that MPOA transplants, and not other brain tissue, enhanced masculine behavior.

2.4. Effects of Feminine Sexual Behavior

Unexpectedly, some male brain tissue transplants increase feminine sexual behavior in the female transplant recipients.[18] Lordosis responsiveness (as measured by the lordosis quotient, LQ) after minimal priming with estradiol benzoate (s.c.; 2 μg/3 days)

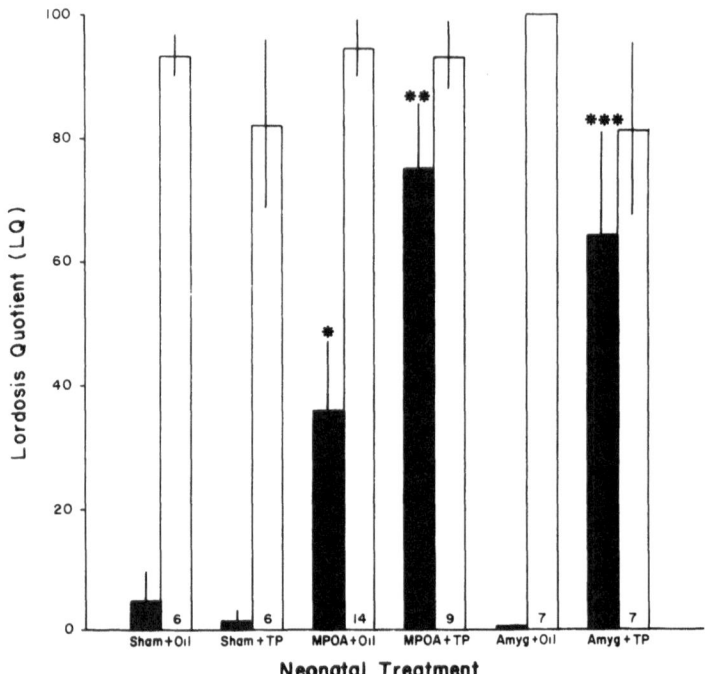

FIGURE 4. Effects of transplanting brain tissue from neonatal male into littermate females on lordotic behavior as adults following minimal estrogen priming (closed bars) or estrogen and progesterone priming (open bers). Vertical lines are S.E. Numbers at the base of the open bars indicate the number of rats in each group. Single asterisk, significantly ($p < 0.05$) different from all other estrogen-only groups treated with oil neonatally; double asterisks, significantly ($p < 0.01$) different from all other estrogen-only groups except the one receiving amygdala plus TP; triple asterisks significantly ($p < 0.01$) different from all other estrogen-only groups except those receiving MPOA transplants. Modified from Ref. 18 with permission.

was markedly enhanced by the neonatal transplantation of male MPOA or amygdala tissue in combination with 8 μg TP neonatally (Fig. 4). A significant, though less dramatic, increase in receptivity after minimal estrogen priming was also observed in females that received MPOA tissue plus oil neonatally (but not amygdala plus oil). These enhancements in female behavior seen in transplant recipients after minimal estrogen priming, although unexpected, are not unreasonable, for estrogen receptors are present in the MPOA and amygdala of male and female rat neonates.[5,20] It is possible that, in the process of transplanting these two brain regions neonatally, a sizeable number of estrogen receptors or estrogen-responsive neurons were also transferred, resulting in an enhanced MPOA estrogen receptor population in the recipient's brain. Interestingly, four of the five MPOA + oil recipients showing the highest levels of male behavior also displayed very high levels of female behavior (LQ ≥ 70) after minimal estrogen priming. Thus, an MPOA transplant-mediated enhancement of both male and female sexual behavior can be observed in the same animal. In contrast to the enhanced lordotic behavior observed in transplant recipients after estrogen treatment alone, progesterone-facilitated lordotic responsiveness was not affected by transplantation or TP treatment (Fig. 4).

3. Transplant Connectivity with Recipient Brain Tissue

The transplant-induced behavioral results presented indicate that mammalian intracerebral transplants can produce behavioral changes in recipient animals that are consistent with a normal function (i.e., sexual behavior) of the transplanted tissue (i.e., MPOA). Indeed, these results may represent the first behavioral modifications induced by mammalian brain tissue transplanted directly into normal recipient brain tissue. Although our histological examination of thionin (Nissl)-stained brain sections indicated that healthy, normal-appearing neurons were present throughout the transplanted tissue (Fig. 2), this should be verified ultrastructurally, and does not prove the existence of connections between the male transplant and the recipient female's brain. However, through a combination Golgi–Nissl histochemical staining procedure, we have been able to identify the neural transplants (via Nissl), and visualize entire transplanted neurons (via Golgi), including their dendritic-axonal processes.* Brains from 1-month-old female recipients, neonatally implanted with male MPOA tissue, were histologically processed for combination Golgi–Nissl staining according to the method of Ramon-Moliner et al.[21] A representative Golgi–Nissl-stained brain section of the interface between a male MPOA transplant and the host female's MPOA is shown in Fig. 5. Dendritic processes can be seen extending from cell bodies on both sides of the transplant–host interface and, most importantly, extensive networks of neuronal processes traverse the interface. Such neurohistological preparations of transplant tissue serve to verify the existence of neuronal connectivity between male transplant tissue and the recipient female brain, thus providing a possible neuroanatomical basis for the transplant-induced behavioral modifications discussed in this chapter.

4. Hormonal Effects Induced by Transplanted Male Brain Tissue

To broaden the challenge of the ability of brain tissue transplants to alter reproductive function, we tested the effects of male neural transplants into neonate females on their hormonal function during adulthood, as revealed by their patterns of vaginal cyclicity and LH secretory responses to the feedback actions of estrogen and progesterone (Arendash, Leung, and Gorski, unpublished data).

The overwhelming majority (90%) of females implanted neonatally with male MPOA or amygdala tissue showed normal vaginal cyclicity after temporally normal vaginal opening. This suggests the presence of basically intact steroid feedback mechanisms in these animals, a premise further substantiated by the observation of inhibitory and facilitatory LH secretory responses to estrogen and progesterone treatment, respectively (Fig. 6; oil-treated groups). Estradiol benzoate treatment (5 μg/100 g body wt)

---

*This neuroanatomical study, which is still in progress, is in collaboration with Dr. R. Hammer of NIMH.

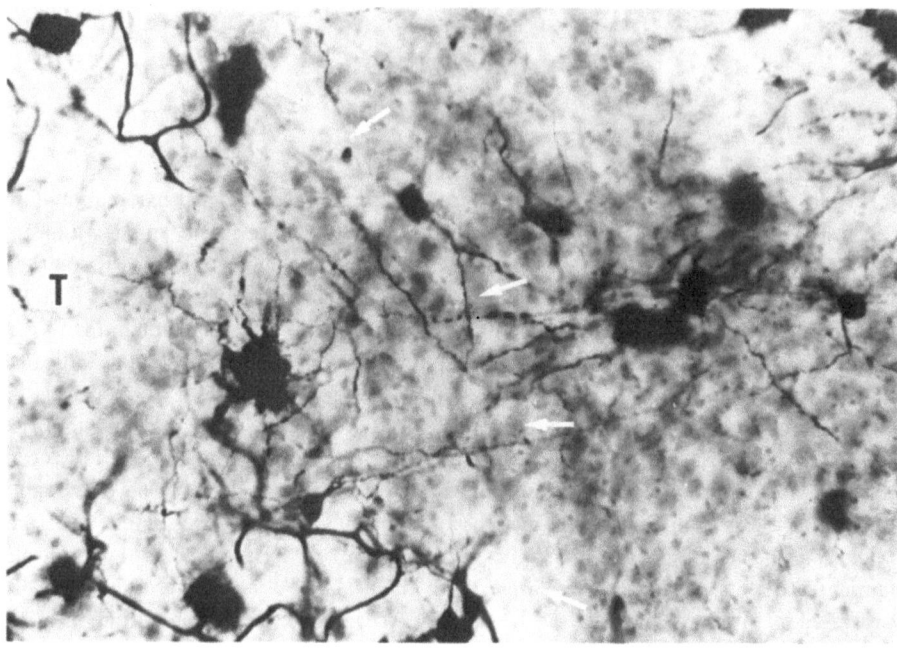

FIGURE 5. The interface between a male MPOA transplant (T) and the host's MPOA stained for both Golgi and Nissl. The white arrows indicate the approximate border between the transplant and host (to the right) tissue. Note neuronal processes traversing the transplant–host interface.

resulted in a significant suppression of blood LH levels at 6 hr that was generally maintained throughout the 64-hr post treatment sampling period. Interestingly, an increase in LH titers (probably reflective of a positive feedback action of estrogen) was observed 12 hr after estradiol benzoate treatment in sham and MPOA-implanted animals given oil neonatally (Fig. 6). However, male amygdala or caudate nucleus implants apparently blocked this early LH elevation in recipient females; this result may suggest a minor transplant-induced masculinization of LH responses to estradiol benzoate, for normal male rats, unlike normal female rats, do not show a facilitation of LH secretion after estrogen (or progesterone) administration.[22]

The administration of 8 μg TP neonatally to females also receiving male tissue transplants or sham surgery induced acyclicity (i.e., constant vaginal cornification and anovulation) either immediately or after a delay. The latter effect of a small dose of TP has previously been termed the *delayed anovulatory syndrome*[1] and suggests that gonadal steroids can have organizational effects on the brain postpubertally. In addition to its above effects on cyclicity, TP (8 μg) administration to neonatal females also receiving male tissue transplants or sham surgery masculinized their LH responses to ovarian steroids. Both the early estradiol benzoate-induced elevation in plasma LH and progesterone-facilitated LH secretion were considerably attenuated or completely eliminated in these animals (Fig. 6; TP-treated groups). This TP-induced masculinization is con-

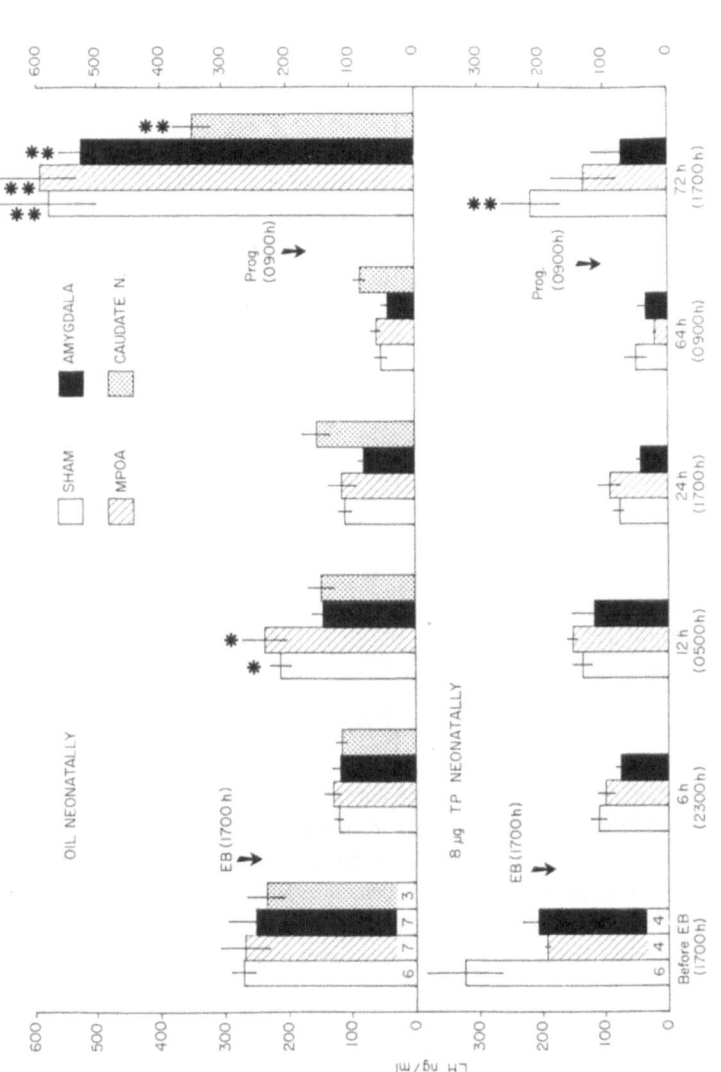

FIGURE 6. Effects of estradiol benzoate (EB) and progesterone (Prog) treatment on plasma LH titers in adult ovariectomized rats that had been implanted neonatally with male brain tissue and given oil or μg testosterone propionate (TP) concurrently with the transplant. Values are mean ± S.E. Number of rats in each group is indicated at the bottom of the first set of bars. Single asterisk, significantly ($p < 0.02$) elevated above those for comparable group at 6 hr after EB; double asterisks, significantly ($p < 0.03$) elevated above those for comparable group at 64 hr after EB. From Arendash, Leung, and Gorski (unpublished).

sistent with previous work involving ovarian steroid feedback mechanisms in such "lightly androgenized" rats.[1,23,24] Furthermore, male MPOA or amygdala transplants appeared to synergize with neonatal TP treatment to eliminate an LH response after progesterone which was only reduced, but not eliminated, by neonatal TP treatment alone (Fig. 6; TP-treated groups). A similar synergism between male tissue transplants and neonatal TP treatment was also apparently responsible for a paradoxical enhancement of female sexual behavior in a number of MPOA and amygdala transplant recipients (see Section 2.4).

To summarize this work, although certain of the male brain tissue transplants (either alone or in combination with TP treatment neonatally) altered LH secretion to a minor degree, intracerebral transplants of male *MPOA* brain tissue into female neonates did not, *by themselves,* greatly modify the negative and positive feedback actions of estrogen and progesterone, respectively, on adult LH secretion in these recipients. However, the fact that these same male MPOA transplants did substantially enhance sexual behavior (masculine and feminine) in female recipients provides a most interesting dichotomy between the dramatic behavioral and negligible hormonal effects of such MPOA transplants, suggestive of selectivity in the functional innervation of the recipient brain by transplanted MPOA tissues.

It must be emphasized that in any test of the functional effects of brain transplants, the specific nature of that test is very important. In the case of steroid modification of hormonal release, for example, the paradigm chosen may not have been the most appropriate to reveal functional activity of the grafts. In this regard, note that graft modification of lordosis responsiveness was not seen in animals primed with both estrogen and progesterone, but only in those primed minimally with estrogen. Designing the most appropriate test for the functional integration of transplanted and recipient neural tissue remains a most critical step.

5. *Effects of Testosterone Treatment of Recipient on Transplant Volume*

Thus far we have shown that the transplantation of the MPOA of the male rat, which includes the SDN-POA, does appear to make functional connections with the brain of the female recipient. This supports our earlier suggestion that cross-sex transplants of the neural substrate for sex-specific behaviors may facilitate the recognition of functional connectivity. However, it is possible that the transplantation of sex-specific and hormone-dependent neural tissue can also be used to approach more general questions related to the survival and differentiation of grafted neural tissues.

As mentioned in Section 1, the exact mechanism whereby gonadal steroids (such as testosterone) "organize" or program the sexual and neuroanatomical differentiation of the brain is currently unknown and remains an important and fundamental question in developmental neurobiology. However, some insight into the CNS mechanism(s) of steroid action has been obtained through the use of *in vitro* culture techniques that show that gonadal steroids accelerate and enhance the outgrowth of neuronal processes from explants of newborn mouse MPOA tissue.[25] Such steroid-induced growth effects could be important for the very survival of neurons, for cells that form only a limited number of synaptic connections (or inappropriate connections) are preferentially eliminated during brain development.[26] Therefore, gonadal steroids could be acting as "neuronotrophic agents" by selectively stabilizing certain neuronal populations through growth-promoting effects and/or the direct prevention of cell death.

We have already mentioned that some female recipients of male MPOA tissue were also injected with TP in an attempt to subject these transplants to a hormonal environment that we felt might be critical for their survival and functional integration. Although no hormonal administration was necessary for transplant-induced behavioral changes, analysis of the volume of the grafts revealed that in the oil-treated females, graft volume $(0.12 \pm 0.02 \text{ mm}^3)$ 6–7 months after transplantation was significantly smaller than the initial volume of the brain punch. In those females given a single injection of 8 μg TP concurrent with the transplant neonatally, the volume of these grafts $(0.19 \pm 0.04 \text{ mm}^3)$ was maintained at the initial value. This observation suggested that the hormonal environment might actually promote grafts survival as initially assumed.

Therefore, in an effort to maximize such an effect, females with transplants of male MPOA and for control purposes, caudate nucleus, were treated with multiple and larger doses of androgen. Female neonates, in addition to receiving bilateral implants of one of these brain regions aimed at their MPOA, received either 200 μg TP or oil (s.c.) on the day of transplantation (postnatal day 1) and on the following 4 days. All recipients were sacrificed at 30 days of age, before any test of functional activity, and their brains serially sectioned and stained with thionin.

The results of an ensuing analysis of transplant volumes are depicted in Fig. 7. MPOA transplants in oil-treated recipients were substantially reduced in size compared with the initial transplant volume of 0.19 mm^3. However, MPOA transplants in recipients treated with 200 μg TP for 5 days actually showed a 79% increase above their initial volume. The resulting five- to sixfold difference in MPOA transplant volume between oil- and TP-treated recipients was highly significant ($p < 0.001$). Figure 8 shows representative coronal sections through the largest extent of MPOA transplants

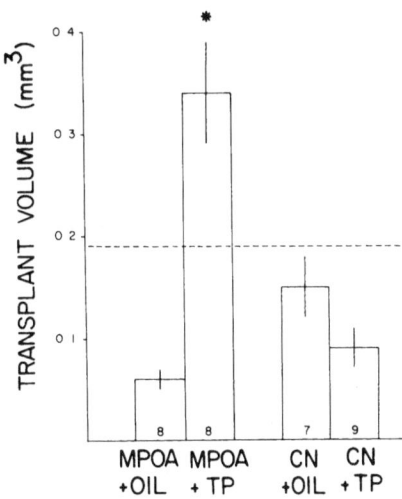

FIGURE 7. Effects of testosterone propionate (TP) treatment for 5 days beginning with the day of transplantation (postnatal day 1) on the volume of medial preoptic area (MPOA) or caudate nucleus (CN) grafts measured 30 days after surgery. Dashed line indicates initial transplant volume, i.e., the size of the punch. Asterisk, significantly different ($p < 0.001$) from MPOA + oil group.

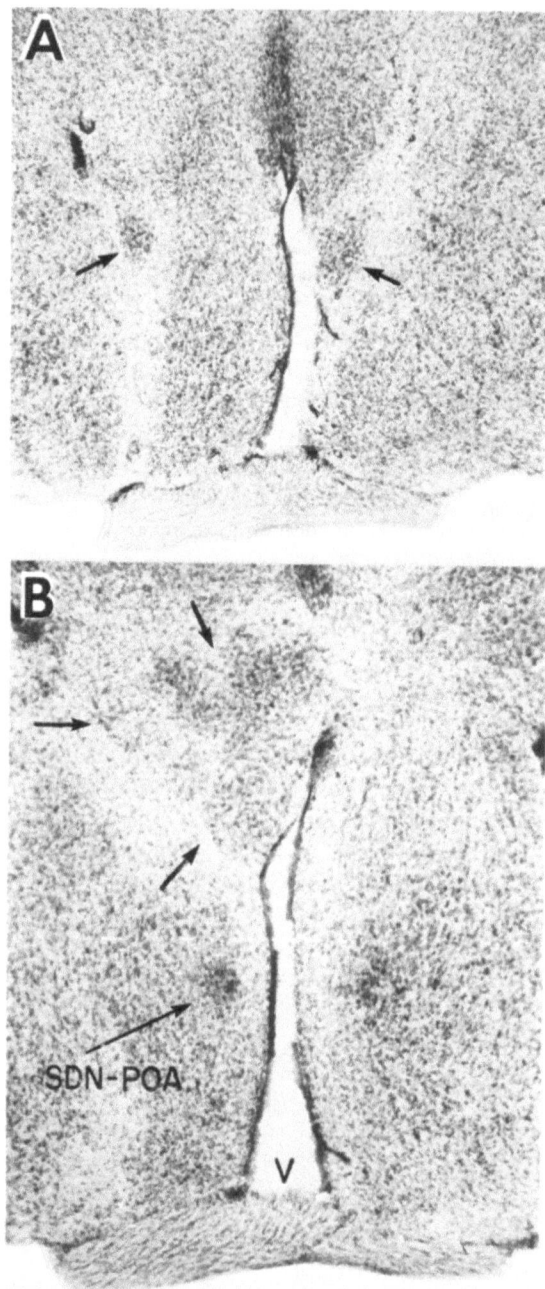

FIGURE 8. Coronal sections through the largest extent of male MPOA transplants in the MPOA of female recipients treated (A) with oil or (B) with testosterone propionate (200 μg/day for 5 days). Both at same magnification. Arrows indicate the transplants [bilateral in (A)]. SDN-POA, sexually dimorphic nucleus of the preoptic area of the host female; V, third ventricle. Thionin stain.

in an oil-treated and a TP-treated recipient. In sharp contrast to the ability of TP to enhance MPOA transplant volume, the same treatment to recipients receiving male caudate nucleus transplants resulted in no such enhancement (Fig. 7). As the MPOA of newborn rats concentrates [³H] testosterone, whereas the caudate nucleus of newborn rats does not,[5] it is not surprising that transplants involving the former, but not the latter area of the brain, respond to TP treatment with an enhancement of volume.

These results provide evidence that even in the transplant paradigm, testosterone is a "neuronotrophic" agent during development that may act specifically on cells within steroid-sensitive brain areas, possibly by a prevention of cell death within these areas through growth-promoting actions. Clearly, steroid action may be involved with the development of sexual dimorphisms in behavior, hormonal secretory patterns, and CNS neuroanatomical structure both *in vivo* and after transplantation. Interestingly, some evidence indicates that steroids may be influencing neural growth and development *indirectly,* by regulating the neuronal microenvironment through actions on glial cells.[27] Thus, glial cells could be involved in the mechanism(s) whereby TP enhances MPOA transplant volume.

6. Transplantation of Radioactively Labeled MPOA Tissue Containing the SDN-POA

Thus far, the behavioral modifications of the female that receives a male MPOA transplant, and the stimulatory effect of steroid treatment on surviving transplant volume, are consistent with the possibility that it is the neurons of the SDN-POA that become functionally integrated with the recipient's brain. As there is a growing list of sex differences in synaptic or dendritic organization of the brain[28-30] and in the gross volume of defined regions[11, 31,32] in several species, the approach of transplanting regions of the brain that are both functionally and structurally dimorphic, may have widespread applicability. Thus, it becomes even more important to identify the specific neurons from a transplant that modify the functional capacity of the recipient. These studies are clearly in their infancy and depend on our ability to identify or characterize neurons in the SDN-POA of the male and then, presumably, in the transplant.

At present, there is one such characteristic. The results of autoradiographic studies following the administration of [³H]thymidine have suggested that a number (approximately 30%) of SDN-POA neurons are born, as late as day 18 of gestation, after neurons in the rest of the preoptic areas have apparently become postmitotic.[33] Thus, neurons within the SDN-POA can be specifically labeled and their development followed after [³H]thymidine exposure on day 18. Importantly, [³H]thymidine exposure 1 day earlier results in a heavily labeled SDN-POA plus lightly labeled surrounding MPOA tissue (Arendash and Gorski, unpublished observations; see Fig. 9).

Utilizing the intraparenchymal transplantation technique, we have been able to punch out and transplant MPOA tissue (including the SDN-POA) from 1-day-old male neonates, previously exposed to [³H]thymidine on day 17 of gestation, into the MPOA of 1-day-old "cold" female recipients from a different litter not exposed to [³H]thymidine. After sacrifice of recipients 1 month later, subsequent autoradiographic

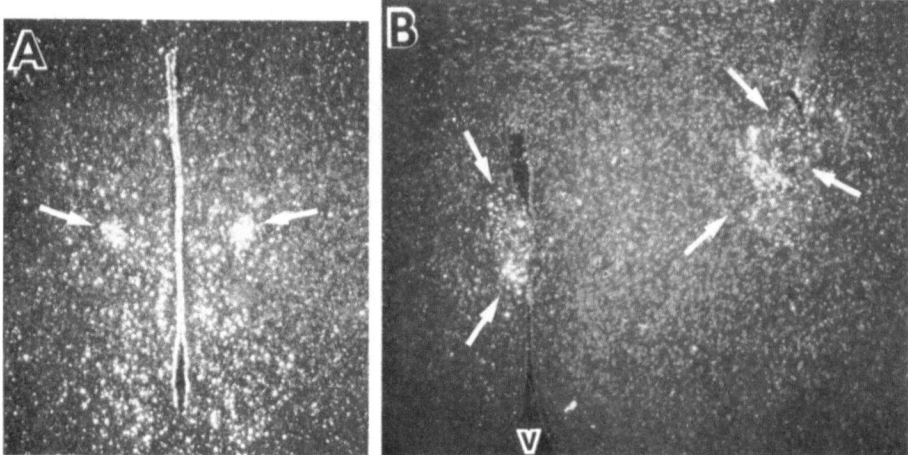

FIGURE 9. (A) Dark-field autoradiograph through the MPOA of a 1-day-old neonatal male rat that was exposed to [³H]thymidine on day 17 of gestation. Arrows indicate the heavily labeled SDN-POA. Less heavy labeling is seen throughout the MPOA. (B) Dark-field autoradiograph of the MPOA of a femele rat 1 month after the transplantation of the radiolabeled MPOA of a 1-day-old male rat [as in (A)]. Arrows indicate the transplants in which a heavily labeled component is clearly visible. V, third ventricle. Both at same magnification.

analysis has revealed that the male SDN-POA can be consistently transplanted *intact*, along with surrounding male MPOA tissue. Figure 9 shows two such autoradiographically labeled MPOA transplants bilaterally present within the MPOA of the recipient's brain. Note the heavily labeled male SDN-POA within each transplant and the lightly labeled male MPOA tissue adjacent to each transplanted SDN-POA.

Therefore, through the combination of radiolabeling techniques with brain tissue transplantation, a specific marker for transplanted male neural tissue (and specific groups of neurons within transplanted tissue, such as the SDN-POA) can be obtained. This may allow one to follow the development of transplanted SDN-POA neurons and to observe possible effects of testosterone treatment on their development and survival.

Conceivably, anatomical studies using other tags may permit us to identify the connections of these [³H]thymidine-labeled transplanted neurons, which would represent an important first step in elucidating the mechanisms by which transplanted neural tissue may assume functional significance. At this point in time, it does appear that the intraparenchymal transplantation of sex-specific and hormonally dependent nervous tissue, such as the SDN-POA, should promide new insight both in the developmental events responsible for sexual dimorphisms in brain structure and in those processes required for successful brain tissue transplantation.

7. Summary

Although brain tissue transplantation has recently been used to investigate the development and potential for functional integration of various transplanted brain regions, the cross-sex transplantation of sex-specific brain regions which can be expected to facilitate the detection of functional integration is only now being cohesively pursued. In this regard, the studies discussed in this chapter have utilized intraparenchymal transplantation methodology (1) to investigate the functional capacity of transplanted male brain tissues within female recipients and (2) to begin an elucidation of the mechanisms whereby gonadal steroids organize or modulate brain development, resulting in the sexual differentiation of the brain. With respect to the first aim, transplantation of the reproductively important MPOA from male neonates into the MPOA of female neonates results in enhanced masculine sexual behavior in recipients during adulthood, indicating that the neonatal male MPOA already has the potential to masculinize behaviorally the neonatal female brain. Such transplants also enhance feminine sexual behavior. Furthermore, MPOA transplants do not, by themselves, greatly modify reproductive hormonal profiles, as neither vaginal cyclicity nor the feedback effect of estrogen and progesterone on LH secretion was significantly altered by transplanted MPOA tissue. An interesting dichotomy would, therefore, appear to be present between the dramatic behavioral and minor hormonal effects of male MPOA transplants, suggestive of a degree of selectivity in the apparent functional integration of transplanted male MPOA tissue into the female recipient's brain.

With respect to the second aim, intraparenchymal transplantation methodology should provide new insight into the mechanisms whereby gonadal steroids organize or modify brain development. The marked enhancement of the volume of transplanted male MPOA tissue by testosterone treatment suggests that testosterone is a "neuronotrophic" agent that acts to stabilize selective neuronal populations within the brain possibly through growth-promoting functions and/or the direct prevention of cell death. Much more information on the CNS mechanisms of steroid action should be obtained in future studies involving specific radiolabeling of cells within the SDN-POA. Study of the development of such labeled cells after their transplantation, and the effects of gonadal steroids on this development and survival of transplanted SDN-POA neurons, should enhance our understanding of the ways in which gonadal steroids induce sexual dimorphisms in behavior, hormonal secretory systems, and neuroanatomy. Therefore, the intraparenchymal transplantation of neural tissue would appear to have immense potential for studying the development of neuronal circuitry involved in reproductive mechanisms and, as such, represents a new tool for research in neuroendocrinology. At the same time, the use and study of sex-specific hormone-dependent neural tissue for transplantation may provide new insight into those factors that govern transplant survival.

ACKNOWLEDGMENT. The authors' research reported herein was supported by NIH Grant HD-01182.

References

1. Harlan, R., Gordon, J., and Gorski, R., 1979, Sexual differentiation of the brain: Implications for neuroscience, in: *Reviews of Neuroscience*, Vol. 4 (D. Schneider, ed.), pp. 31–71, Raven Press, New York.

2. Gorski, R. A., 1983, Steroid-induced sexual characteristics in the brain, in: *Neuroendocrine Perspectives*, Vol. 2 (E. E. Müller and R. M. MacLeod, eds.), Elsevier, Amsterdam, p. 35.

3. Ward, I., 1969, Differential effect of pre- and postnatal androgen on the sexual behavior of intact and spayed rats, *Horm. Behav.* 1:25.

4. MacLusky, N., and Naftolin, F., 1981, Sexual differentiation of the central nervous system, *Science* 211:1294.

5. Sheridan, P., Sar, W., and Stumpf, E., 1975, in: *Anatomical Neuroendocrinology* (W. Stumpf and L. Grant, eds.), p. 134, Karger, Basal.

6. Christensen, L., and Gorski, R., 1978, Independent masculinization of neuroendocrine systems by intracerebral implants of testosterone or estradiol in the neonatal female rat, *Brain Res.* 146:325.

7. Heimer, L., and Larsson, K., 1966-1967, Impairment of mating behavior in male rats following lesions in the preoptic-anterior hypothalamic continuum, *Brain Res.* 3:248.

8. Arendash, G. W., and Gorski, R. A., 1983, Effects of discrete lesions of the sexually dimorphic nucleus of the preoptic area or other medial preoptic regions on the sexual behavior of male rats, *Brain Res. Bull.* 10:147.

9. Barraclough, C. A., and Gorski, R. A., 1961, Evidence that the hypothalamus is responsible for androgen-induced sterility in the female rat, *Endocrinology* 68:68.

10. Clemens, J., Smalstig, E., and Sawyer, C., 1976, Studies on the role of the preoptic area in the control of reproductive function in the rat, *Endocrinology* 99:728.

11. Gorski, R., Harlan, R., Jacobson, C., Shryne, J., and Southam, A., 1980, Evidence for the existence of a sexually dimorphic nucleus in the preoptic area of the rat, *J. Comp. Neurol.* 193:529.

12. Gorski, R., Gordon, J., Shryne, J., and Southam, A., 1978, Evidence for a morphological sex difference within the medial preoptic area of the rat brain, *Brain Res.* 148:333.

13. Döhler, K. D., Coquelin, A., Davis, F., Hines, M., Shryne, J. E., and Gorski, R. A., 1982, Differentiation of the sexually dimorphic nucleus in the preoptic area of the rat brain is determined by the perinatal hormone environment, *Neurosci. Lett.* 33:295.

14. Jacobson, C., Shryne, F., Shapiro, J., and Gorski, R., 1980, Ontogeny of the sexually dimorphic nucleus of the preoptic area in the rat, *J. Comp. Neurol.* 196:519.

15. Schmidt, R. H., Björklund, A., and Stenevi, U., 1981, Intracerebral grafting of dissociated CNS tissue suspensions: A new approach for neuronal transplantation to deep brain sites, *Brain Res.* 218:347.

16. Reier, P., Cullen, M., Froelich, J., and Rothchild, I., 1977, The ultrastructure of the developing medial preoptic nucleus in the postnatal rat, *Brain Res.* 122:415.

17. Lawrence, J., and Raisman, G., 1980, Ontogeny of synapses in a sexually dimorphic part of the preoptic area in the rat, *Brain Res.* 183:466.

18. Arendash, G., and Gorski, R., 1982, Enhancement of sexual behavior in femele rats by neonatal transplantation of brain tissue from males, *Science* 217:1276.

19. Weisz, J., and Ward, I., 1980, Plasma testosterone and progesterone titers of pregnant rats, their male and female fetuses, and neonatal offspring, *Endocrinology* 106:306.

20. MacLusky, N., Leiberburg, I., and McEwen, B., 1979, The development of estrogen receptor systems in the rat brain: Perinatal development, *Brain Res.* 178:129.

21. Ramon-Moliner, E., Vane, M., and Fletcher, G., 1964, in: *Stain Technology*, Vol. 39, p. 65.

22. Brown-Grant, K., 1974, Steroid hormone administration and gonadotropin secretion in the gonadectomized rat, *J. Endocrinol.* 62:319.

23. Mennin, S., and Gorski, R., 1975, Effects of ovarian steroids on plasma LH in normal and persistent estrous adult female rats, *Endocrinology* 96:486.

24. Harlan, R., and Gorski, R., 1978, Effects of postpubertal ovarian steroids on reproductive function and sexual differentiation of lightly androgenized rats, *Endocrinology* 102:1716.

25. Toran-Allerand, C., 1976, Sex steroids and the development of the newborn mouse hypothalamus and preoptic area *in vitro*: Implications for sexual differentiation, *Brain Res.* 106:407.

26. Hamburger, V., and Oppenheim, R. W., 1982, Naturally occurring neuronal death in vertebrates, *Neurosci. Comment.* 1:39.

27. Vernadakis, A., Culver, B., and Nides, R., 1978, Actions of steroid hormones on neural growth in culture: Role of glial cells, *Psychoneuroendocrinology*, 3:47.

28. Raisman, G., and Field, P., 1973, Sexual dimorphism in the neuropil of the preoptic area of the rat and its dependence on neonatal androgen, *Brain Res.* 54:1.

29. Greenough, W., Carter, C., Steerman, C., and DeVoogd, T., 1977, Sex differences in dendritic patterns in hamster preoptic area, *Brain Res.* **126**:63.

30. Ayoub, D., Greenough, W., and Juraska, J., 1983, Sex differences in dendritic structure in the preoptic area of the juvenile macaque monkey brain, *Science* **219**:197.

31. Nottebohm, F., and Arnold, A. P., 1976, Sexual dimorphism in vocal control areas of the songbird brain, *Science* **194**:211.

32. Breedlove, S., and Arnold, A., 1980, Hormone accumulation in a sexually dimorphic motor nucleus of the rat spinal cord, *Science* **210**:564.

33. Jacobson, C., and Gorski, R., 1981, Neurogenesis of the sexually dimorphic nucleus of the preoptic area in the rat, *J. Comp. Neurol.* **196**:519.

10

Morphological and Functional Properties of Transplanted Vasopressin Neurons

John R. Sladek, Jr., and Don M. Gash

1. Introduction

Three basic principles in neural transplantation have been demonstrated by studies conducted since 1970. Building on the earlier experiments of Dunn,[1] May,[2] and Le Gros Clark,[3] a number of investigators[4-8] have shown that fetal CNS neurons survive and develop anatomically normal features in the host brain. Lund and Hauschka[9] provided some of the first evidence that grafted neural tissue becomes structurally integrated with the parenchyma of the host nervous system by demonstrating that fiber projections are established between the host and the donor. An extensive literature now exists to support this concept that grafted neurons readily send efferent fibers into the host brain and in turn receive afferents from host neurons (see Chapter 4). Finally, the ability of transplanted neurons to synthesize and release neurohormones and neurotransmitters in an appropriate manner to effect host behavior has been documented in at least three different model systems.[10-14] Thus, the principles of (1) graft viability, (2) structural integration, and (3) appropriate function have been described. At issue now is the precise definition of the limits and important variables of transplant development and function.

We have developed a model system for analyzing the properties of transplanted neurons.[8,11-13,15,16] Our method involves placement of vasopressin neurons from the hypothalamic neurosecretory system of normal fetal donors into Brattleboro stain rats, which congenitally lack vasopressin-produced neurons.[17,18] The grafted vasopressin neurons and their axonal and derndritic projections can be unequivocally identified in the

John R. Sladek, Jr., and Don M. Gash • Department of Anatomy, University of Rochester School of Medicine and Dentistry, Rochester, New York 14642.

host brain by immunohistochemical staining for vasopressin or vasopressin-specific neurophysin. Amelioration of the symptoms of diabetes insipidus in the host is evaluated as a correlate of transplant function. Due to the absence of vasopressin, Brattleboro rats exhibit a pronounced polydipsia and polyuria.[17,18] Vasopressin administration decreases water consumption and increases urine osmolality in a dose-dependent fashion.[16]

Since our model utilizes the hypothalamic neurosecretory system, it will be appropriate to first describe the normal features of this neuroendocrine system before discussing experiments with vasopressin neural transplants.

2. The Hypothalamic Neurosecretory System

This system of large-sized, hypothalamic neurons has been well characterized anatomically, biochemically, and physiologicaly. The pioneering work of Scharrer and Scharrer[19] and Bargmann [20,21] clearly established the principles of neurosecretion. They proposed that neurons of the supraoptic and paraventricular nuclei synthesized vasopressin and oxytocin and transported these hormones to the posterior pituitary for release into the systemic circulation. Subsequent immunohistochemical studies verified the existence of each hormone within separate neurons of each hypothalamic nucleus and further demonstrated neuroanatomical projections to sites other than the pituitary.[22]

2.1. Vasopressin

Oxytocin and vasopressin were the first neurohormones to be characterized chemically and their structures verified by synthesis.[23] Vasopressin, which differs from oxytocin by two amino acids, is a nonapeptide (MW 1228) that is synthesized *in vivo* as part of a much larger precursor molecule, propressophysin, of about 20,000 daltons. Vasopressin-specific neurophysin and a 39-amino-acid glycopolypeptide are other major components of propressophysin.[24]

Vasopressin has several important functions. It is the antidiuretic hormone and thus serves a vital role in regulating fluid and ionic homeostasis.[25] At high physiological titers, vasopressin has a significant pressor effect and may be involved in the peripheral regulation of blood pressure.[26] In recent years there has been increasing evidence that vasopressin, perhaps through its extrahypothalamic projections in the CNS, has other effects including modulating emotional–motivational (arousal) and temperamental aspects of behavior.[27]

2.2. Anatomy

Vasopressin is synthesized by large (20–35 μm in diameter) neurons located in the supraoptic, paraventricular, and accessory nuclei of the hypothalamus and in parvocel-

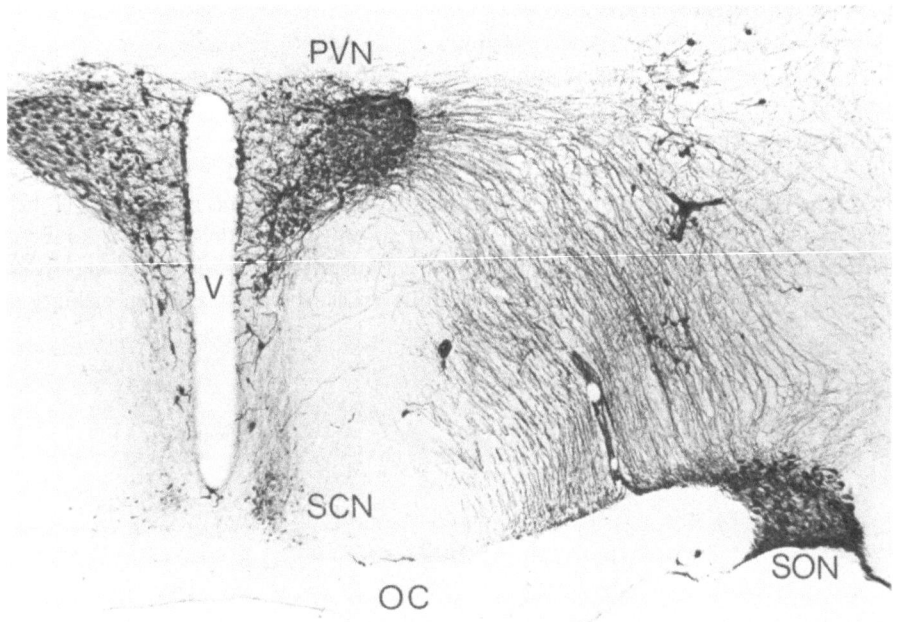

FIGURE 1. The hypothalamo–neurohypophyseal system is illustrated following neurophysin immunohisto-chemistry. Landmarks such as the third ventricle (V) and optic chiasm (OC) are indicated. The magnocellular system consists of peptidergic neurons of the paraventricular nucleus (PVN) and supraoptic nucleus (SON). Internuclear groups also exist as seen lateral to the PVN. Beaded axonal fibers ramify from the PVN and SON in a ventral course to eventually reach the neurohypophysis via passage through the pituitary stalk. The suprachiasmatic nucleus (SCN) contains vasopressin neurons of a parvicellular type which do not contribute to the neurosecretory pathway. × 40.

lular (10–14 µm in diameter) neurons of the suprachiasmatic nucleus (Fig. 1). Immunohistochemical and tract-tracing studies have revealed that there are significant extra-hypothalamic vasopressinergic projections in addition to the classical neuroendocrine projections to the posterior pituitary.[28,29] We now know that the paraventricular nucleus is an exceptionally global nucleus, projecting to widespread portions of the neuraxis including the brain stem, septum, hippocampus, cingulum, and autonomic centers in the spinal cord, to cite a few locations. It is also clear that the release of vasopressin is regulated by a number of different transmitters and modulators including noradrenaline, acetylcholine, angiotensin II, and certain opiate peptides.[30] The classic hormones also are known to coexist with other peptides; dynorphin has been discovered within vasopressin neurons[31] and CCK-8 has been colocalized to oxytocin neurons.[32] Evidence also points to the coexistence of enkephalin[33] and possibly angiotensin II[34] within magnocellular neurons. Thus, our knowledge of this classical neurosecretory system has expanded to include the concept that vasopressin and oxytocin project to numerous nonhypothalamic sites and therein may play a role in phenomena other than water balance.

3. The Transplantation Procedure

Our procedures for transplanting neural tissue have been taken, with some modifications, from classical techniques employed in grafting endocrine tissues.[35] In our neurotransplantation studies, the donor tissue is derived from timed-pregnant rats. For the initial experiments, inbred albino Wistar/Lewis rats were used for the tissue donors. Recently, we have changed and begun routinely using Long–Evans rats because of their close genetic relationship to the Brattleboro rat strain (the Brattleboro rat is a mutant derived from the Long–Evans strain in 1960). Both Wistar/Lewis and Long–Evans seem to serve well with no apparent strain differences between transplant viability or

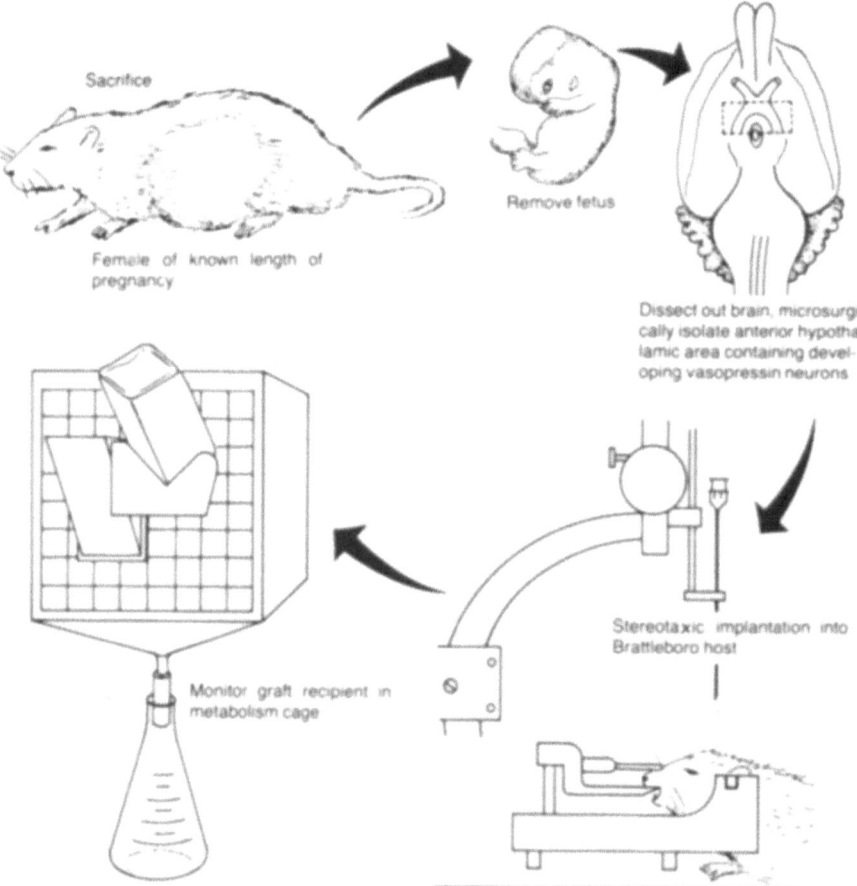

FIGURE 2. The transplant protocol is depicted beginning with removal of fetal rat pups from 17 to 19 day gestational pregnant rats. Following brain removal, a small portion of the hypothalamus is dissected and stereotaxically implanted into the third ventricle of the host Brattleboro rat. Following surgery, animals are monitored in metabolism cages to determine the effect of the grafting procedure on water balance. × 90. From Ref. 11.

function. Fetal age is determined by the presence of sperm in vaginal smears; the initial detection is considered day 0 of pregnancy.

Dissection procedures depend on the age of the donor. For embryos ranging from 11 to 15 days postcoitus, the entire ventral diencephalon is removed. For 16-day and older fetuses, it is possible to dissect out the anlagen of the supraoptic and paraventricular nuclei (see Ref. 36; Boer, Dick, and Gash, unpublished data). In all cases, the uterine horns containing the concepti are removed from a timed-pregnant dam and placed in ice-cold culture media (Fig. 2). The prenatal brains are removed for transplantation purposes within 90 min after uterine horn isolation. Tissue blocks are dissected free under a dissecting microscope with care being taken to keep the brain well hydrated with Eagles minimum essential medium. In all the experiments discussed presently, the neural tissue used for grafting included all of the anterior-ventral hypothalamus, i.e., the anlagen of the supraoptic and suprachiasmatic nuclei and associated neuropil. It also contained the periventricular regions and the arcuate nucleus with its accompanying peptidergic and dopaminergic neurons. Tissue viability is high following these procedures, routinely ranging from 97 to 99% (Notter and Gash, unpublished data).

Each block is dissected into 8–12 smaller components that are drawn together into a 20-gauge spinal needle. This needle is inserted stereotaxically into the third ventricle of an adult Brattleboro-strain rat that has been anesthetized with intraperitoneal sodium brevital (40 mg/kg). This cannula is lowered to a position immediately dorsal to the median eminence, which comprises the floor of the third ventricle. The tissue blocks are then deposited by a downward movement of the cannula onto the dorsum of the median eminence.

Following surgery, the animals are returned to metabolism cages for the continuation of water balance monitoring. The hosts are killed for histological examination at 20 or 40 days subsequent to surgery. A variety of procedures are employed for tissue analysis including electron microscopy, monoamine histochemistry, and peptide immunohistochemistry. Additionally, a combined approach for the concurrent demonstration of peptides and monoamines has been utilized to analyze the potential for connectivity between host and graft neurons.[37]

4. Anatomical Characteristics

4.1. General Features

The hypothalamic neural graft usually occupies a position dorsal to the floor of the third ventricle in contiguity with the underlying median eminence (Fig. 9, 15). Its dorsal–ventral extent is considerably greater than its medial–lateral width; nevertheless, the third ventricle appears expanded especially ventral to the hypothalamic sulcus. Usually, the transplant is separated from the host tissue laterally by the ependymal lining of the third ventricle. On occasion, when the ependyma is denuded, the graft and host neuropil appear continuous. Furthermore, the dorsal aspect of the transplant often appears integrated with the host hypothalamus and thalamus (Fig. 8). The rostral–caudal extent of

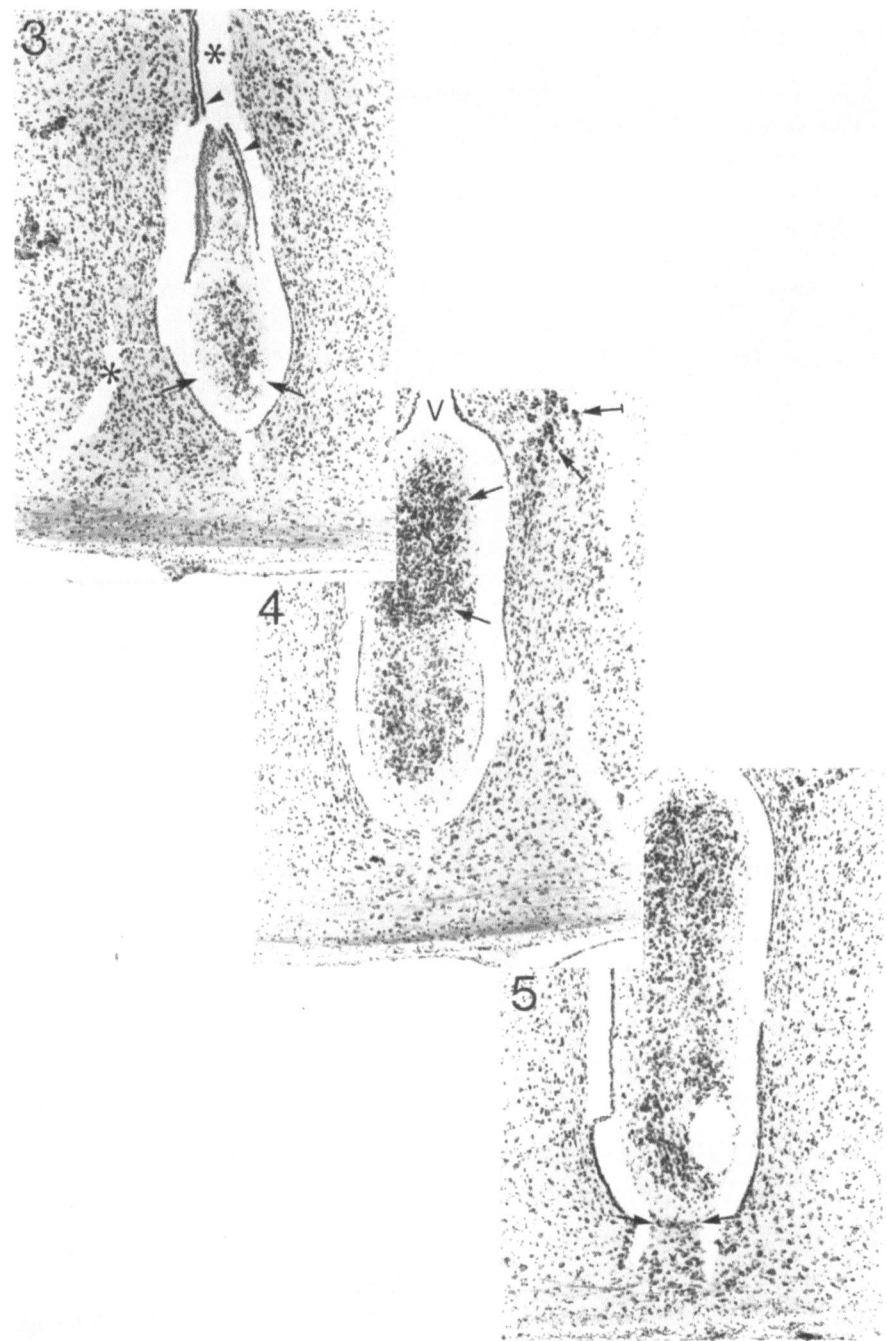

FIGURES 3–14. This series represents frontal, 10-μm sections from a functional transplant following staining with cresyl violet and luxol fast blue. The sections are 200 μm part and are sequential from rostral (Fig. 3) through caudal (Fig. 14) levels of the graft. In general, the graft begins at its rostral and caudal poles as a small piece of tissue seen within the third ventricle (V). It is larger in dimensions at its ventral extent where the median eminence has expanded to its fullest width. Highlights of some of these figures follow. In Fig.

the transplant is such that grafted tissue often is seen in the infundibular recess caudally at levels of the mammillary bodies and pituitary stalk (Figs. 13, 14, 42, 43) and rostrally is seen as far anterior as the preoptic area (Fig. 3). The overall shape is usually that of a wedge of tissue, somewhat pyramidal with a base slightly narrower than the underlying median eminence (Fig. 15). In cresyl violet-stained sections, the graft is seen to contain an abundant neuropil consisting of small and large neurons, neuroglia, and a rich vascular arbor (Figs. 3–14). Scar tissue and areas of prominent gliosis generally are absent. Even with fluorescence microscopy, which shows glial scarring and tissue wounds as a medium to bright yellow fluorescence, there is a general absence of this fluorescent index at the interface between the host and the graft.

4.2. Vascularization

The transplants appear to be richly vascularized as a result of growth of blood vessels from the underlying median eminence. Elements of the arterial as well as venous system appear to ramify to a greater than normal extent in the host median eminence (Figs. 16, 28, 33, 54) and further to extend dorsally through the interface and into the transplanted hypothalamic tissue (Figs. 16–18, 22–24). Arteries as well as arterioles appear to penetrate this interface and then to richly arborize at all levels of the transplant. There is usually a tuft of vessels that extends in a dorsal direction at rostral and middle levels of the transplant (Figs. 16, 17, 19). Blood vessels also appear to cross the interface laterally at points of denuded ependyma and dorsally at the dorsal hypothalamic and ventral thalamic junctions. Although it is not possible to determine the directionality of these latter vessels, they are seen also in transplants that solely occupy the

---

3 the rostral pole of the graft is encapsulated by an ependymal lining dorsally (arrowheads) and appears somewhat wider at its base (arrows). These sections were prepared from tissue that had been freeze-dried to accommodate Falck–Hillarp histofluorescence studies; therefore, tissue cracks (∗) occur and should not be interpreted to represent either necrosis or ventricular cavities. In Fig. 4, a marked cellular density is seen in the dorsal one-half of the graft (arrows). These cells are considerably smaller than those of the host paraventricular nucleus (crossed arrow). Subsequent analysis revealed this area to contain parvicellular, suprachiasmatic-like, vasopressin neurons. Figure 5 illustrates the rostral point of contiguity between the host and the graft (arrows). An ependymal lining, which is prominent on the walls of the third ventricle, is absent at this point of interface. Figures 6 and 7 illustrate the increased surface area of the interface. Moreover, another area of increased, parvicellular density is seen in Fig. 7 (arrow) as well as a vacuolated portion of the optic chiasm (∗). This dorsally placed cell density persists in Fig. 8. Here the graft appears to be in contiguity with the host brain as an ependymal lining in this area is absent (arrowhead). The ventral portion of the graft has expanded (arrows) to fill the floor of the third ventricle. In Fig. 9, a blood vessel (arrows), presumably from the underlying median eminence, appears to course in the lower right quadrant of the graft. This apparent vascularity is seen again in Fig. 10 (crossed arrows), which is depicted at a more advantageous magnification in Fig. 18. Numerous large neurons occupy a position at the interface between the graft and the host (arrows) and are common features of the ventral one-third of the graft. In Fig. 13 and 14, the caudal portion of the graft begins to diminish in size as the ventricle transforms into the infundibular recess. A continuation of graft tissue into the underlying pituitary stalk (arrows) is seen in Fig. 13 and 14; in the latter, large neurons are evident within the stalk (arrow). × 45.

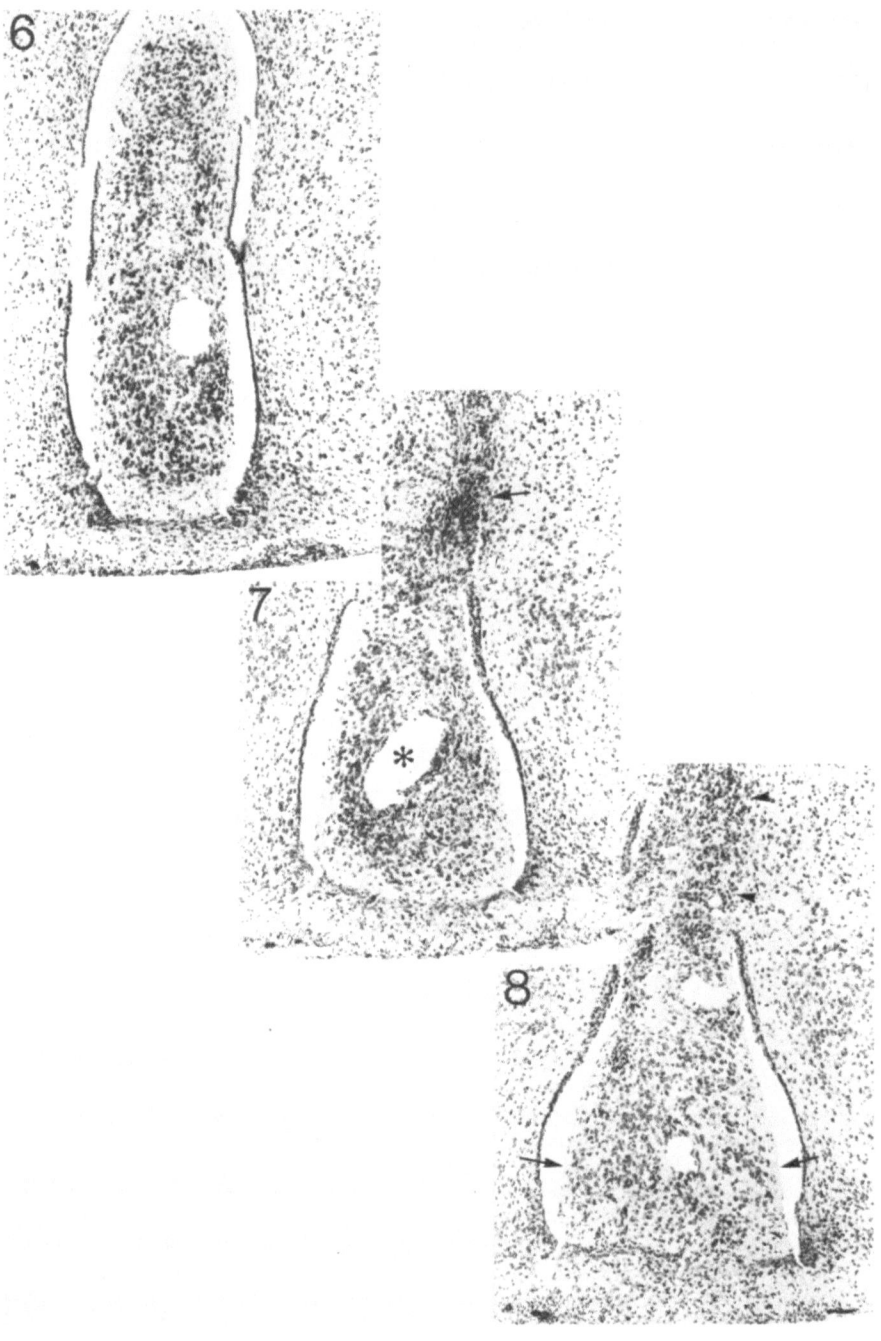

FIGURES 3–14 (*continued*)

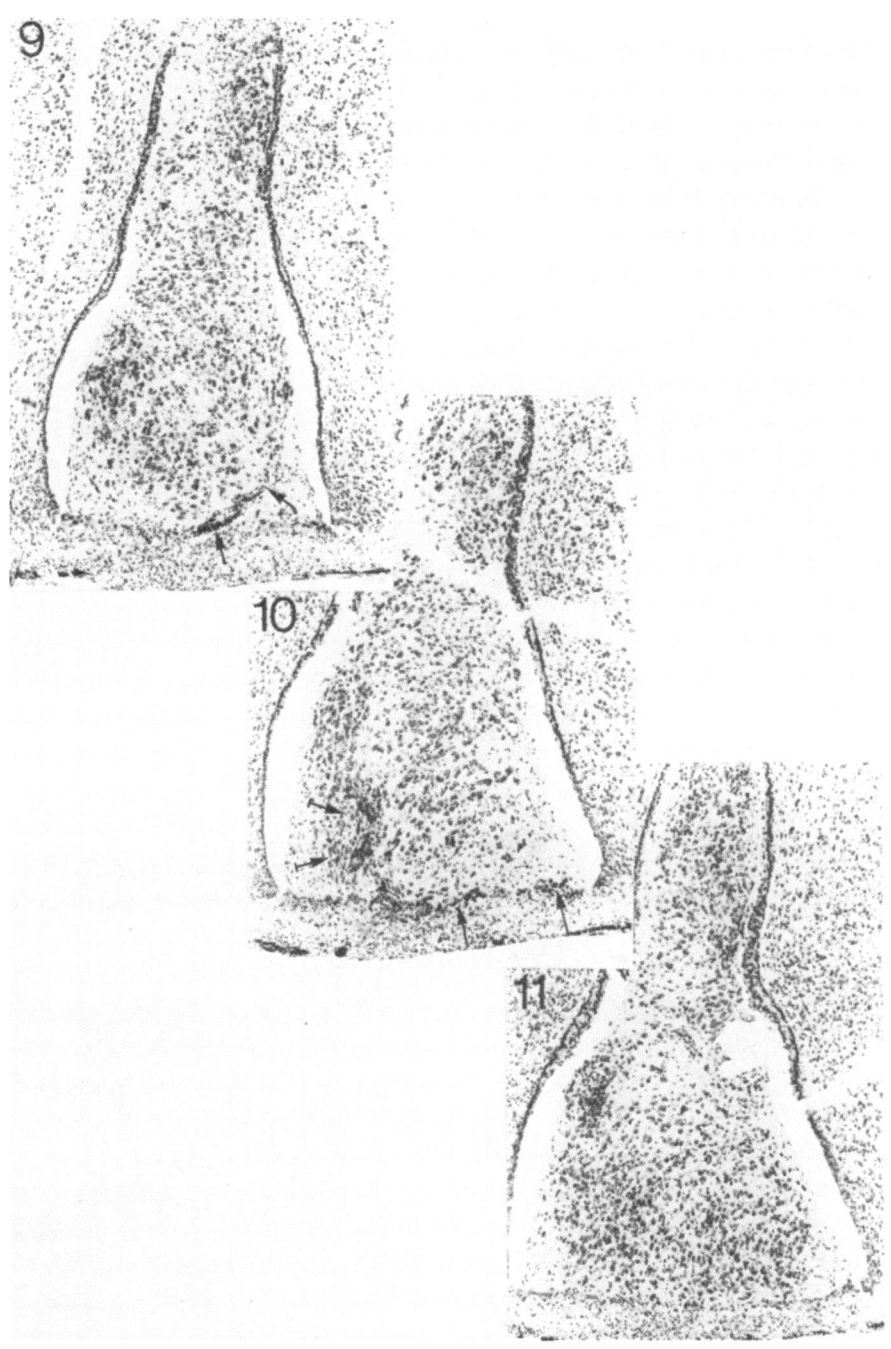

FIGURES 3–14 (*continued*)

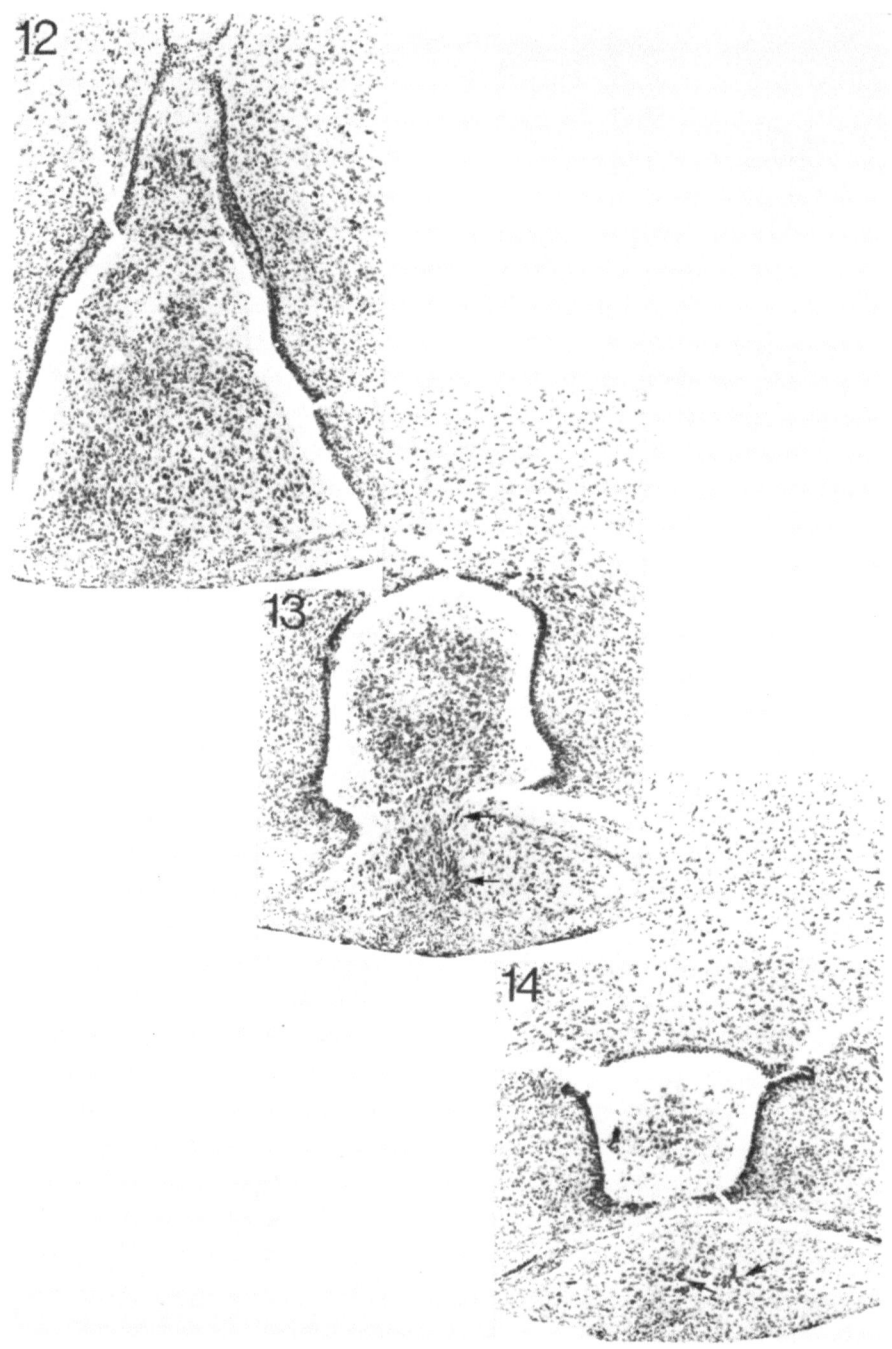

FIGURES 3–14 (*continued*)

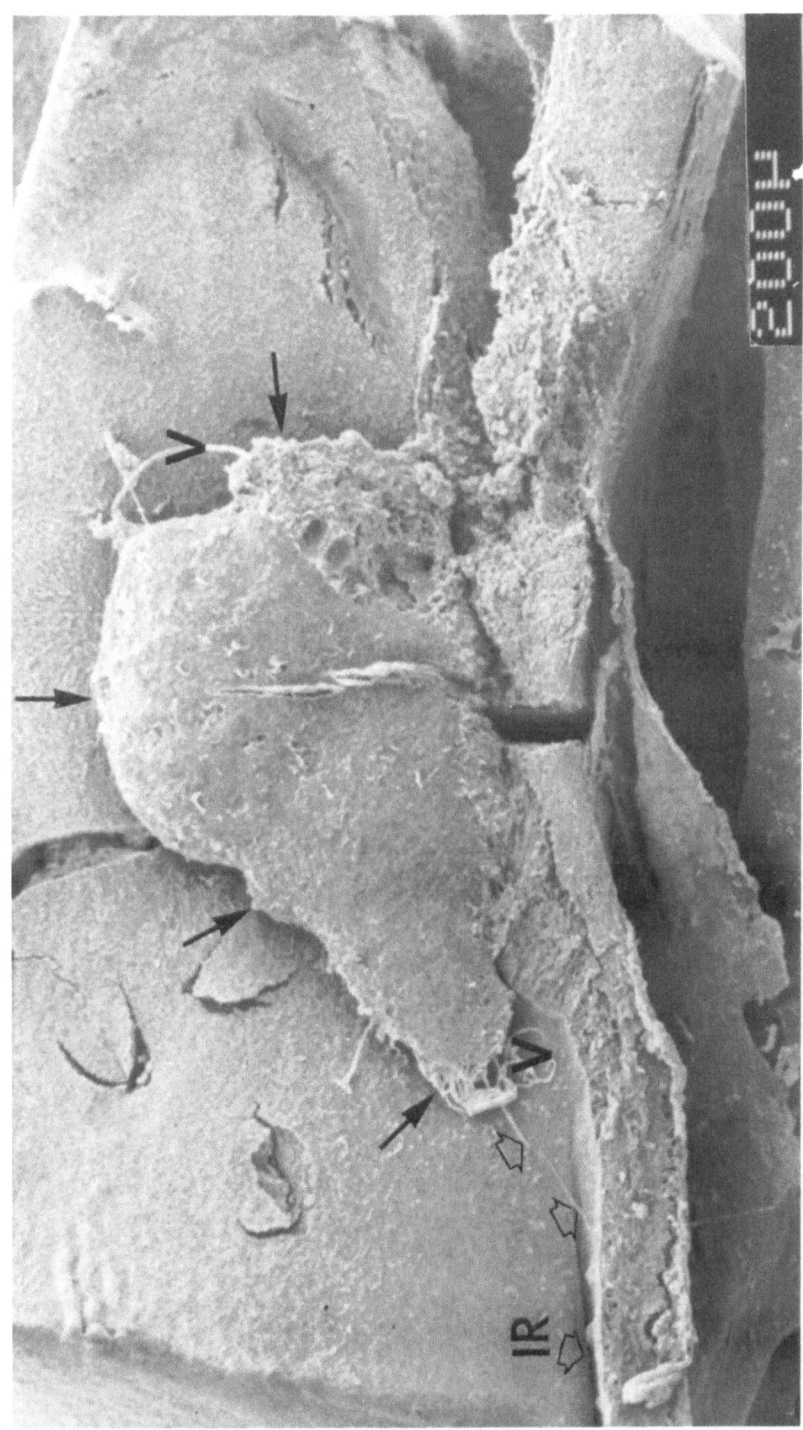

FIGURE 15. A low-power scanning electron micrograph of the lateral wall of the third ventricle of a host Brattleboro rat. Rostral is to the right. The graft (arrows) occupies a central position in the ventricle (V) and extends caudally toward the infundibular recess (IR). Processes from supraependymal cells (open arrows) appear to interconnect the graft and host brain. From Ref. 54.

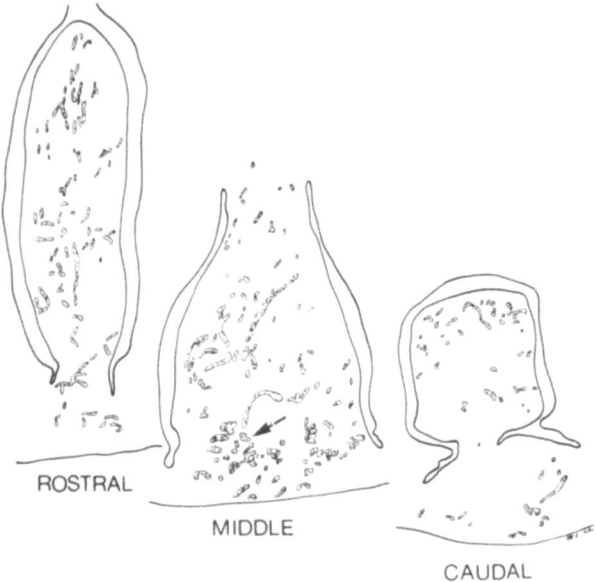

FIGURE 16. This schematic illustrates the vascularization of a functional graft. These drawings were each composed from 7 to 10 serial sections with the aid of a camera lucida. Sections were 10 μm in thickness. Blood vessels of the host median eminence appear to grow dorsally into the graft to provide an extensive vascular network. They also appear to extend in rostral to caudal directions and at caudal levels do not appear continuous with the underlying pituitary stalk. At middle levels, a large tuft of vessels often is seen coursing through the host–graft interface (arrow). This phenomenon is seen histologically in Figs. 17 and 18.

dorsum of the third ventricle. Thus, it is probable that vessels grow into the transplant from the adjacent periventricular stratum as well as the richly vascularized median eminence. This intratransplant network is extensive and appears to arborize evenly throughout the dorsal–ventral extent of the graft. In some instances, the caudal extremes are less well-vascularized by this plexus. This plexus is present at 20 days posttransplantation, but the earliest posttransplantation time at which the vessels grow into the grafted tissue is not known.

In cresyl violet-stained sections, magnocellular neurons often appear in juxtaposition to the vasulature (Figs. 17–19). This is especially prominent where large vessels have coursed through the substance of the graft and also at the interface of the graft and the median eminence (Fig. 19).

Recent analysis of hypothalamic transplants employing microfil injection of the vascular system has confirmed the presence of this extensive vascular plexus throughout the transplants (Figs. 20, 21). Correlative transmission electron microscopy has resulted in the identification of numerous fenestrated capillaries within the transplant also (Scott and Gash, unpublished data).

4.3. Peptidergic Neurons

Vasopressin and oxytocin neurons can be identified separately by the use of immunohistochemistry with specific antisera to their respective neurophysins or collectively with antisera that recognize both peptides. The majority of antineurophysin staining, employed presently, reveals the presence of both oxytocin- and vasopressin-containing neurons. Additional staining for the specific hormones, vasopressin and oxytocin, also has been performed as well as for the specific neurophysins. The following neuronal

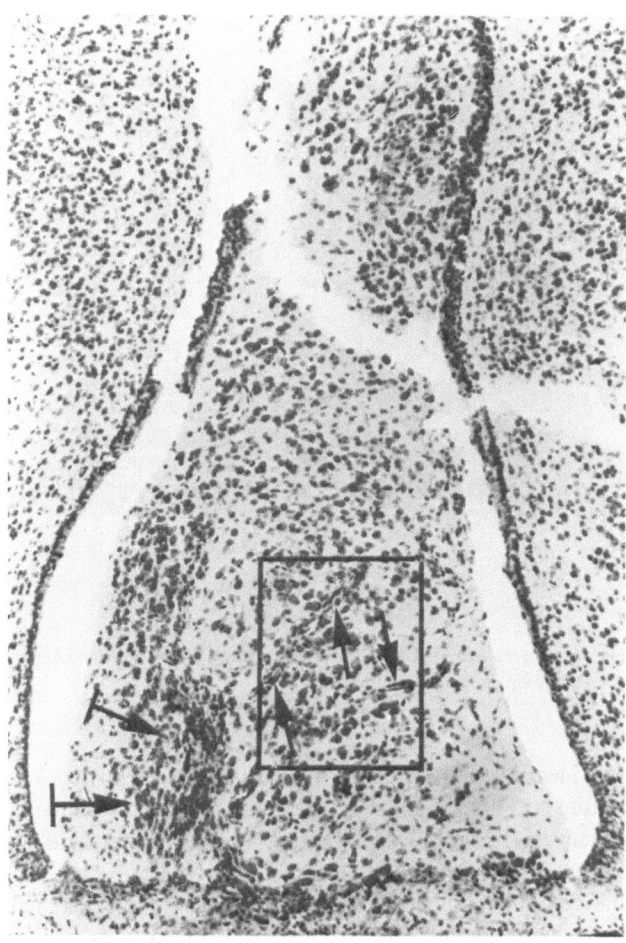

FIGURE 17. This low-power view of a cresyl violet–luxol fast blue-stained section depicts a large tuft of vessels in the lower left quadrant of the graft (crossed arrows). Smaller vessels appear to branch from this incoming source (arrows). The area outlined in the rectangle is illustrated at higher magnification in Fig. 18. × 75.

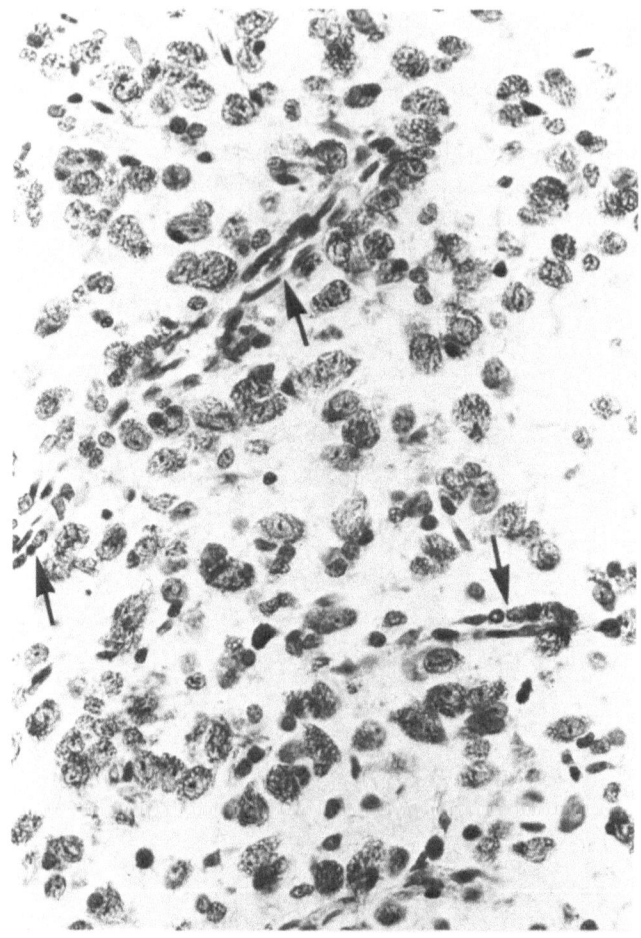

FIGURE 18. The same vessels seen in Fig. 17 are indicated (arrows). They are characterized by elongated nuclei of endothelial cells and by an identifiable lumen. × 525.

elements have been identified in transplants: parvicellular elements that usually appear adjacent to a structure reminiscent of the optic chiasm, and magnocellular elements that appear in various locations (Figs. 30–41). The parvicellular components stain intensely for vasopressin as well as neurophysin and appear as part of a complex of small peri-karya and abundant immunopositive fibers (Figs. 25, 26, 31, 35). This, in all likelihood, represents the suprachiasmatic nucleus, which often is included in the tissue block of transplanted hypothalamus. Interestingly, vasopressin as well as neurophysin staining is equally strong in parvicellular perikarya. Magnocellular perikarya usually occur pri-marily in three locations: (1) within the ventral one-third of third ventricular grafts (Fig. 37), (2) at the interface of the graft and host (Figs. 19, 27–29), and (3) within the median eminence and pituitary stalk of the host brain (Figs. 42–44). Intense immuno-

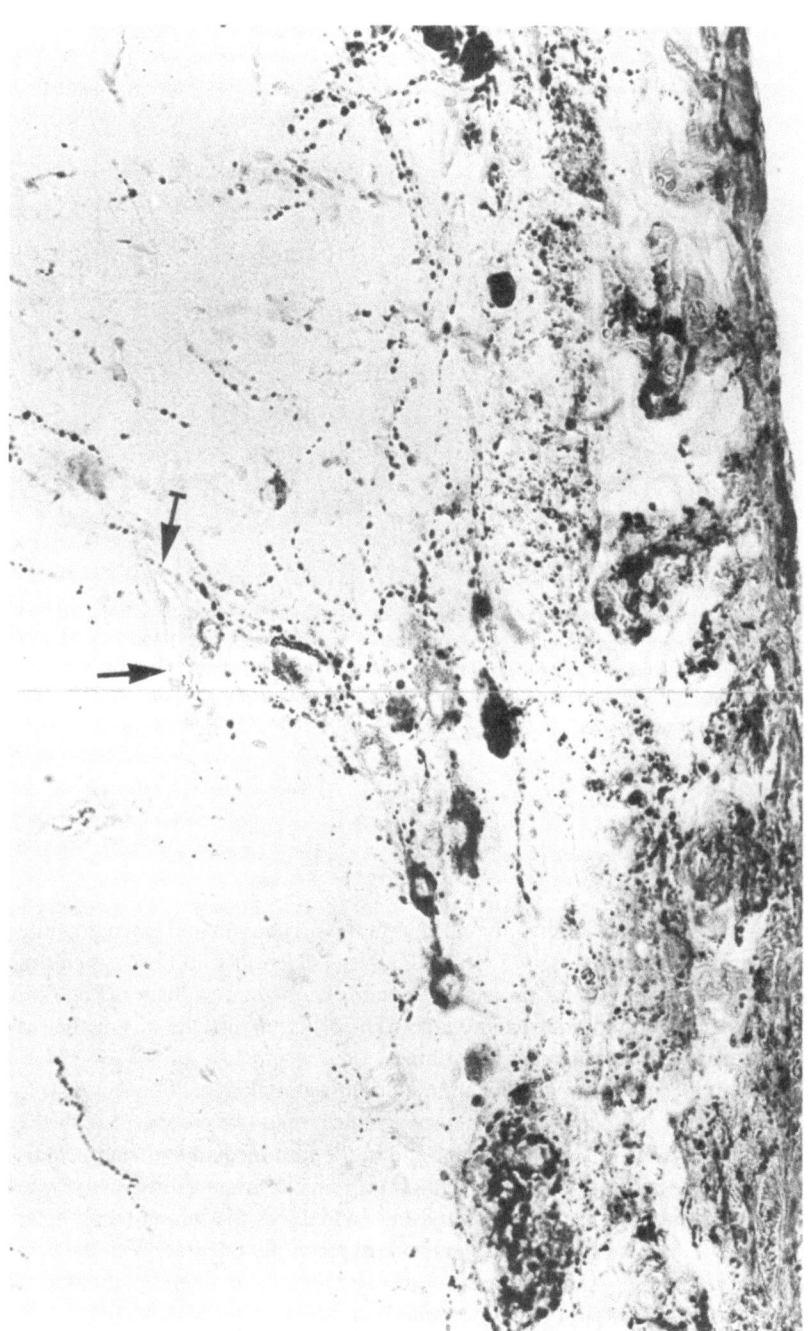

FIGURE 19. In a section 70 μm caudal to the level illustrated in Figs. 17 and 18, blood vessels again are seen in the vicinity of those depicted above (arrow). Numerous fibers, stained for neurophysin (crossed arrow), appear in the vicinity of this vascular component. The underlying host median eminence also is characterized by dense neurophysin staining in association with blood vessels. × 330.

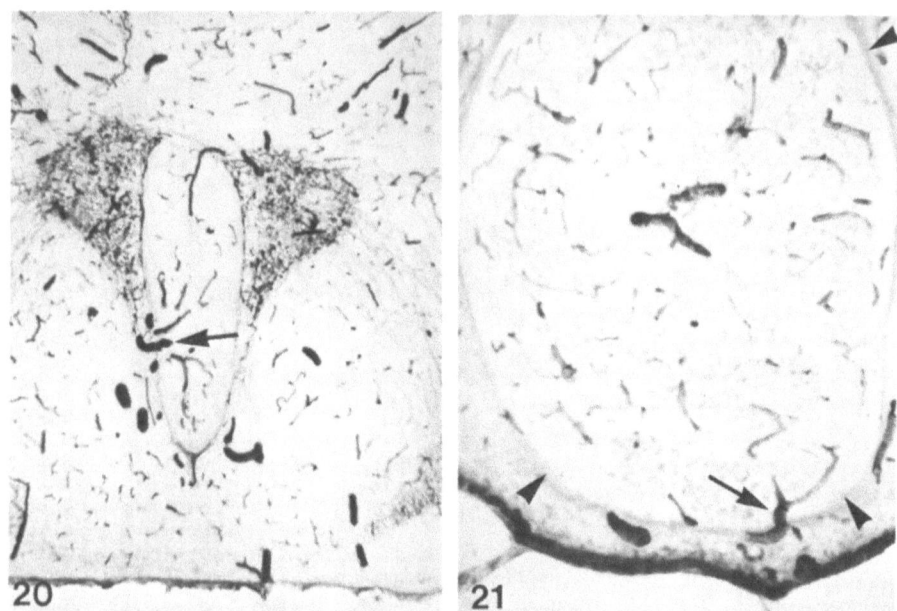

FIGURE 20. This low-power view of a frontal section through the hypothalamus depicts a neural graft that fills the cavity of the third ventricle. Within it a large number of blood vessels (arrow) appear dark as a result of microfil injection into the vascular system. The normal vasculature of the host brain also is evident and is especially prominent in the paraventricular nucleus immediately adjacent to the third ventricle. The section is counterstained for neurophysin. × 50. From Scott and Gash, unpublished data.

reactive precipitate is seen surrounding the nucleus of the perikaryon and extending into axonal and dendritic processes (Figs. 45, 46). Stain also is identified in fiber systems of these perikarya, which usually show a polarity toward the median eminence (Figs. 19, 37). Although vasopressin immunoreactivity is seen in perikarya, usually it is neither as dense nor as extensive as the neurophysin staining.

Fibers appear to emanate from the immunopositive perikarya. This suggests a rapid growth process by these developing neurons, which would be consistent with the extensive fiber networks seen in the transplant and in the host median eminence. Additionally, magnocellular perikarya appear to migrate from the transplanted tissue into the host median eminence and have been seen as far caudally as the infundibular stalk. Without exception, third ventricular transplants that successfully integrate with the host median eminence do not show neurophysin-positive perikarya or fibers in the dorsal extremes of the transplant except for parvicellular, suprachiasmatic components.

The fiber system of magnocellular neurons ramifies extensively throughout the host median eminence to produce a denser than normal fiber pattern in both the internal and the external laminae (Figs. 22, 24, 27–29, 47–49). This extensive ramification of fibers is absent in nonfunctional transplants (Figs. 50, 51) and furthermore is absent in the normal, unimplanted Brattleboro rat, which contains neurophysin-positive fibers in the *internal* lamina of the median eminence; these fibers consist of oxytocin-containing axons en route to the neurohypophysis.[15] Functional neural transplants, however, result in the ramification of neurophysin- and vasopressin-positive fibers throughout all laminae of the median eminence including the external zone, which contains a dense network of fibers particularly around portal vessels. Vasopressin staining in the median eminence is dense, but again not quite as extensive as neurophysin staining.

4.4. Catecholaminergic Neurons

Catecholaminergic neurons also occur within these hypothalamic grafts. Their cell bodies appear as small, rounded perikarya that are reminiscent of A12 dopaminergic neurons of the tuberoinfundibular system. They appear scattered throughout the graft with no particular orientation or aggregation. Catecholamine-containing varicosities also are common features of the graft and can appear throughout all levels. They may arise from the endogenous dopaminergic neurons or via invasion of the graft from at least three routes. The first is the underlying median eminence wherein dopaminergic fibers appear to follow the ingrowth of portal vessels, which normally are heavily rimmed by a fine-sized dopaminergic plexus of tuberoinfundibular axons (Figs. 54, 55). These often extend several millimeters dorsally into the graft and can appear as punctate varicosities in the graft neuropil. A second route from the median eminence may involve noradrenergic fibers that arise from the reticuloinfundibular, noradrenergic system. These too can follow the ingrowth of blood vessels in the accompanying neuropil and appear as linear profiles of varicosities (Fig. 56). A third route of ingrowth of catecholamine fibers involves the adjacent periventricular stratum, which is richly innervated by the ascending, noradrenergic periventricular system as well as probable inputs from the locus coeruleus through the medial forebrain bundle. In instances where the ependyma is denuded and there is contiguity between the ventricular wall and the transplant, linear profiles of catecholamine varicosities have been seen to cross the host–transplant interface (Figs. 52, 53). Some of these ramify throughout the dorsal and middorsal region of the graft.

Simultaneous analysis of catecholamine histofluorescence and peptide immunohistochemistry reveals the presence of catecholamine varicosities in juxtaposition to magnocellular neurons (Figs. 57–59). This is especially prominent for those magnocellular neurons that occupy the ventral extreme of the transplant at the interface with the host median eminence. Comparator bridge superimposition revealed some contact between these two chemically identified elements. Although this pattern appeared less extensive than the normal innervation of magnocellular neurons by catecholamine varicosities[38,39], it is consistent with the general avidity of noradrenaline and vasopressin.

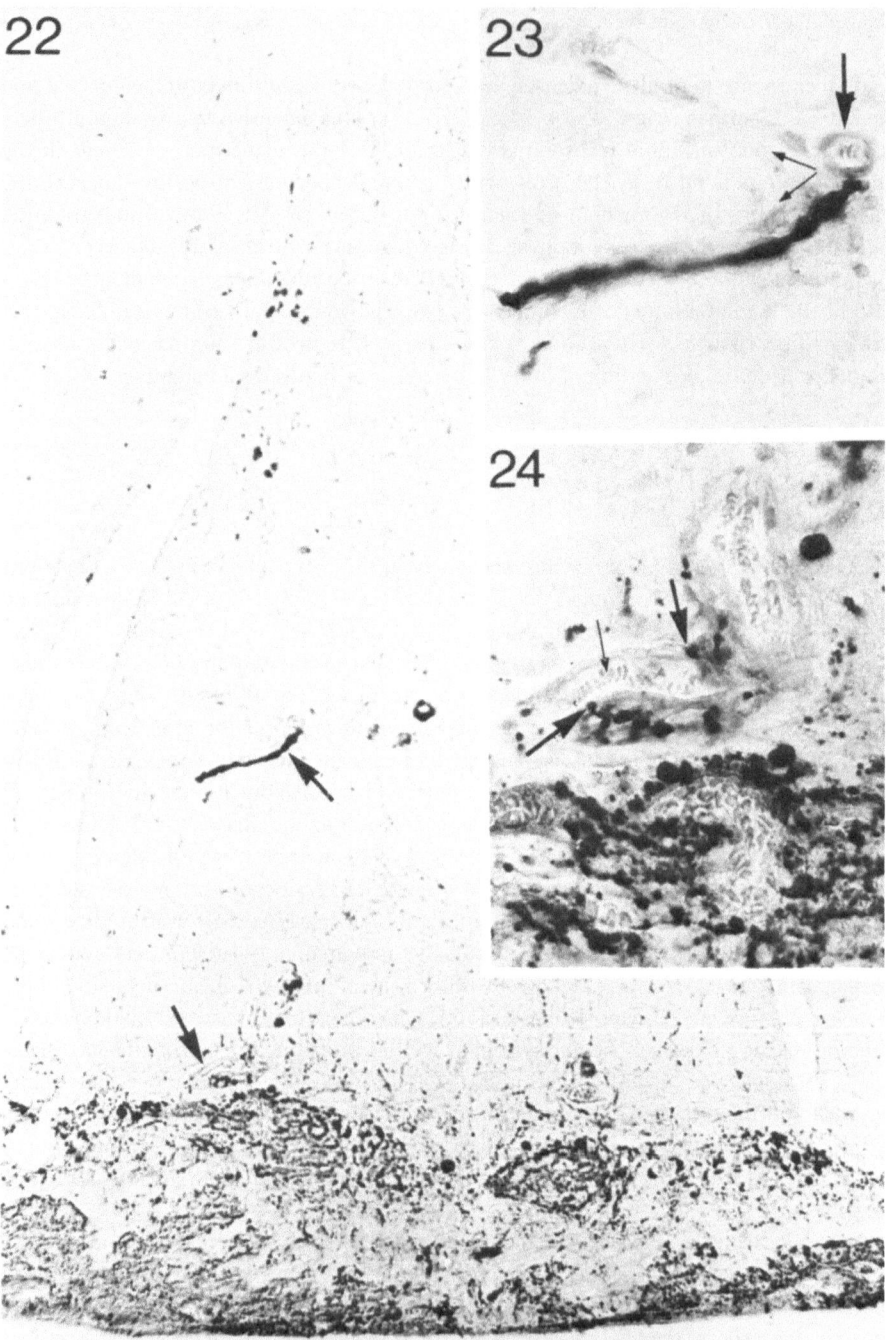

FIGURE 22. The association between neurophysin-positive fibers and the new vascular tree of the graft is depicted. Two blood vessels (arrows) are prominent features of the graft and are seen at higher magnifications in Figs. 23 and 24. Also note the extensive network of neurophysin-positive fibers in the underlying host median eminence. × 180.

5. Functional Considerations and Discussion

5.1. Transplants into Young Adult Hosts

Successful transplants (i.e., those that result in a significant change in water balance characteristics) appear to have several common anatomical characteristics. These include (1) a rich, vascular invasion from the underlying median eminence, (2) the presence of magnocellular perikarya stained positively for neurophysin and vasopressin, (a) within the transplant, (b) at the interface of the transplant and the host, and (c) in the host brain, and (3) an apparent association between immunopositive fiber systems and the portal vasculature of the host median eminence (Fig. 63). It is probable that these three elements combine to promote the appropriate release of vasopressin and subsequent delivery to the target cells of the kidney tubules. In contrast, nonfunctional grafts may contain magnocellular, vasopressin neurons that fail to contact a sufficient number of portal blood vessels due to a lack of anatomical proximity to the median eminence (Fig. 64). Because the presurgical polydipsia is replaced, as early as 3–4 days postsurgery, with a reduced fluid intake, it is probable that the vasopressin neurons are receiving some signal, possibly osmotic, that coupled with the fact that they are removed from their noradrenergic inhibitory input[40] could account for the rapid shift in fluid homeostasis. Moreover, it is possible that the tissue being transplanted may contain some degree of neurological control of vasopressin release perhaps mediated through a cholinergic facilitory input. It has recently been shown that cholinergic neurons exist in the immediate vicinity of the supraoptic nucleus[41] and therefore could be a component of the transplanted tissue. Although their viability is undetermined at present, it is possible that they may continue to influence the release of vasopressin. Excellent evidence to this effect has been generated with an *in vitro* explant that contains many of the same anatomical characteristics of the transplants, i.e., viable supraoptic neurons and probable neuronal input to these neurons.[42] Alternately or in concert with this, the noradrenergic system has been denervated, initially, by removal of the vasopressin neurons from their hypothalamic locus. It is well established that the noradrenergic input to the supraoptic nucleus arises from brain stem reticular neurons in the ventrolateral medulla and to some extent dorsomedial medulla.[43–45] Hypothalamic sources have not been identified as afferent to the supraoptic nucleus. Physiologically, this input is thought to exert an inhibitory, perhaps modulatory effect on the cholinergic-induced release of vasopressin from magnocellular neurons.[40] Thus, the mere deafferentation of this inhibitory link might result in a supersensitive magnocellular neuron that responds maximally to osmotic as well as other potential stimuli.

FIGURE 23. Two blood vessels appear to branch (in the direction of the small arrows) from a small arteriole (arrow) seen in cross-section. The more ventral branch appears associated with a large neurophysin-positive fiber. × 625.

FIGURE 24. This vessel is further identified by the presence of red blood cells (small arrow) within its lumen. Numerous beaded fibers (large arrows), containing neurophysin, are seen in association with this blood vessel. Moreover, the underlying median eminence is richly supplied with neorophysin-positive fibers especially in association with the portal vasculature. × 625.

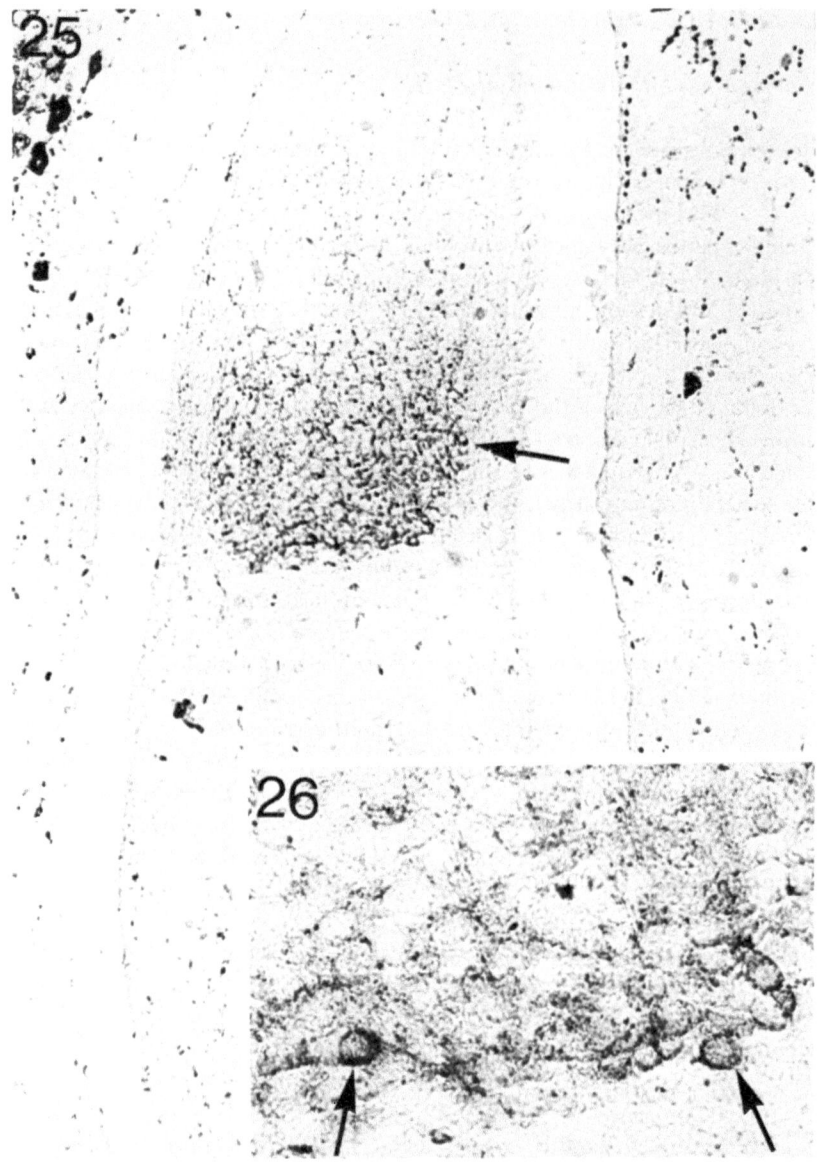

FIGURE 25. Densely packed neurophysin staining is seen within the dorsum of a graft (arrow). It is reminiscent of the suprachiasmatic nucleus and is composed of a fibrous network as well as parvicellular neurons. ×165.

FIGURE 26. Suprachiasmatic-like neurons are seen to advantage at higher magnification following neurophysin staining (arrows). Unlike magnocellular elements, these neurons do not appear to migrate to the interface of the host and graft and also are present in nonfunctional grafts that lack magnocellular elements. × 760.

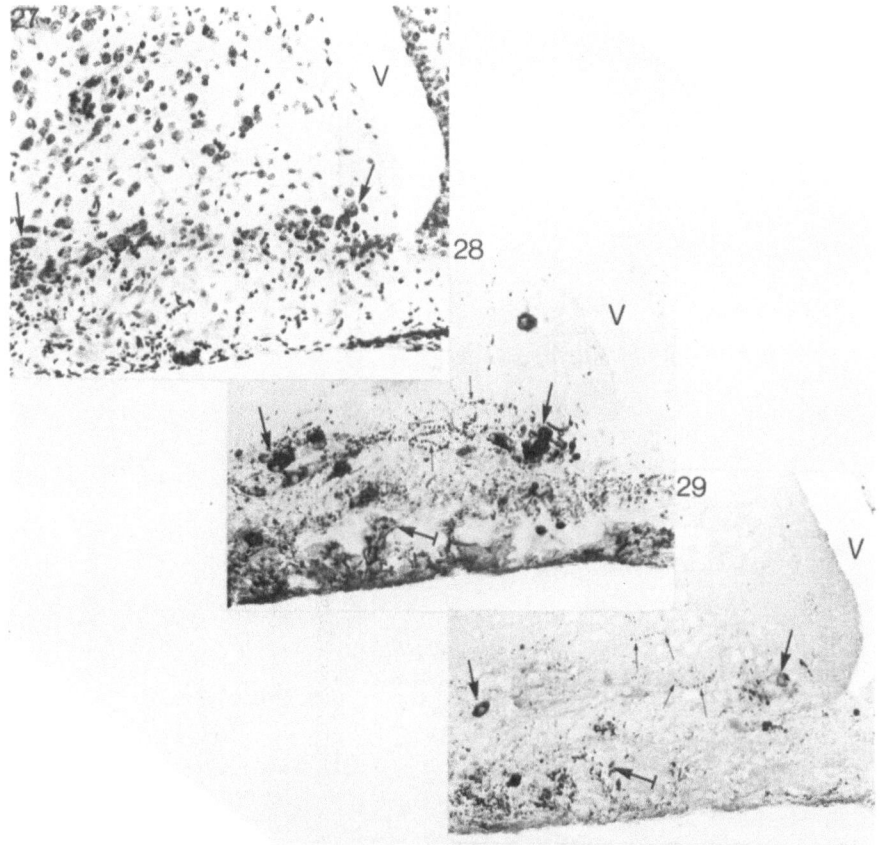

FIGURES 27-29. These alternate sections at 20-μm intervals depict the lower right portion of a graft. Figure 27 has been stained with cresyl violet and luxol fast blue, Fig. 28 for rat neurophysin, and Fig. 29 for vasopressin. The third ventricle (V) is indicated as a landmark. The graft is situated on the dorsum of the median eminence and is characterized by the presence of magnocellular neurons that stain positively for Nissl, neurophysin, and vasopressin (large arrows). These neurons are common features of functional grafts and frequently are located at the interface between the graft and host brain. They are continuous with a fiber network in the graft that extends into the host brain and is seen in association with portal vessels (crossed arrows). × 115.

The observation that vasopressin stains less densely than does neurophysin for *magnocellular* but not *parvicellular* neurons may indicate that the magnocellular neurons are releasing vasopressin at capacity. The parvicellular, suprachiasmatic-like neurons are not believed to play a role in water balance, which may account for their consistent staining, immunohistochemically. It is logical that a significant facilitory drive for vasopressin release exists in the grafted, Brattleboro rat and that the immunohistochemical staining may reflect minimal storage of vasopressin within magnocellular perikarya. It should be remembered that the perikaryon is the site of vasopressin synthesis.[24] Thus,

FIGURES 30–41. This series is comparable to that illustrated in Figs. 3–14 and is stained for neurophysin. Because the neurophysin antiserum recognizes both oxytocin- and vasopressin-associated neurophysin, positive staining also is seen in the host brain and represents oxytocin neurons and fibers. Positive staining in

in the chronically dehydrated rat, axonal transport and terminal release may exceed the rate of replenishment via biosynthetic mechanisms. This postulated increase in peptide turnover then could be reflected by the strong neurophysin and moderate vasopressin staining seen presently.

About 20–25% of hosts that contain hypothalamic tissue grafts exhibit a dramatic shift in fluid balance. As in control grafts (Fig. 60) described below, hypothalamic graft hosts (Fig. 61) displayed a period of transient adipsia for 48 hr following surgery. Unlike controls, the decrease in water consumption continued past the 48-hr recovery period in functional grafts. Although water intake fluctuated, it established a new, reduced level between 0.1 and 0.3 ml H_2O/g body wt. Urine osmolality also changed inversely; it was common to find animals that responded by concentrating urine to 800–1000 mOsm/liter H_2O. Animals such as these, exhibiting significant drops in water intake and increases in urine osmolality, are considered "functionally responsive." In contrast, nonfunctional transplants are characterized by a lack of change of these parameters. Several types of control implants have been performed; these include grafts of (1) occipital cortex into the third ventricle and (2) hypothalamus into the lateral ventricle. In each instance, surgery is followed by a period of transient adipsia, but polydipsia quickly develops again and reaches presurgical levels.

Another possibility for the early change in fluid intake is that vasopressin may be released from degenerating nerve cells. While this may occur, it is clear that a substantial number of magnocellular neurons survive the transplant procedure and integrate with the host brain. Because these neurons stain positively for neurophysin and vasopressin at 40 days after surgery, it is probable that they are actively synthesizing these neurohypophyseal principles and releasing vasopressin in response to an undefined stimulus. In order to fully address this issue, it would be necessary to challenge the host with physiological or pharmacological stimuli and to measure subsequent alterations in serum vasopressin levels.

---

the graft brain, therefore, can represent either neurohypophyseal peptide. Vasopressin staining as depicted in Fig. 29 has confirmed that a substantial amount of neurophysin staining within the graft is due to the presence of vasopressin neurons. The ventricle (V) is indicated. In Fig. 30, at the rostral pole of the graft, a dense staining appears and is continuous with the larger, more densely aggregated mass seen in Fig. 31 (arrow), which represents the suprachiasmatic nucleus. Oxytocin neurons of the caudal portion of the host paraventricular nucleus are seen in the dorsal portion of the field. In Fig. 33, fiber staining in the graft and host becomes more apparent. A large beaded fiber (small arrow) courses in the ventral portion of the graft. In the host median eminence, normal fibers of the internal lamina (crossed arrow) are seen. However, the external lamina is filled with neurophysin-positive material (arrow), which is atypical for the unimplanted rat. Figure 34 depicts a dorsally placed area of neurophysin staining equivalent to the suprachiasmatic nucleus contralateral to that seen in Fig. 31 (arrows). In Fig. 35, additional beads of immunopositive material are seen within the graft (arrow). The suprachiasmatic nuclear staining continues in Fig. 36 and is seen in association with a vacuolated area (*), which may represent a resorbed optic chiasm. Magnocellular neurons are prominent (arrows) and beaded fibers are seen in association with blood vessels (arrowheads; see Fig. 23). In Fig. 37, an increased number of magnocellular neurons is seen in the ventral third of the graft and at the graft–host interface (arrows). Figures 39–41 show a gradual decrease in the size of the graft as the infundibular recess is approached caudally. Neurophysin-positive staining occurs in the pituitary stalk (S) seen to advantage in Figs. 42–44. × 50.

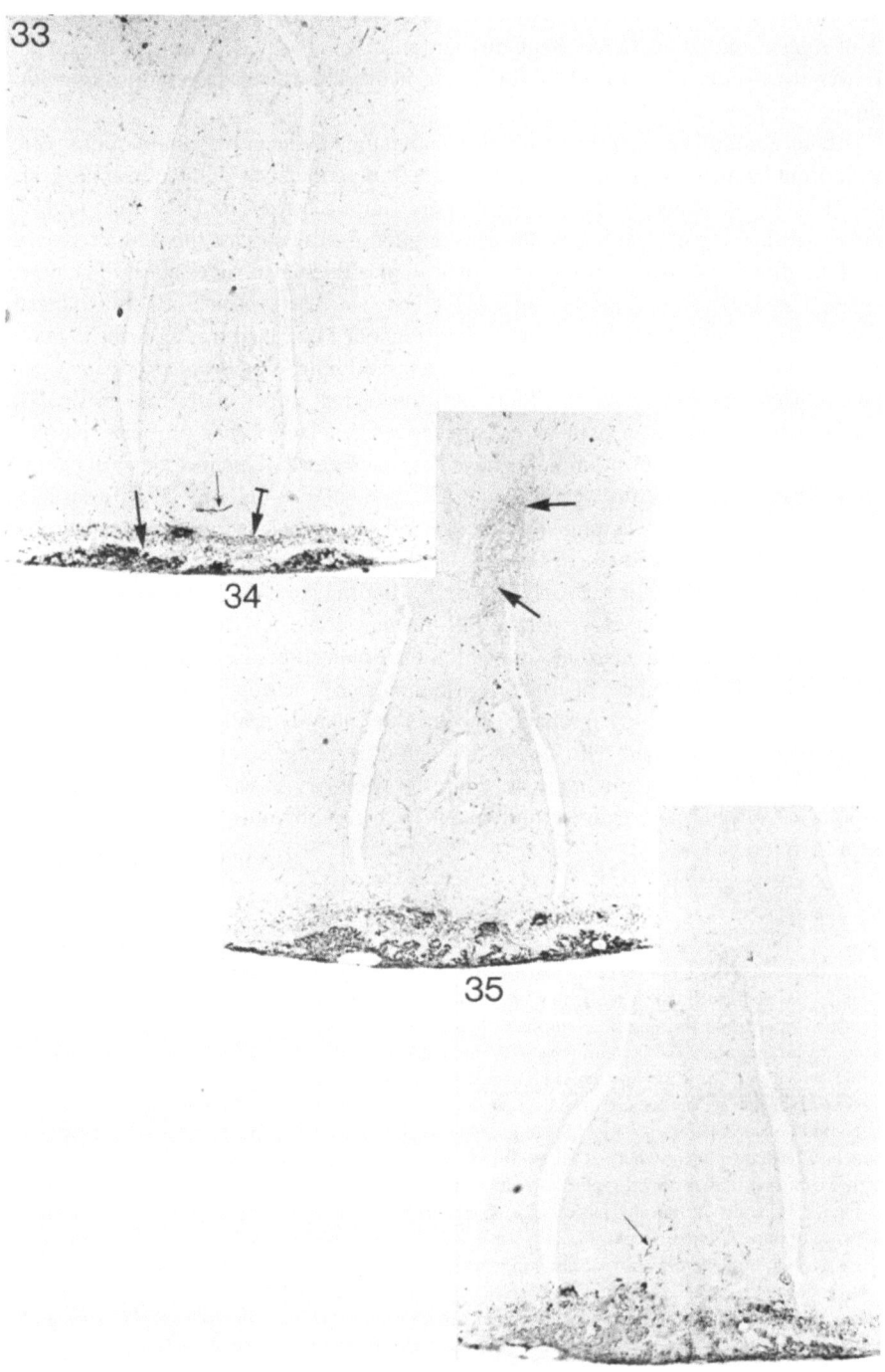

FIGURES 30–41 (*continued*)

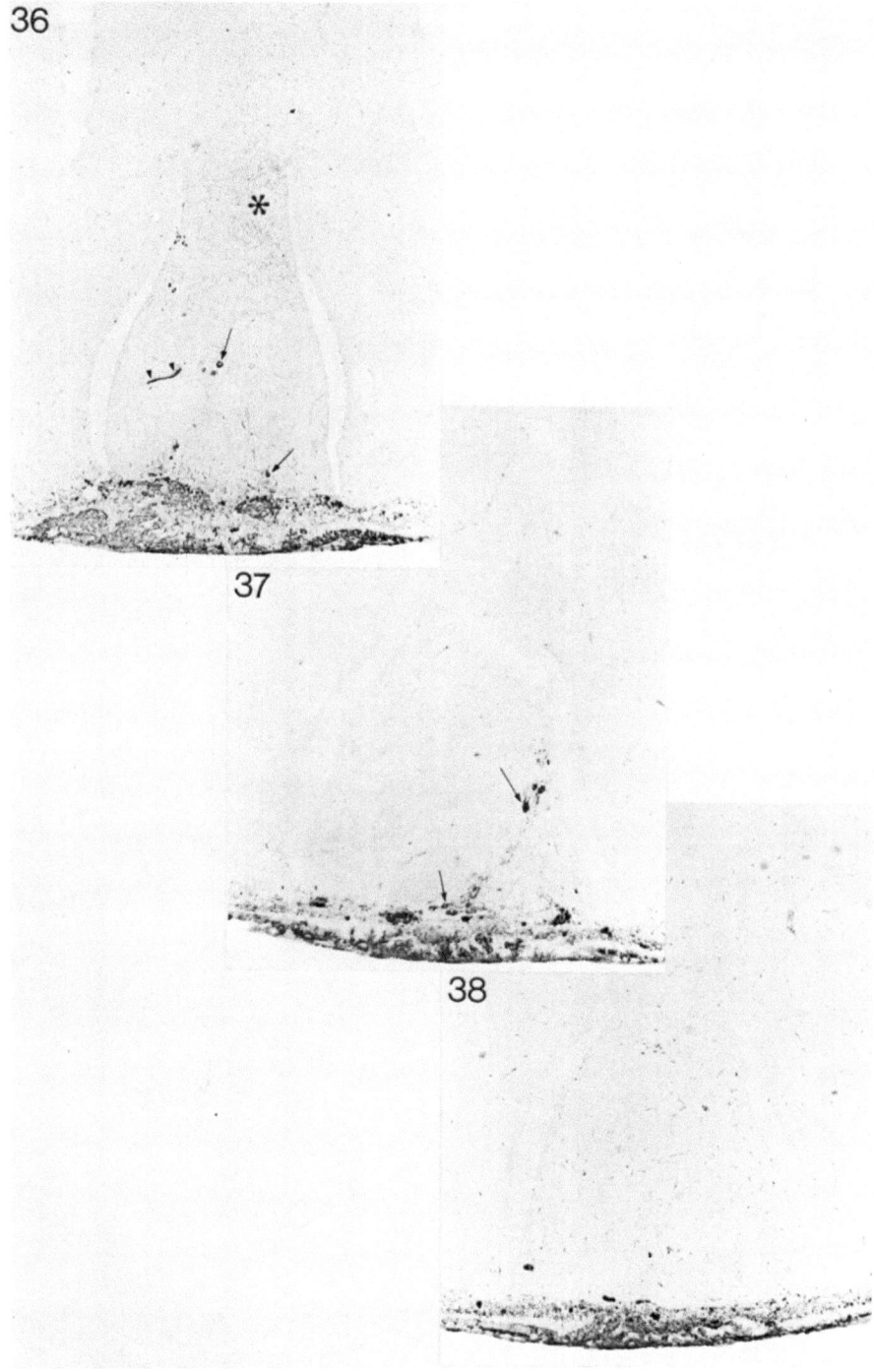

FIGURES 30–41 (continued)

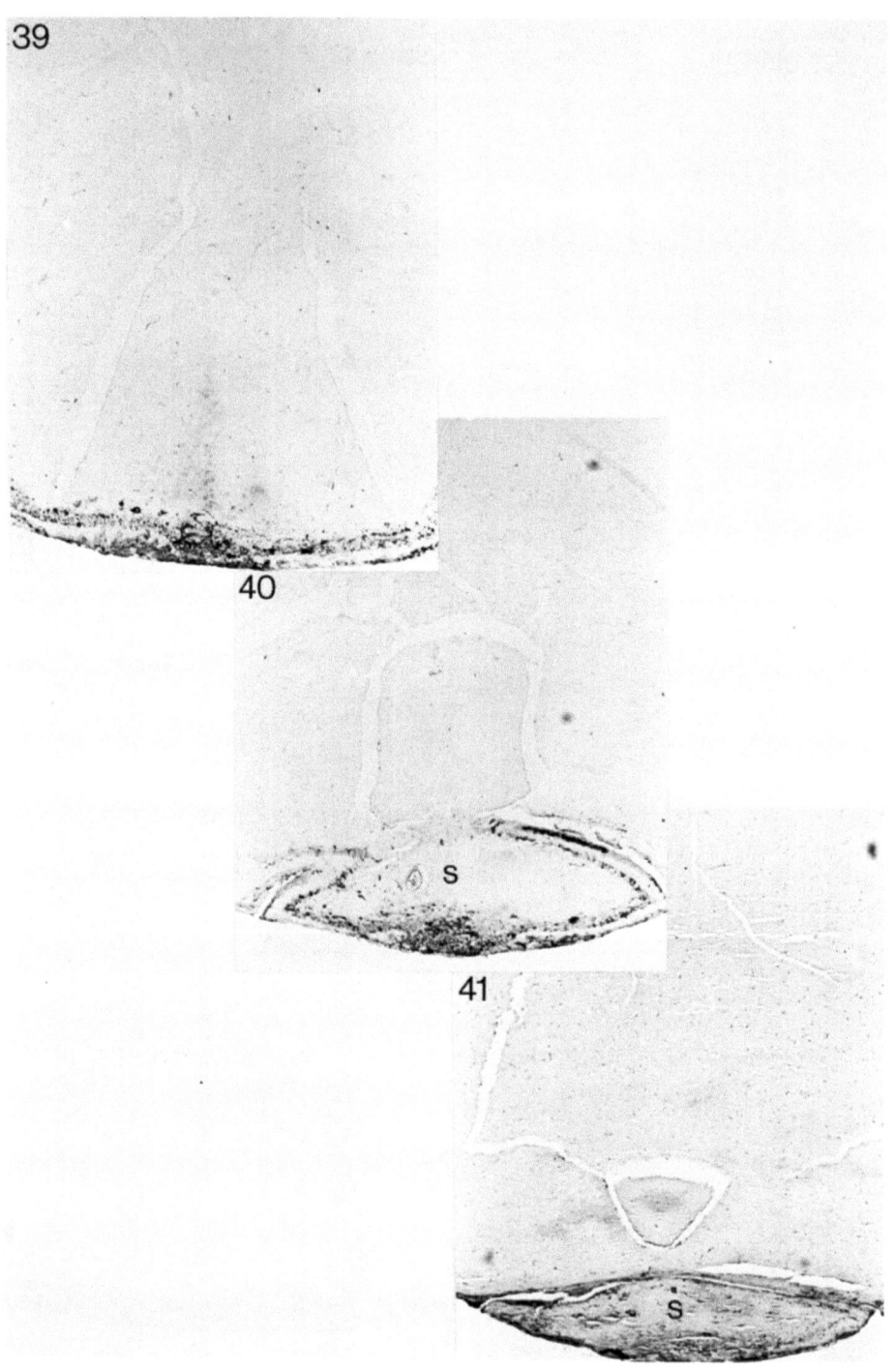

FIGURES 30–41 (*continued*)

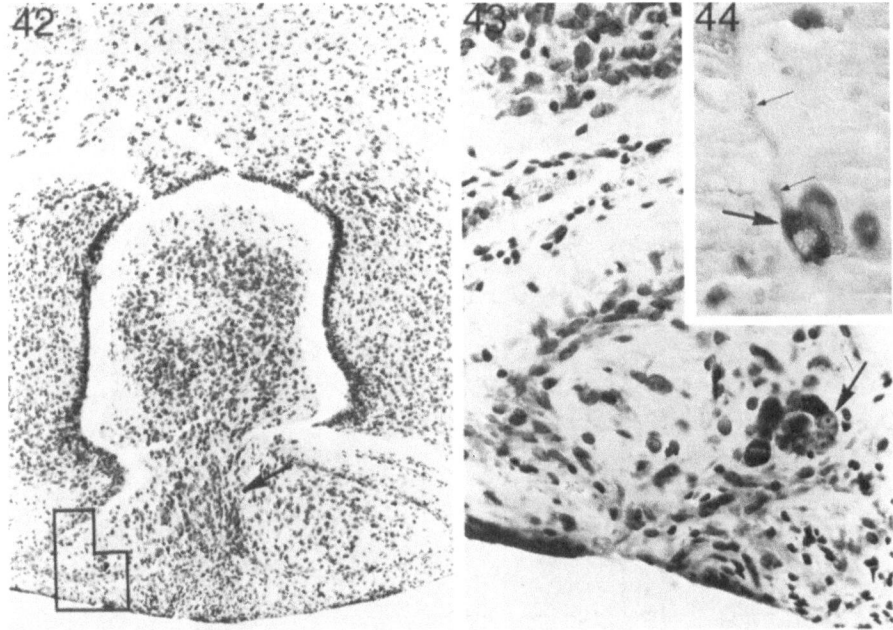

FIGURE 42. The infundibular recess of the third ventricle is seen at this caudal level of the graft. The graft appears to show a continuation of cellularity (arrow) ventrally into the underlying pituitary stalk. A portion of the stalk (outlined area) contains magnocellular neurons that are seen to advantage in Fig. 43. × 80.

FIGURE 43. Magnocellular neurons, staining with cresyl violet, are seen within the pituitary stalk (arrow). They apparently have migrated from the graft tissue. × 220.

FIGURE 44. The large neurons depicted in Fig. 42 and 43 appear to contain neurophysin (large arrow). One of these neurons is continuous with a process that extends dorsally (small arrows) in a different plane of focus than the perikaryon. × 300.

5.2. Transplants into Aged Hosts

Considerable interest has focused on normal characteristics of aged neurons. Many neural systems are known to undergo certain declines in function during the aging process. Dopamine is often deficient in aged humans and has been shown to be depressed in content within the substantia nigra of aged monkeys.[46] Sharp declines in acetylcholine accompany age-related memory loss,[47] and vasopressin responsiveness to dehydration is diminished in aged rats.[48] Neuronal loss also occurs in some brain loci.[49] Moreover, cell loss occurs for certain areas that is exaggerated in pathological conditions such as senile dementia of the Alzheimer type. Transplantation of neurons may present an opportunity to restore either lost cell numbers or function that often accompany the aging process.

We have tested the ability of aged brain to successfully host transplanted neurons in preliminary experiments that parallel those described above, i.e., fetal vasopressin neurons into aged, host third ventricle. Host Brattleboro rats were either 12 or 25 months of age at the time of receipt of fetal tissue. Most animals were monitored for 40 days for changes in water balance after which time the brains were processed for cate-

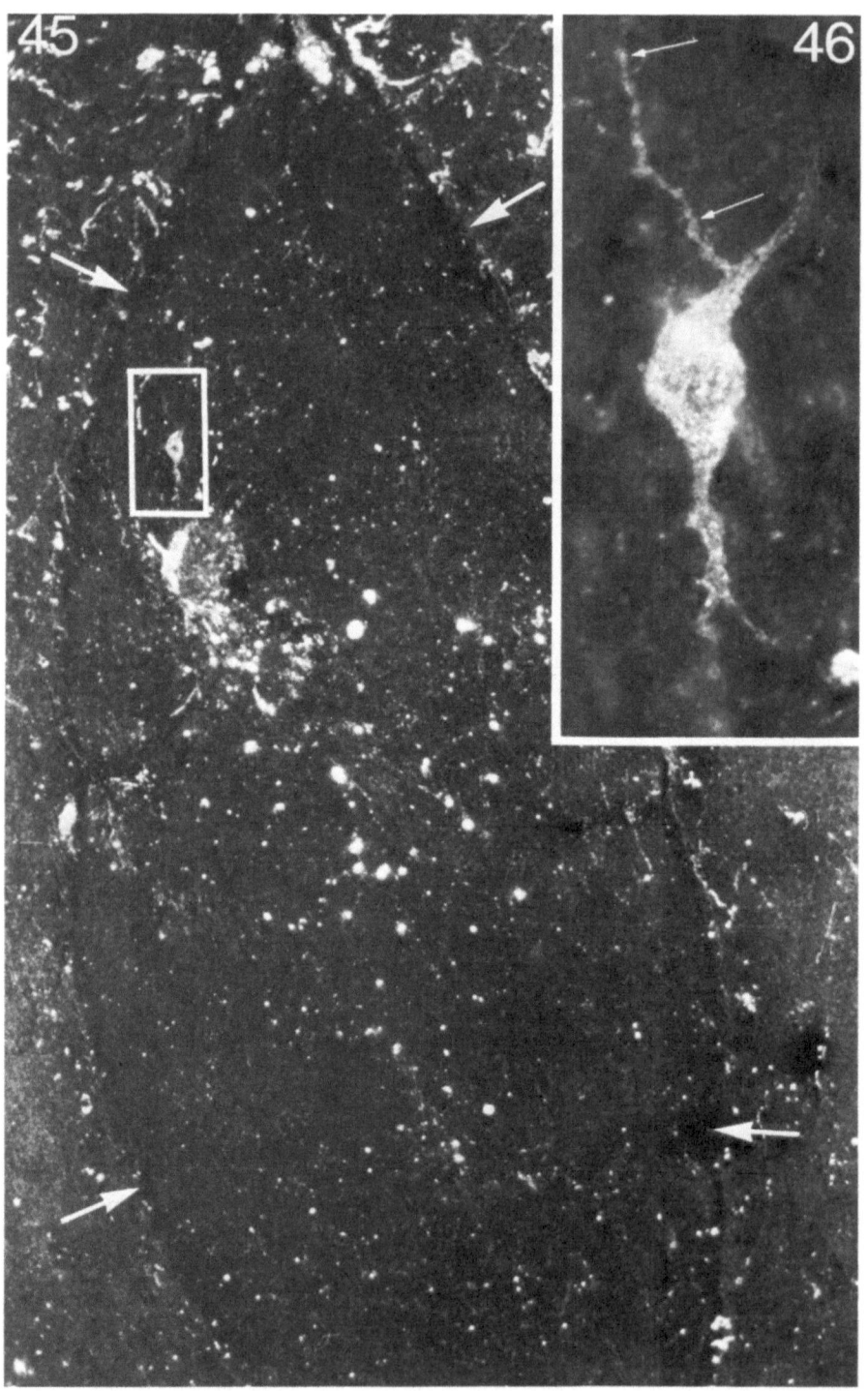

cholamine fluorescence and neurophysin immunohistochemistry; a few of the 12-month-old hosts were killed after 56 days. Although the grafts did not appear to affect water intake, many grafts appeared to contain parvi- and magnocellular hypothalamic neurons that stained positive for neurophysin (Figs. 45, 46). Catecholaminergic neurons also were prominent features. Beaded processes extended from peptidergic and catecholaminergic neurons; however, they did not show evidence of growth across the graft–host interface. Exceptionally fine, smooth processes, which are reminiscent of axons, appeared to emanate from perikarya and proximal dendrites of neurophysin-positive neurons found within the grafted tissue.

These findings are suggestive that the age of the host tissue may not be an obstacle to vigorous growth of grafted neurons and are consistent with the finding that serotoninergic neurons can survive transplantation into the hippocampus of 24-month-old mice.[50] A number of characteristics paralleled those seen in our younger graft recipients including the presence of magno- and parvicellular neurophysin-positive neurons, the occurrence of a rich vascularization, and the absence of tissue rejection or a scar tissue barrier phenomenon. Although the diabetes insipidus of the aged hosts was not significantly altered in these grafts, it is possible that this reflects the critical nature of the graft sites, as many of the aged host grafts were found either in hypothalamic areas adjacent to the third ventricle or dorsally in the overlying thalamus. One potentially significant difference between grafts into young and old hosts is the absence of growth of axons from magnocellular vasopressin neurons into the host median eminence. Although the positioning of the grafts did not always favor this in the old hosts, it is possible that a reduced stimulus for directed growth into this vascularized bed existed in the aged brain through a loss of some type of growth or neuronotrophic factor.

5.3. Transplants into Neurohypophysectomized Rats

The consequences of surgically removing the neural lobe of the pituitary have been extensively investigated.[51–53] Following neurohypophysectomy, rats exhibit a severe diabetes insipidus that, over a course of several months, is gradually alleviated. Histological studies have found[51,53] that the infundibular stalk, proximal to the lesion, develops into a miniature posterior lobe approximately one-quarter the size of the normal neurohypophysis. While the diabetes insipidus was ameliorated, water intake and urine output are still about twice as high as normal 1 to 2 months after stalk sectioning,[52] indicating that while there is functional recovery, it is incomplete.

We (Gash, Rohrer, and Marciano, studies in progress) have been examining the ability of hypothalamic grafts to assist in the functional recovery of the hypothalamic

FIGURE 45. A graft (arrows) appears to occupy a large portion of the third ventricle in a 12-month-old Brattleboro rat host. A magnocellular neuron (outlined in the rectangle) is a prominent feature and is seen to advantage in Fig. 46. × 240.

FIGURE 46. This neurophysin-positive magnocellular neuron appears to project two large primary dendrites at dorsal and ventral poles; the dorsal process appears continuous with a beaded axon and is exemplary of vigorous growth of grafted neurons (arrows). × 990.

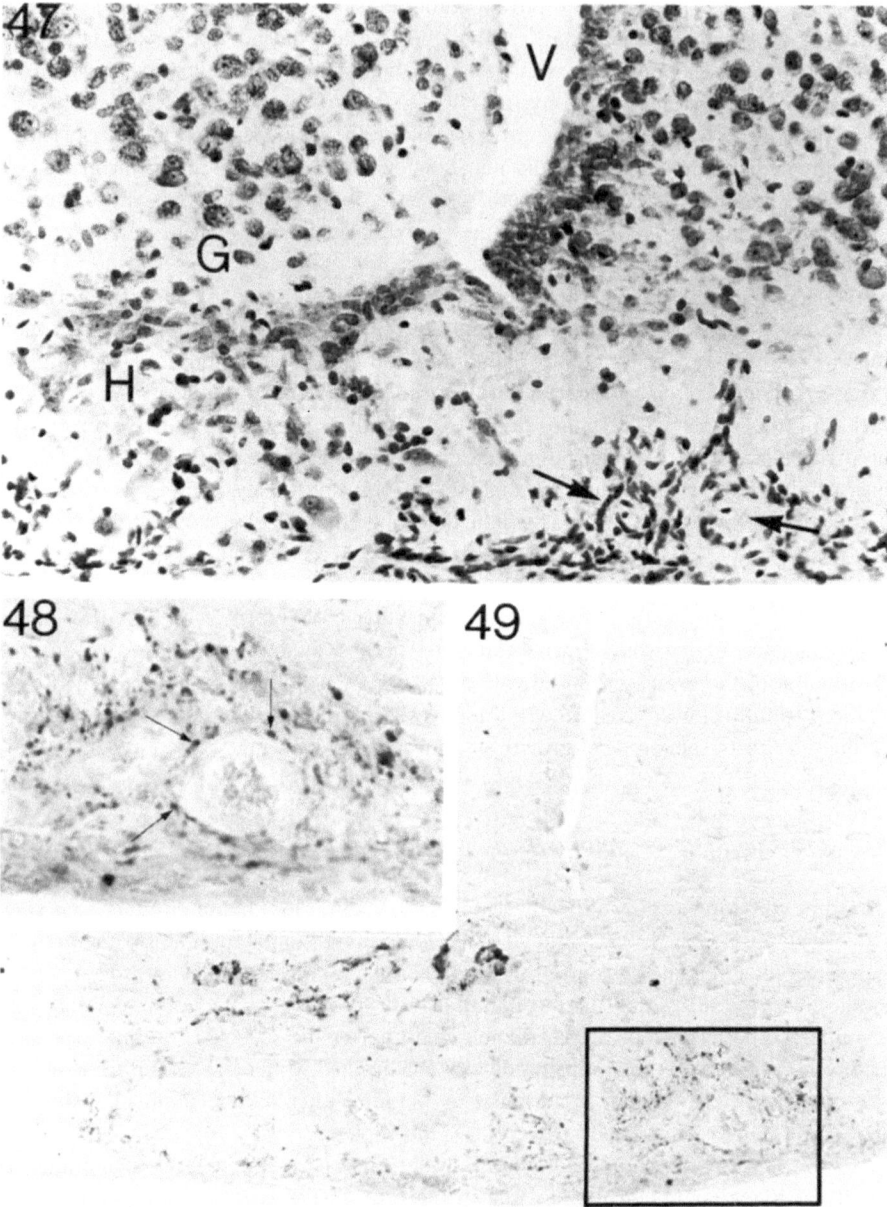

FIGURE 47. Cresyl violet staining reveals the presence of portal blood vessels in the external lamina of the median eminence (arrows). The relative position of the graft (G) and host (H) is indicated. The blood vessels are seen following vasopressin staining in Figs. 48 and 49. × 280.

FIGURES 48 AND 49. Figure 48 is a high-power insert of the field illustrated by the rectangle in Fig. 49. × 670. Figure 49 serves to verify the position of these blood vessels in relation to those identified with cresyl violet staining. Numerous vasopressin-positive fibers are seen adjacent to the tunica adventitia of the portal vessel. The lumen is characterized by the presence of blood cells. × 280.

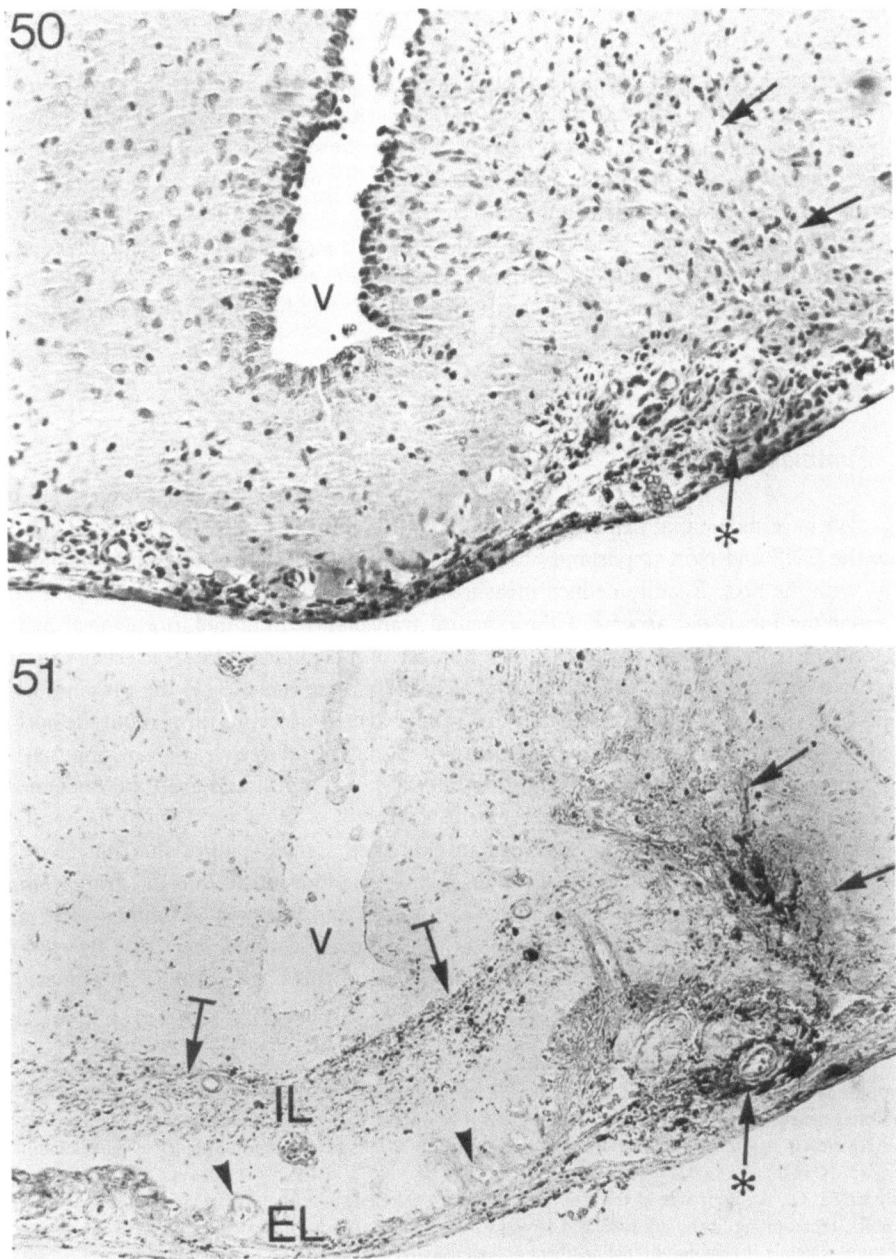

FIGURE 50. A nonfunctional graft. The position of the graft is marked (arrows) as well as the position of a large blood vessel (arrow with *). The third ventricle is indicated (V). × 180.

FIGURE 51. In this section, adjacent to Fig. 50, neurophysin immunochemistry reveals the presence of immunoreactive fibers within the misplaced graft in a position lateral to the third ventricle (V). Some neurophysin fibers are seen in association with a blood vessel (arrow with *). The host median eminence appears normal for a Brattleboro rat, i.e., neurophysin-positive fibers (crossed arrows) are seen within the internal lamina (IL), but neurophysin immunoreactivity (arrowheads) does not occur in the external lamina (EL). This is in distinction to functional grafts, which have an abundance of immunopositive material in association with blood vessels of the external as well as the internal lamina. × 180.

neurosecretory system following neurohypophysectomy. Young adult male Long–Evans rats have been used in this study. Our initial results (Fig. 62) indicate that 17 days post-coitus fetal hypothalamic grafts significantly alleviate the symptoms of diabetes insipidus in rats that were neurohypophysectomized 6 days prior to transplantation. Functional recovery in these graft recipients mimics the recovery seen in Brattleboro rats following successful transplantation. Evidence of graft function is seen within the first 6 days following transplantation. Daily fluctuations in antidiuretic activity also are seen that may reflect upon the development of host–graft interactions including innervation of the graft by host fibers.

6. Summary

We have shown that peptidergic, hypothalamic neurons can survive transplantation into the CNS and most importantly can integrate, both morphologically and functionally, with the host. Readily defined measures such as fluid homeostasis can be used to monitor the functional aspects of these neural transplants. Immunohistochemical and histofluorescence analysis have revealed a number of morphological characteristics that appear to be common features of functional grafts. These include (1) the presence of well-differentiated vasopressin neurons that ramify axonal processes throughout the host median eminence in the immediate vicinity of portal blood vessels, (2) an apparent migration of magnocellular neurons to the interface of the host and graft and to some extent into the host brain, and (3) the extensive growth of blood vessels into the graft, especially from the underlying host median eminence. Other features that may have functional implications include the growth of catecholamine fibers into the graft from various sources including the periventricular stratum and the median eminence. It is conceivable that some degree of influence is exerted over vasopressin release by catecholamines as the juxtaposition of catecholamine varicosities and magnocellular peri-

---→

FIGURE 52. The dorsal portion of a third ventricular graft A line separating the graft (G) and host (H) indicates the relative position of the interface, which is bridged by catecholamine fibers from the adjacent periventricular stratum. The third ventricle is indicated (V). This field is seen in higher magnification in Fig. 53. × 165.

FIGURE 53. On the graft side of the interface, numerous catecholamine-containing varicosities and linear profiles, representing fibers, are indicated (arrows). These appear to have grown from the periventricular stratum, which is rich in noradrenaline fibers as well as other aminergic and peptidergic elements. × 500.

FIGURE 54. The ventral portion of a graft (G) is seen dorsal to an atypical median eminence. The median eminence is characterized by an unusually large number of portal vessels which appear quite large (∗). These vessels are surrounded by a dense plexus of dopamine fibers from the A-12, tuberoinfundibular dopamine neurons. × 170.

FIGURE 55. In this high-power view of the ventral portion of a functional graft, a proliferation of blood vessels (∗) carrying dopaminergic fibers is seen to extend from the host (H) into the graft tissue. The relative position of the graft–host interface is indicated by a dashed line. Dopaminergic fibers (long arrows) course dorsally into the graft to yield a plexus of fine-size varicosities (short arrows) throughout the ventral portion of the graft (G). × 300.

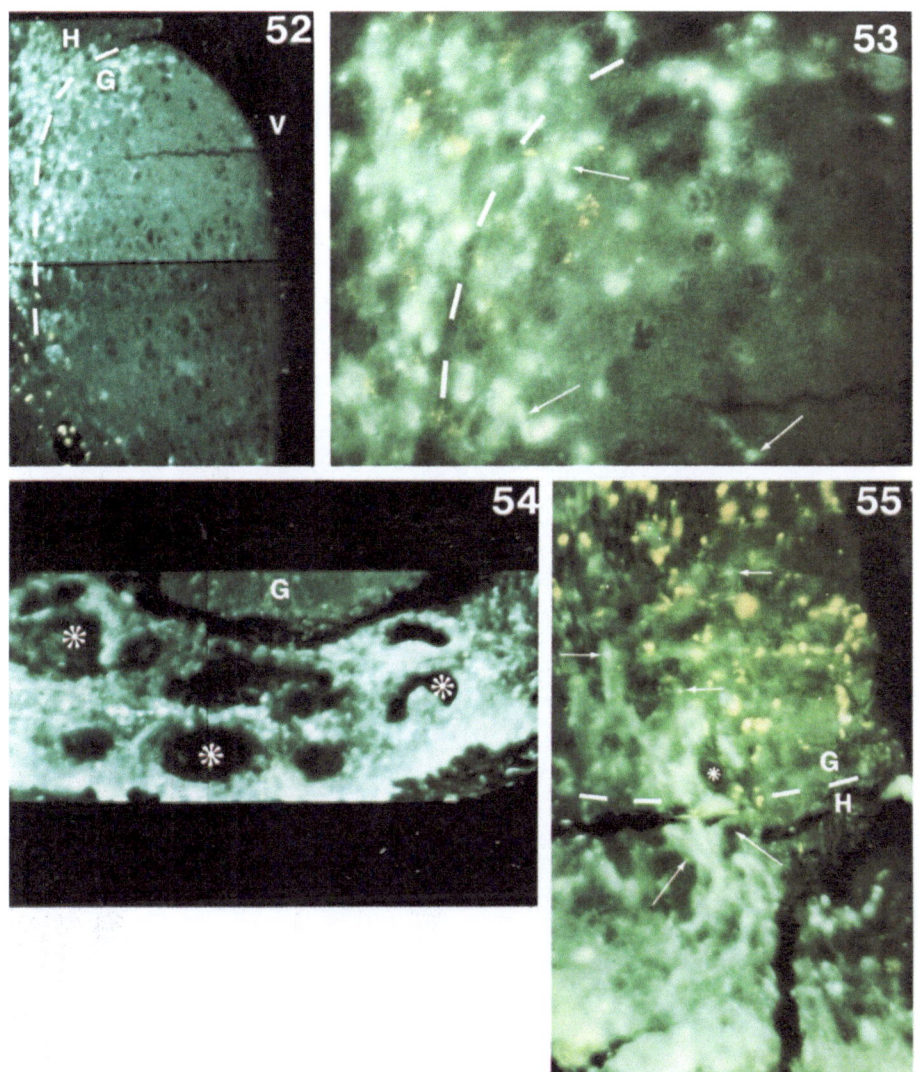

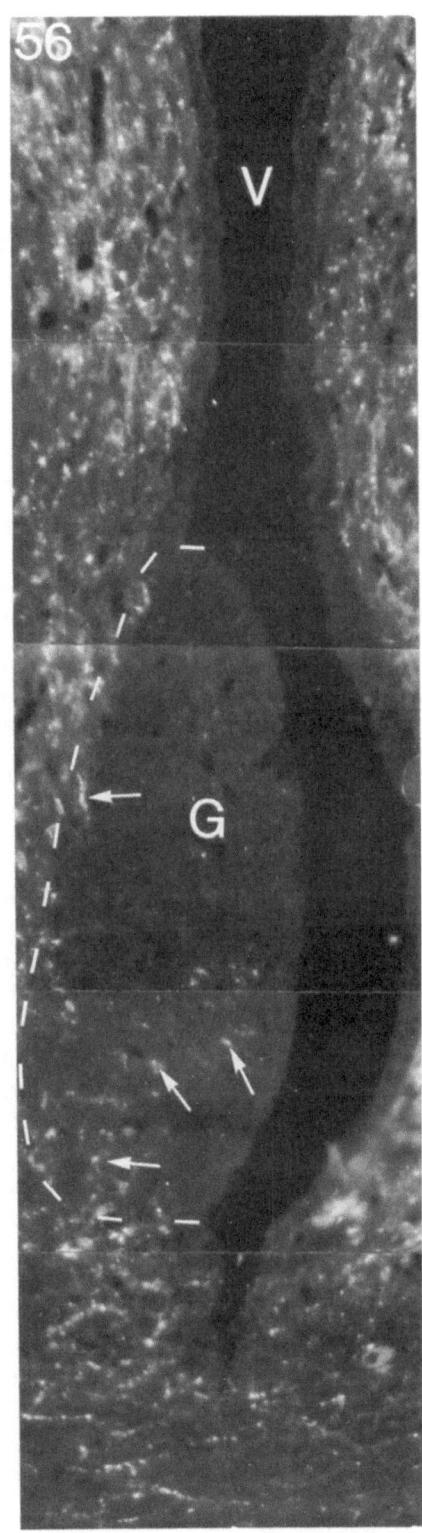

FIGURE 56. This photomontage illustrates the ingrowth of reticuloinfundibular noradrenergic fibers from the underlying floor of the third ventricle. Although the ingrowth by these fibers is rather modest, it does represent a route of entry of noradrenergic fibers (arrows) in contast to the dopaminergic fibers which follow portal vessels. The catecholamine-rich periventricular stratum is adjacent to the lateral walls of the third ventricle. × 285.

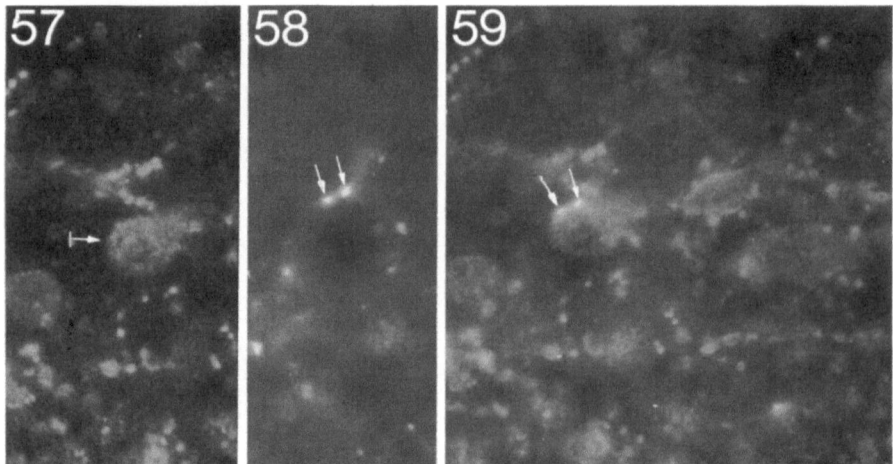

FIGURES 57–59. These photomicrographs depict comparator bridge analysis of peptides and catecholamines in a functional graft. In Fig. 57, a 10-μm section has been stained for neurophysin and reveals the presence of a magnocellular neuron (crossed arrow) as seen with dark-field illumination. It is neurophysin positive. The adjacent, serial tissue section, as seen with catecholamine histofluorescence, is depicted in Fig. 58; intensely fluorescent catecholamine variosities (arrows) are adjacent to the nonfluorescent perikaryon which was seen to stain for neurophysin in the previous section. Superimposed, comparator bridge analysis is seen in Fig. 59. Here, the catecholamine varicositis appear in apposition to the neurophysin-positive neuron, but are not as intense as in Fig. 58 because of fading due to repetitive photographic exposures. × 550.

karya suggests a neuronal interaction; this would be particularly effective in long-term grafts as the 20- and 40-day survivial times examined presently are somewhat shorter than optimal for innervation.

It is of interest that fluid intake is altered quickly following transplantation. It is probable that this rapid response to hypothalamic tissue transplants, seen in functionally responsive hosts, may represent an initial, i.e., nonneurologically controlled, release of vasopressin. Magnocellular neurons are characterized by spontaneous electrical activity which could result in the release of vasopressin. In addition, the removal of these neurons from the developing hypothalamus of the fetus may result in the elimination of an inhibitory input, such as the ascending noradrenaline neurons, which physiologically appear to modulate vasopressin release. This then could augment facilitory input from neurons endogenous to the grafted tissue. It is of interest in this regard that acetylcholine is known to be a powerful stimulus to vasopressin release and that recent immunohistochemical studies of choline acetyltransferase localization clearly place cholinergic neurons within the anatomical areas dissected for these transplant purposes. As osmoregulation appears to be mediated through a cholinergic mechanism, it would follow that the strong osmotic signal present in the Brattleboro rat host could result in stimulation of grafted magnocellular neurons to release vasopressin. This release presumably is into blood vessels of the pituitary portal system, which rapidly grow into the transplant from the underlying median eminence; grafts into the lateral ventricle do not appear to alter

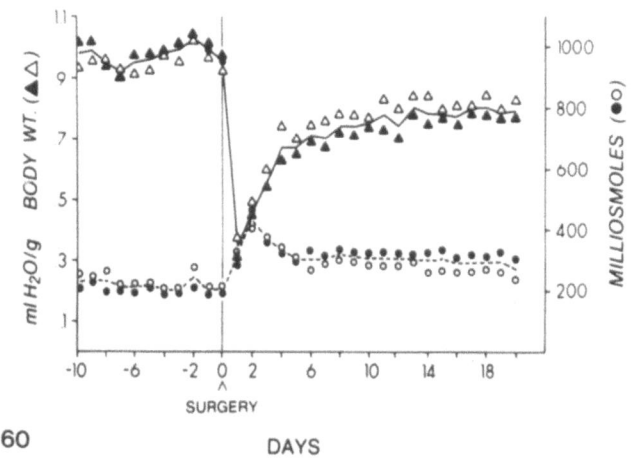

60

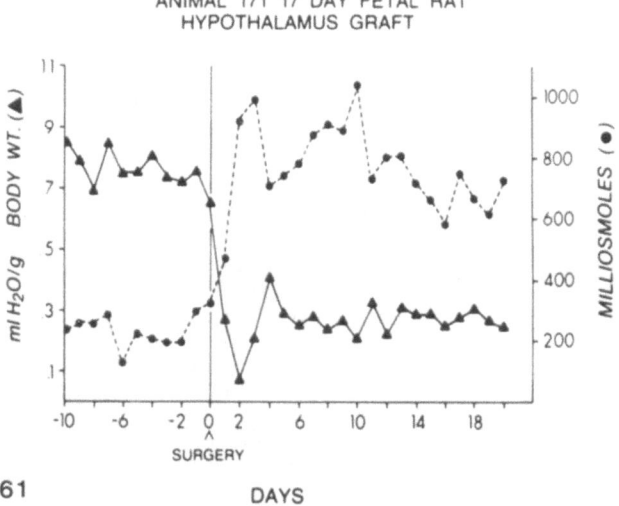

ANIMAL 171 17 DAY FETAL RAT
HYPOTHALAMUS GRAFT

61

FIGURES 60 AND 61. Effects of surgery and transplantation in control (Fig. 60) and experimental (Fig. 61) Brattleboro rats. The average daily consumption of water by the sham-operated rats ($\triangle$, $N = 11$) and the control tissue-implanted rats ($\blacktriangle$, $N = 18$) is indicated by the solid line. The average osmolality of the urine is plotted on the same graph. Because we did not measure osmolality for all the controls, the sample size is smaller; $N = 18$ for the sham-operated rats ($\bigcirc$) and $N = 10$ for the rats with occipital cortex implants ($\bullet$). The standard error for each daily average was ± 0.04 ml for water consumption and ± 43 mOsm for urine osmolality. In some of the experimental animals (a rat that received vasopressin neurons from a 17-day-old fetus), there were sustained decreases in water consumption and sustained increases in urine osmolality. Changes of these magnitudes were never observed in control animals. Ref. 13.

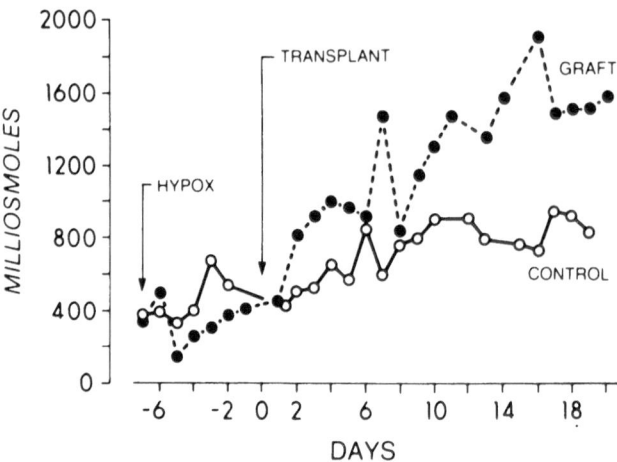

RECOVERY FROM NEUROHYPOPHYSECTOMY

FIGURE 62. Effects of surgery and transplantation in neurohypophysectomized rats. The average daily urine osmolality is plotted from the time of hypophysectomy (hypox) through transplantation and during the 20 days postsurgery. A matched sham-transplanted control animal (O) was examined concurrently with each graft recipient (●).

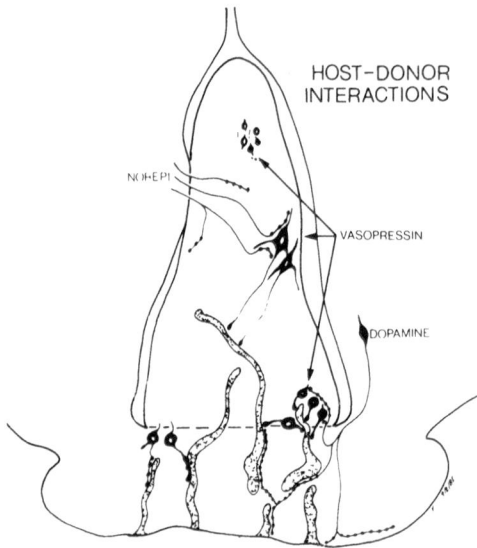

HOST–DONOR INTERACTIONS

FIGURE 63. This schematic illustrates common features of functional vasopressin grafts. They include the presence of a well-differentiated piece of neural tissue within the third ventricle situated on the floor of the median eminence. It is often continuous with the walls of the ventricle especially if the ependymal lining has been denuded. Blood vessels of the underlying median eminence are prominent features of the graft and appear to result from dorsal growth into the grafted tissue. Parvicellular and magnocellular vasopressin-containing neurons, seen both with neurophysin and vasopressin staining, are common features of the graft as are oxytocin-containing neurons. Parvicellular neurons do not appear to migrate to the ventral, portal-vessel-rich portion of the graft and host, but magnocellular neurons do. These larger vasopressin neurons also appear to ramify processes toward the median eminence and in the vicinity of portal vessels. These neurons also appear to migrate to the interface of the host and graft and further to extend into the host

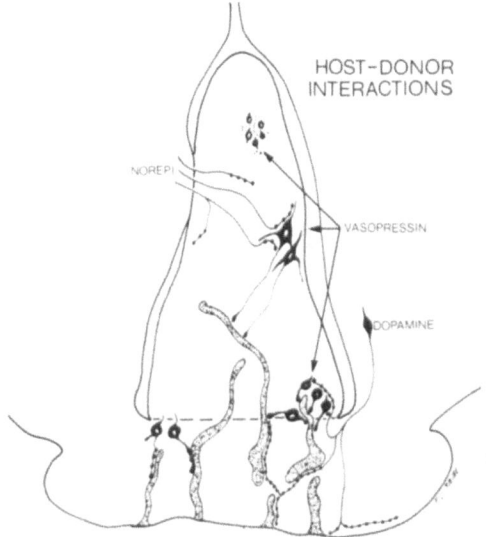

FIGURE 64. This schematic illustrates three types of nonfunctional, vasopressin grafts. Animal 86 (illustrated in Fig. 51) demonstrated magnocellular neurons in the graft as well as some association of fibers with blood vessels lateral to the host median eminence. However, it did not display a rich plexus of fibers in the host median eminence. Animal 174 displayed a dorsally placed graft with a suprachiasmatic-like plexus of fibers. Animal 1134 contained a dorsal hypothalamic–ventral thalamic graft containing magnocellular neurons and fibers which did not appear in association with blood vessels.

fluid balance, which would appear to lend important supporting evidence for this hypothesis.

Neural grafting presents itself as an approach of major benefit for the study of neuronal growth and development. It also poses numerous possibilities for the study of disease states and offers some promise for application to neurological disorders such as Parkinson's disease and others as discussed elsewhere in this book. While the complexity of the nervous system may prohibit reconstruction of complete systems in a morphological sense, it is clear that knowledge of transmitters employed by a system permits placement of the proper neurochemical messenger in appropriate neural loci for restitution of function. With the current expansion of our understanding of neurotransmitter candidates to include over 30 different peptides, it is likely that continued study of the nervous system will provide needed information about the role of these endogenous neural substances in normal function and disease. This knowledge therefore could lead to rational therapies to address certain neurological disorders through the use of neural grafting procedures.

median eminence and pituitary stalk. They provide a rich plexus of vasopressin fibers intimately associated with portal vessels. Noradrenaline-containing fibers appear to enter from two routes, the periventricular stratum and the underlying (not illustrated) reticuloinfundibular system. Dopaminergic fibers appear to follow portal vessels into the graft.

ACKNOWLEDGMENTS. The authors are appreciative of support from the USPHS (Grants NS 15109, NS 15816, and AG 00847) and the NSF (Grant BNS 82 06061). Excellent technical support was provided by Leslie Dick, Barbara Blanchard, and Julie Fields, and excellent secretarial-editorial service was provided by Mary Beth Horan and Joyce Goodberlet.

References

1. Dunn, E. H., 1917, Primary and secondary findings in a series of attempts to transplant cerebral cortex in the albino rat, *J. Comp. Neurol.* **27**:565.

2. May, R. M., 1930, La greffe dans l'oeil de rat blanc adulte du tissu cerebral de rat nouveau-ne, *Arch. Anat. Microsc.* **26**:433.

3. Le Gros Clark, W. E., 1940, Neuronal differentiation in implanted foetal cortical tissue, *J. Neurol. Psychiatry* **3**:263.

4. Das, G. D., and Altman, J., 1972, Studies on the transplantation of developing neural tissue in the mammalian brain. I. Transplantation of cerebellar slabs into the cerebellum of neonate rats, *Brain Res.* **38**:232.

5. Das, G. D., 1974, Transplantation of embryonic neural tissue in the mammalian brain. I. Growth and differentiation of neuroblasts from various regions of the embryonic brain in the cerebellum of neonatal rats, *TITJ. Life Sci.* **4**:93.

6. Seiger, A., and Olson, L., 1977, Quantitation of fiber growth in transplanted central monoamine neurons, *Cell Tissue Res.* **179**:285.

7. Stenevi, U., Björklund, A., and Svendgaard, N. A., 1976, Transplantation of central and peripheral monoamine neurons to the adult rat brain: Techniques and conditions for survival, *Brain Res.* **114**:1.

8. Gash, D., and Sladek, J. R., Jr., 1980, Vasopressin neurons grafted into Brattleboro rats: Viability and activity, *Peptides* **1**: 11.

9. Lund, R. A., and Hauschka, S. D., 1976, Transplanted neural tissue develops connections with host rat brain, *Science* **193**:582.

10. Perlow, M. J., Freed, W. J., Hoffer, B. J., Seiger, Å., Olson, L., and Wyatt, R. J., 1979, Brain grafts reduce motor abnormalities produced by destruction of nigrostriatal dopamine system, *Science* **204**:643.

11. Cash, D., Sladek, C. D., and Sladek, J. R., Jr., 1980, A model system for analyzing functional development of transplanted peptidergic neurons, *Peptides* **1**(Suppl. 1) :125.

12. Gash, D. M., Boer, G. J., Notter, M. F. D. and Sladek, J. R., Jr., 1983 Transplanted vasopressin neurons and central nervous system effects of vasopressin, *Prog. Brain Res.* **60**:189–201.

13. Gash, D., Sladek, J. R., Jr., and Sladek, C. D., 1980, Functional development of grafted vasopressin neurons, *Science* **201**:1367.

14. Krieger, D. T., Perlow, M. J., Gibson, M. J., Davies, T. F., Zimmerman, E. A., Ferrin, M., and Charlton, H. M., 1982, Brain grafts reverse hypogonadism of gonadotropin releasing hormone deficiency, *Nature (London)* **298**:468.

15. Sladek, J. R., Jr., Schöler, J., Notter, M. F. D., and Gash, D. M., 1982, Immunohistochemical analysis of vasopressin neurons transplanted into the Brattleboro rat, *Ann. N.Y. Acad. Sci.* **394**:102.

16. Gash, D. M., Warren, P. H., Dick, L. B., Sladek, J. R., Jr., and Ison, J. R., 1982, Behavioral modification in Brattleboro rats due to vasopressin administration and neural transplantation, *Ann. N.Y. Acad. Sci.* **394**:672.

17. Sokol, E. W., Zimmerman, E. A., Sawyer, W. H., and Robinson, A. G., 1976, The hypothalamic-neurohypophyseal system of the rat: Localization and quantitation of neurophysin by light microscopic immunocytochemistry in normal rats and in Brattleboro rats deficient in vasopressin and a neurophysin, *Endocrinology* **98**:1176.

18. Valtin, H., 1982, The discovery of the Brattleboro rat, recommended nomenclature, and the question of proper controls, *Ann. N.Y. Acad. Sci.* **394**:1.

19. Scharrer, E., and Scharrer, B., 1954, Hormones produced by neurosecretory cells, *Recent Prog. Horm. Res.* **10**:183.

20. Bargmann, W., 1949, Über der neurosekretorische vernupfung vonhypothalamus and neurohypophyse, *Z. Zellforsch. Mikrosk. Anat.* **34**:610.

21. Bargmann, W., and Scharrer, E., 1951, The site of origin of the hormones of the posterior pituitary, *Am. Sci.* **39**:255.

22. Silverman, A. J., and Zimmerman, E. A., 1983, Magnocellular neurosecretory system, *Annu. Rev. Neurosci.* **6**:357.

23. Land, H., Schutz, G., Schmale, H., and Richter, D., 1982, Nucleotide sequence of cloned cDNA encoding bovine arginine vasopressin-neurophysin II precursor, *Nature (London)* **45**:299.

24. Brownstein, M. J., 1983, Biosynthesis of vasopressin and oxytocin, *Annu. Rev. Physiol.* **45**:129.

25. Verney, E. B., 1947, The antidiuretic hormone and the factors which determine its release, *Proc. R. Soc. London* **135**:25.

26. Kimura, T., Share, L., Wang, B. C., and Crofton, J. T., 1981, The role of central adrenoreceptors in the control of vasopressin release and blood pressure, *Endocrinology* **108**:1829.

27. Gash, D. M., and Thomas, G. J., 1983, What is the importance of vasopressin in memory processes?, *Trends Neurosci.* **6**:197.

28. Swanson, L. W., and Kuypers, H. B. J. M., 1980, The paraventricular nucleus of the hypothalamus: Cytoarchitectonic subdivision and organization of projections to the pituitary, dorsal vagal complex, and spinal cord as demonstrated by retrograde fluorescence double-labeling methods, *J. Comp. Neurol.* **194**:555.

29. Nilaver, G., Zimmerman, E. A., Wilkins, J., Michaels, J., Hoffman, D., and Silverman, A. J., 1980, Magnocellular hypothalamic projections to the lower brain stem and spinal cord of the rat, *Neuroendocrinology* **30**:150.

30. Sladek, C. D., 1983, Regulation of vasopressin release by neurotransmitters, neuropeptides and osmotic stimuli, *Prog. Brain Res.* **60**:71–90.

31. Watson, S. J., Akil, H., Fischli, W., Goldstein, A., Zimmerman, E., Nilaver, G., and van Wimersma Greidanus, T. B., 1982, Dynorphin and vasopressin: Common localization in magnocellular neurons, *Science* **216**:85.

32. Vanderhagen, J. J., Lofstra, F., Vandersande, F., and Dierickx, K., 1981, Coexistence of cholecystokinin and oxytocin-neurophysin in some magnocellular hypothalamo-hypophyseal neurons, *Cell Tissue Res.* **221**:227.

33. Rossier, J., Battenberg, E., Pittman, Q., Bayon, A., Koda, L., Miller, R., Guillemin, R., and Bloom, F., 1979, Hypothalamic enkephalin neurons may regulate the neurohypophysis, *Nature (London)* **277**:653.

34. Phillips, I., Weyhenmeyer, J., Felix, J., Ganten, D., and Hoffman, W. E., 1979, Evidence for an endogenous brain renin–angiotensin system, *Fed. Proc.* **38**:2260.

35. Gash, D. M., Roos, T. B., and Chambers, W. F., 1975, Development of Rathke's pouch transplanted into adult hypophysectomized female rats, *Neuroendocrinology* **19**:214.

36. Gash, D. M., Boer, G. J., Notter, M. F. D., and Sladek, J. R., Jr., 1983, Development and function of transplanted vasopressin neurons, *Psychopharmacol. Bull.* **19**:308.

37. McNeill T. H., and Sladek, J. R., Jr., 1980, Simultaneous monoamine histofluorescence and neuropeptide immunocytochemistry. V. A methodology for examining correlative monoamine–neuropeptide neuroanatomy, *Brain Res. Bull.* **5**:599.

38. McNeill, T. H., and Sladek, J. R., Jr., 1980, Simultaneous monoamine histofluorescence and neuropeptide immunocytochemistry. II. Correlative distribution of catecholamine varicosities and magnocellular neurosecretory neurons in the rat supraoptic and paraventricular nuclei *J. Comp. Neurol.* **193**:1023.

39. Sladek, J. R., Jr., Khachaturian, H., Hoffman, G. E., and Scholer, J., 1980, Aging of central endocrine neurons and their aminergic afferents, *Peptides* **1**(Suppl. 1):141.

40. Armstrong, W. E., Sladek, C. D., and Sladek, J. R., Jr., 1982, Characterization of noradrenergic control of vasopressin release by the organ-cultured rat hypothalamo-neurohypophyseal system, *Endocrinology* **111**:273.

41. Armstrong, D. M., Saper, C. B., Levey, A. I., Wainer, B. H., and Terry, R. D., 1983, Distribution of cholinergic neurons in rat brain: Demonstration by the immunocytochemical localization of choline acetyltransferase *J Comp. Neurol.* **216**:53.

42. Sladek, C. D., and Joynt, R. J., 1979, Characterization of cholinergic control of vasopressin release by the organ-cultured rat hypothalamo-neurohypophyseal system, *Endocrinotogy* **104**:659.
43. Ungerstedt, U., 1971, Stereotaxic mapping of the monoamine pathways in the rat brain, *Acta Physiol. Scand. Suppl.* **367**:1.
44. Sawchenko, P., and Swanson, L. W., 1982, Central noradrenergic pathways for the integration of hypothalamic neuroendocrine and automatic responses, *Science* **214**:685.
45. McKellar, S., and Loewy, A. D., 1981, Organization of some brain stem afferents to the paraventricular nucleus of the hypothalamus in the rat, *Brain Res.* **217**:351.
46. Sladek, J. R., Jr., McNeill, T. H., Walker, P., and Sladek, C. D., 1979, Age-related alterations in monoamine and neurophysin systems in primate brain, in: *Aging in Non-human Primates* (D. M. Bowden, ed.), pp. 1–37, Van Nostrand–Reinhold, Princeton, N. J.
47. Davies, P., 1979, Neurotransmitter-related enzymes in senile dementia of the Alzheimer type, *Brain Res.* **171**:319.
48. Sladek, C. D., McNeill, T. H., Gregg, C. M., Blair, M. L., and Baggs, R. B., 1981, Vasopressin and renin response to dehydration in aged rats, *Neurobiol. Aging* **2**:293.
49. Vijayashankar, N., and Brody, H., 1979, A quantitative study of the pigmented neurons in the nuclei locus coeruleus and subcoeruleus in man as related to aging, *J. Neuropathol. Exp. Neurol.* **38**: 490.
50. Azmitia E. C., Perlow, M. J., Brennon, M. J., and Lauder, J. M., 1981, Fetal raphe and hippocampal transplants into adult and aged C57BT/6N mice: A preliminary immunocytochemical study, *Brain Res. Bull.* **7**:703.
51. Moll, J., and de Wied, D., 1962, Observations on the hypothalamo-posthypophyseal system of the posterior lobectomized rat, *Gen. Comp. Endocrinol.* **2**:215.
52. Laszlo, F. A., and de Wied, D., 1966, Antidiuretic hormone content of the hypothalamo-neurohypophysial system and urinary excretion of antidiuretic hormone during the development of diabetes insipidus after lesions in the pituitary stalk, *J. Endocrinol.* **36**:125.
53. Dellmann, H. D., 1973, Degeneration and Regeneration of neurosecretory systems, *Int. Rev. Cytol.* **36**:215.
54. Gash, D. M., and Scott, D. E., 1980, Fetal hypothalamic transplants in the third ventricle of the adult rat brain, *Cell Tissue Res.* **211**:191.

11

The Use of Fetal Hypothalamic Transplants in Developmental Neuroendocrinology

CHARLES M. PADEN, ANN-JUDITH SILVERMAN, ULF STENEVI, AND BRUCE S. MCEWEN

1. Introduction

Developmental neuroendocrinology is a two-way street. We wish to know not only how the hypothalamic–pituitary complex comes to govern the activity of the endocrine glands, but also to understand the roles endocrine secretions may play in neural development. The introduction of techniques that permit isolated pieces of embryonic brain to survive, grow, and differentiate following transplantation into adult host brains has provided a powerful new approach to these problems. We have been transplanting embryonic rat hypothalamus and preoptic area onto the choroidal pia of adult hosts in order to study the influence of hormones on differentiation of these areas and their acquisition of neuroendocrine function. In addition, we have begun to explore the ability of fetal hypothalamic transplants to modify the neuroendocrine physiology and behavior of the adult recipients.

2. Differentiation of Embryonic Hypothalamic Transplants

In our initial studies,[1,2] we wished to determine whether or not the embryonic hypothalamic primordium would survive and develop differentiated characteristics fol-

CHARLES M. PADEN ● Department of Biology, Montana State University, Bozeman, Montana 59717.   ANN-JUDITH SILVERMAN ● Department of Anatomy, Columbia University, College of Physicians and Surgeons, New York, New York 10027.   ULF STENEVI ● Departments of Histology and Opthalmology, University of Lund, Lund, Sweden.   BRUCE S. MCEWEN ● Laboratory of Neuroendocrinology, The Rockefeller University, New York, New York 10021.

lowing transplantation into the adult brain. Thus, we chose characteristics for study that were either absent or at the limit of detection in the fetal tissue at the time of transplantation between gestational days 16 and 18.[3-13] These included dopamine histofluorescence, immunocytochemical demonstration of the neuropeptides luteinizing hormone-releasing hormone (LHRH), somatostatin, and neurophysin, and autoradiography of estrogen-concentrating cells. The differentiation of all these cellular characteristics is also of particular interest because of their important neuroendocrine functions.

2.1. Dopamine Histofluorescence

The great majority of our initial series of hypothalamic transplants survived and increased in size.[1] Fourteen of twenty-five transplants examined exhibited a distinctive median eminence-like region in association with a ventricle-like, ependymal-lined cavity that appeared to be in direct contact with the cerebrospinal fluid of the host brain. The similarity to a normal median eminence was most clearly revealed by fluorescence histochemistry for dopamine (Fig. 1). This median eminence-like structure appeared as a differentiated part of the ventricular cavity of the trannsplant, facing the vascular, superficial pia of the host colliculus. Indeed, the outer part of the structure had a palisade terminal arrangement characteristic of the normal median eminence (Fig. 2b); dopaminergic terminals appeared to end close to blood vessels on the pial surface or near pial vessels that made loops into the palisade region (Fig. 2b). Cell soma exhibiting dopamine histofluorescence were found near the median eminence-like structure (Fig. 2a). These findings suggest that the tuberoinfundibular dopaminergic system underwent extensive differentiation and development following transplantation of the fetal hypothalamus. Transplants also proved capable of retaining these characteristics indefinitely, as the median eminence-like structure was found in two of three transplants examined after 19 months survivial (Fig. 2b).

2.2. Neuropeptide Immunocytochemistry

Further evidence for the differentiated character of the transplants was provided by the presence of positive immunocytochemical staining for LHRH, somatostatin, and neurophysin.[1] In our initial series of experiments, LHRH staining was found in each of 5 transplants examined, somatostatin in 3 of 4, and neurophysin in 12 of 15. Beaded processes, presumably axons, were stained positively for each of the three peptides. Cell bodies positive for LHRH were observed occasionally (Fig. 3), and cells positive for neurophysin were found frequently. While neurophysin-containing cells often were found in clusters scattered throughout the transplant, nuclear groupings similar to the normal paraventricular (PVN) or supraoptic (SON) nuclei were not observed. This pattern has persisted in our more recent transplants,[2] as shown in Fig. 4. The neurophysin-immunoreactive perikarya range from 12 to 25 μm in diameter and are similar in size and appearance to magnocellular neurons of the normal PVN and SON (Fig.

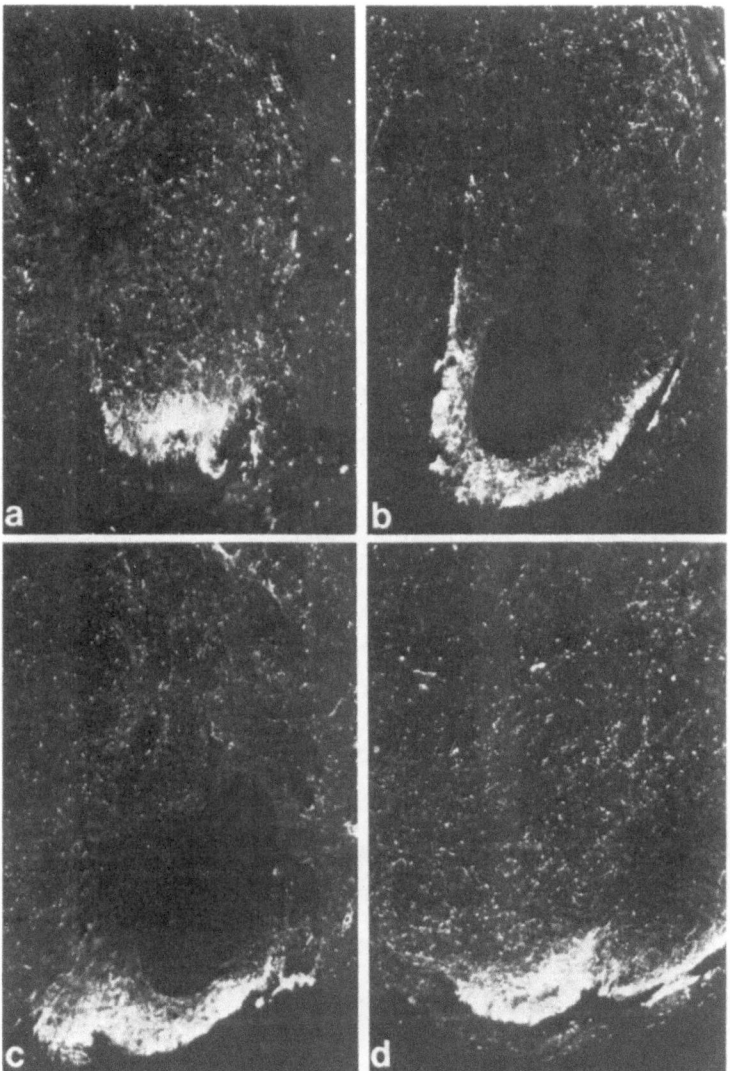

FIGURE 1. Fluorescence histochemistry for dopamine in a hypothalamic transplant. Transplant (8 weeks survival) taken from a 16-mm embryo: (a)–(d) illustrate four levels through the dopamine-innervated median eminence-like structure that forms the ventral wall of the ventricle-like cavity in the transplant. × 120. Reprinted from Ref. 1 with permission.

5). A second, smaller type of neurophysin-immunoreactive cell, approximately 10 μm in diameter also has been observed in a number of transplants. These cells typically are found in large clusters in association with a dense neurophysin fiber complex (Fig. 6). The appearance of these cell clusters is quite similar to those of vasopressin- and neurophysin-containing parvicellular neurons that have been described in the rat supra-

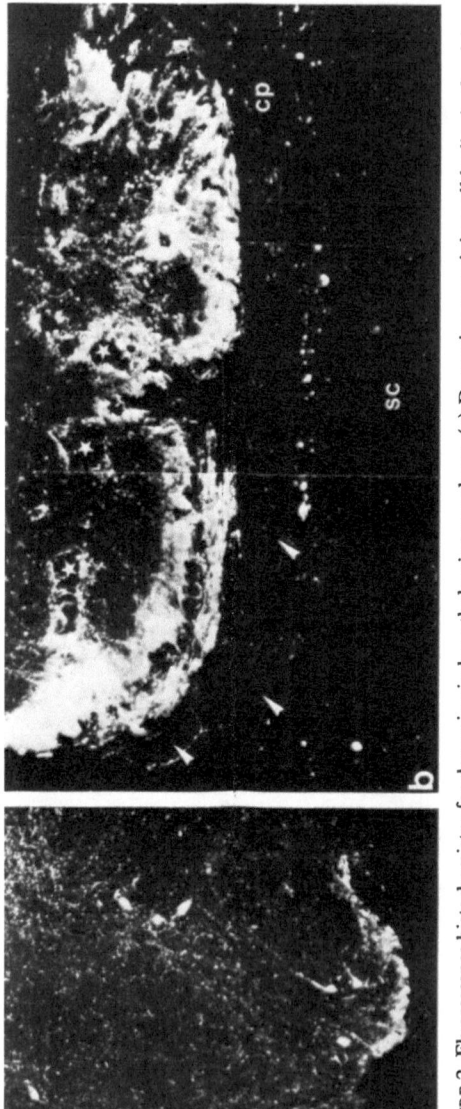

Figure 2. Fluorescence histochemistry for dopamine in hypothalamic transplants. (a) Dopamine-containing cell bodies in the vicinity of the median eminence-like region in an 8-week hypothalamic transplant from a 16-mm embryo. (b) Detail of the median eminence-like structure in a hypothalamic transplant from a 20-mm embryo, 19 months' survival. The vessels (arrowheads) in the choroidal pia (cp) have formed loops into the palisade layer (stars), surrounded by dopamine-fluorescent fibers. The median eminence was found in the ventral part of the transplant, facing the superior colliculus (sc). × 135. Reprinted from Ref. 1 with permission.

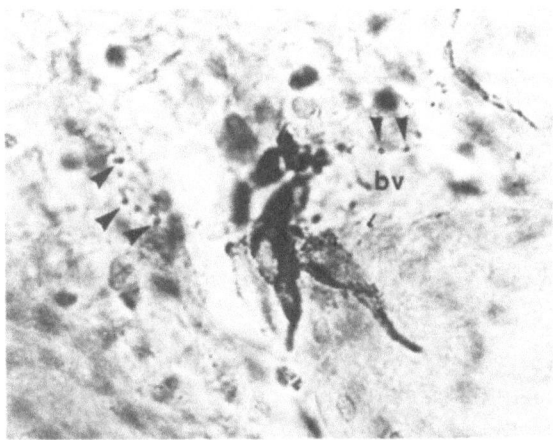

FIGURE 3. Immunocytochemistry (PAP unlabeled antibody technique) for LHRH in a hypothalamic transplant, showing a small cluster of LHRH-containing neurons. Some of the positive cells lie outside of the plane of focus. Note the presence of beaded axons (arrowheads); one, near a blood vessel (bv), is partially outside the plane of focus. × 490. Reprinted from Ref. 1 with permission.

chiasmatic nucleus.[14] Others have observed similar cells in fetal hypothalamic grafts placed in the third ventricle.[15]

Neuropeptide-containing fibers were found throughout the transplants, indicating that peptidergic cells had elaborated extensive processes following transplantation. Fibers frequently were seen in proximity to blood vessels (Figs. 3, 4b, 7). Processes stained for neurophysin were the most abundant, and they often were concentrated along the edge of the transplant tissue bordering the cerebrospinal fluid of the subarachnoid space (Fig. 4). Dense granular clumps of staining, reminiscent of the neurohypophysis, frequently were present in these areas. The general pattern of immunoreactive fibers suggests that these processes sought targets during their differentiation in the transplanted hypothalamus that could act as sites of neuropeptide release. However, in marked distinction to the pattern of dopaminergic fibers noted above, palisade arrangement of peptidergic processes as normally seen in the median eminence was not evident.

2.3. Estradiol Autoradiography

The final measure of neuronal differentiation employed in our initial studies[1] was the ability of cells to concentrate [^{3}H] estradiol. Autoradiography of brains from transplant-bearing animals given [^{3}H]estradiol i.v. prior to death revealed clusters of estrophilic cells within every transplant examined (Figs. 8a–c). Both the intensity of labeling of cells and the density of labeled cells were comparable to those found in the hypothalamus of the host animals (Fig. 8d).

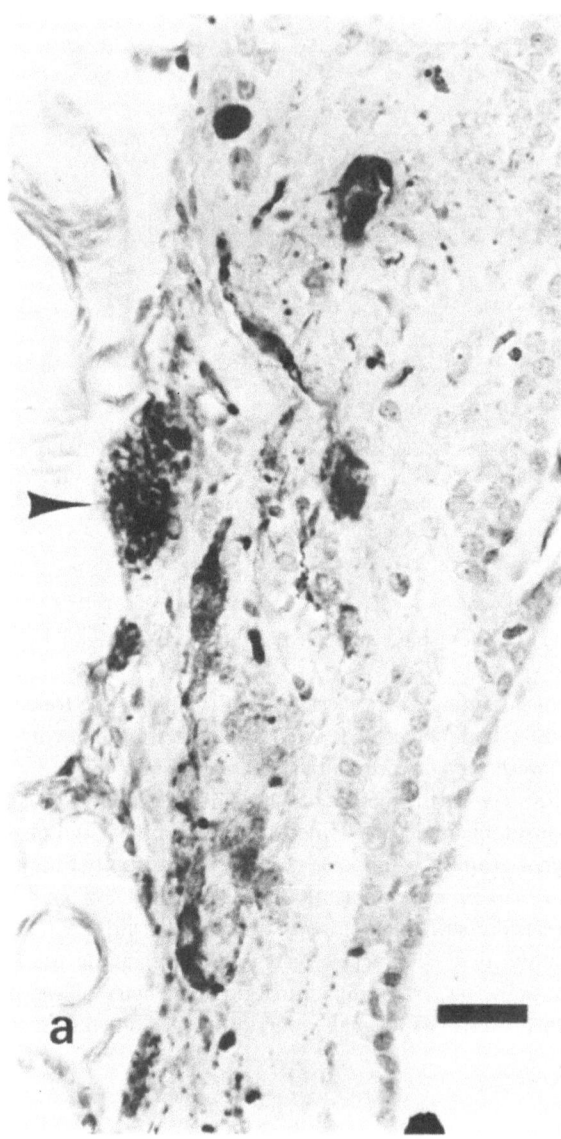

FIGURE 4. Immunocytochemistry (PAP unlabeled antibody technique) for neurophysin in hypothalamic transplants. (a) Neurophysin-positive cell bodies and fibers along the ventral edge (left side of micrograph) of a transplant in a hormonally intact female host. Cells tended to be found in small groups. Fibers and terminals are frequently observed in clumps (arrowhead). Bar = 25 μm. (b) Example of a cluster of large (> 10 μm) neurophysin-positive cells with fibers coursing near blood vessels (*). Bar = 25 μm. Reprinted from Ref. 2 with permission.

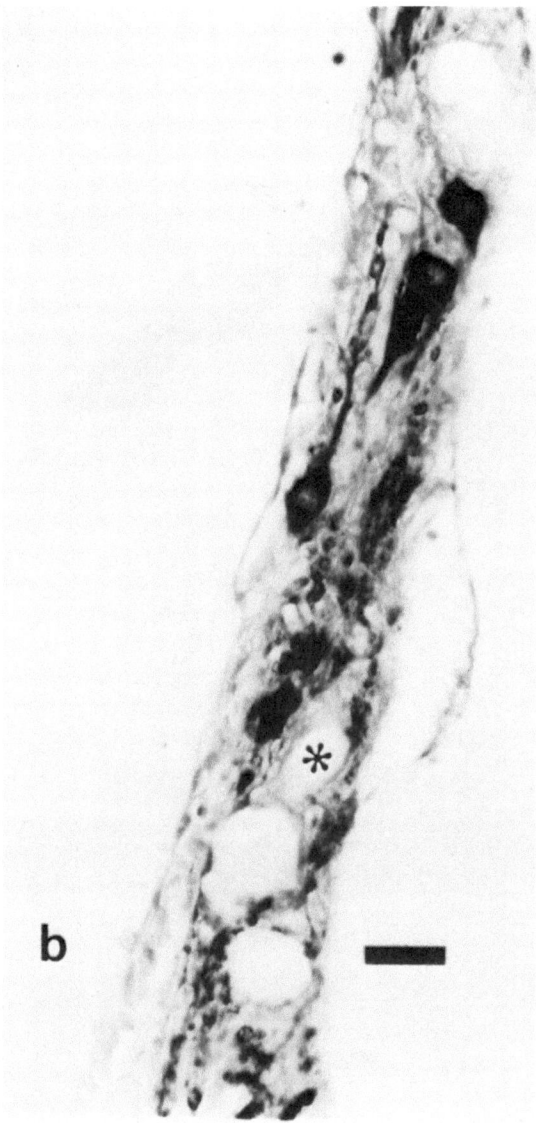

FIGURE 4 (*continued*)

2.4. Summary

These initial studies revealed that a considerable degree of neuronal differentiation occurs during the first 2 to 10 weeks after transplantation of the fetal hypothalamus. Each of the cell types studied survived and elaborated extensive processes containing characteristic products; dopamine, LHRH, somatostatin, and neurophysin. The pres-

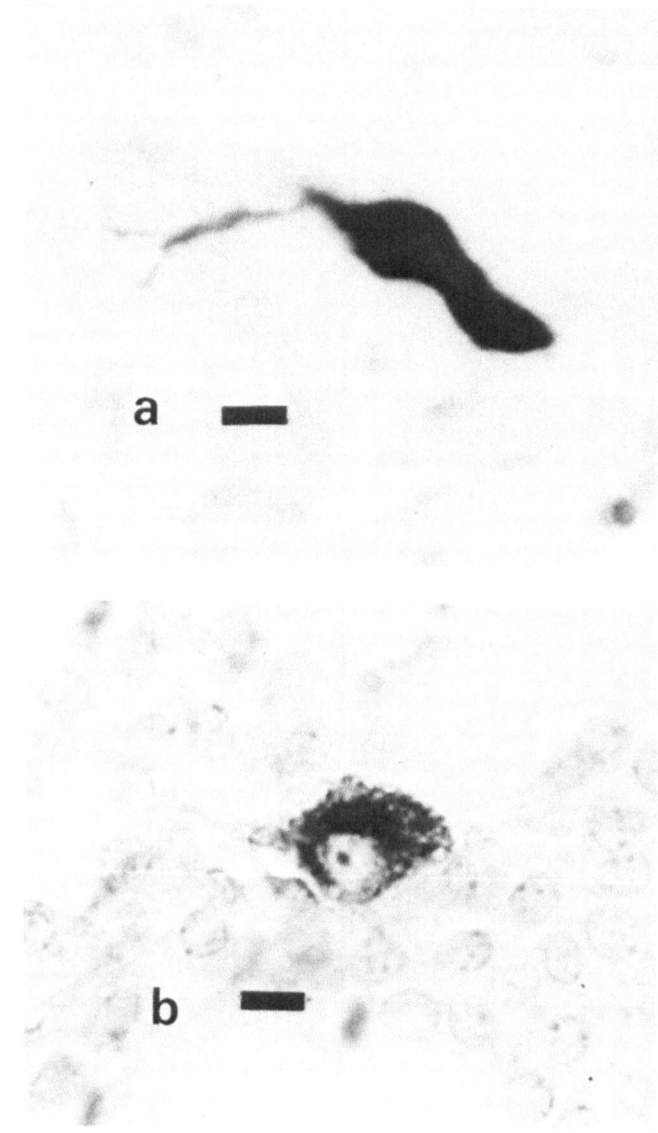

FIGURE 5. High magnification of large neurophysin-positive cells in a hypothalamic transplant. (a) Note the immunoreactive processes, presumably dendritic, extending from the cell. (b) Example of a cell that markedly resembles the magnocellular neurophysin cells of the host supraoptic and paraventricular nuclei. Bar = 10 μm. Reprinted from Ref. 2 with permission.

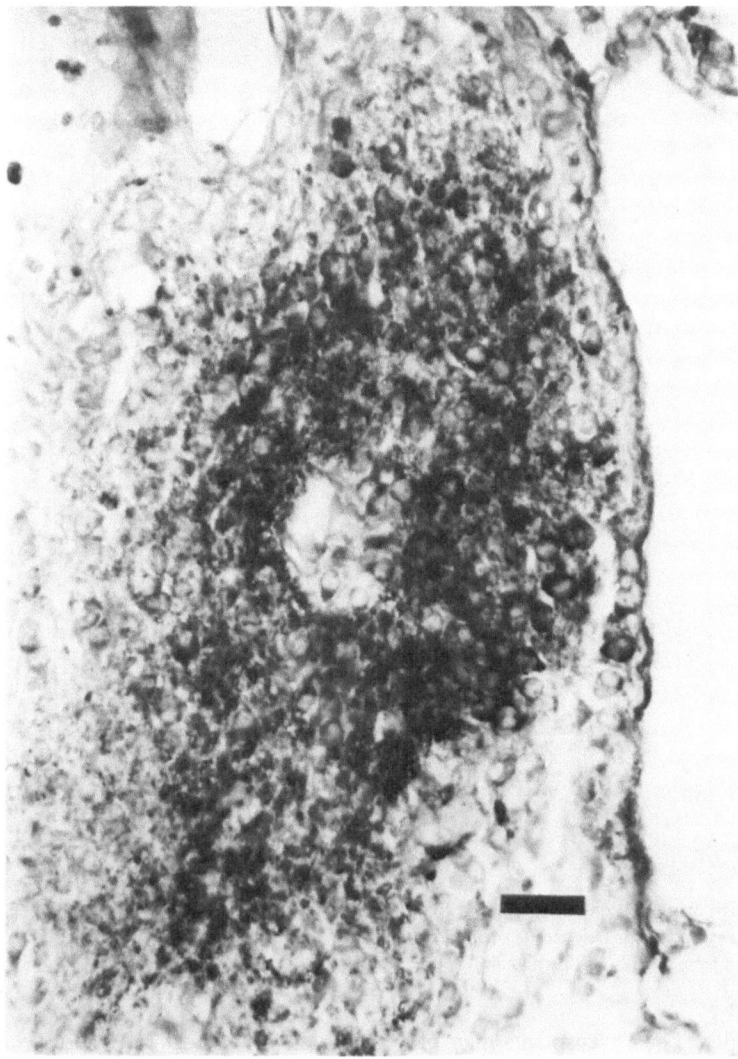

FIGURE 6. Large cluster of small (< 10 μm) neurophysin-positive cells in a transplant which resemble the neurophysin cells of the suprachiasmatic nucleus. These positive cells are typically found in the transplant embedded in a dense immunoreactive fiber plexus. Bar = 25 μm. Reprinted from Ref. 2 with permission.

ence of estrophilic cells was a further indication of normal differentiation. Fetal transplants therefore appear to be an excellent preparation for studying development of specific hypothalamic cell types.

Nuclear groupings of neuropeptide or dopaminergic cells were absent. Instead, these cells usually were found in small clusters. However, the frequent presence of a well-defined median eminence-like palisade zone of dopaminergic fibers clearly showed that fetal transplantation does not preclude the development of normal morphological

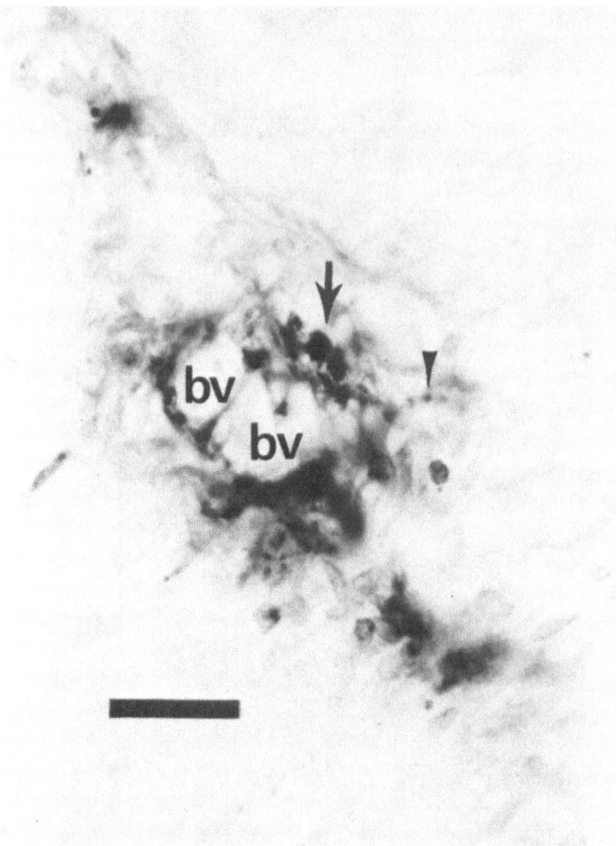

FIGURE 7. High magnification of neurophysin fibers and terminals near blood vessels (bv) within a transplant. Arrowhead points to a slender beaded axon; the arrow indicates a presumptive terminal. Bar = 25 μm. Reprinted from Ref. 2 with permission.

characteristics. Thus, transplants may prove to be useful in studying the role various factors may play in the structural development of the hypothalamus.

3. The Use of Transplants to Study the Role of Steroid Hormones in Hypothalamic Development

Having established that hypothalamic neurons with neuroendocrine functions differentiate and survive in transplants, we have begun using the transplantation technique to investigate the role that steroid hormones may play in neuronal differentiation. The developing fetus is exposed to a wealth of steroid hormones, both gonadal[16,17] and adrenal.[18-20] These hormones are produced by fetal as well as maternal glands. Because it is impossible to remove all these sources of prenatal steroids and at the same time permit the fetus to develop *in situ,* there are very few data regarding the possible effects of endogenous steroids on prenatal neural ontogeny. Explant cultures of fetal hypothalamus have been used to demonstrate a profound stimulatory effect of estrogens on early outgrowth of processes from hypothalamic neurons, but this approach is hampered by the necessity of using steroid-containing sera to maintain viable cultures.[21] One of the reasons why fetal transplantation is such a promising technique in the area of developmental neuroendocrinology is that the steroid environment in the host animal can be controlled easily.

3.1. Sexual Differentiation and the Development of Estrogen Receptors

The great majority of studies to date on the organizational effects (i.e., irreversible actions exerted during development) of steroids on neural ontogeny have been concerned

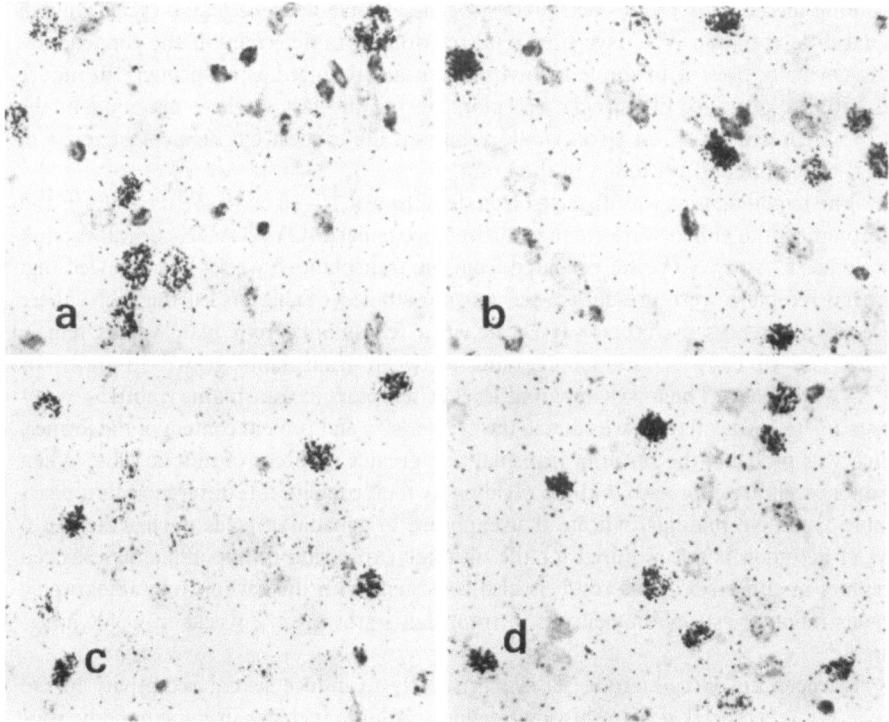

FIGURE 8. [³H]-Estradiol autoradiography of hypothalamic transplants and host hypothalamus. (a,b) [³H]-Estradiol-concentrating cells in a hypothalamic transplant, showing concentrations of silver grains. The host adult female was ovariectomized 4 days before sacrifice, which occurred 15 days after transplantation. Unfixed, unembedded frozen sections were exposed for 170 days and stained lightly with cresyl violet. × 66. (c) [³H]-Estradiol-concentrating cells in a hypothalamic transplant from a host female ovariectomized at the time of transplantation. Methods as in (a) and (b). × 166. (d) [³H]-Estradiol-concentrating cells in the arcuate nucleus of the hypothalamus of the host for the transplant shown in (a) and (b). Methods as in (a) and (b). × 620. Reprinted from Ref. 1 with permission

with the sexual differentiation of the brain (for reviews see Refs. 22, 23). In the rat, the classic defeminizing actions of gonadal steroids (induction of acyclic gonadotropin secretion and decreased propensity for female sexual behavior) are exerted during a postnatal critical period in the first week of life. It is apparent, however, that masculinizing actions of gonadal steroids on sexual and aggressive behaviors occur both pre- and postnatally in rodents.[24-27] Most recently, Gorski and colleagues have described a prenatal effect of gonadal steroids in increasing the size of a cluster of cells, termed the *sexually dimorphic nucleus,* in the preoptic area of male rats.[28] Thus, there is good reason to believe that crucial early events in sexual differentiation of the brain occur before birth in this species.

Interestingly, evidence has been accumulating over the last few years that the defeminizing actions of testicular steroids actually are mediated by the "female" hormone estradiol, formed by aromatization of testosterone within the developing rat brain.[22,29-31] Additionally, it has been argued that the complete development of female sexual behavior is not passive, but requires exposure to low levels of estrogens.[32,33] It is becoming increasingly clear, then, that estrogen-sensitive neurons play a crucial role in sexual differentiation. We have used fetal transplants to determine if the gonadal steroids normally present in the fetal environment are required as a "primer" to induce the initial appearance of estrogen receptors during the last week of gestation in the rat.[11,12] If present, such an effect would represent the earliest biochemical correlate of sexual differentiation yet found in the brain.

The hypothalamus was dissected from fetal rats of 15–16 or 17–18 days of age and transplanted into either ovariectomized–adrenalectomized (OVX–ADX) or intact adult female hosts. Assay of cytosol prepared from the transplants 8 weeks later showed that estrogen receptors were present in every group studied (Table I). Furthermore, there were no differences in receptor levels between transplants from male versus female fetuses (17–18 days gestational age) nor between transplants grown in intact or OVX–ADX hosts. The lower apparent level of receptors in transplants from 15- to 16- versus 17- to 18-day fetuses was due to the greater size and protein content of the former, which was probably the result of transplanting greater numbers of mitotic cells. When receptor levels are expressed without dividing by total protein, this difference disappears (Table I). These findings indicate that exposure to gonadal steroids during the last 6 days of gestation is not required for the ultimate expression of hypothalamic estrogen receptors in either sex. This result is also consistent with the earlier autoradiographic demonstration of estrophilic neurons in transplants grown for 2 weeks in OVX hosts[1] (Fig. 8c).

Estrogen and progesterone act synergistically to induce sexual receptivity in the female rat.[34] A crucial step in this joint action is a large increase in hypothalamic progestin receptors on the day of proestrus, induced by the earlier rise in circulating estrogens.[35-37] Sequential administration of estradiol and progesterone will mimic this natural sequence in OVX rats.[38] In order to determine if the estrophilic neurons present in hypothalamic transplants are capable of inducing progestin receptors, host animals were given two injections of either 10 μg estradiol benzoate or oil vehicle 72 and 48 hr prior to sacrifice. A significantly higher level of saturable binding of the synthetic progestin ^{3}H-R5020 was found in the transplant cytosol from estrogen-treated hosts: 7.52 + 0.96

TABLE I

Estrogen Receptor Levels in Cytosol from Fetal Hypothalamic Transplants Grown under
Different Hormonal Conditions

Adult host	Gestational age and sex of fetal donor			
	Day 15–16 Sexes pooled	Day 17–18		Day 17–18 Sexes pooled
		Male	female	
	Mean ± S.E.M. (n)			
Intact	1.30 ± 0.27 (7)[a]	4.56 ± 0.65 (4)	3.28 ± 0.30 (3)	3.77 ± 0.47
female	0.70 ± 0.24[b]	0.88 ± 0.14	0 78± 0.16	0.89 ± 0.15
OVX–ADX	2.03 ± 0.71 (5)	4.53 ± 1.63 (3)	4.72 ± 0.91 (4)	4.47 ± 0.81
female	0.91 ± 0.34	0.75 ± 0.18	1.22 ± 0.27	0.88 ± 0.14

[a]The upper row in each group are estrogen receptor levels expressed as femtomoles of estradiol bound per milligram protein A 2 × 3 analysis of variance using an unweighted means procedure revealed that there are no difference in receptor levels between transplants grown in intact versus OVX–ADX female hosts, nor between transplants from day 17–18 male versus female fetuses. Levels in transplants from day 15–16 fetuses (sexes not determined) are significantly lower than either those from day 17–18 male or female fetuses ($p < 0.01$) when expressed in this manner

[b]The lower row in each group are estrogen receptor levels expressed as femtomoles of estradiol bound per transplant There are no significant differences between any groups when receptor levels are expressed in this manner.

($n = 7$) fmoles/mg protein (mean + S.E.M.) versus 4.58 + 1.00 ($n = 4$) in oil-treated animals (Student's $t = 2.12$, $df = 9$, $p < 0.05$). This induction shows that the transplant estrogen receptors are biochemically functional. Induction was observed in transplants from both intact (OVX prior to treatment) and OVX–ADX hosts, indicating that exposure to gonadal steroids during hypothalamic development is not required for the appearance of the estrogen-responsive mechanism of progestin receptor induction.

Our results to date have demonstrated the utility of the transplantation procedure for studying the early events in sexual differentiation of the brain. Specifically, it appears that biochemically functional estrogen receptors appear during development in both sexes independently of the perinatal steroid environment. Other important biochemical aspects of sexual differentiation, such as the development of androgen aromatase activity, can be investigated using the fetal transplantation technique in future studies.

3.2. Glucocorticoid Influences on the Differentiation of Magnocellular Neurons

During the course of the estrogen receptor studies, we made the fortuitous discovery that markedly fewer magnocellular neurons immunoreactive for neurophysin were visible in transplants grown in OVX–ADX versus intact hosts[2] (Table II). Preliminary studies using low levels of corticosterone replacement suggest that this effect may be caused by the absence of glucocorticoids (Table II). An additional highly unusual immunocytochemical observation is shown in Fig. 9. In this case, a dissociation between immunostaining for oxytocin and vasopressin and that of neurophysin was observed in

TABLE II
Number of Large Perikarya Immunoreactive for Neurophysin in Hypothalamic Transplants[a]

Host	Number of perikarya	Percent of sections stained[b]
Intact female	34	11.7
	56	16.3
OVX–ADX female	10	11.5
	6	25.0
OVX–ADX female + corticosterone	11	10.3
	24	10.4

[a]Reprinted from Ref. 2 with permission.
[b]Each transplant was sectioned throughout its entire length. This figure is the percentage of the total number of sections that were taken for neurophysin immunocytochemistry.

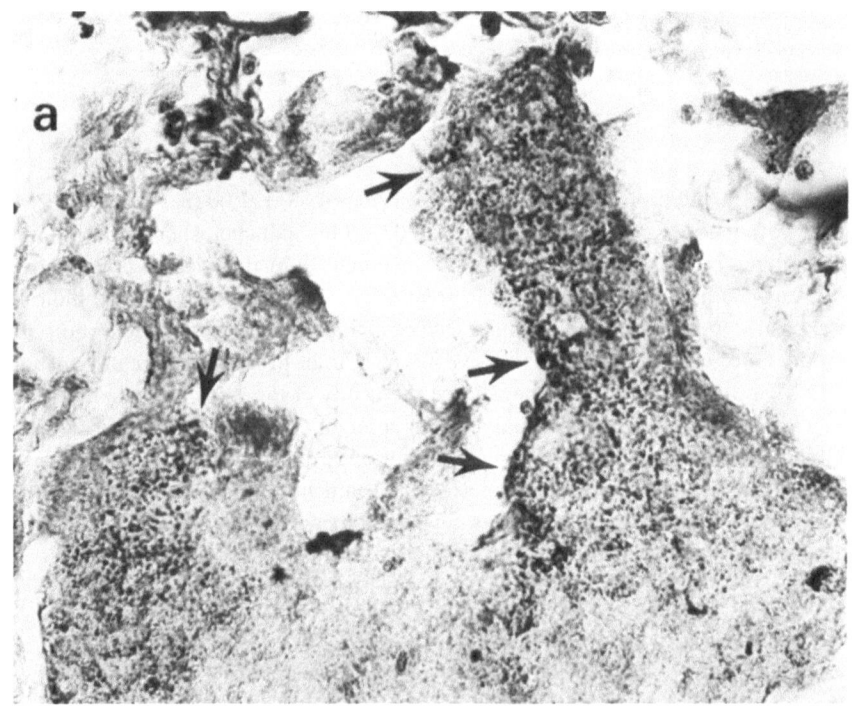

FIGURE 9. Immunocytochemistry (PAP unlabeled antibody technique) for oxytocin, vasopressin, and neurophysin in a hypothalamic transplant grown for 8 weeks in a glucocorticoid-free adult host. Granular staining (arrows) of beaded fibers and presumptive terminals along the dorsal edge of the transplant is apparent when antisera to oxytocin (panel a) or vasopressin (panel b) are employed, but is totally absent when stained with antisera to neurophysin (panel c).

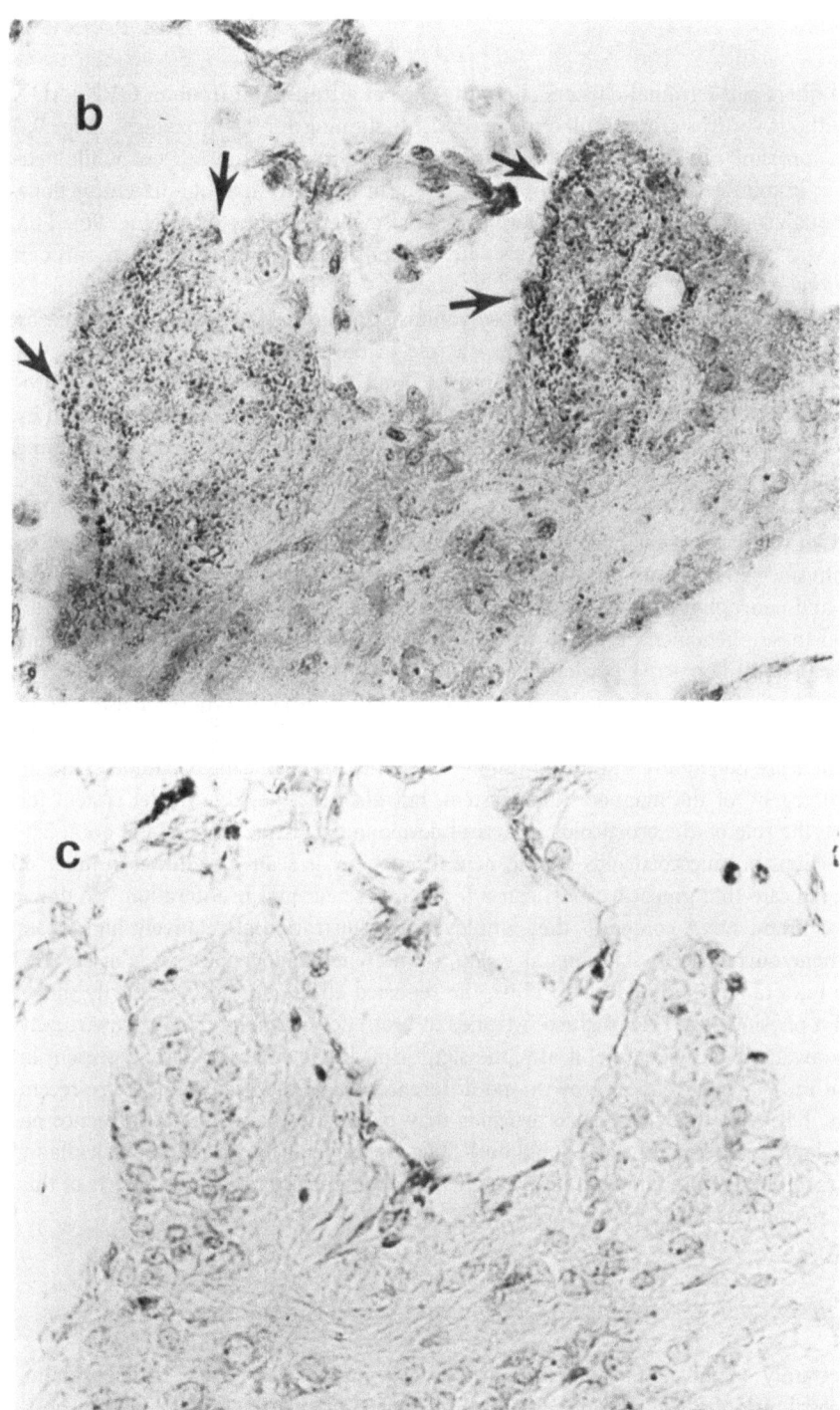

FIGURE 9 (*continued*)

beaded fibers and terminal clusters along the edge of a transplant from an OVX-ADX host without corticosterone replacement. Positive staining for both oxytocin (Fig. 9a) and vasopressin (Fig. 9b) was observed in adjacent 10-μm paraffin sections while neurophysin immunoreactivity was completely absent in the same area of adjacent sections, even though immunostained fibers were present elsewhere in the section (Fig. 9c). This finding was confirmed by staining slides with each of the three antisera, in turn, through 12 sections.

To our knowledge, both of these observations are unique. There is, however, reason to suspect that glucocorticoids could play a role in the differentiation and function of magnocellular neurons. Adrenalectomy results in an increase in the number of neurophysin fibers in the median eminence of adult rats,[39] an effect that can be prevented by glucocorticoid but not mineralocorticoid treatment.[39,40] There is also evidence suggesting that while oxytocin and vasopressin as well as their "carrier" protein neurophysins are believed to be cleaved from a common larger precursor,[41] this relationship is not obligatory. Oat cell carcinomas have been shown to synthesize vasopressin without making neurophysin.[42] In addition, large ontogenetic fluctuations in the molar ratios of the hormones and neurophysins have been observed in the rat brain.[43,44] The influence of steroids on these phenomena is unknown. We are currently working to determine if the decrease in immunoreactive perikarya and the unusual staining pattern in fibers are true developmental effects and to define their precise relationship to the hormonal milieu of the transplant.

These are potentially significant findings, not only for increasing our understanding of the ontogeny of the magnocellular system, but also as a possible model system for studying the role of glucocorticoids in neural development. Knowledge of the organizational actions of glucocorticoids during neural ontogeny is scanty. While a number of studies indicate that glucocorticoids can affect central neuronal proliferation,[45,46] tissue weight,[45,46] and DNA content,[47] they employed administration of relatively high doses of hormone during the first postnatal week, a time when endogenous levels of corticosterone have fallen to low levels.[48] Thus, the reported effects are not necessarily indicative of a physiological role of glucocorticoids in brain development. There are virtually no data available on what role, if any, the significant levels of corticosterone present in the fetal rat[18-20] may play in growth and differentiation of the brain. There are recent findings, however, that fetal glucocorticoids may exert an organizational influence on the developing superior cervical ganglion.[49] The use of fetal hypothalamic transplants promises to provide fundamental new insights into possible organizational effects of this class of steroid hormones in the CNS.

4. Neuroendocrine Function of Hypothalamic Transplants

Certainly one of the most exciting aspects of working with fetal brain transplants is the possibility of alleviating deficits in the behavior or physiology of the adult host. We have previously reported that cotransplants of fetal hypothalamus and pituitary can partially overcome the lack of growth hormone secretion caused by hypophysectomy.[2] In addition, we have begun investigating whether or not the estrogen-sensitive neurons that develop in fetal transplants are capable of affecting sex behavior in host animals with electrolytic lesions of the ventromedial nucleus of the hypothalamus (VMN). Such lesions drastically lower the incidence of estrogen–progesterone-induced female sexual behavior in the rat.[50] This deficit is thought to be the result of destruction of a relatively small population of steroid-sensitive neurons whose activity influences cells in the midbrain central gray that are part of the reflex pathway of the lordosis response.[51-54] Because this terminal field of VMN projections is only a few millimeters ventral to the transplantation site on the choroidal pia, we reasoned that estrophilic neurons in the transplanted hypothalamus might be capable of sending axons to the denervated midbrain target areas of host animals with VMN lesions. Transplants of fetal substantia nigra, for example, will send dopaminergic projections several millimeters into denervated host striatum.[55] As we have demonstrated that estrophilic cells are capable of responding to circulating estrogens in the host by inducing progestin receptors, it seemed plausible that they could also act to facilitate lordosis in response to estrogen and progesterone treatment.

The results of our initial experiment are shown in Fig. 10. A significantly greater lordosis response was observed over a period of months in female rats that received fetal hypothalamic transplants 1 week after bilateral VMN lesions as compared to similar animals given sham transplant surgery. To our surprise, however, surgical removal of the transplants after animals had regained high lordosis scores did not cause a new deficit in sex behavior (Fig. 10). The animals that had borne transplants continued to score significantly higher than controls until some 3 months later, when differences were diminished by recovery of the sham-operated group (Fig. 10).

Although we cannot completely exclude the possibility that small pieces of transplant tissue were not removed and continued to mediate hormone-induced lordosis, we feel that this possibility is unlikely. Further experiments are in progress to test our current hypothesis that the presence of fetal brain tissue in the transplant group somehow stimulates the normal recovery process that occurs following VMN lesions. The mechanism of this recovery is not known, but is assumed to involve reorganization of the connections of surviving hormone-sensitive cells in or near the VMN.

A variety of other dramatic behavioral effects of neural transplants are described elsewhere in this volume. They are each believed to result from growth of neuronal processes from the transplant into the host tissue. To our knowledge, these results with VMN-lesioned hosts are the first to suggest a trophic influence of fetal transplants on spontaneous recovery from lesions of the CNS. These findings add yet another tantalizing aspect to the growing body of work with neural transplants.

5. Summary

Our studies have clearly demonstrated a marked degree of normal cellular differentiation in fetal hypothalamic transplants. Other aspects of hypothalamic morphology do not develop normally following transplantation of the isolated hypothalamus, but for that reason the technique may prove useful in determining the role of various factors in the acquisition of mature morphology. Transplantation is a powerful tool for investi-

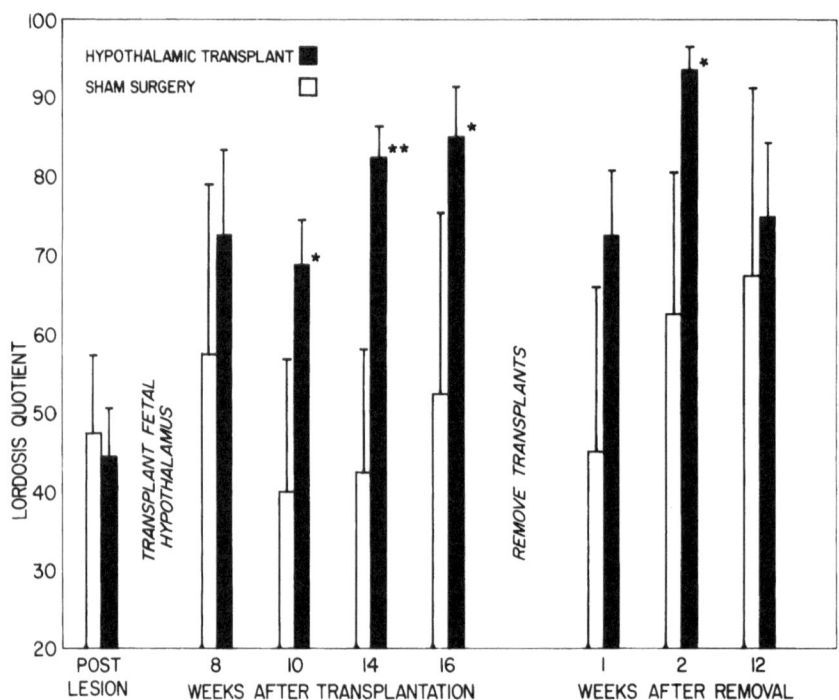

FIGURE 10. Effect of fetal hypothalamic transplants on female sexual behavior of adult hosts with lesions of the ventromedial hypothalamic nucleus (VMN). Ovariectomized adult female (200–250 g) CD rats received bilateral DC lesions of the VMN using an iron electrode (1 mA for 30 sec, anodal; coordinates 0.2 mm posterior to bregma, lateral 0.7 mm, ventral 8.1 mm from dura with the incisor bar at +5.0 mm). Behavior tests were performed following three daily injections of 10 μg estradiol benzoate in oil s.c. and a single injection of 1 mg progesterone in oil s.c. approximately 5 hrs. prior to testing. The percentage of lordosis responses to 10 mounts by stud males is shown as the lordosis quotient (LQ) and its S.E.M. Initial LQs (post lesion) of transplant ($n = 8$) and sham surgery ($n = 4$) groups were not different 4 days after the lesions. Ten days after lesioning, hypothalami from 22-mm embryos were transplanted on to the choroidal pia as previously described[1] or the transplant site was prepared and closed without transfer of fetal tissue (sham surgery). Multivariate analysis (University of Rochester Weighted ANOVA System, Program BMDP4V, copyright 1981 Regents of the University of California) using the initial postlesion scores as a covariable revealed a significantly greater mean LQ in the transplant group versus sham surgery controls between 8 and 16 weeks after surgery ($F = 7.41$, $p < 0.05$). The transplant group scored significantly higher on the specific test dates indicated by * ($p < 0.05$) and ** ($p < 0.01$) as determined by one-way ANOVA with the initial postlesion scores as covariates. Surgical removal of transplants 18 weeks after insertion did not return LQs to postlesion levels. While the transplant group did not quite score significantly higher than sham controls 1 week after transplant removal ($p < 0.07$), the difference was again significant at 2 weeks ($p < 0.024$). By 12 weeks after removal, control scores had increased and were not significantly different from those of the transplant group.

gating perinatal influence of hormones on neural development because the endocrine environment of the adult host can be manipulated easily. Ablation of hormones and controlled replacement already have begun to yield new insights into the mechanisms of sexual differentiation as well as the possible influence of glucocorticoids on hypothalamic development. Finally, fetal hypothalamic transplants appear to be capable of alleviating deficits in the neuroendocrine functions of the adult recipient in ways we have only begun to explore.

ACKNOWLEDGMENTS. The authors thank Dr. Gaj Nilaver for generous supplies of PAP complex and antisera to oxytocin and vasopressin; Dr. Allen Robinson for the gift of antisera to bovine neurophysin I (prepared under NIH grant AM 16166); Dr. Marilyn McGinnis for assistance with lordosis measurements; and Claude Chaptal, Valerie DeGroff, John Gerlach, Marie Nicolas, Gunilla Lundquist, and Gertrude Stridsberg for expert technical assistant. This research was supported by grants from the NIH (NS-07080 to B.S.M.; NS-17974 to C.M.P.; HD-10665 to A.-J.S.) and from the Swedish MRC (O4X-3874 to U.S.). B.S.M. received support from an institutional grant (RF 70095) from the Rockefeller Foundation for research in reproductive biology. A.-J.S. is an Alfred P. Sloan Foundation Fellow and is the recipient of an Irma T. Hirschl Career Scientist Award. C.M.P. was a recipient of an NIH Postdoctoral Fellowship (MH 07493).

References

1. Stenevi, U., Björklund, A., Kromer, L. F., Paden, C. M., Gerlach, J. L., McEwen, B. S., and Silverman, A.-J., 1980, Differentiation of embryonic hypothalamic transplants cultured on the choroidal pia in brains of adult rats, *Cell Tissue Res.* **205**:217.
2. Paden, C. M., Silverman, A.-J., McEwen, B. S., Stenevi, U., Björklund, A., and Thorngren, K. G., 1980, Hormonal effects on development of transplanted embryonic hypothalamus, *Peptides* **1**(Suppl. 1):117.
3. Glydon, R. S. J., 1957, The development of the blood supply of the pituitary in the albino rat with special reference to the portal vessels, *J. Anat.* **91**:237.
4. Enemar, A., 1961, The structure and development of the hypophysial portal system in the laboratory mouse, with particular regard to the primary plexus, *Arch. Zool.* **13**:203.
5. Björklund, A., Enemar, A., and Falck, B., 1968, Monoamines in the hypothalamohypophyseal system of the mouse with special reference to the ontogenetic aspects, *Z. Zellforsch. Mikrosk. Anat.* **89**:590.
6. Hyyppä, M., 1969, A histochemical study of the primary catecholamines in the hypothalamic neurons of the rat in relation to the ontogenetic and sexual differentiation, *Z. Zellforsch. Mikrosk, Anat.* **98**:550.
7. Araki, S., Tòran-Allerand, C. D., Ferin, M., and Vande Wiele, R. L., 1979, Immunoreactive gonadotropin-releasing hormone (Gn-RH) during maturation in the rat: ontogeny of regional hypothalamic differences, *Endocrinology* **97**:693.
8. Chiappa, S. A., and Fink, G., 1977, Releasing factor and hormone changes in the hypothalamic–pituitary-gonadotrophin and adrenocorticotrophin systems before and after birth and puberty in male, female and androgenized female rats, *J. Endocrinol.* **72**:211.
9. Daikoku, S., Kawano, H., and Matsumura, H., 1978, In vivo and in vitro studies on the appearance of LHRH neurons in the hypothalamus of perinatal rats, *Cell Tissue Res.* **194**:433.
10. Silverman, A.-J., Goldstein, R., and Gadde, C., 1980, The ontogenesis of neurophysin-containing neurons in the mouse hypothalamus, *Peptides* **1**(Suppl. 1):27.
11. Vito, C. C., and Fox, T. O., 1979, Embryonic rodent brain contains estrogen receptors, *Science* **204**:517.

12. MacLusky, N. J., Lieberburg, I., and McEwen, B. S., 1979, The development of estrogen receptor systems in the rat brain: Perinatal development, *Brain Res.* **178**:129.

13. Hoffman, G. E., Dick, L. B., and Gash, D., 1980, Development of somatostatin neurons: Examination by the technique of combined autoradiography and immunocytochemistry, *Peptides* **1**(Suppl. 1):79.

14. Vandesande, F., Dierickx, K., and DeMey, J., 1975, Identification of the vasopressin–neurophysin producing neurons of the rat suprachiasmatic nuclei, *Cell Tissue Res.* **156**:377.

15. Gash, D., Sladek, C. D., and Sladek, J. R., Jr., 1980, A model system for analyzing functional development of transplanted peptidergic neurons, *Peptides* **1**(Suppl. 1):125.

16. Ward, I., and Weisz, J. M., 1980, Maternal stress alters plasma testosterone in fetal males, *Science* **207**:328.

17. Weisz, J., and Ward, I. L., 1980, Plasma testosterone and progesterone titers of pregnant rats, their male and female fetuses, and neonatal offspring, *Endocrinology* **106**:306.

18. Chatelain, A., Dupouy, J.-P., and Allaume, P., 1980, Fetal–maternal adrenocorticotropin and corticosterone relationships in the rat: Effects of maternal adrenalectomy, *Endocrinology* **106**:1297.

19. Milkovic, K., Paunovic, J., Kniewald, Z., and Milkovic, S., 1973, Maintenance of the plasma corticosterone concentration of adrenalectomized rat by the fetal adrenal glands, *Endocrinology* **93**: 115.

20. Negellen-Perchellet, E., and Cohen, A., 1975, Effect of ether inhalation by adrenalectomized pregnant rats on the adrenal corticosterone concentration in normal, decapitated, and encephalectomized fetuses, *Neuroendocrinology* **17**:225.

21. Toran-Allerand, C. D., 1980, Sex steroids and the development of the newborn mouse hypothalamus and preoptic area in vitro. II. Morphological correlates and hormonal specificity, *Brain Res.* **189**:413.

22. McEwen, B. S., Biegon, A., Davis, P. G., Krey, L. C., Luine, V. N., McGinnis, M. Y., Paden, C. M., Parsons, B., and Rainbow, T. C., 1982, Steroid hormones: Humoral signals which alter brain cell properties and functions, in: *Recent Progress in Hormone Research*, Vol. 38 (R. O. Greep, ed.), pp. 41–83, Academic Press, New York.

23. Goy, R. W., and McEwen, B. S. (eds.) 1980, *Sexual Differentiation of the Brain*, MIT Press, Cambridge, Mass.

24. Gerall, A. A., and Ward, I. L., 1966, Effects of prenatal exogenous androgen on the sexual behavior of the female albino rat, *J. Comp. Physiol. Psychol.* **62**:370.

25. Grady, K. L., Phoenix, C. H., and Young, W. C., 1965, Role of the developing rat testis in differentiation of the neural tissues mediating mating behavior, *J. Comp. Physiol. Psychol.* **59**: 176.

26. vom Saal, F. S., and Bronson, F. H., 1978, *In utero* proximity of female mouse fetuses to males: Effect on reproductive performance during later life, *Biol. Reprod.* **19**:842.

27. Ward, I. L., 1971, Prenatal stress feminizes and demasculinizes the behavior of males, *Science* **175**:82.

28. Gorski, R. A., Harlan, R. E., Jacobson, C. D., Shryne, J. E., and Southam, A. M., 1980, Evidence for the existence of a sexually dimorphic nucleus in the preoptic area of the rat brain, *J. Comp. Neurol.* **193**:529.

29. Reddy, V. V. R., Naftolin, F., and Ryan, K. J., 1974, Conversion of androstenedione to estrone by neural tissues from fetal and neonatal rats, *Endocrinology* **94**:117.

30. Lieberburg, I., Wallach, G., and McEwen, B. S., 1977, The effects of an inhibitor of aromatization (1,4,6-androstatriene-3,17-dione) and an antiestrogen (CI628) on in vivo formed testosterone metabolites recovered from neonatal rat brain tissues and purified cell nuclei: Implications for sexual differentiation of the rat brain, *Brain Res.* **128**:176.

31. Davis, P. G., Chaptal, C. V., and McEwen, B. S., 1979, Independence of the differentiation of masculine and feminine sexual behavior in rats, *Horm. Behav.* **12**:12.

32. Döhler, K. D., 1978, Is female sexual differentiation hormone-mediated?, *Trends Neurosci.* **1**:138.

33. Toran-Allerand, C. D., 1981, Gonadal steroids and brain development—In vitro veritas?, *Trends Neurosci.* **4**:118.

34. Beach, F. A., 1976, Sexual attractivity, proceptivity and receptivity in female animals, *Horm. Behav.* **7**:105.

35. Moguilewsky, M., and Raynaud, J.-P., 1977, Progestin binding sites in the rat hypothalamus, pituitary and uterus, *Steroids* **30**: 99.

36. MacLusky, N. J., and McEwen, B. S., 1978, Oestrogen modulates progestin receptor concentrations in some brain regions but not in others, *Nature (London)* **274**:276.

37. Blaustein, J. D., and Wade, G. N., 1978, Progestin binding by brain and pituitary cell nuclei and female rat sexual behavior, *Brain Res.* **140**:360.
38. Parsons, B., MacLusky, N. J., Krey, L., Pfaff, D. W., and McEwen, B. S., 1980, The temporal relationship between estrogen-inducible progestin receptors in the female rat brain and the time course of estrogen activation of mating behavior, *Endocrinology* **107**:774.
39. Stillman, M. A., Recht, L. D., Rosario, S. L., Seif, S. M., Robinson, A. G., and Zimmerman, E. A., 1977, The effect of adrenalectomy and glucocorticoid replacement on vasopressin and vasopressin-neurophysin in the zona externa of the rat, *Endocrinology* **101**:42.
40. Silverman, A.-J., Hoffman, D., Gadde, C. A., Krey, L. C., and Zimmerman, E. A., 1981, Andrenal steroid inhibition of the vasopressin-neurophysin neurosecretory system to the median eminence of the rat: Differential effects of corticosterone and deoxycorti costerone administration after adrenalectomy, *Neuroendocrinology* **32**:129.
41. Brownstein, M. J., Russell, J. T., and Gainer, H., 1980, Synthesis, transport, and release of posterior pituitary hormones, *Science* **207**:373.
42. Pettengill, O. S., Faulkner, C. S., Wurster-Hill, D. H., Maurer, L. H., Sorenson, G. D., Robinson, A. G. and Zimmerman, E. A., 1977, Isolation and characterization of a hormone-producing cell line from human small cell anaplastic carcinoma of the lung, *J. Natl. Cancer Inst.* **58**:511.
43. Sinding, C., Robinson, A. G., Seif, S. M., and Schmid, P. G., 1980, Neurohypophyseal peptides in the developing rat fetus, *Brain Res.* **195**:177.
44. Sinding, C., Czernichow, P., Seif, S. M., and Robinson, A. G., 1980, Quantitative changes in neurohypophyseal peptides in the developing brain, *Peptides* **1**(Suppl. 1):45.
45. Cotterrell, M., Balázs, R., and Johnson, A. L., 1972, Effects of corticosteroids on the biochemical maturation of rat brain: Postnatal cell formation, *J. Neurochem.* **19**:2151.
46. Bohn, M. C., and Lauder, J. M., 1978, The effects of neonatal hydrocortisone on rat cerebellar development, *Dev. Neurosci.* **1**:250.
47. Howard, E., 1973, Hormonal effects on the growth and DNA content of the developing brain, in: *Biochemistry of the Developing Brain*, Vol. 2 (A. Himwich, ed.), pp. 1–68, Dekker, New York.
48. Sze, P. Y., 1976, Glucocorticoid regulation of the serotonergic system of the brain, in: *Advances in Biochemical Psychopharmacology* Vol. 15 (E. Costa, E. Giacobini, and R. Paoletti eds.), pp. 251–265, Raven Press, New York.
49. McLennan, I. S., Hill, C. E., and Hendry, I. A., 1980, Gluco-cortico-steroids modulate transmitter choice in developing superior cervical ganglion, *Nature (London)* **283**:206.
50. Mathews, D., and Edwards, D. A., 1977, Involvement of the ventromedial and anterior hypothalamic nuclei in the hormonal induction of receptivity in the female rat, *Physiol. Behav.* **19**:319.
51. Morrell, J. I., and Pfaff, D. W., 1982, Characterization of estrogen-concentrating hypothalamic neurons by their axonal projections, *Science* **217**:1273.
52. Sakuma, Y., and Pfaff, D. W., 1979, Facilitation of female reproductive behavior from mesencephalic central grey in the rat, *Am. J. Physiol.* **237**(5):R278.
53. Sakuma, Y., and Pfaff, D. W., 1979, Mesencephalic mechanisms for integration of female reproductive behavior in the rat, *Am. J. Physiol.* **237**(5):R285.
54. Pfaff, D. W., 1980, *Estrogens and Brain Function*, Springer-Verlag, Berlin.
55. Björklund, A., and Stenevi, U., 1979, Reconstruction of the nigrostriatal dopamine pathway by intracerebral nigral transplants, *Brain Res.* **177**:555.

12

Specificity of Termination Fields Formed in the Developing Hippocampus by Fibers from Transplants

Carl W. Cotman

1. Introduction

One of the major unsolved problems in neurobiology is understanding the mechanisms specifying the development of specific synapses and their plasticity after injury. A variety of mechanisms have been postulated to account for the development and plasticity of specific connections between central neurons, i.e., mechanical or contact guidance, chemotaxis, afferent competition, temporal matching, chemospecificity, etc. Transplantation provides a means for evaluating the potential role of at least some of these mechanisms. It is relatively simple, for example, to vary the position of the afferent source and alter the temporal sequence of events during the development of a specific afferent projection. Experiments varying such parameters have proven instrumental in analyzing the mechanism of synapse formation in nonmammalian species such as the retinotectal field in amphibians.

In this chapter I will describe our results on brain transplants aimed at understanding the normal development and injury-induced reorganization of circuitry in the immature dentate gyrus. The objective is to describe the types of mechanisms involved ultimately at a biochemical level of analysis. We have examined the ability of septal, striatal,

Carl W. Cotman ● Department of Psychobiology, University of California, Irvine, California 92717 and raphe tissues, implanted into the entorhinal cortex of a neonatal rat at various times after lesion, to grow into the hippocampal formation and form their characteristic terminal fields.

Transplanted fetal neurons will form appropriate terminal fields in the developing hippocampus despite being forced to grow from abnormal positions and at abnormal times. The host hippocampus apparently has sufficient indigenous cues for fetal cells to direct their fibers to their proper targets. In the partially damaged hippocampus, transplanted cells create terminal fields that replicate the pattern of the native fibers. This takes place against a background of biochemical events associated with injury that are only beginning to be understood. In response to injury, the brain appears to produce an increase in neurotropic factors that probably aid in the maintenance of existing structures and facilitate the regrowth of new circuitry. The increased concentration of these factors can be used to enhance the survival and growth of certain difficult-to-transplant neurons by placing the transplant in an injured area several days after the injury when the neurotropic factor concentration is maximal.

2. Normal Organization of the Dentate Gyrus

We have selected the hippocampal formation for study because of its defined and relatively simple structure and because of the wealth of existing data on the development and plasticity of its connections.[9] The hippocampal formation consists of two major divisions, the dentate gyrus and the hippocampus proper with its subfields CA1, CA2, CA3, and CA4 (Fig. 1). In the dentate gyrus, the major cell type (granule cells) is located in a layer with their dendrites ramifying into the molecular layer and their axons projecting to areas CA3, and CA4. The major inputs to the molecular layer are those from the entorhinal cortex and CA4 hippocampal neurons. Entorhinal fibers project to the outer three-fourths of the molecular layer while CA4 neurons project to the inner one-fourth. Septohippocampal fibers, a prominent cholinergic pathway, coinhabit the molecular layer along with serotoninergic fibers and several other minor inputs. A supragranular band of septal fibers is found just beneath CA4 fibers, while a moderately dense outer zone coexists along with entorhinal fibers. The CA4 zone contains very few, if any, septal fibers.

The septal system can be readily followed by histochemical methods that stain for acetylcholinesterase activity (AChE) (Fig. 2A). AChE histochemistry reveals a trilaminar pattern: a dark supragranular band immediately above the granule cell layer followed by a very lightly stained zone above it. The outer three-fourths of the molecular layer stains at an intermediate intensity. Microchemical analyses of choline acetyltransferase activity and immunocytochemistry with choline acetyltransferase antibodies also show a trilaminar pattern.[11,16,38]

Raphe fibers form a relatively dense band directly beneath the granule cell bodies (the infragranular band). In the molecular layer, raphe fibers are diffusely and uniformly distributed.[30]

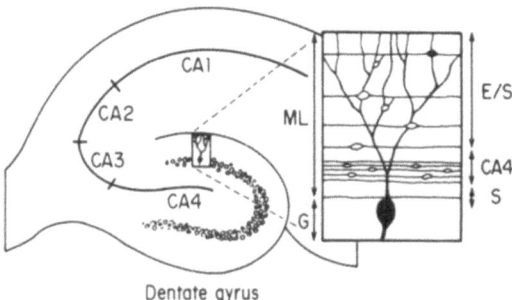

Dentate gyrus

FIGURE 1. Organization of afferent inputs to the molecular layer (ML) of the dentate gyrus. Granule cells, the major cell type of the dentate, are arranged in a layer (G) with their dendrites extending into the molecular layer. Ipsillateral associational fibers (A) and contralateral commissural fibers (C) from CA4 pyramidal cell innervate the inner one-fourth of the molecular layer. The outer three-fourths of the molecular layer consists primarily of fibers from the entorhinal cortex (E). Septohippocampal fibers (S) form a relatively dense supragranular innervation immediately above the granular cell layer: a less dense input is intermixed with entorhinal fibers. Septohippocampal fibers are absent from the A/C zone.

3. Restoration of Septal Lamination by Septal Tissues Transplanted into the Entorhinal/Occipital Cortex

A cavity was made in the occipital cortex of 3-day-old rat pups and embryonic septal tissues were implanted in it. In animals that received fimbrial transections on postnatal day 3, no AChE staining was present in the dentate molecular layer or hippocampal regio inferior (Fig. 2B). Residual AChE staining was present in the hilus and in stratum oriens and radiatum of CA1 adjacent to the subiculum, and superficial to the hippocampal fissure.[28] When embryonic septal tissue was implanted into the occipital cortex, the AChE staining pattern in the hippocampal formation was strikingly similar to that seen in the normal brain (Fig. 2C). The only substantial differences in AChE staining were quantitative, and these differences appeared to depend upon the number of cells surviving in the implant. In the dentate gyrus, the laminar distribution of AChE reaction product closely paralleled the pattern in control sections. The trilaminar pattern normally observed in the outer molecular layer of the dendate gyrus was recreated in animals with septal transplants.[21]

In a second group of experiments, septal tissue was implanted into the entorhinal cortex of the host animal in order to study the effects of a different implant site, and partial deafferentation of the pattern of implant growth into the host hippocampal formation (Figs. 2C, D). In the hippocampus proper, the distribution of AChE stain was the same as that previously described for occipital implants. In the dentate gyrus, however, the laminar pattern of AChE activity was different in that the pale-staining CA4 zone was wider. The AChE staining was confined to the outermost one-half to one-fourth of the outer molecular layer. The increased width of the CA4 zone in our entorhinal implant condition (20 μm on the average) correlated well with the width increase (4–25 μm) previously reported in animals with entorhinal lesions made at 11–16 days

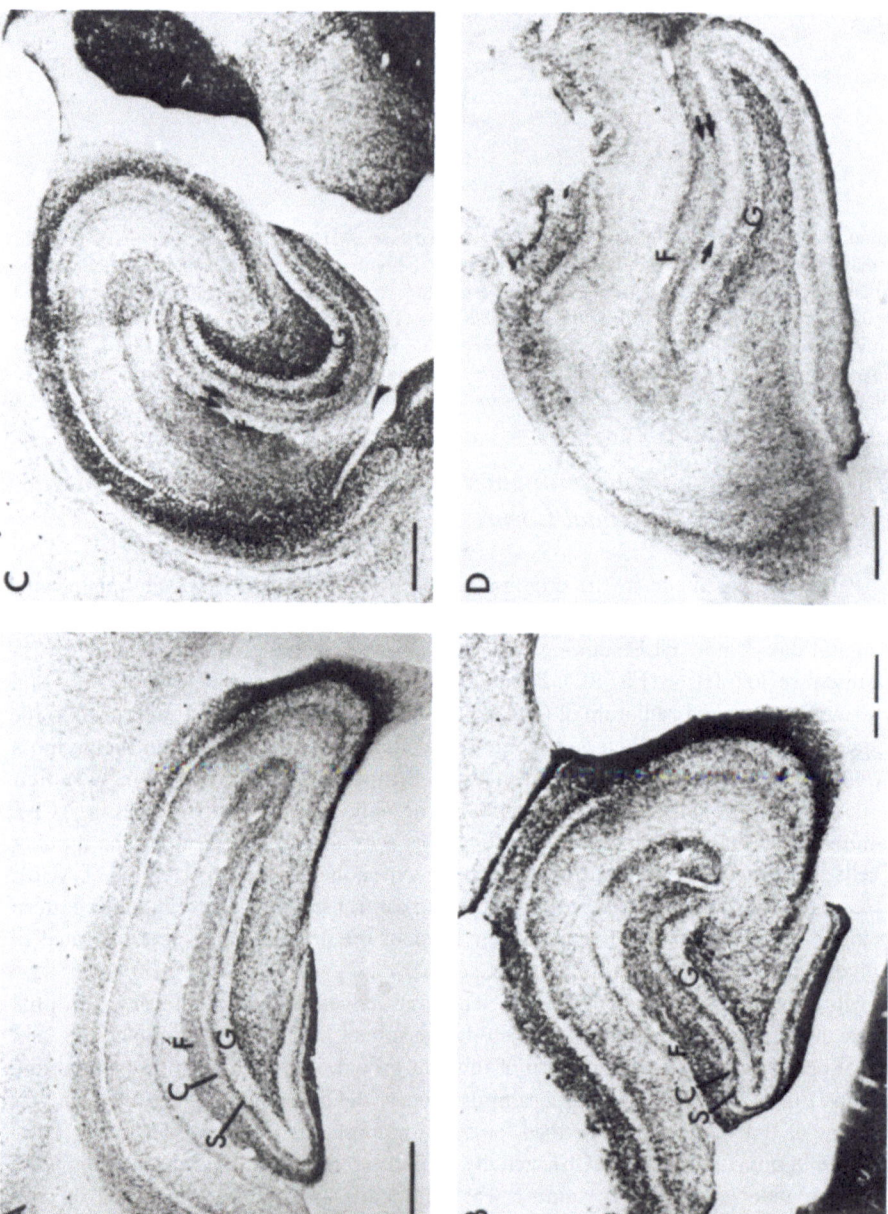

FIGURE 2. Comparison of control hippocampus (A) and hippocampus innervated by septal implant (B) as seen in coronal section. In (B), septal tissue was implanted into the occipital cortex dorsal to the hippocampus on postnatal day 3 and the animal was sacrificed 30 days after implant. In (C) and (D), control hippocampus (C) and hippocampus innervated by septal implant (D) in horizontal section are compared. Septal tissue was placed in a cavity in the ento-rhinal cortex on postnatal day 3 and the animal was sacrificed 30 days after implant. Note the expansion of the pale-

of age.[31] This result illustrates that septal tissues assume the reorganized pattern produced by an entorhinal lesion.[21]

Axons of septal neurons implanted on postnatal day 3 were first detectable in the dentate molecular layer between postnatal days 10 and 12.[17] This was 4 to 6 days later than the time that native septal axons arrived but still before the period of peak synaptogenesis in the dentate molecular layer. Furthermore, tissues grafted on postnatal day 6 or 9 produced the same pattern of AChE staining in the target area as day 3 implants despite their arrival during the period of active synapse formation.

The ability of septal fibers from the implant to produce a pattern of innervation quite similar to that produced by native septal efferents, despite their anomalous position in the occipital or entorhinal cortex, suggests that the location of the presynaptic cell is not critical to the mechanisms that guide these fibers to their target cells in the hippocampal formation. The developmental disparity between implanted (3 days prenatal) and host tissue (3 days postnatal) suggests that the outgrowth of axons from implanted cells should be delayed by approximately 1 week relative to the time of arrival of native septal efferents, which occurs around day 6 postnatal.[27,32,35] This delay does not appear to disrupt the laminar distribution of fibers from implanted cells. A mechanism of septohippocampal lamination that relies on a strict temporal correspondence between pre- and postsynaptic elements is inconsistent with these findings.

4. Absence of Guidance by Degenerative Debris

As a consequence of the transplantation surgery, neurons in the developing entorhinal cortex are ablated and it is possible that axons from transplanted neurons may be guided by degenerating afferent fibers in the host. In order to evaluate this possibility, we characterized the time course for the removal of degeneration products and made transplants at times when degeneration had been removed. The growth of septal implants could then be examined in a host brain devoid of observable degenerative debris.

The notion that growing axons might be attracted by degenerating nerves seems to have originated with Ramón y Cajal.[34] In the amphibian visual system, guidance via myelin debris has been postulated to account for the distribution of axons regenerating to the tectum.[22] A similar phenomenon was hypothesized in the mammalian CNS by Björklund et al.[4] to explain the growth of transplanted monoaminergic neurons along the lesioned perforant path into the hippocampal formation of adult rat recipients.

Three-day-old rat pups received lesions of the entorhinal cortex and were sacrificed 24 or 72 hr after the operation. The presence of degeneration products was monitored by electron microscopy and by a silver stain specifically modified for use in developing animals (DeOlmes method). At 24 hr postoperation, silver deposits indicative of degeneration products were found in zones corresponding to normal entorhinal input. At 72 hr postsurgery, however, no degeneration argyrophilia was found in the dentate gyrus. Electron microscopic analysis of the molecular layer confirmed the absence of residual degenerative elements at 72 hr.

Accordingly, septal implants were placed in the entorhinal cortex on postnatal day 6, 3 days after the implant cavity was created. The grafts implanted according to the two-stage surgery were significantly larger than those placed according to the one-stage protocol and they showed greater vigor in their ability to send axons into the neuropil.

The pattern formed in the hippocampus was essentially identical to that in the normal rat hippocampal formation and in the hippocampi of rats implanted with septal tissues according to the one-stage surgery.[21] In the hippocampus proper, AChE staining was dense in the lamina external to the pyramidal cells (stratum oriens) and of moderate intensity internal to the pyramidal cell layer and along the hippocampal fissure. The hilus also stained darkly for AChE. A distinct bilaminar distribution of AChE reaction product was present in the dentate gyri of both normal and implanted rats. Thus, the specific growth of axons from implanted neurons does not depend upon cues derived from the presence of degeneration.[18]

5. Patterns Formed by Serotonin Fibers from Raphe Transplants

It is of interest to compare this ingrowth of raphe fibers to those of septal fibers. Can raphe cells placed in the entorhinal cortex late in development locate and negotiate the appropriate cues in the neuropil to re-form their correct pattern? Do these fibers form their normal pattern or a septal-like pattern?

The pattern of reinnervation of the hippocampal formation by 5-HT and noradrenergic neurons has been investigated previously in adult animals. Azmitia et al.[1] have shown that mouse fetal raphe tissue implanted into adult mice formed a pattern of 5-HT innervation similar to that found in normal animals, although the total density of 5-HT fibers was substantially increased compared to the contralateral hippocampus. However, Björklund et al.[3-5] have reported that 5-HT neurons implanted in the retrosplenial cortex or in the hippocampal fimbria of adult rats grow into the dentate gyrus and reinnervate not only the infrapyramidalzone, where the densest serotoninergic innervation normally is observed,[29,30,36] but also into the outer part of the dentate gyrus molecular layer where the entorhinal perforant path fibers normally terminate.

In order to study transplant fibers in the absence of normal raphe input, the native 5-HT input was removed by a fimbrial lesion and entorhinal cortex ablation. In these animals, essentially all 5-HT fibers were absent in the hippocampus. Raphe tissues were placed into the entorhinal cavity made in 3-day-old pups 3 days after the lesions. This two-stage procedure improved the growth and survival of the raphe tissues over that of the one-stage procedure and allowed the outgrowth of raphe fibers to be studied in the absence of degeneration.

At 7 days postimplant, raphe neurons appeared rounded and had not yet formed processes. By 14 days, the cells displayed fiber processes and sent axons into the hippocampal formation. The 5-HT fibers were long with many beaded varicosities, and coursed in parallel from the implant to the hippocampal formation. The fibers had reached the CA3 area of the hippocampus and followed the hippocampal fissure to the dentate gyrus. From this point, the fibers passed into the dentate gyrus molecular layer and coursed along either side of the granule cells into the supra- and infragranular layers. Some of the 5-HT fibers grew into the hilus of the dentate gyrus where a moderately dense innervation was observed (Fig. 3). In the molecular layer, fibers were

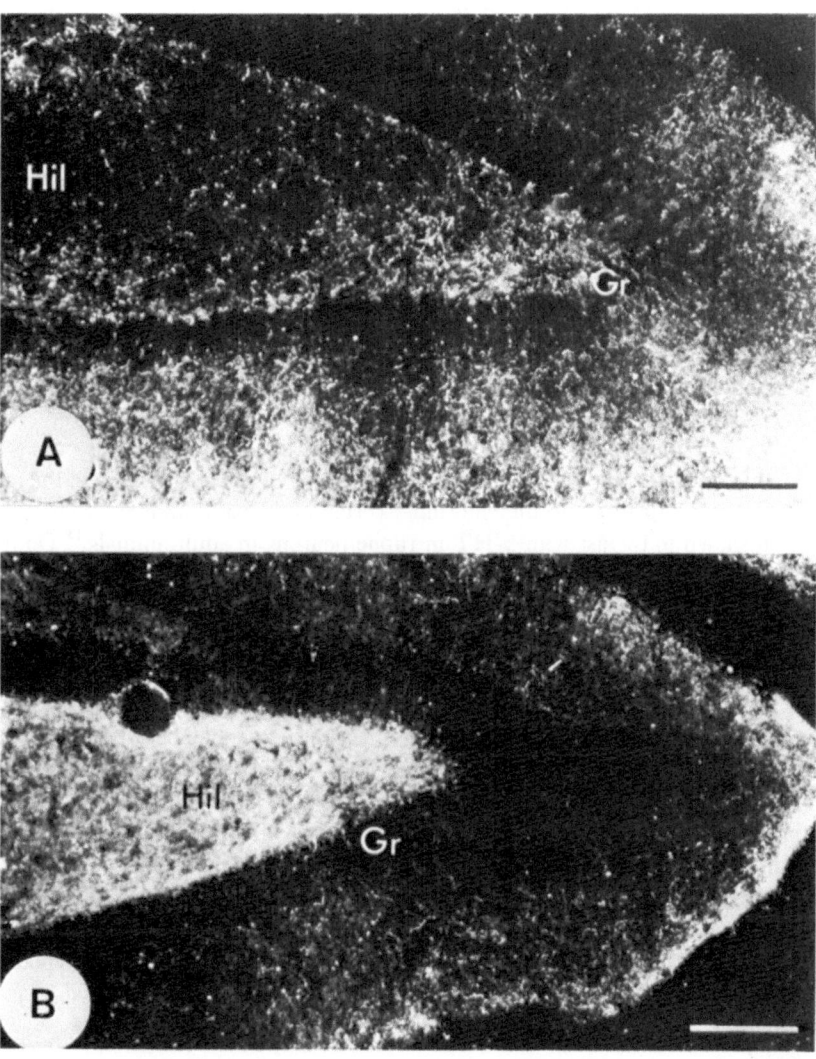

FIGURE 3. Photomicrographs of the hippocampal formation following raphe implants. A) darkfield photomicrograph of the dentate gyrus (in coronal section) from an animal 21 days after a raphe implant. B) darkfield photomicrograph of the dentate 30 days after a raphe implant. In this animal the hilus is very densely innervated. Antisera to 5-HT were used to visualize the raphe fibers. Hil, hilus; Gr, granule cells. Bar = 50 μ. From Ref. 15.

diffuse and seemingly randomly distributed in contrast to the strict laminar appearance of septal fibers. Both the hippocampus and the dentate gyrus appeared hyperinnervated when compared to normal rats. Raphe neurons in the transplant could be retrogradely labeled following injections of fast blue into the hippocampus or dentate gyrus, establishing that raphe neurons in the transplant had in fact given rise to the new 5-HT hippocampal fibers. The majority of retrogradely labeled neurons contained 5-HT; however, a significant number did not.

The finding that the earliest transplant fibers that arrived in the dentate molecular layer were diffusely distributed is similar to their normal developmental sequence. In control animals, 5-HT fibers were evident by 7 days after birth at which time they were diffusely organized with no discernible pattern. The major change during development was an increase in fiber density. Thus, transplant fibers recapitulate their developmental sequence and final organization in the molecular layer even though they have grown from an abnormal position and at a delayed time during development.[15]

6. Possible Coexistence of Transmitters in Raphe Transplants

Methionine-enkephalin (ME) and substance P (SP) were also studied in the raphe nuclei following implants into the entorhinal cortex of neonatal animals. ME and SP have been shown to coexist with 5-HT in raphe neurons in adult animals.[14] Do these transmitters coexist in the implanted raphe neurons?

SP and ME fibers were not observed in the hippocampus although SP and ME neurons were present in the implant (Fig. 4). Both SP and ME-immunoreactive neurons contained 5-HT immunoreactivity, but these neurons were not those retrogradely labeled. It was unclear whether their fibers stayed within the transplant or projected to other sites in the host brain and therefore were not retrogradely labeled.

The coexistence of transmitters in these transplanted neurons indicates that the ability to express transmitter coexistence does not require that the raphe neurons differentiate in the brain stem and receive their normal afferents. However, it may be important that the peptidergic neurons are placed in their proper site in order for the normal development of their complete afferent projections. Other peptidergic neurons have been shown to reinnervate the host brain following implantation in their normal site. Vasopressin-containing neurons of the paraventricular nucleus of the hypothalamus, for example, will reinnervate the median eminence and alleviate polydipsia and polyuria in the adult Brattleboro rat.[12,13]

7. Comparison Between Septal and Striatal Transplants

Is the pattern generated by septal fibers shared by other neurons perhaps sharing transmitter type or is the pattern unique to septal neurons? Striatal tissues contain a population of cholinergic neurons that can be grafted into the entorhinal cortex and examined to determine if they can replicate the laminated pattern of septal neurons. Cholinergic striatal neurons are short-axon neurons and these do not normally innervate the hippocampus.

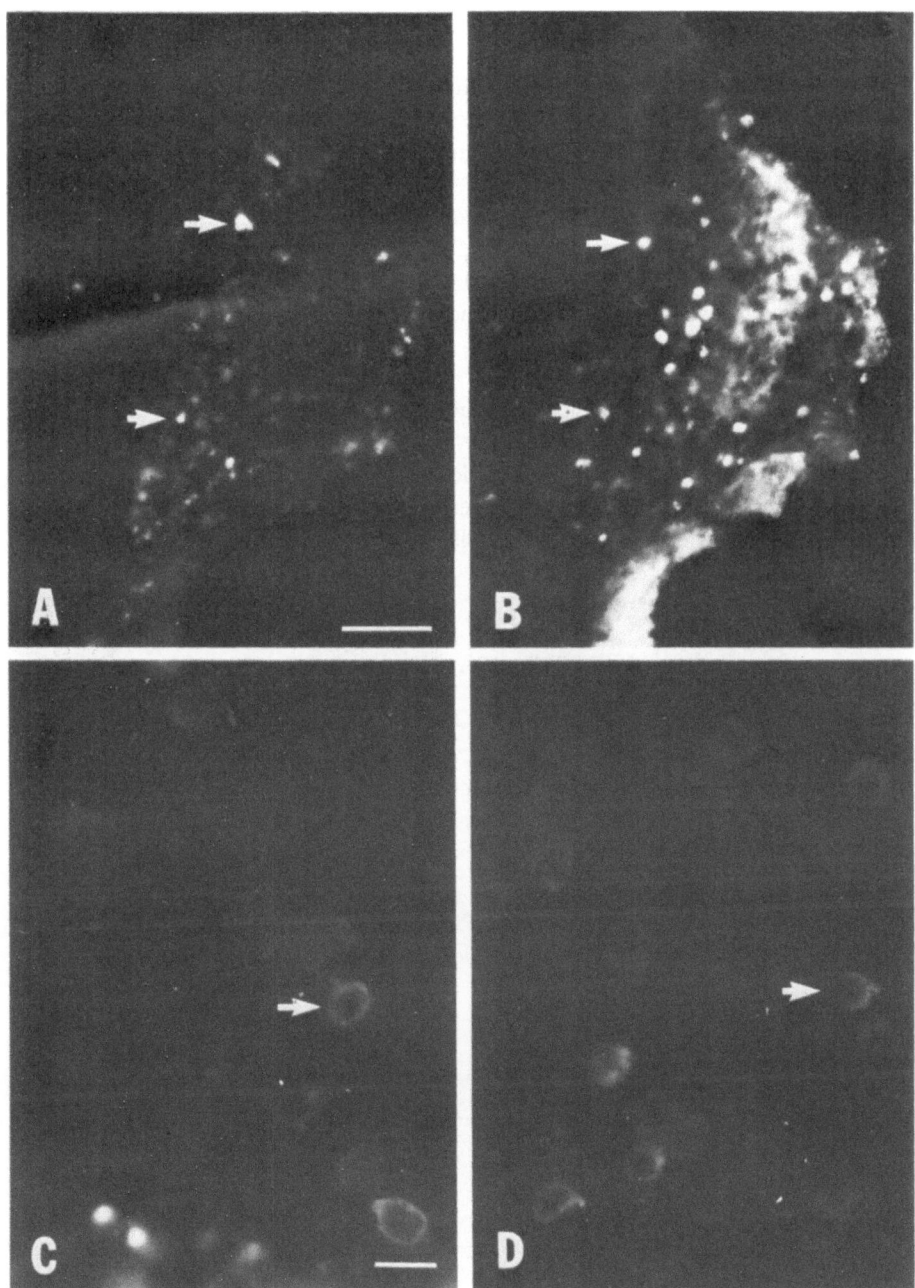

FIGURE 4. Fluorescence (A) and immunofluorescence photomicrographs (B–D) of a raphe implant 30 days postimplant incubated with 5-HT antisera (A, B, D) and SP antisera (C), (A, B) and (C, D) represent the same section, respectively. In (A), retrogradely labeled fast blue neurons can be seen (arrows). Some of the retrogradely labeled cells exhibit a bright fluorescence (arrows), whereas others are only lightly labeled. Some of the 5-HT-immunoreactive neurons which are also retrogradely labeled are shown (arrows) in (B). SP-immunoreactive neurons are observed in the raphe implant (arrow, C), some of which also contain 5-HT immunoreactivity (arrow, D). Bars = 50 μm. From Ref. 15.

The two-stage transplantation procedure was used for striatal tissues so that a direct comparison to septal and raphe transplants could be made and because striatal tissues were particularly difficult to transplant. The delay appeared to enhance the survival of striatal neurons.

Figure 5 compares the AChE staining in the two implant conditions with the hippocampal formation from a control rat (Fig. 5A) and a deafferented hippocampal formation from a control animal that received a fimbrial transection on postnatal day 3 (Fig. 5B). The distribution of AChE reaction product in the hippocampal formation ipsilateral to a striated tissue implant (Fig. 5C) was essentially identical to that seen following septal transplants (Fig. 5D).[20]

The ability of striatal implants to produce AChE staining in the appropriate "cholinergic" terminal laminae of the immature host hippocampal formation suggests that an identity of neurotransmitter characteristics encourages similar types of interactions between developing afferent fibers and target neurons. As the raphe transplants give a different pattern, this eliminates the possibility that any central tissue implanted adja-

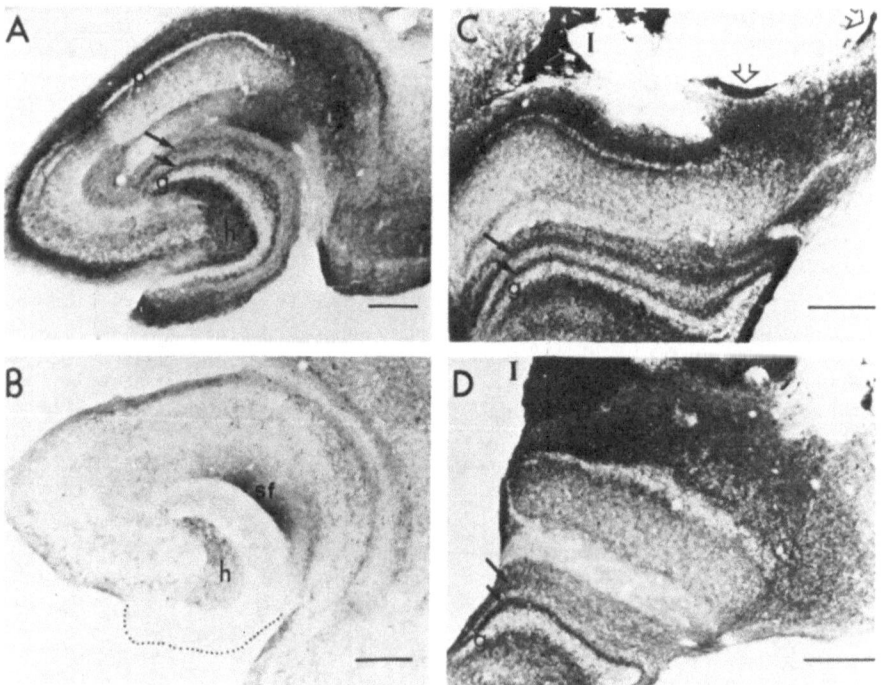

FIGURE 5. Comparison of AChE staining pattern in the hippocampus in control animals (A), animals with a fimbrial transection (B) and with a transplant of striatal tissues (C), and animals with a transplant of septal tissue (D). The striatal implant occupies the anteromedial portion of a cavity of the host entorhinal cortex. Implant-associated AChE staining is intense in the subiculum and on either side of the hippocampal pyramidal cells. In the molecular layer, bands of stain are present suprajacent to the granule cells and the outer portion of the molecular layer. Bars = 150 μm, From Ref. 20.

cent to an immature hippocampal formation would produce the same "septal-like" pattern of innervation. Previous experiments also argue against a general effect. In the adult hippocampus, implants of monoaminergic tissues produce patterns of innervation unlike the AChE pattern produced by septal grafts. Dopaminergic and serotoninergic axons from tissues grafted into the entorhinal area accumulate in hippocampal zones that have been deafferented by the entorhinal cortex lesions. Dispersed fetal serotoninergic neurons injected into the mature hippocampus form their normal monolaminated pattern in the molecular layer.[1] Similarly, the distribution of implanted noradrenergic fibers is similar to their normal pattern of innervation.[4] Finally, native sympathetic noradrenergic axons, sprouting in response to fimbrial transections, show the highest density in the hilus and generally resemble the normal pattern of noradrenergic innervation. [23,24] Thus, the pattern of septal and striatal tissue grafts is unlike that formed by cells that employ other transmitters.

8. Competition between Native and Transplanted Cholinergic Fibers

Can native septohippocampal fibers outcompete cholinergic fibers from the transplants? Numerous intrinsic and extrinsic fibers are making synapses during development which can compete with ingrowing transplant fibers. Does a specific competition exist between related inputs that can restrict the ingrowth of transplant fibers? Mechanisms must exist that limit the abundance of particular types of inputs and it would appear that examining the interaction between transplant and native fibers may provide information on this process.

Accordingly, native septal afferents were left intact during the period of implant growth.[19] Septal or striatal tissues were transplanted into the entorhinal cortices of neonatal rats by the two-stage procedure. Thirty days after implantation and five days prior to sacrifice, recipient rats received a transection of their native septohippocampal fibers. This allowed examination of implant-associated fibers at a time when native cholinergic fibers had disappeared.

Septal implants appeared to survive effectively and fibers were evident in residual subicular tissue in the strata oriens and radiatum of CA1 and along the hippocampal fissure (Fig. 6). The dentate molecular layer, in contrast, contained a small amount of staining that barely exceeded background levels. A band of transplant fibers was evident along the hippocampal fissure which ran for several hundred micrometers.[19] Thus, it appears that fibers from the implanted neurons approach the target cells but fail to proliferate within the terminal laminae. They are outcompeted for terminal space by developing septal afferents in the host brain. After the fimbrial transection, no additional regions of staining nor intensification of fiber growth occurred with increasing time up to 1 month. Thus, the transplant fibers did not appear to reinitiate growth once the native fibers were removed.[19]

In view of the immature state of the dentate molecular layer, it seems all the more surprising that processes from implanted cells are unable to gain access to postsynaptic target zones. We have not carried out a developmental analysis to determine whether

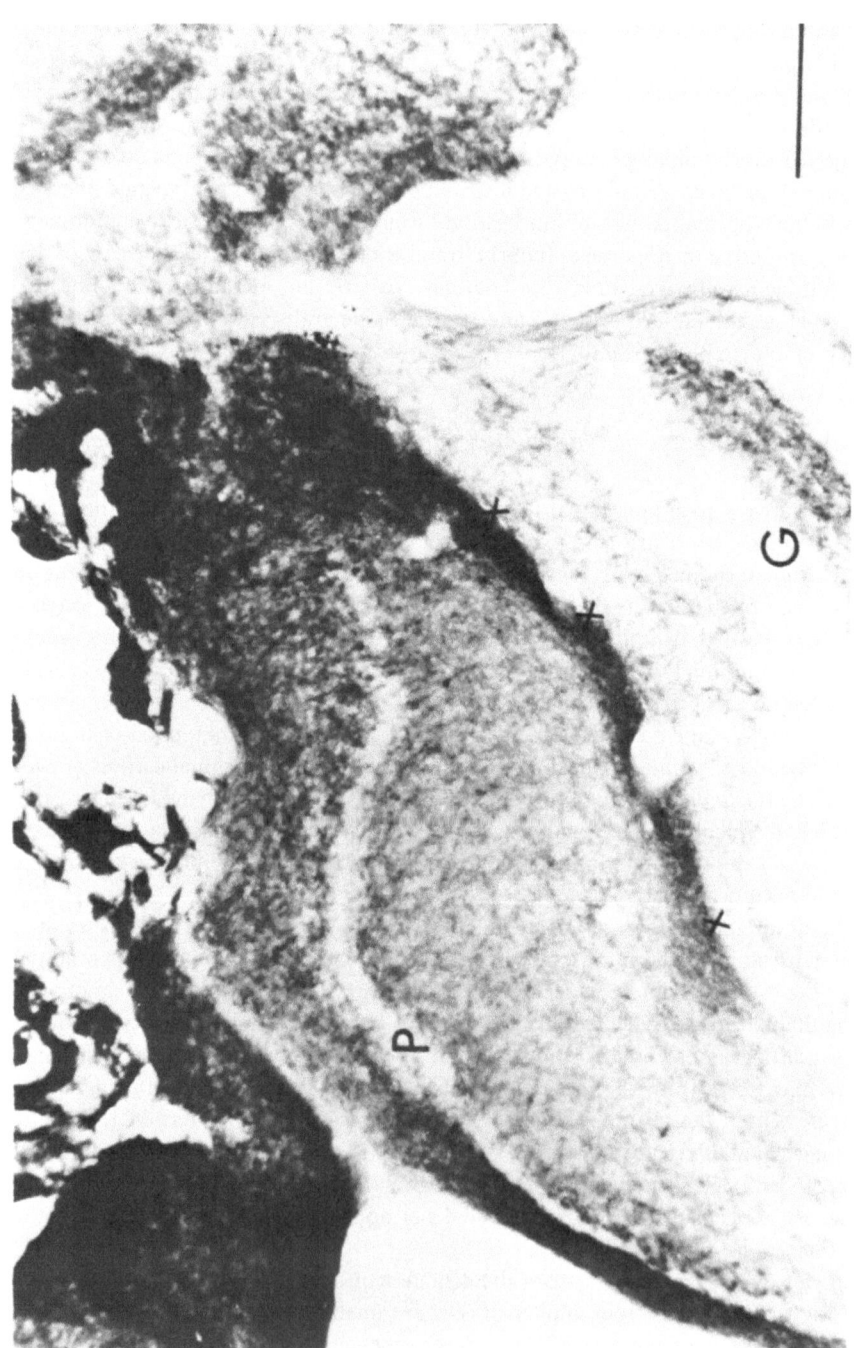

FIGURE 6. A septal implant made with fimbria intact 1 one week prior to the time of sacrifice. Internal AChE staining is present across the subiculum and along the hippocampal fissure but it does not enter the dentate molecular layer. Bar = 250 μm. From Ref. 19.

axons of cells in the septal grafts initially enter the target area and are then displaced by maturing native septal afferents. In either case, however, it appears that there may be a limited supply of termination sites in the dentate molecular layer in that a hyper-innervation by the supernumerary septal afferents from the grafts does not occur. The possibility that developing granule cells in the dentate gyrus may quantitatively restrict their afferent input has been postulated also with respect to the perforant path projection from the entorhinal cortex.[17] Such afferent-specific restriction has not been previously reported. It suggests highly specific biochemical mechanisms such as the release of selective growth-inhibiting factors or affinity for a limited quantity of "septal" postsynaptic sites. Similar experiments have been carried out in the adult hippocampus. In adult animals, fibers from septal grafts placed in the entorhinal cortex are restricted to terminal laminae which are deafferented during implant surgery when the septal input to the target hippocampus is intact.[2] Our data demonstrate that the exclusion is more complete in developing animals.

9. Factors Involved in Lamination of Cholinergic Fibers in the Dentate Gyrus

A variety of mechanisms all working in concert participate in organizing terminal fields in the dentate. What factors might account for the terminal distribution of axons from septal or striatal grafts? As argued in the previous section, it does not appear that initial position of the neurons is critical nor is the exact time of arrival of the fibers in the target area. Previous results on lesion-induced organization of the septal system provide an important clue and a possible determinant to cholinergic organization in the dentate gyrus. These studies indicate that the CA4 afferent system plays a critical role in establishing septal lamination. It appears that CA4 fibers can selectively displace or exclude cholinergic fibers. This hypothesis appears to explain the data on the normal and lesion-induced reorganization of septal afferents. Septohippocampal fibers are always sparse in areas rich in commissural associational fibers (Fig. 7), as if these fibers exclude septohippocampal ones. For example, the expansion of the AChE clear zone following an entorhinal lesion corresponds very closely to the widening of the CA4 fiber plexus. As noted above, this exclusion also appears to pertain to fibers from transplants.

The capacity to exclude cholinergic fibers appears specific to CA4 fibers. Thus, for example, a partial entorhinal lesion causes septal reactivity and a proliferation of contralateral entorhinal and probably other fibers in that zone, but it does not lead to a loss of septal innervation.

It appears that the distribution of a fiber system can establish that of another: CA4 fibers appear to serve as an afferent that controls the pattern of both septal growth and reorganization. A hierarchy of interactions exists probably established during development and retained throughout life. We have called this model the *critical afferent theory of lamination;* it is a special type of competition mechanism where a particular afferent can dictate the pattern of select other ones.[6,7,17] In excluding septal afferents, CA4 fibers have the same capacity that native septal afferents have for transplanted ones.

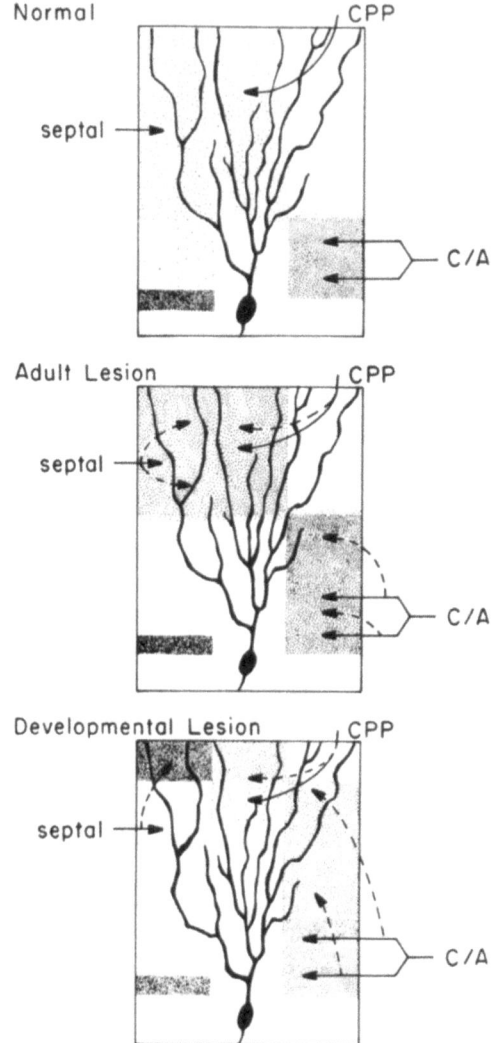

FIGURE 7. Schematic showing the organization of inputs in the dentate gyrus of normal animals and those with entorhinal lesions made during development or adulthood. Septohippocampal fibers are absent in the CA4 zone (C/A). From Ref. 8.

10. Possible Basis for Enhanced Survival in the Two-Stage Procedure

It appeared that the two-stage procedure enhanced the survival and growth of neurons placed in the entorhinal cavity. This was particularly critical for striatal neurons, which showed very limited survival when transplanted at the time of injury (Fig. 8). Septal and raphe tissues survived in the one-stage procedure but the size of the transplant and vigor of outgrowth appeared better in the two-stage procedure.[15,18,26]

Survival of transplants depends on at least the following variables: (1) adequate oxygen and nutrient supply, i.e., vascularization; (2) absence of lethal concentrations of toxic substances; (3) availability of trophic substances. Graft survival was associated by Stenevi et al.[37] with implantation adjacent to a highly vascularized surface. Certainly, vascularization of the tissue adjacent to the original cavity accounts at least in part for the enhancement of implant survival when implantation is delayed. However, it is also possible that toxic substances may be released from the injury area and these may decrease with time. Various hydrolytic enzymes and microglia, for example, may injure or destroy transplanted neurons particularly immediately after injury.

Another possibility is that injury to the CNS causes an increase over time in the concentration or activity of factors promoting neuron survival and growth. Neural development is dependent on neurotrophic factors that regulate neuron survival, growth, and differentiation. Such factors may also be released as a sequel to injury in the developing and adult CNS. Factors that promote neuron survival have recently been shown to appear following peripheral nerve injury.[10,25] However, the long suspected possibility that brain injury induces the release of trophic factors has only recently been directly tested. We hypothesized that injury to the CNS was accompanied by an increase of factors promoting neuron survival and that their appearance was a major cause of the enhanced transplant survival when transplantation was carried out several days after the injury.

In order to test this possibility, fractions were collected from the area of injury and assayed in cell culture for neurotropic activity. A wound was made in the entorhinal cortex of developing rats (3 days old) and the lesion cavity was filled with a fragment of saline-moistened Gelfoam. At various times postlesion, the animals were sacrificed and the Gelfoam fragment and the tissue surrounding the wound cavity were analyzed for their content of neurotropic activity in cell culture. A survey of the neurotropic activity in the wound fluid was carried out on various neurons for its ability to support the survival of these cells (Fig. 9). It was found that neurotrophic activity clearly accumulates in the wound cavity. These trophic factors supported the survival of neurons in monolayer cultures of chick embyro spinal cord, ciliary ganglion, and sympathetic ganglion as well as mouse dorsal root ganglion. Trophic activity was low in noninjured

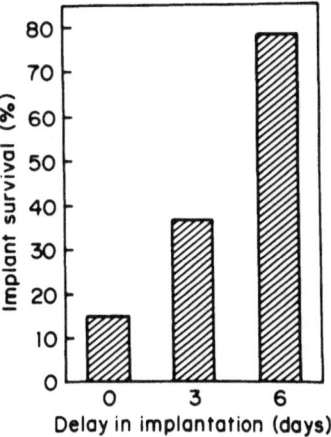

FIGURE 8. Survival of corpus striatum grafts into host cortex as a function of the delay in implantation. Implants of embryonic corpus striatum survived much better if they were placed in the host cortical cavity 3 or 6 days after this was made. The delay also dramatically increased the ability of the surviving implants to innervate the host hippocampus. From Ref. 26.

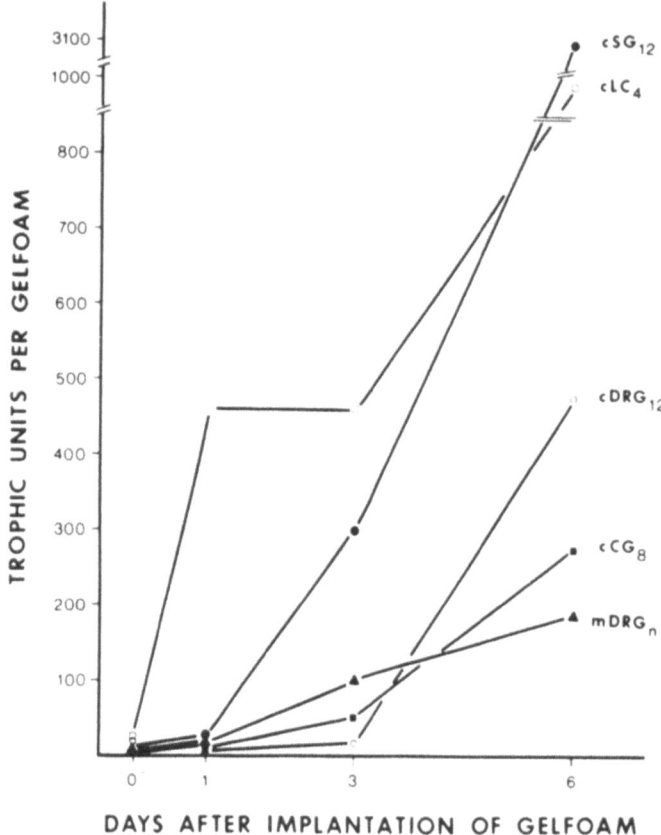

DAYS AFTER IMPLANTATION OF GELFOAM

FIGURE 9. Time course of neurotropic activity in Gelfoam fragments implanted for increasing periods. Extracts derived from ablated entorhinal tissue of 3-day rat pups (0 time) as well as from Gelfoam fragments implanted for 1, 3, or 6 days in the resulting cavity were assayed for neurotropic activity toward the five indicated neuronal cell types. Chick embryo tissues: 12-day sympathetic (cSG_{12}) and dorsal root ganglia ($cDRG_{12}$), 4-day lumbar cord (cLC_4), 8-day ciliary ganglion (cCG_8). Neonatal mouse dorsal root ganglion ($mDRG_n$). From Ref. 26.

brain tissue and in the wound cavity 1 day postlesion. Activity increased 4- to 30-fold during the subsequent 2–5 days. The magnitude of the increase depended on the target tissue used in the assay. Activity was highest for sympathetic ganglion and spinal cord neurons. It was lowest for mouse dorsal root ganglion neurons. The neurotrophic factors appeared to be proteins immunologically distinct from mouse submaxillary nerve growth factors.[16,33]

If the hypothesis that higher levels of neurotrophic factors improve the survival of transplants were correct, then exogenous supply of these factors should enhance survival when the implant is placed in the cavity with no postsurgical delay or even when the delays are too long for sufficient endogenous production. Recently, we have found that the survival of striatal tissues implanted at the time of making the wound can be significantly improved by providing a supply of exogenous factors (M. Nieto-Sampedro and C. Cotman, unpublished observations). This raises the possibility that such factors once purified may prove therapeutically useful as a treatment for cell loss.

11. Summary and Conclusion

We have focused our efforts on attempting to understand the principles involved in the development of two minor but important inputs to the dentate gyrus, septal and raphe inputs. Transplants have proven to be valuable tools for exploring the properties of the developing hippocampus because they allow manipulations otherwise prohibitive. Transplants of septal, raphe, or striatal tissues were placed into a cavity made in the entorhinal or occipital cortex of a developing rat pup. Transplantation was carried out in two stages. In the first stage, the septum was removed and a cavity made in the occipital or entorhinal cortex of 3-day-old rats. In the second stage, the transplant was placed in the cavity 3 to 6 days after it was made. This allowed time for the removal of degeneration products and improved the survival of transplanted neurons as well as the vigor of their outgrowth. Fibers from septal transplants enter in the hippocampus approximately 11 to 14 days after transplantation, which is at least 1 week after native septal fibers arrive in normal animals. When placed in the occipital cortex, sparing the entorhinal cortex, septal fibers recreate the pattern of an uninjured hippocampus. Thus, fibers from these transplanted septal neurons can locate their appropriate terminal fields even though they grow into the hippocampus from the opposite direction and at a later time. Transplants were also made in the entorhinal cortex which partially denervates the hippocampus. In animals with entorhinal lesions, native septal fibers sprout and develop an anomalous organization confined to the outermost part of the molecular layer. This pattern is largely replicated by transplant fibers. When placed in the entorhinal cortex, septal fibers appear to create the pattern of an injured nervous system as opposed to repairing circuitry to normal. The ability to create a septal-like pattern is not unique to septal neurons but is shared by striatal cholinergic neurons. Thus, it appears that a property related to transmitter type may help code for locating the proper terminal fields in the hippocampus. Septal or striatal transplant fibers will not outcompete native septal fibers if they are left intact. Septal fibers do enter the hippocampus along the hippocampal fissure but are not found in the neuropil. The septal input most likely accounts for only a small fraction of the total input to the dentate. It can compete effectively with other intrinsic and extrinsic fibers but apparently not native septal fibers. This illustrates the existence of highly specific selection mechanisms that may operate to set the total cholinergic input to the hippocampus during development. Removal of the native input after transplant fibers have grown along the hippocampal fissure does not trigger their regrowth into the neuropil. Once "mature," septal transplant fibers will not reinitiate growth.

Raphe neurons survive quite well when transplanted into the entorhinal cortex, and they send fibers into the hippocampus and the dentate gyrus. In the dentate gyrus, raphe fibers are diffusely and apparently randomly dispersed throughout the molecular layer. This pattern is highly distinct from the septal pattern. The fibers appear to be more abundant than in normal animals, particularly in the hilar zone.

Why do cholinergic neurons form one type of pattern and serotoninergic neurons another? The simultaneous transplantation of both septal and raphe tissues in the entorhinal cortex results in patterns unique to each cell type in the same animal. The terrain is virtually equivalent so that the fibers must monitor different cues in the neuropil and/or have different inherent cellular properties. One critical issue is whether synapses are formed and of what types. Recently, we have found that neither native or transplant raphe fibers form synaptic junctions. Instead, 5-HT fibers form numerous varicosities throughout the molecular layer (Holets and Cotman, unpublished observations). This may explain the absence of a discrete organization and the tendency to hyperinnervate the molecular layer. Raphe fibers may not have the ability to make junctions or they may not experience restrictive local interactions. Cholinergic fibers, on the other hand, may benefit more from local interactions. One distinctive feature of septal and striatal fibers is that they are controlled by CA4 fibers which can exclude them from CA4 fiber laminae. Raphe fibers freely intermix with the CA4 zone. The distinct boundary between cholinergic fibers and CA4 fibers suggests that CA4 fibers employ specific local properties to exert their control. Possibly, for example, they make the target cell membrane incompatible with cholinergic input. It is not known at present, however, if they form synaptic junctions. It is clear, though, that areas of the molecular layer outside CA4 are permissive to cholinergic growth whereas the CA4 zone is restrictive.

Successful transplants depend on adequate cell survival. Indeed, the normal maintenance and optimal return of function after injury depends on neuron survival. We have found that transplants survive better if the transplant is made several days after the initial injury. Part of the enhanced success appears to be due to a time-dependent increase in neurotrophic factors at the injury site. These factors may be an important mechanism in the maintenance and repair of injured circuits. Once fully characterized and available in pure form, they may be useful aids in transplantation experiments as well as in maintaining and repairing residual circuitry after injury. As the study of transplants progresses, it is likely that fundamental knowledge about the injured nervous system will lead to discoveries and treatments that will ultimately, it is hoped, replace the therapeutic need for transplants themselves in repairing nerve injury.

References

1. Azmitia, E. C., Perlow, M. J., Brennan, M. J., and Lauder, J. M., 1981, Fetal raphe and hippocampal transplants into adult and aged C57BL/6N mice: A preliminary immunocytochemical study, *Brain Res.* **7**:703.
2. Björklund, A., and Stenevi, U., 1977, Reformation of the severed septohippocampal cholinergic pathway in the adult rat by transplanted septal neurons, *Cell Tissue Res.* **185**:289.
3. Björklund, A., and Stenevi, U., 1979, Regeneration of monoaminergic and cholinergic neurons in the mammalian central nervous system, *Physiol. Rev.* **59**:62.
4. Björklund, A., Stenevi, U., and Svendgaard, N. A., 1976, Growth of transplanted monoaminergic neurons into the adult hippocampus along the perforant path, *Nature (London)* **262**:787.
5. Björklund, A., Wiklund, L., and Descarries, L., 1981, Regeneration and plasticity of central serotoninergic neurons: A review, *J. Physiol. (Paris)* **77**:247.
6. Cotman, C. W., 1979, Specificity of synaptic growth in brain: Remodeling induced by kainic acid lesions, in: *Development and Chemical Specificity of Neurons* (M. M. Cuenod, G. W. Kreutzberg, and F. E. Bloom, eds.), pp. 203–215, Elsevier/North-Holland, Amsterdam.

7. Cotman, C. W., Lewis, E. R., Hand, D., 1981, The critical afferent theory: A mechanism to account for septohippocampal development and plasticity, in: *Proceedings in the Life Sciences* (H. Flohr and W. Precht, eds.), pp. 13–26, Springer-Verlag, Berlin.

8. Cotman, C. W., and Nadler, J. V., 1978, Reactive synaptogenesis in the hippocampus, in: *Neuronal Plasticity,* (C. W. Cotman, ed.), pp. 227–271, Raven Press, New York.

9. Cotman, C. W., and Scheff, S. W., 1979, Synaptic growth in aged animals, in: *Aging,* Vol. 8 (E. Cherkin, C. E. Finch, N. Kharash, T. Makinodan, F. L. Scott, and B. L. Strehler, eds.), pp. 109–120, Raven Press, New York.

10. Ebendal, T., Olson, L., Seiger, A., and Hedlun, K. O., 1980, Nerve growth factors in the rat iris, *Nature (London)*, **286**:25.

11. Fonnum, F., 1970, Topographical and subcellular localization of choline acetyltransferase in rat hippocampal region, *J. Neurochem.* **17**:1029.

12. Gash, D., and Sladek, J. R., Jr., 1980, Vasopressin neurons grafted into Brattleboro rats: Viability and activity, *Peptides* **1**:11.

13. Gash, D., Sladek, J. R., Jr., and Sladek, C. D., 1980, Functional development of grafted vasopressin neurons, *Science* **210**:1367.

14. Hokfelt, T., Ljungdahl, A., Steinbusch, H., Verhofstad, A., Nilsson, G., Brodin, G., Pernow, B., and Goldstein, M., 1978, Immunohistochemical evidence of substance P-like immunoreactivity in some 5-hydroxytryptamine containing neurons in the rat central nervous system, *Neuroscience* **3**:517.

15. Holets, V. R., and Cotman, C. W., 1984, Postnatal development of the serotonin innervation of the hippocampus and dentate gyrus: Normal development and reinnervation following raphe implants, *J. Comp. Neurol.,* in press.

16. Houser, G., Crawford, G., Anderson, L., Barber, R., Salvaterra, P. M., and Vaughn, J. E., 1982, Immunocytochemical localization of cholinergic neurons with a monoclonal antibody to choline acetyltransferase, *Soc. Neurosci. Abstr.* **8**:662.

17. Lewis, E. R., and Cotman, C. W., 1980, Factors specifying the development of synapse number in the rat dentate gyrus: Effects of partial target loss, *Brain Res.* **191**:35.

18. Lewis, E. R., and Cotman, C. W., 1982, Mechanisms of septal lamination in the developing hippocampus revealed by outgrowth of fibers from septal implants. II. Absence of guidance by degenerative debris, *J. Neurosci.* **2**:66.

19. Lewis, E. R., and Cotman, C. W., 1982, Mechanisms of septal lamination in the developing hippocampus revealed by outgrowth of fibers from septal implants. III. Competitive interactions, *Brain Res.* **233**:29.

20. Lewis, E. R., and Cotman, C. W., 1983, Neurotransmitter characteristics of brain grafts: Striatal and septal tissues form the same laminated input to hippocampus, *Neuroscience* **8**:57.

21. Lewis, E. R., Mueller, J. C., and Cotman, C. W., 1980, Neonatal septal implants: Development of afferent lamination in the rat dentate gyrus, *Brain Res. Bull.* **5**:217.

22. Lo, R. Y. S., and Levine, R. L., 1980, Time course and pattern of optic fiber regeneration following tectal lobe removal in the goldfish, *J. Comp. Neurol.* **191**:295.

23. Loy, R., Milner, T. A., and Moore, R. Y., 1980, Sprouting of sympathetic axons in the hippocampal formation: Conditions necessary to elicit growth, *Exp. Neurol.* **67**:399.

24. Loy, R., Koziell, J. E., and Moore, R. Y., 1980, Noradrenergic innervation of adult rat hippocampal formation, *J. Comp. Neurol.* **189**:699.

25. Lundborg, G., Longo, F. M., and Varon, S., 1982, Nerve regeneration model and trophic factors *in vivo, Brain Res.* **232**: 157.

26. Manthorpe, M., Nieto-Sampedro, M., Skaper, S. D., Lewis, E. R., Barbin, G., Longo, F. M., Cotman, C. W., and Varon, S., 1983, Neuronotrophic activity in brain wounds of the developing rat: Correlations with implant survival in the wound cavity, *Brain Res.* **267**:47.

27. Matthews, D. A., Nadler, J. V., Lynch, G. S., and Cotman, C. W., 1974, Development of cholinergic innervation in the hippocampal formation of the rat, *Dev. Biol.* **36**:130.

28. Mellgren, S. I., and Srebro, B., 1973, Changes in acetylcholinesterase and distribution of degenerating fibers in the hippocampal region after septal lesions in the rat, *Brain Res.* **52**:19.

29. Moore, R. Y., 1975, Monoamine neurons innervating the hippocampal formation and septum: Organization and response to injury, in: *The Hippocampus,* Vol. I (R. L. Isaacson and K. H. Pribram. eds), pp. 215–237, Plenum Press, New York.

30. Moore, R. Y., and Halaris, A. E., 1975, Hippocampal innervation by serotonin neurons of the midbrain raphe in the rat, *J. Comp. Neurol.* **164**:171.
31. Nadler, J. V., Cotman, C. W., and Lynch, G. S., 1977, Histochemical evidence of altered development of cholinergic fibers in the rat dentate gyrus following lesions. I. Time course after complete unilateral entorhinal lesion at various ages, *J. Comp. Neurol.* **171**:561.
32. Nadler, J. V., Mathews, D. A., Cotman, C. W., and Lynch, G. S., 1974, Development of cholinergic innervation in the hippocampal formation of the rat. II. Quantitative changes in choline acetyltransferase and acetylcholinesterase activity, *Dev. Biol.* **36**: 142.
33. Nieto-Sampedro, M., Lewis, E. R., Cotman, C. W., Manthorpe, M., Skaper, S. D., Barbin, G., Longo, F. M., and Varon, S., 1982, Brain injury causes a time-dependent increase in neuronotrophic activity at the lesion site, *Science* **217**:860.
34. Ramón y Cajal, S., 1959, *Degeneration and Regeneration of the Nervous System* (R. M. May, translator), Hafner, New York.
35. Shelton, D. L., Nadler, J. V., and Cotman, C. W., 1979, Development of high affinity choline uptake and associated acetylcholine synthesis in the rat fascia dentata, *Brain Res.* **163**:263.
36. Steinbusch, H. W. M., 1981, Distribution of serotonin-immunoreactivity in the central nervous system of the rat—Cell bodies and terminals, *Neuroscience* **6**:557.
37. Stenevi, U. A., Björklund, A., and Svendgaard, N. A., 1976, Transplants of central and peripheral monoamine neurons in the adult rat brain: Techniques and conditions for survival, *Brain Res.* **114**:1.
38. Storm-Mathisen, J., 1970, Quantitative histochemistry of acetylcholinesterase in rat hippocampal region correlated to histochemical staining, *J. Neurochem.* **17**:739.

13

Use of CNS Implants to Promote Regeneration of Central Axons across Denervating Lesions in the Adult Rat Brain

ULF STENEVI, ANDERS BJÖRKLUND, AND
LAWRENCE F. KROMER

1. Introduction

Following lesions that cause substantial tissue damage, such as spinal cord transection or large hemorrhagic necroses in the brain, mammals exhibit very little evidence of regeneration of the severed axons across the necrosis (see Refs. 1–4 for reviews). This holds true also for those nonmyelinated or finely myelinated systems, such as the monoaminergic and cholinergic neurons, that have been shown to possess a very pronounced regenerative capacity (see Ref. 5 for review). Early this century, Tello[6,7] and Ramón y Cajal[8] described this regenerative failure as due to a lack of neurotropic mechanisms and growth pathways in the adult mammalian CNS. Based on this notion, these authors were the first to attempt to promote regeneration of lesioned central axons by using implants of peripheral nerve.

Predegenerated peripheral nerve segments, which are known to stimulate peripheral nerve regeneration, did, according to Tello's and Ramón y Cajal's observations, exert positive neurotropic effects also on sprouting central axons. The idea of using transplants to bridge lesions in the CNS, and particularly in the spinal cord, has subsequently been tested in a number of studies,[9–22] which have been reviewed in some

ULF STENEVI AND ANDERS BJÖRKLUND ● Departments of Histology and Ophthalmology, University of Lund, Lund, Sweden.   LAWRENCE F. KROMER ● Department of Anatomy and Neurobiology, College of Medicine, University of Vermont, Burlington, Vermont 05405.

detail in Chapter 17 in this volume. The results of the earlier studies are, however, conspicuously inconsistent, variable, and often contradictory. An important reason for this is the difficulty of studying axonal growth phenomena in the CNS when using nonselective staining methods, particularly the silver stains. As evident, e.g., from the critical analysis of Le Gros Clark,[18] the interpretational problems inherent in the use of nonselective methods have prevented safe conclusions from being drawn and hence created controversy over the extent to which any central regeneration had occurred in these experiments.

In the present review, we describe a series of experiments using the more recently available selective histochemical tracing methods in combination with biochemical assays of specific neurochemical markers. There are significant advantages in using selective methods in that one defined axonal system can be visualized in its entirety in the microscope, and the axonal regrowth can be monitored biochemically through assays of transmitter content or transmitter-related enzymes. In the present experiments, we have studied the influence of adult peripheral tissue or embryonic CNS tissue on the regenerative sprouting of central cholinergic and monoaminergic neuron systems, using the thiocholine method for acetylcholine esterase (AChE) and the formaldehyde histofluorescence method for monoamines, respectively. In particular, we have been interested in determining whether regeneration of central cholinergic and adrenergic axons can be stimulated by transplants of peripheral tissue normally innervated by cholinergic and adrenergic neurons, or alternatively, by transplants taken from embryonic CNS regions that would receive central cholinergic and adrenergic innervations during development. The results show that such implants possess sufficiently strong attractive forces, which stimulate the sprouting central axons to grow not only across the developing scar into the transplant, but also across the transplant into the initially denervated area.

2. Techniques

In our experimental model (Fig. 1), the connections between the septum and the hippocampus have been severed through a lesion of the fimbria–fornix system. Transplants of the iris or embryonic hippocampus have then been inserted into the lesion cavity. Some results obtained from other intracerebral transplantation sites will also be discussed. In all cases, the experiments have been carried out on young female Sprague-Dawley rats, weighing, 180–200 g at the time of surgery. In the first series of experiments,[23,24] autologous transplants of the iris were pierced through the fimbria–fornix on one side, using a thin flat glass rod, as illustrated in Fig. 1A. This operation lesioned part of the septohippocampal connections and left the iris in contact with the transected axons. In the second series of experiments,[25–27] the fimbria–fornix system on one side was completely transected by making a suction cavity (approximately 3 × 3 mm) through the parietal and cingulate cortices and the hippocampal fimbria, as described by Stenevi *et al.*[28] In some of the animals, the cavity was filled with Gelfoam alone; in others, a transplant, consisting either of two irides (taken from the recipient's own eyes) or of embryonic hippocampus (taken from 7- to 40-mm rat fetuses), was placed as a bridge between the septum and the hippocampus at the bottom of the cavity, as illustrated in Fig. 1B.

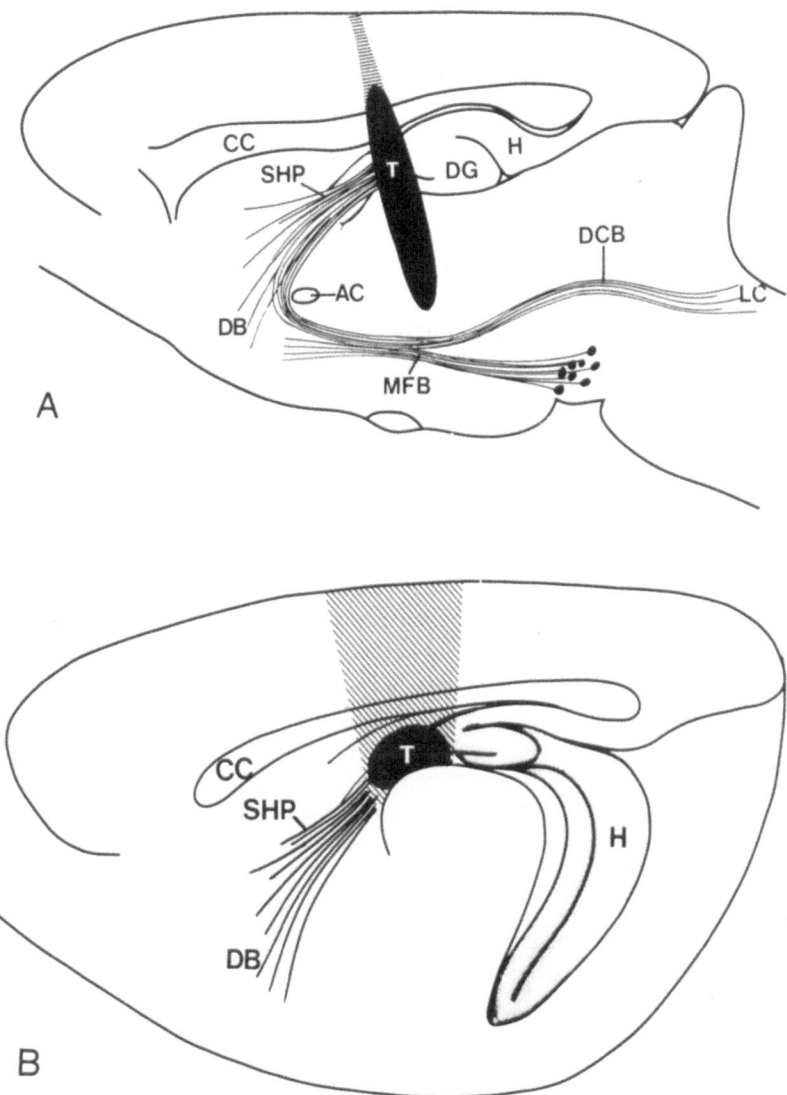

FIGURE 1. Schematic representation of the two types of transplantation experiments carried out in the septohippocampal system. (A) The position of the autologous iris transplant (T) in contact with the lesioned cholinergic axons of the septohippocampal pathway (SHP). (B) The position of the embryonic hippocampal transplant (T) at the bottom of the fimbria–fornix cavity (hatched area). AC, anterior commissure; CC, corpus callosum; DB, nucleus of the diagonal band; DCB, dorsal tegmental catecholamine bundle; DG, dentate gyrus; H, hippocampus; LC, locus coeruleus; MFB, medial forebrain bundle.

After survival times varying between 3 days and 14 months, the animals were sacrificed and processed for AChE histochemistry and, in the case of the iris transplants, some specimens were taken for monoamine fluorescence histochemistry as well. Other groups of rats were taken for biochemical analysis of the acetylcholine-synthesizing enzyme, choline acetyltransferase (ChAT), or the catecholamine-synthesizing enzymes, dopa decarboxylase (DDC) and tyrosine hydroxylase (TH). For details on the experimental material and the methods used, see Svedgaard *et al.*[24,29] Emson *et al.*,[23] and Kromer *et al.*[26,27]

3. Survival of the Implants

Good survival of the intracerebral implants depends primarily on their rapid and efficient revascularization from pial vessels and, to a lesser extent, from vessels in the brain parenchyma. In the study of Svendgaard *et al.*[29] on irides transplanted to the caudal diencephalon, the revascularization started within the first 24 hr, an approximately half-normal vascular density was observed at 5 days, and a fairly normal vascular supply was reached by the end of the third postoperative week. A similar time course of revascularization also has been seen with respect to grafts of embryonic CNS tissue (Beebe, Møllgård, Björklund, and Stenevi, unpublished observations).

The survival rate of iris grafts inserted into the brain parenchyma or placed in the fimbrial cavity was over 90%, [24,29] The survival of embryonic hippocampal implants, placed on the vessel-rich thalamic surface at the bottom of the fimbrial cavity, was 80–90%.[26] In the iris transplants, which were fully mature at the time of transplantation, the size of the graft, as measured by wet weight or protein content, was reduced by 50–60% during the first 8 days and remained at this level for at least 6 weeks.[23,24] In the hippocampal transplants, which were taken at an early stage of development, the average weight increased about fivefold during the 3-month survival period, probably due to the continued proliferation of glial and neuronal elements in the graft.[26]

The hippocampal implants frequently showed a high degree of intrinsic differentiation.[30,31] Thus, the dentate gyrus often developed its normal laminated crescent shape, and hippocampal and subicular subdivisions were discernible. Likewise, the iris transplants retained their normal structural integrity with easily recognizable sphincter, dilator, and ciliary body regions.[29] In fact, *in vitro* experiments demonstrated that intracerebral transplants of portal vein that contained smooth muscle retained their normal contractile properties.[32]

4. Regeneration of Central Adrenergic and Cholinergic Axons into the Implants

4.1. Transplants of Peripheral Tissues

Autologous iris transplants, as well as certain other smooth muscle-containing grafts, became extensively reinnervated within 3–4 weeks by both adrenergic and cholinergic central axons, provided that the graft was placed in the vicinity of the sprouting transected axons.[2,24,29] After implantation into the ascending pathway of the *noradre-nergic axons* from the locus coeruleus, the sprouting adrenergic axons were seen to reach the transplant by traveling straight through necroses that were up to 1 mm in width. Once the sprouts reached the transplant, the growth was markedly accelerated. Thus, whereas it took 2–3 days for the new sprouts to reach across a necrotic zone of less than 0.25 mm, they extended several millimeters within the transplant during the subsequent 4 days. From their observations in whole-mount preparations of iris transplants, Svend-gaard *et al.*[29] have estimated the minimum mean growth rate of the regenerating locus coeruleus axons in the iris at between 0.5 and 0.8 mm/day. This value is about 3–4 times less than that calculated by Olson[33] for peripheral noradrenergic axons regener-ating along the rat sciatic nerve (2.8 mm/day during the second week after compression of the nerve), and by Ramón y Cajal[8] for the overall axonal population in the same nerve (2.5 mm/day). It is greater, though, than that reported by Ramón y Cajal for the growth of regenerating peripheral axons across a scar (about 0.2–0.3 mm/day).

The first noradrenergic axon sprouts were observed in the iris transplant 5 days after axotomy. During the next few days, the sprouts grew rapidly in bundles out over the iris, covering about a third of the iris surface by 9 days and reaching across the whole iris by 15 days after operation. During the second postoperative week, the regrowth was expressed primarily in the extensive branching of the bundles of sprouting fibers and in the beginning of the development of fiber plexuses in the dilator and the sphincter. Dur-ing the third and fourth week, the regenerated terminal plexuses expanded gradually to cover practically all vital areas of the iris. At this stage, the arrangement and patterning of the regenerated central noradrenergic fibers mimicked in several respects the normal sympathetic nerve supply (an innervation that was totally removed at transplantation).

Regeneration of *cholinergic neurons* into intracerebral iris transplants has been studied in the septohippocampal system and the striatum of adult rats.[23,24] In these experiments, the irides were implanted into either a lesion transecting the septohippo-campal cholinergic pathway or into the striatum in a position that lesioned the axons of intrinsic cholinergic interneurons within the caudate nucleus and the axons of the cho-linergic basal forebrain neurons projecting through the striatum to the neocortex. All these systems appeared to produce axons that efficiently regenerated into the iris trans-plant. Moreover, these cholinergic fibers formed organotypic terminal plexuses within 2–4 weeks, very much in the same manner as the noradrenergic locus coeruleus neurons. In parallel, there was a marked recovery in ChAT within the initially denervated implant (Fig. 2A). When irides were placed so that both the cholinergic telencephalic neurons and the adrenergic locus coeruleus axons could reinnervate the target simulta-neously, the two systems reinnervated the same areas of the transplant. In fact, the fibers were shown to run close together in the terminal plexuses just as in the normal auto-nomic ground plexus of the intact iris.[24] Field stimulation experiments *in vitro*[32] have shown that the reinnervating central adrenergic and cholinergic axons formed normal functional connections with the smooth muscle in the graft.

Schwann cells are known to play an important role in the guidance of regenerating peripheral nerves (e.g., see Refs. 8 and 34). From the results obtained with intracerebral iris implants, it seems highly probable that the Schwann cells of the autonomic ground plexus can interact in the very same manner with both central and peripheral adrenergic axon sprouts. As during regeneration of peripheral adrenergic axons, it seems likely,

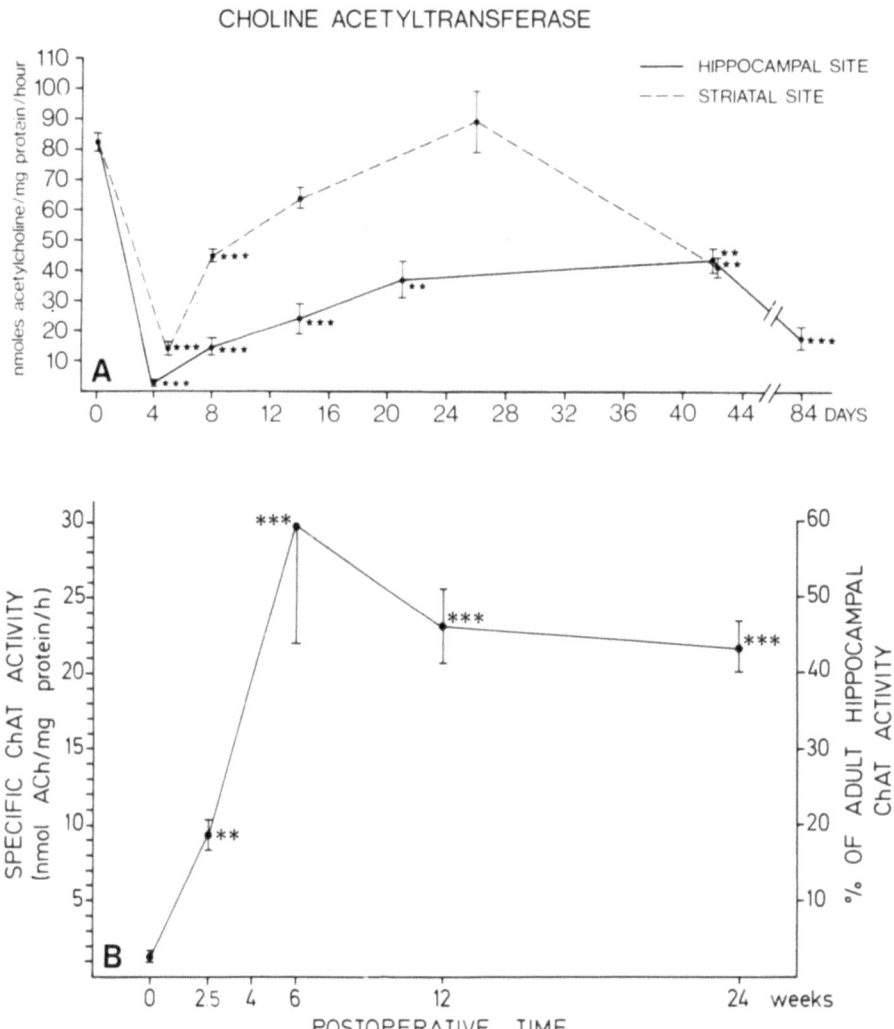

FIGURE 2. Time course in the appearance of choline acetyltransferase (ChAT) in iris transplants placed in the hippocampal fimbria or in the striatum (A), and in embryonic hippocampal transplants in the fimbria–fornix cavity (B). In (A), the asterisks denote differences from normal iris values, and in (B), differences from pretransplant values: ***$p < 0.001$, **$0.001 < p < 0.01$, From Refs. 23 and 26.

therefore, that the attracting and organizing influences exerted by the denervated target tissue on the central sprouting axons resulted from an activity of the neural sheath elements which also normally occurred during regeneration of peripheral adrenergic axons.

4.2. Hippocampal Transplants

In these experiments,[26] a piece of embryonic hippocampus was placed into a 3 × 3-mm-wide cavity made through the fimbria-fornix. The lesion shown in Fig. 1B produced a complete and highly reproducible transection of the septohippocampal cholinergic pathway. After a survival time of 6 weeks or more, possible afferent and efferent connections of the transplant were traced using small injections of HRP or [³H]proline into the transplant. The regeneration of the septal cholinergic system was monitored by utilizing AChE staining and the ChAT assay.

In most cases, the hippocampal implants fused with the cut surface of the ipsilateral host hippocampus, and often formed attachments with the septum, caudate-putamen, or neocortex of the host (Fig. 3). Silver and myelin stains revealed numerous fibers interconnecting the implant and the host septum and between the implant and the host hippocampus. The HRP and [³H]proline injections showed that the fibers penetrating the host–transplant borders represented both efferent and afferent connections between the implant and the septum or hippocampus of the host. This indicated that reciprocal septal-hippocampal and hippocampal-hippocampal connections indeed could be established with the hippocampal graft.

The AChE histochemistry and ChAT determinations provided evidence that the cholinergic neurons in the medial septum–diagonal band of the host regenerated efficiently into the implant. In addition, the AChE staining suggested that, in some cases, fibers from the caudate-putamen and the supracallosal striae contributed to the AChE plexus as well. The axons of the AChE-positive septohippocampal cholinergic system sprouted into the implant and gave rise to a new AChE-stained neuropil that, with time, grew to cover the entire implant. In parallel, there was a progressive appearance of ChAT in the implant, reaching between 40 and 60% of the normal hippocampal level by 6–24 weeks after transplantation (Fig. 2B). Interestingly, the ChAT level in the hippocampal implants was quite similar to that found in irides transplanted to the fimbria (Fig. 2A). The septal origin of the ingrowing cholinergic axons was supported by experiments with electrolytic lesions and HRP injections. Thus, the major portion of the cholinergic afferents to the implant were found to disappear after electrolytic destruction of the ipsilateral medial septum–diagonal band area, and neurons in the ipsilateral medial septal nucleus and the nucleus of the diagonal band became labeled after small HRP injections confined to the hippocampal transplant.

These results provide strong support for the notion that regeneration of axotomized central neurons can indeed be prompted either by a piece of deafferented embryonic CNS tissue or by an appropriate denervated peripheral tissue. This effect may be a fairly specific one, providing a stimulus for directed sprouting into the denervated target by the types of axons (e.g., cholinergic or adrenergic) that would normally innervate the implanted tissue.

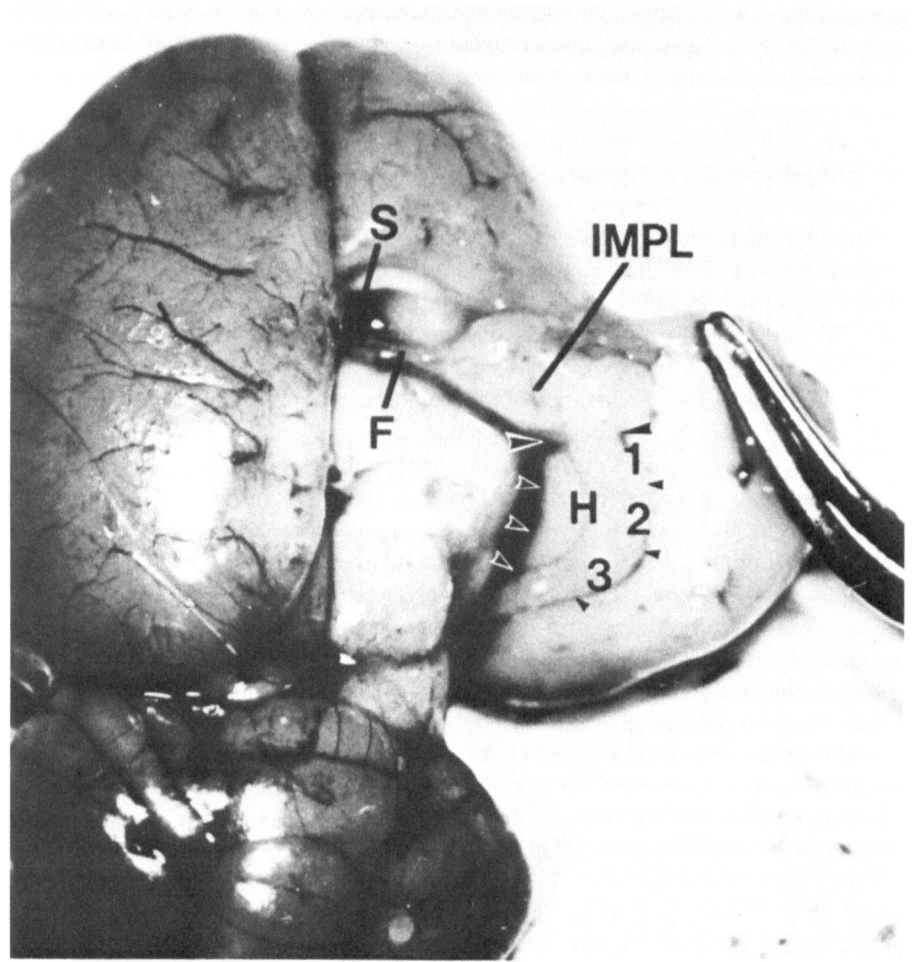

FIGURE 3. Appearance of a hippocampal implant (IMPL) that forms a bridge across a lesion of the host fimbria–fornix (6 months' survival). There is extensive fusion (demarcated by the large arrowheads) between the transplant and the hippocampal formation (H) of the recipient, and a thin fimbria-like connection (F) between the transplant and the host septum (S). The small arrowheads indicate the regions of the host hippocampus (1, 2, and 3) dissected for the ChAT assay (results illustrated in Fig. 5). Adapted from Ref. 27.

5. Bridging of Regenerating Central Cholinergic Axons across Lesions in the Hippocampal Fimbria

Several studies[27,35–37] have demonstrated that suction lesions of the hippocampal fimbria–fornix, made under visual control, reliably produced a complete axotomy of the cholinergic septohippocampal system innervating the dorsal hippocampus. Although some sprouting AChE-positive fibers were seen to grow across the cavity in some specimens in the absence of any transplant, they reached at most a very narrow zone of the denervated hippocampus, bordering on the cavity. By contrast, in adult rats that received a transplant of embryonic hippocampus that was placed in the fimbrial cavity at the time of operation, the denervated hippocampus became cholinergically reinnervated via the transplant.[27] Due to the axotomy caused by the transplantation cavity, the AChE-positive innervation of the dorsal part of the hippocampus was initially totally removed and the ChAT level was reduced to about 5% of normal. By 4–6 weeks after transplantation, new AChE-positive fibers appeared in parts of the subiculum, hippocampus, and/or dentate gyrus immediately bordering on the implant. By 3 months, the new fibers expanded about 2.5–3.5 mm further caudally to cover the entire dorsal hippocampal formation, as shown in Fig. 4. In parallel, there was a reappearance of ChAT activity (not seen in control rats without implants) that averaged 30% of the ChAT level normally present in the dorsal hippocampus. In the most successful cases, a recovery of as much as 50% of normal ChAT activity has been recorded (filled circles in Fig. 5).

Interestingly, a very similar reinnervation of the denervated hippocampus has been demonstrated by AChE histochemistry also following implantation of autologous grafts of iris into the cavity (previously unpublished observations). In this operation, two irides were placed in parallel (in the shape of collapsed tissue bands) at the bottom of the fimbrial cavity, with the two ends touching the cut fimbrial and hippocampal surfaces, respectively. The AChE staining revealed abundant fiber growth into the irides and, presumably via the irides, into the dorsal parts of the initially denervated hippocampus and dentate gyrus. The cholinergic regrowth was revealed also in the ChAT levels, as shown by the triangles in Fig. 5. Within 3–12 months after transplantation, the acetyl-choline-forming enzyme had thus recovered to the same degree as in the rats bearing transplants of embryonic hippocampus.

In animals with hippocampal bridge grafts, there is evidence that the regenerating septal cholinergic neurons indeed establish functional synaptic contacts. Segal et al.[38] have thus reported that stimulation of the host medial septum produced evoked responses from neurons within the hippocampal implant, as well as from neurons within the rein-nervated part of the host hippocampus. The cholinergic nature of these synaptic contacts was supported by the ability of the cholinergic blocker atropine to counteract these excit-atory effects.

The above observations from animals with hippocampal implants[27,38] provide strong evidence that the cholinergic fibers reinnervating the hippocampus indeed are of central origin and that they mainly arise from a regeneration of the initially axotomized neurons in the medial septum–diagonal band area. The evidence for this is the following:

1. The recovery of ChAT levels and the reappearance of AChE-positive fibers in the denervated hippocampus did not occur in animals with cavity lesions alone. This recovery was directly dependent on how well the implant had survived and established contacts across the cavity.

2. The time course of reappearance of the cholinergic enzyme markers, as well as the rostrocaudal gradient in the final ChAT content in the implanted rats, strongly suggest that the regrowing axons first reach the implant and then the host hippocampus, and that they extend within the hippocampus in a rostral to caudal direction.

3. The newly formed AChE-positive fibers in the hippocampus in most cases could

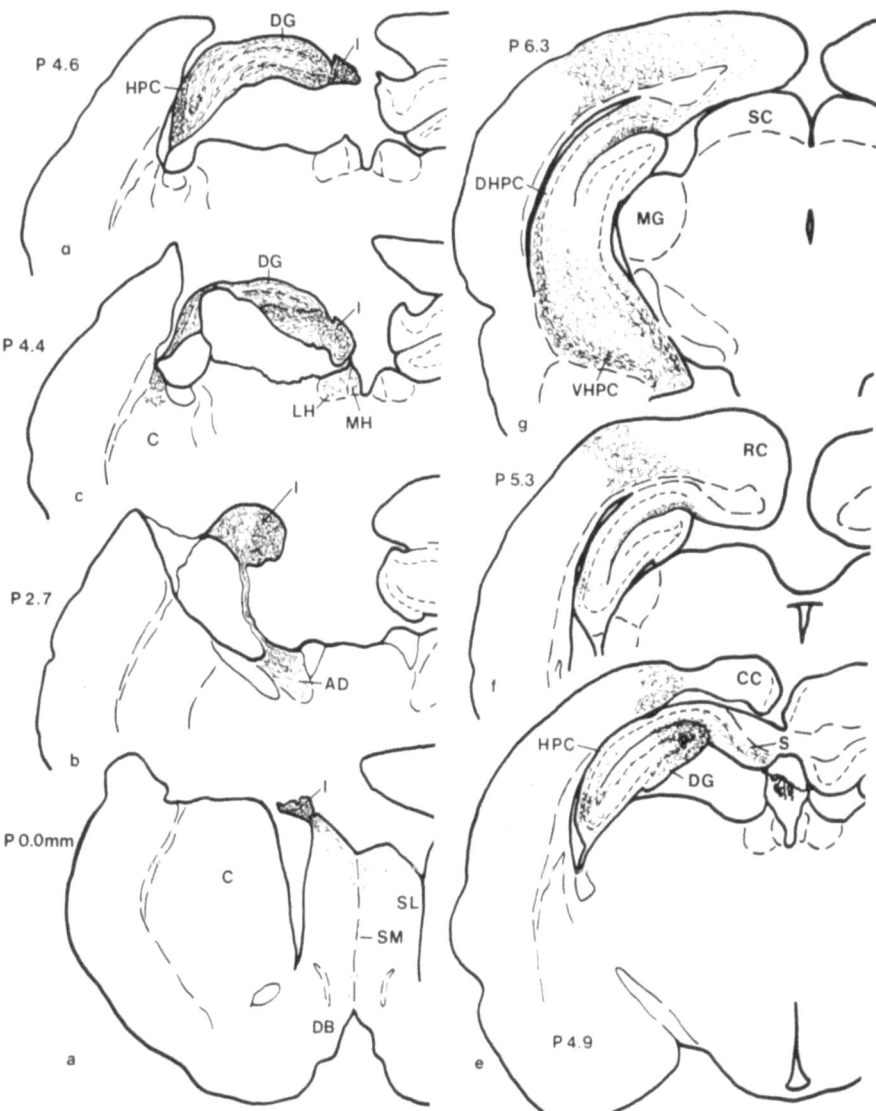

FIGURE 4. Growth of AChE-positive fibers across an implant of embryonic hippocampus (I) into the initially denervated hippocampus (HPC), subiculum (S), and dentate gyrus (DG) of the host rat. The seven levels (a–g) represent successively more caudal frontal planes. In the subiculum and the hippocampal CA1 area, fibers are seen to have extended into the most caudal level (g), 6.3 mm behind the septum (level a), whereas there is a slightly less extensive reinnervation of the dentate gyrus.

AD, anterodorsal thalamic nucleus; C, caudate-putamen; CC, cingulate cortex; DB, nucleus of the diagonal band; DHPC, dorsal hippocampal formation; LH, lateral habenular nucleus; RC, retrosplenial cortex; SC, superior colliculus; SL, lateral septum; SM, medial septum; VHPC, ventral hippocampal formation, MG, medial geniculate body; MH, medial habenular nucleus. From Ref. 27.

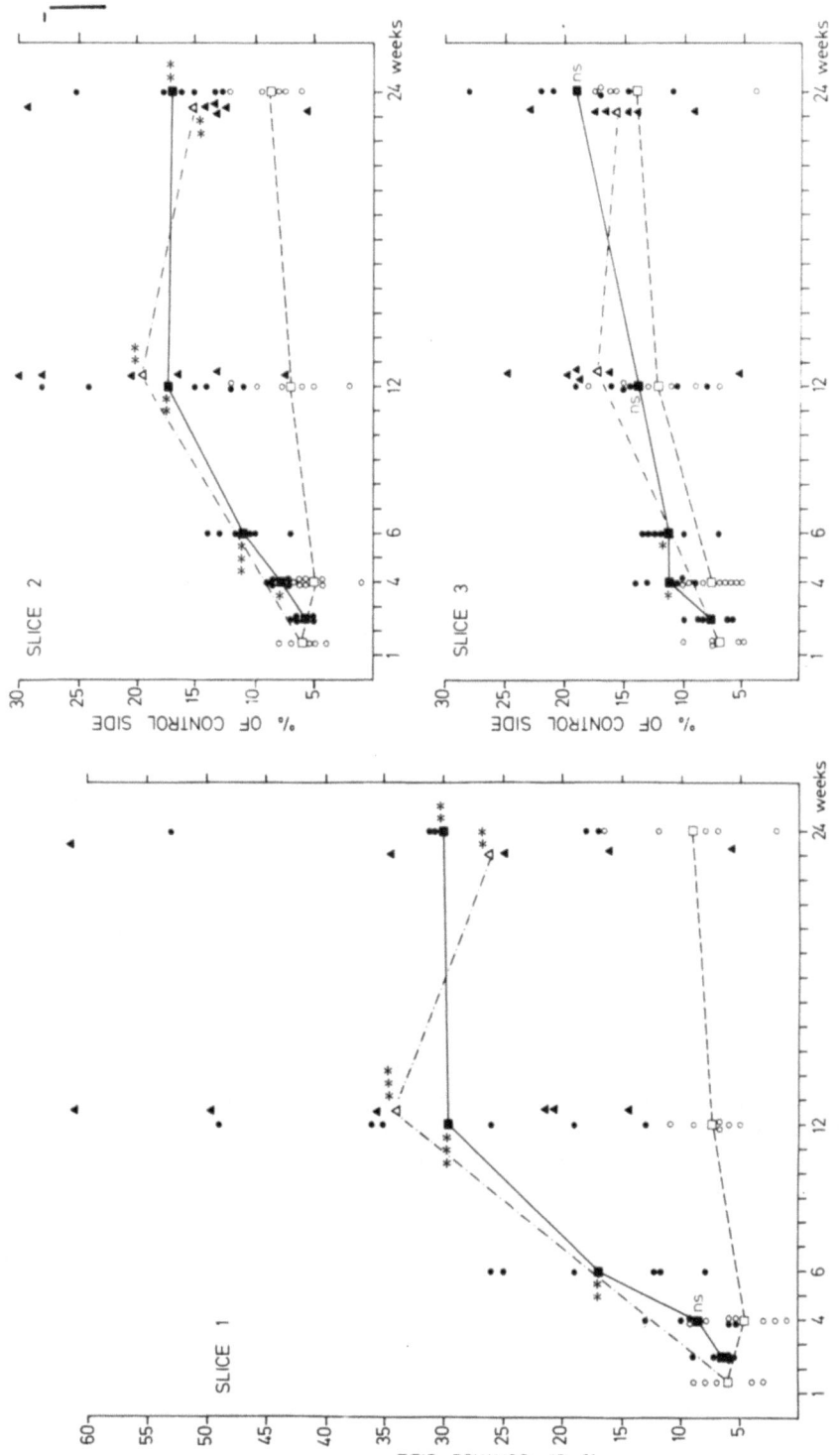

FIGURE 5. Comparison of AChE staining pattern in the hippocampus in control animals (A), animals with a fimbrial transection (B) and with a transplant of striatal tissues (C), and animals with a transplant of septal tissue (D). The striatal implant occupies the anteromedial portion of a cavity of the host entorhinal cortex. Implant-associated AChE staining is intense in the subiculum and on either side of the hippocampal pyramidal cells. In the molecular layer, bands of stain are present suprajacent to the granule cells and the outer portion of the molecular layer. Bars = 150 μm, From Ref. 20.

readily be traced back to the implant, and AChE-positive axons were seen to enter the implant in large numbers from the host brain.

4. The vast majority of the AChE-positive axons in the implant and the dorsal hippocampus were removed following a large lesion of the septal–diagonal band area ipsilateral to the implant.

5. Injections of HRP into the dorsal hippocampus, caudal to the implant, produced retrogradely labeled neurons in the ipsilateral medial septum and diagonal band. Similar injections into specimens without implants produced no labeled cells in this region.

6. Electrical stimulation of the medial septal area produced atropine-sensitive excitatory responses in the reinnervated part of the hippocampus, as well as in the implant itself.

The HRP experiments indicated that only a small proportion of the septal nucleus neurons managed to regenerate axons that reached the hippocampus. This is explained by the fact that the vast majority of these neurons undergo retrograde degeneration within several weeks after fimbrial transection.[39,40] Severe retrograde cell loss also seemed to occur in the ipsilateral septal–diagonal band area even in animals with successful implants. The apparently limited number of septal–diagonal band neurons that regenerated axons into the hippocampus in the implanted rats could thus, by itself, explain why the reinnervation process ceases before the entire denervated region has been reached by the regrowing axons. Previous studies[35,37] on the reinnervation of the cholinergically denervated hippocampus by embryonic septal implants suggest, in fact, that the entire denervated hippocampus may be innervated by regrowing septal neurons provided their number is sufficiently large.

6. Comments

Adverse scarring has often been regarded as a major obstacle to successful regeneration in the mammalian CNS. In the present experimental system utilizing embryonic CNS transplants, glial scarring appeared to be minimal.[41] It is likely that the tight fusion between the graft and the host brain tissue, and the absence of any prominent glial barriers, is an important reason for the efficient extension of the regenerating central cholinergic axons into and across the embryonic CNS implant. In a recent series of elegant experiments, Aguayo, Richardson, and collaborators[10,13,19,20] have shown that many types of axons from CNS neurons also will sprout into and grow along segments of peripheral nerve inserted into the brain or the spinal cord. This indicates that the ability of intracerebral implants to promote axonal regrowth is not limited to aminergic and cholinergic neurons but is of a more general nature. Both our own experiments and those of Aguayo *et al.* further indicate that the role of the implant in axonal regeneration is not a passive one, but that the implanted tissue exerts a direct stimulatory or attractive influence on sprouting central axons. Moreover, the relative precision by which the septal cholinergic axons can grow for long distances through the denervated adult hippocampus in order to establish their normal innervation pattern in this structure suggests that growth-regulating mechanisms can be quite specific and operate also within the adult host CNS.

In the absence of interfering scars, there is substantial evidence that at least certain types of nonmyelinated or weakly myelinated systems, notably the monoaminergic ones, also can regenerate efficiently within the adult CNS environment.[42-46] The bulbospinal noradrenergic and serotoninergic neurons which form long descending projections from the pons and medulla oblongata to all segments of the spinal cord, showed extensive regeneration after selective chemical axotomy, induced by the monoamine neurotoxins. The most clear-cut case was provided after axotomy of the bulbospinal serotonin system by 5,6-dihydroxytryptamine (5,6-DHT).[42-44] After an intraventricular injection of 75 μg 5,6-DHT, this system was completely axotomized at a high cervical level, while other axonal systems, as well as the general tissue architecture of the spinal cord, were undamaged. During a protracted regrowth process, the lesioned serotonin neurons regenerated down the cord within 8 months and reinnervated the initially denervated gray matter. The most extensive reinnervation occurred in the cervical cord, but there was evidence that some axons reached all the way down into the lumbar segments, a distance of at least 5 cm in the rat. Interestingly, although the newly formed serotoninergic terminal plexus never reached normal density in the lower segments of the cord, the parallel biochemical data[43] indicated that this is at least partly compensated for by an increased turnover of the transmitter in the hypoinnervated segments.

These results suggest that efficient regeneration of CNS pathways is possible under favorable conditions, and that intracerebral transplants can successfully be used to promote bridging of regenerating central axons across lesions. To what extent such stimulated regeneration can promote behavioral recovery in rats with brain or spinal cord damage remains, however, to be established.

ACKNOWLEDGMENTS. The studies summarized in this review have been supported by the Swedish MRC, the Syskonen Svensson and the Kock Foundations, NIH Grant NS-18126, and the A.P. Sloan Foundation.

References

1. Windle, W. F., Clemente, C. D., and Chambers, W. W., 1952, Inhibition of formation of a glial barrier as a means of permitting a peripheral nerve to grow into the brain, *J. Comp. Neurol.* **96**:359.
2. Björklund, A., and Stenevi, U., 1971, Growth of central catecholamine neurones into smooth muscle grafts in the rat mesencephalon, *Brain Res.* **31**:1.
3. Clemente, C. D., 1964, Regeneration in the vertebrate central nervous system, *Int. Rev. Neurobiol.* **6**:257.
4. Puchala, E., and Windle, W. F., 1977, The possibility of structural and functional restitution after spinal cord injury: A review, *Exp. Neurol.* **55**:1.
5. Björklund, A., and Stenevi, U., 1979, Regeneration of monoaminergic and cholinergic neurons in the mammalian central nervous system, *Physiol. Rev.* **59**:62.
6. Tello, J. F., 1911, La influencia del neurotropismo en la generacion de los centros nervioso, *Trab. Lab. Invest. Biol.* **9**:123.
7. Tello, J. F., 1923, Gegen wärtige Anschaungen über den Neurotropismus, *Vortr. Entwicklungsmech. Org.* **33**:1.
8. Ramón y Cajal, S., 1928, *Degeneration and Regeneration of the Nervous System,* Oxford University Press, London.

9. Aihara, H., 1970, Autotransplantation of the cultured cerebellar cortex for spinal cord reconstruction, *Brain Nerve* **22**:769.
10. Benfey, M., and Aguayo, A. J., 1982, Extensive elongation of axons from rat brain into peripheral nerve grafts, *Nature (London)* **296**:150.
11. Brown, J. D., and McCouch, G. P., 1947, Abortive regeneration of the transected spinal cord, *J. Comp. Neurol.* **87**:131.
12. Campbell, J. B., Bassett, C. A. L., Thulin, C. A., and Feringa, E. R., 1960, The use of nerve grafts to orient axonal regeneration in transected spinal cord, *Anat. Rec.* **136**:174.
13. David, S., and Aguayo, A. J., 1981, Axonal elongation into peripheral nervous system "bridges" after central nervous system injury in adult rats, *Science* **214**:931.
14. Feigin, I., Geller, E. H., and Wolf, A., 1951, Absence of regeneration in the spinal cord of the young rat, *J. Neuropathol. Exp. Neurol.* **10**:420.
15. Glees, P., 1955, Studies on cortical regeneration with special reference to central implants, in: *Regeneration in the Central Nervous System* (W. F. Windle, ed.), pp. 94–111, Thomas, Springfield, Ill.
16. Kao, C. C., 1974, Comparison of healing process in transected spinal cords grafted with autogenous brain tissue, sciatic nerve, and nodose ganglion, *Exp. Neurol.* **44**:424.
17. Kao, C. C., Chang, L. W., and Bloodworth, J. M. B., Jr., 1977, Axonal regeneration across transected mammalian spinal cords: An electron microscopic study of delayed microsurgical nerve grafting, *Exp. Neurol.* **54**:591.
18. Le Gros Clark, W. E., 1942, The problem of neuronal regeneration in the central nervous system. 1. The influence of spinal ganglia and nerve fragments grafted in the brain, *J. Anat.* **77**:20.
19. Richardson, P. M., McGuinness, U. M., and Aguayo, A. J., 1980, Axons from CNS neurones regenerate into PNS grafts, *Nature (London)* **284**:264.
20. Richardson, P. M., McGuinness, U. M., and Aguayo, A. J., 1982, Peripheral nerve autografts to the rat spinal cord: Studies with axonal tracing methods, *Brain Res.* **237**:147.
21. Sugar, O., and Gerard, R. W., 1940, Spinal cord regeneration in the rat, *J. Neurophysiol.* **3**:1.
22. Woolsey, D., Minkler, J., Rezende, N., and Klemme, R., 1944, Human spinal cord transplant, *Exp. Med. Surg.* **2**:93.
23. Emson, P. C., Björklund, A., and Stenevi, U., 1977, Evaluation of the regenerative capacity of central dopaminergic, noradrenergic and cholinergic neurones using iris implants as targets, *Brain Res.* **135**:87.
24. Svendgaard, N.-A., Björklund, A., and Stenevi, U., 1976, Regeneration of central cholinergic neurons in the adult rat brain, *Brain Res.* **102**:1.
25. Kromer, L. F., Björklund, A., and Stenevi, U., 1978, Development of specific connection between embryonic brain implants and the adult rat brain. 1. Establishment of a hippocamposeptal pathway by implanted hippocampal neurons, *Neurosci. Lett. Suppl.* **1**:S33.
26. Kromer, L. F., Björklund, A., and Stenevi, U., 1981, Innervation of embryonic hippocampal implants by regenerating axons of cholinergic septal neurons in the adult rat, *Brain Res.* **210**:153.
27. Kromer, L. F., Björklund, A., and Stenevi, U., 1981, Regeneration of the septohippocampal pathways in adult rats is promoted by utilizing embryonic hippocampal implants, as bridges, *Brain Res.* **210**:173.
28. Stenevi, U., Björklund, A., and Svendgaard, N.-A., 1976, Transplantation of central and peripheral monoamine neurons to the adult rat brain: Techniques and conditions for survival, *Brain Res.* **114**:1.
29. Svendgaard, N.-A., Björklund, A., and Stenevi, U., 1975, Regenerative properties of central monoamine neurons, *Adv. Anat. Embryol. Cell Biol.* **51**:1.
30. Kromer, L. F., Björklund, A., and Stenevi, U., 1979, Intracephalic implants: A technique for studying neuronal interactions, *Science* **204**:1117.
31. Kromer, L. F., Björklund, A., and Stenevi, U., 1983, Intracephalic embryonic neural implants in the adult rat brain. I. Growth and mature organization of brainstem, cerebellar, and hippocampal implants, *J. Comp. Neurol.* **218**:433.
32. Björklund, A., Johansson, B., Stenevi, U., and Svendgaard, N.-A., 1975, Re-establishment of functional connections by regenerating central adrenergic and cholinergic axons, *Nature (London)* **253**:446.
33. Olson, L., 1969, Intact and regenerating sympathetic norepinephrine axons in the rat sciatic nerve, *Histochemie* **17**:349.
34. Speidel, C. C., 1964, In vivo studies of myelinated nerve fibers, *Int. Rev. Cytol.* **16**:173.

35. Björklund, A., Kromer, L. F., and Stenevi, U., 1979, Cholinergic reinnervation of the rat hippocampus by septal implants is stimulated by perforant path lesion, *Brain Res.* **173**:57.
36. Björklund, A., and Stenevi, U., 1977, Reformation of the severed septohippocampal cholinergic pathway in the adult rat by transplanted septal neurons, *Cell Tissue Res.* **185**:289.
37. Kromer, L. F., 1982, Cholinergic axons from delayed septal implants and sympathetic fibers co-exist in the denervated dentate gyrus, *Brain Res. Bull.* **9**:539.
38. Segal, M., Stenevi, U., and Björklund, A., 1981, Reformation in adult rats of functional septo-hippo-campal connections by septal neurons regenerating across an embryonic hippocampal tissue bridge, *Neurosci. Lett.* **27**:7.
39. Daitz, H. M., and Powell, T. P. S., 1954, Studies of the connections of the fornix system, *J. Neurol. Neurosurg. Psychiatry* **17**:75.
40. McLardy, T., 1955, Observations on the fornix of the monkey. 1. Cell studies, *J. Comp. Neurol.* **103**:305.
41. Kromer, L. F., 1980, Glial scar formation in the brain of adult rats is inhibited by implants of embryonic CNS tissue, *Soc. Neurosci. Abstr.* **6**:688.
42. Björklund, A., and Lindvall, O., 1979, Regeneration of normal terminal innervation patterns by central noradrenergic neurons after 5, 6-dihydroxytryptamine-induced axotomy in the adult rat, *Brain Res.* **171**:271.
43. Björklund, A., and Wiklund, L., 1980, Mechanism of regrowth of the bulbospinal serotonin system following 5,6-dihydroxytryptamine induced axotomy. I. Biochemical correlates, *Brain Res.* **191**:109.
44. Nobin, A., Baumgarten, A. H. G., Björklund, A., Lachenmayer, L., and Stenevi, U., 1973, Axonal degeneration and regeneration of the bulbospinal indolamine neurons after 5,6-dihydroxytryptamine treatment, *Brain Res.* **56**:1.
45. Nygren, L.-G., and Olson, L., 1977, Intracisternal neurotoxins and monoamine neurons innervating the spinal cord: Acute and chronic effects on cell and axon counts and nerve terminal densities, *Histochemistry* **52**:281.
46. Wiklund, L., and Björklund, A., 1980, Mechanisms of regrowth in the bulbospinal serotonin system following 5,6-dihydroxytryptamine induced axotomy. II. Fluorescence histochemical observations, *Brain Res.* **191**:129.

14

Intracephalic Embryonic Transplants
A New Experimental Preparation for Developmental Neurobiology

LAWRENCE F. KROMER

1. Introduction

The purpose of the present review is to explore the possible use of "intracephalic" embryonic neural transplants to neonatal and adult recipients as model systems for analyzing various aspects of CNS development. Transplants of embryonic brain tissue into the CNS milieu of immature or adult recipients provide a unique opportunity to analyze both the subsequent development of the embryonic CNS tissue and the interaction of its immature constituent cells with cellular elements and products present in the CNS environment and circulatory system of the host animal.

The *"in cerebro"* implantation technique offers several unique advantages for studying cellular interactions that are necessary to produce the characteristic neuronal morphology and three-dimensional organization and orientation observed in various brain regions, such as the cerebellum and hippocampus. Because the embryonic transplant is rapidly incorporated into the blood and cerebrospinal circulation of the host, it is provided with an excellent environment that permits the development, maturation, and long-term survival of the implant. A large volume of embryonic tissue can develop within the transplantation cavity, providing an excellent environment for studying the three-dimensional organization of the transplant. Embryonic implants of differing volumes and constituent cell types can be dissected from fetuses at a variety of gestational stages. This permits analysis of the role of various cell types and the temporal sequences of cellular interaction *in utero* that are important for the normal cytoarchitectural development of the tissue after transplantation. Furthermore, the effects of alterations in the

LAWRENCE F. KROMER ● Department of Anatomy and Neurobiology, College of Medicine, University of Vermont, Burlington, Vermont 05405.

environment of the implanted tissue upon its *in cerebro* development can be studied by varying the age of the host or by injecting hormones or drugs into the circulatory system or implantation cavity of the recipient.

A considerable amount of literature already has accumulated indicating that embryonic and early neonatal tissue from a variety of regions in the mammalian CNS can survive transplantation either to the anterior eye chamber,[1–12] to the brain of neonatal mammals,[13–29] or to the adult mammalian CNS.[30–42] Several of these studies have attempted to analyze "intrinsic" and "extrinsic" factors that may influence the ultimate growth and survival of the transplants and their cytoarchitectural organization.

2. Transplant Growth and Survival

Several studies using intraocular grafts of embryonic brain stem and hippocampal tissue[4,5,8] or intracephalic transplants of telencephalic, brain stem, or cerebellar tissue[20–22,31,36,37,40,41] have reported that embryonic tissue dissected from early gestational fetuses produces transplants with the greatest final volume. Das *et al.*[22] and Kromer and Björklund,[36] postulated that the increased growth of early gestational implants was due to an increased number of undifferentiated neuroepithelial cells. Quantitative measurements of the volume increases for transplants of various gestational ages placed in the adult CNS (Table I) provide additional evidence that there is a differential growth of the transplants that is correlated with the gestational age of the donor.[40,41] Recent data by Kromer *et al.*[40] further indicate that embryonic transplants dissected from different regions of the neuraxis at the same gestational stage exhibit different growth potentials that appear to correlate with the *in vivo* rate of neurogenesis of medium and large neurons. Thus, cerebellar tissue, which has a protracted production of small interneurons

TABLE I
Volume of Neural Transplants to Adult Recipients

CNS region (donor)	Gestational age (donor)	CNS site (host)	Final transplant volume, mean (mm³)	N	Percent change in size
Brain stem[a]	E13–14	Intracephalic cavity	3.3 ± 0.8[c]	6	+381
	E17–19		0.7 ± 0.2[c]	4	−4
Hippocampus[a]	E13–14	Intracephalic cavity	27.4 ± 4.8[c]	6	+3877
	E17–20		2.4 ± 0.6[c]	6	+249
Cerebellum[a]	E12–13	Intracephalic cavity	2.7 ± 1.2[c]	6	+294
	E17–19		0.3 ± 0.1[c]	6	−62
Neocortex[b]	E14	Cerebellar injection	162.0 ± 6.0[d]	6	+1925
	E17		31.5 ± 2.1[d]	6	+294
Neocortex[b]	E14	Forebrain injection	54.0 ± 4.0[d]	6	+575
	E17		21.0 ± 2.2[d]	6	+162

[a]Adapted from Ref. 40.
[b]Adapted from Ref. 41.
[c]Initial volume of transplant approx. 0.7 mm³.
[d]Initial volume of transplant approx. 8.0 mm³.

(E17–P30) and a short generation period (E12–14) for Purkinje cells and the neurons in the deep cerebellar nuclei,[43] exhibits less growth than brain stem tissue, which has a more prolonged neurogenesis (E11–16) of medium and large cells.[44–46] Telencephalic tissues from the hippocampus[40] or neocortex[41] in which pyramidal neurons are generated over a considerable gestational period (E14–20)[56] produce transplants with the most extensive volume increases.

A second factor that may influence the growth of CNS transplants, particularly from late gestational fetuses, is posttransplantation cell death.[20,21,31,37] This is substantiated by the observation that pyknotic cells are present in the implant shortly following its transplantation to the adult CNS.[48] Postmitotic neuroblasts in certain CNS regions may be more susceptible to the traumatic effects of dissection from the fetus and the potential initial anoxia encountered in the host CNS until vascularization of the transplant occurs. This may be especially true for late gestational cerebellar tissue, which actually decreases in size after transplantation.[40]

Transplant survival also may be influenced by the location of the transplant within the host CNS,[31,41,49] and to a lesser degree, by the age of the host.[41] Das[49] has reported that "extraparenchymal" transplants that are located in ventricular cavities or on the surface of the host brain survive transplantation only for short periods. In contrast, Kromer et al.[40] demonstrate that several types of transplants differentiate and survive for long posttransplantation periods (at least 14 months) when placed in an "extraparenchymal" intracephalic cavity. The difference in the survival of "extraparenchymal" transplants in these two studies could be due to differences in the transplantation procedures employed. Das[49] injected the embryonic tissue into the host ventricular system, whereas Kromer et al.[40] prepared a cavity within the host telencephalon and positioned the embryonic tissue on the well-vascularized surface of the brain stem or choroid fissure. Thus, in the latter study, the embryonic implant could be rapidly vascularized and had ample room for unhindered cell proliferation within the intracephalic cavity.

3. Transplant Organization

Observations of the intrinsic cellular organization of the embryonic transplants after long-term survival in cerebro suggest that several factors may influence the final cytoarchitecture of the implants. Differences in the organization of transplants from the same region of the neuraxis may be due to variations in the dissection procedure and original size of the implant.[20,21,31,37] Mechanical disruption of the tissue during transplantation[15,20,21,37] and the loss of specific extrinsic afferents[16,29,30,40] also may influence implant development and maturation. The gestational age of the donor fetus (and the corresponding developmental state of the CNS region that is dissected) is another possible factor that may influence the posttransplantational development and organization of the implanted tissue.[17,36,40] In order to determine further which of these factors most influences the three-dimensional organization and final cellular morphology of the transplants, it is important to consider the initial developmental state of the dissected tissue when analyzing the data. Moreover, the differentiation observed in intracephalic transplants of a specific CNS region should be correlated with that observed in other

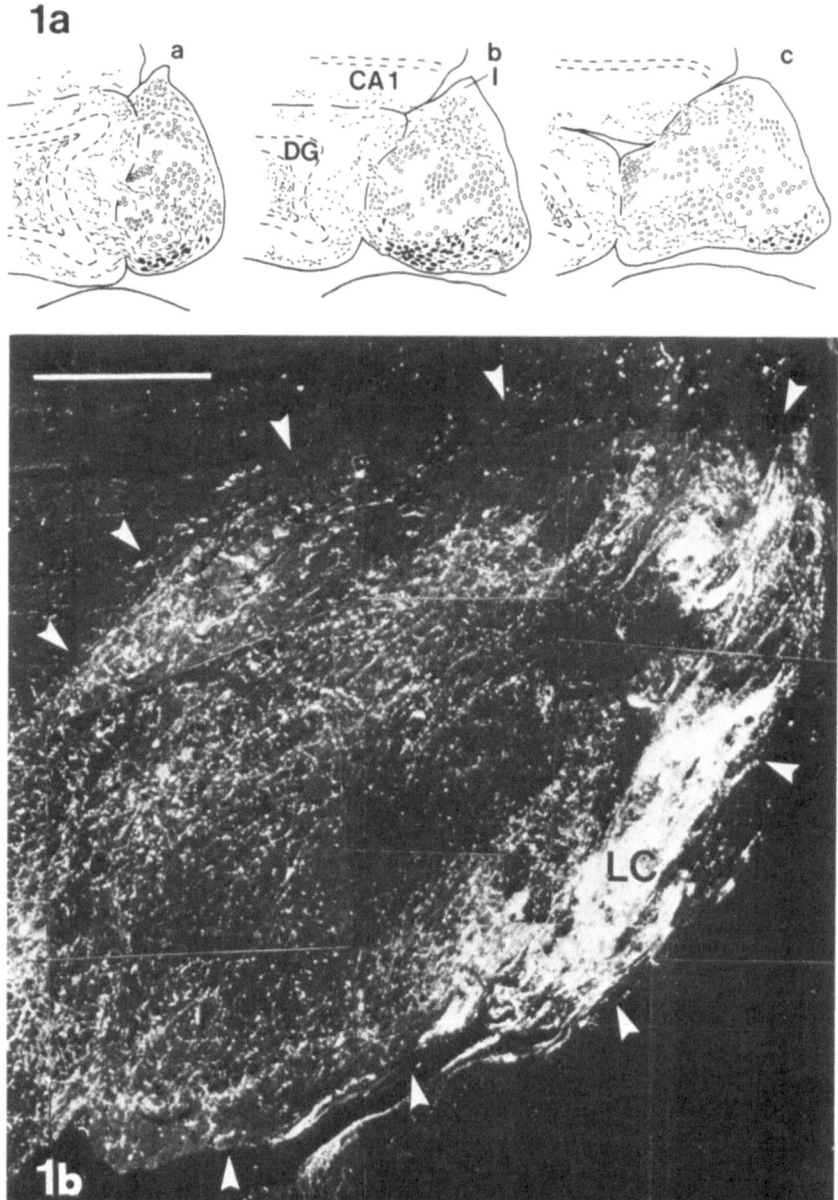

FIGURE 1. (a) Schematic reconstruction of three sagittal planes a–c (medial to lateral) through a brain stem implant (E15; 5 months' survival) prepared for catecholamine histofluorescence and counterstained with cresyl violet. Within the implant (I), cells of various sizes (open symbols) aggregate into nuclear-like configurations. This is particularly evident for the catecholamine neurons (solid shapes) along the ventral surface of the implant, which coalesce to form an oblong nuclear configuration similar to the *in vivo* locus coeruleus. The varicose axons that emanate from these neurons innervate localized regions of the implant and invade the host hippocampus where they form a dense plexus in the host dentate gyrus. This fiber distribution is similar to the normal locus coeruleus innervation, which was removed by an intraventricular injection of 5,7-dihydroxytryptamine (150 μg) prior to transplantation of the brain stem tissue. From Ref. 40. (b) Photomontage of a sagittal section from another brain stem specimen that was prepared for cate-

experimental preparations, such as *in vitro* explants and aggregates and *in oculo* grafts. Thus, a discussion of the organization of several types of embryonic transplants is presented below to illustrate why this technique provides a useful new approach for studying neurodevelopment.

3.1. Brain Stem Transplants

A characteristic feature of brain stem tissue transplanted to the CNS of neonatal[15,27] and adult hosts[32,35,40] or to the anterior eye chamber[4] is the nuclear organization of many of its constituent cells. This is particularly evident when specific cell types can be identified, such as the catecholamine neurons of the substantia nigra or locus coeruleus (Fig. 1). Brain stem tissue dissected from both early and late gestational fetuses possesses this general organization even though the implants are at different stages of development at the time of the initial dissection and exhibit different degrees of posttransplantation growth (Table I and Ref. 40). Since late gestational implants (E17–19) contain predominantly postmitotic neuroblasts which already form a locus coeruleus nuclear configuration,[50–52] the results indicate that the preexisting nuclear organization is maintained after transplantation. However, considerable posttransplantational cell death of the postmitotic catecholamine neurons does occur, for only 10–30% of the normal *in vivo* number of locus coeruleus neurons are present.[33,40]

Early gestational brain stem specimens (E12–14) contain mitotic neuroepithelium and no identifiable nuclear subdivisions at the time of the dissection,[44–47] although the formation of postmitotic catecholamine neuroblasts is beginning.[50–52] Therefore, the nuclear organization present in these specimens must arise subsequent to transplantation. This ability of the developing transplant neurons to migrate and selectively coalesce within the developing implants also is reported to occur in *in vitro* aggregates.[53,54] Thus, when Levitt *et al.*[53] dissociated and reaggregated cells from the region of the substantia nigra, there was a gradual reformation of elongated nuclear bands of catecholamine neurons over a 4-day period. It appears likely that similar cellular interactions could be involved in the formation of selective cell affinities both in culture and in early gestational transplants to the CNS.

Another "intrinsic" property of brain stem transplants is the development of myelinated axons.[35,37,40] Though myelinated fibers cover extensive areas of the implant (Fig. 2a), they also form discrete fiber bundles that interconnect various neuronal subnuclei.[40] In addition, there often is a preferential distribution of the fiber bundles along the external regions of the implant immediately beneath the pial covering (Fig. 2b). Many long descending brain stem projections to the spinal cord also possess a similar

cholamine histofluorescence. The host animal was sympathectomized and denervated of its normal hippocampal monoamine innervation by an intraventricular injection of 5,7-dihydroxytryptamine prior to receiving a brain stem implant from the region of the developing locus coeruleus and vestibular nuclei (E15; 3 months' postimplantation). This implant (arrowheads) contains a dense oblong aggregation of catecholamine-containing neurons from the locus coeruleus (LC) that are positioned along the ventral-posterior surface of the implant. Fluorescent axons from these neurons form a dense plexus within various regions of the implant. Bar = 250 μm. From Ref. 40.

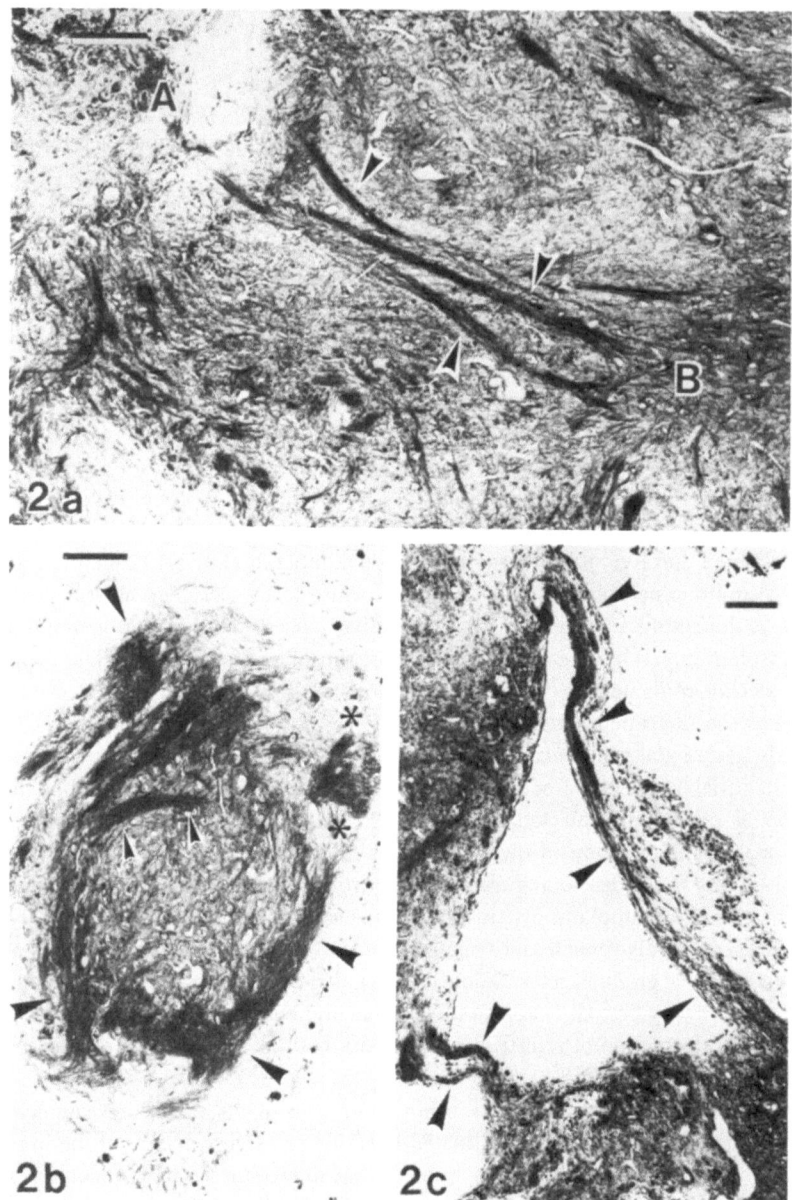

FIGURE 2. (a) Long myelinated fiber tracts (arrowheads) interconnecting two distant regions (A, B) within early gestational implant, BS-12. Klüver–Barrera stain, bar = 100 μm. (b) Dense accumulation of myelinated fibers (large arrowheads) along the exterior surface of a brain stem implant. These fibers bundles are oriented both parallel and perpendicular (asterisks) to the plane of section. Throughout the implant, bundles of myelinated fibers (small arrowheads) interconnect the surface fiber tracts with neuronal clusters in various regions of the implant. Early gestational implant, Klüver–Barrera stain, bar = 100 μm. (c) Two myelinated fiber bridges (arrowheads) interconnecting separate lobes of implant BS-6. These bridges are often quite slender and rarely contain neuronal perikarya. Early gestational implant, Klüver–Barrera stain, bar = 100 μm. From Ref. 40.

subpial location *in vivo*. The myelinated axons within the transplants also possess a remarkable growth capacity, as do axons in tissue culture.[55] Thus, these axons are even capable of interconnecting lobes of the implant that are physically separated provided there is a glial–meningeal substrate that bridges the gap between the two lobes (Fig. 2c). These observations suggest that the embryonic brain stem implants should provide useful model systems for analyzing basic mechanisms of axonal guidance that are responsible for determining the location and orientation of long-projecting axonal pathways in the mammalian CNS.

3.2. Hippocampal Transplants

The posttransplantational organization of embryonic and early neonatal tissue dissected from the hippocampal formation (subiculum, Ammon's horn, and dentate gyrus) has been studied following transplantation to the CNS[18,28,37,40] and to the anterior eye chamber.[10–12] This region of the CNS is especially suited for studying factors that influence the three-dimensional organization of the developing mammalian CNS, for both the hippocampus (Ammon's horn) and the dentate gyrus possess characteristic cell types that form highly organized cellular laminae. Moreover, neurogenesis and the temporal sequences of lamina formation are well characterized in the hippocampal formation.[56–58] Pyramidal neurons in the hippocampus are generated prenatally (E14–18) and a pyramidal cell layer begins forming on E17.[56,57] Granule cells in the dentate gyrus are generated from late embryonic through early neonatal stages (E16–P20). They form a laminated crescent-shaped structure containing a suprapyramidal limb that develops prenatally and an infrapyramidal limb that forms postnatally.[56–58]

Observations of hippocampal transplants from late embryonic and early neonatal donors indicate that there is preservation of the general laminar organization of the large CA3 and smaller CA1 pyramidal neurons of Ammon's horn following transplantation to an intraocular[10,11] or intracephalic site.[18,28,37,40] In all these studies, there is some disruption of the lamina into short segments and some dispersion of the pyramidal cells (Figs. 3 and 4). This spreading of the normally densely packed pyramidal cell layer is most noticeable in hippocampal transplants that are placed in large intracephalic cavities within adult recipients.[40] In these specimens, there is an increased space for expansion of the transplant. Thus, the decreased density of pyramidal cells within a lamina could be due to the loss of some normal *in vivo* physical forces that usually influence the spatial density of the laminae. This phenomenon of a loss of dense pyramidal cell packing also is observed in explants of neonatal hippocampus in which there is flattening of the cultures and dispersion of the pyramidal cells.[59]

A prominent crescent-shaped dentate gyrus has been observed in hippocampal transplants (Fig. 3–5) dissected from late embryonic[12,37,40] and early neonatal rats.[18,28] However, a well-formed C-shaped granule cell lamina appears to require fairly large numbers of granule cells, as many transplants possess only short laminated segments or cell clusters. At the time of transplantation, these specimens already contain a partially formed suprapyramidal blade. Therefore, only the infrapyramidal blade is generated

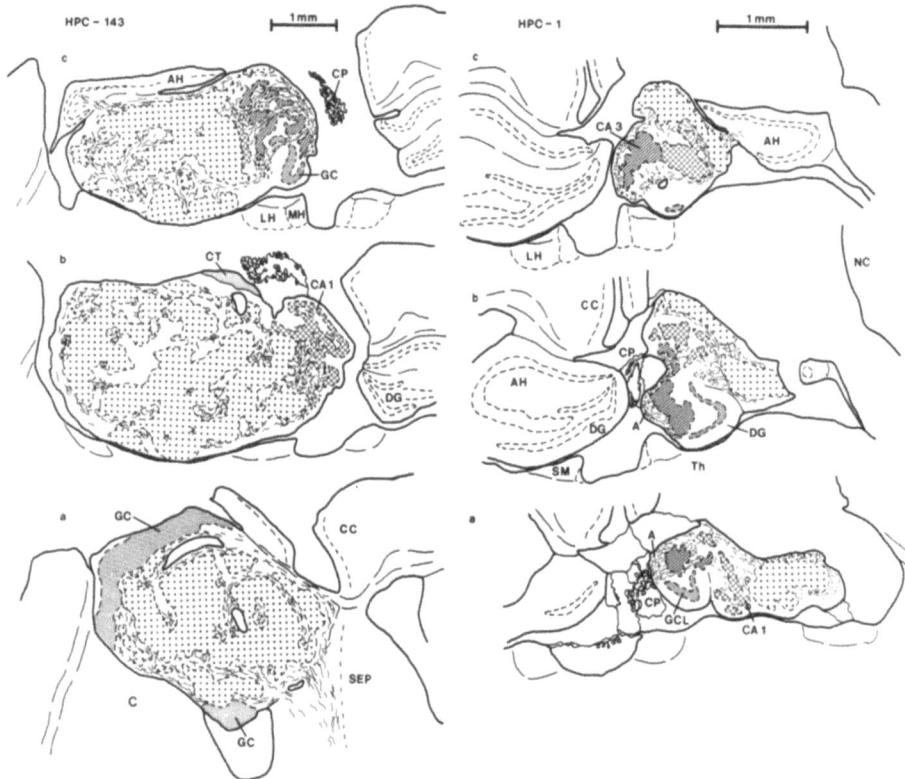

posttransplantation. This development of an infrapyramidal blade of granule cells is rarely observed in hippocampal explants from neonatal mice or rats.[59,60] Because of these observations, LaVail and Wolf[59] and Beach *et al.*[60] speculated that granule cells may require the presence of an entorhinal cortex input in order to continue to form laminae in culture. However, the transplantation studies demonstrate that the formation of the infrapyramidal blade does not require the presence of an input from the entorhinal cortex, for grafts to the anterior eye chamber[12] and transplants to the adult CNS that lack an entorhinal input[40] are capable of generating a crescent-shaped dentate gyrus.

The recent study of Kromer *et al.*[40] also has analyzed the ability of neuroependymal cells in early gestational hippocampal transplants (E13–14) to generate well-organized pyramidal cell and granule cell laminae. In this study, CA1 pyramidal cells sometimes form well-organized laminae but the CA3 neurons usually are arranged in clusters or loose laminar patterns (Fig. 4). This inability of most pyramidal cells to form well-defined laminae could be due to the extensive proliferation of early gestational implant tissue that occurs *in cerebro* (Table I). The massive cell proliferation could easily cause a misorientation of the radial glial cells that are necessary for the proper migration and alignment of the pyramidal neurons.[61] Temporal factors also could influence lamina formation. This is suggested by the *in vitro* reaggregation study of DeLong[62] in which pyramidal cells only form a lamina within the reaggregates when the tissue is dissected at a very specific developmental stage (E18.5). If the hippocampus is removed from fetuses at earlier or later gestational stages, clustering of the pyramidal neurons occurs.

← ———

FIGURE 3. Schematic illustration of coronal sections (a–c; rostral to caudal) through early gestational implant, HPC-143, and late gestational implant, HPC 1. HPC-143 (E14; survival: 5½ months) is a large implant that fills most of the transplantation cavity and is attached to many regions of the host CNS. Myelinated and unmyelinated axons (short wavy lines) interconnect the implant with the host septum and hippocampus. Most of the volume of the transplant is occupied by loosely arranged clusters of medium and large pyramidal cells that are demarcated by the dotted areas. The granule cells (regions with the horizontal lines) are organized in large clusters at the anterior tip of the implant, but develop laminae in the posterior pole. The pyramidal cells also begin to segregate into aggregates of similar-sized neurons in the more posterior portion of the implant. Medium CA1 pyramidal neurons form loose and densely packed laminae (cross-hatched area), and the large CA3 neurons (regions with thicker diagonal lines) are organized mainly into small aggregates or short loose layers. This implant possesses a choroid plexus (CP) and is contaminated by a small region of connective tissue (CT). HPC-1 (E19; survival: 4 months) is subdivided into: (1) a well-developed dentate gyrus (DG) containing a typical crescent-shaped granule cell layer (GCL), which is indicated by the area of thin horizontal lines, an external molecular layer, and hilar region of large neurons; (2) an Ammon's horn region with loosely laminated large CA3 pyramidal neurons (thicker diagonal lines) and medium CA1 pyramidal cells (cross-hatched area); and (3) an area containing a more diffuse subicular-like arrangement of medium pyramidal neurons (dotted region). Myelinated fibers (indicated by the fine short lines) form a dense myelinated alveus-like (A) pathway along the medial aspect of the implant that interconnects regions containing pyramidal neurons. There also is a very extensive plexus of unmyelinated axons (not indicated) that interconnect various implant areas and, in particular, form a mossy fiber-like projection between the granule cells and the CA3 neurons. In addition to these intrinsic connections, some myelinated and unmyelinated axons interconnect the implant and the recipient's Ammon's horn (AH). C, caudate nucleus; CC, cingulate cortex; CP, choroid plexus; DG, dentate gyrus; GC, granule cells; LH, lateral habenula; MH, medial habenula; NC, neocortex; SEP, septum; SM, stria medularis; Th, thalamus. Modified from Ref. 40.

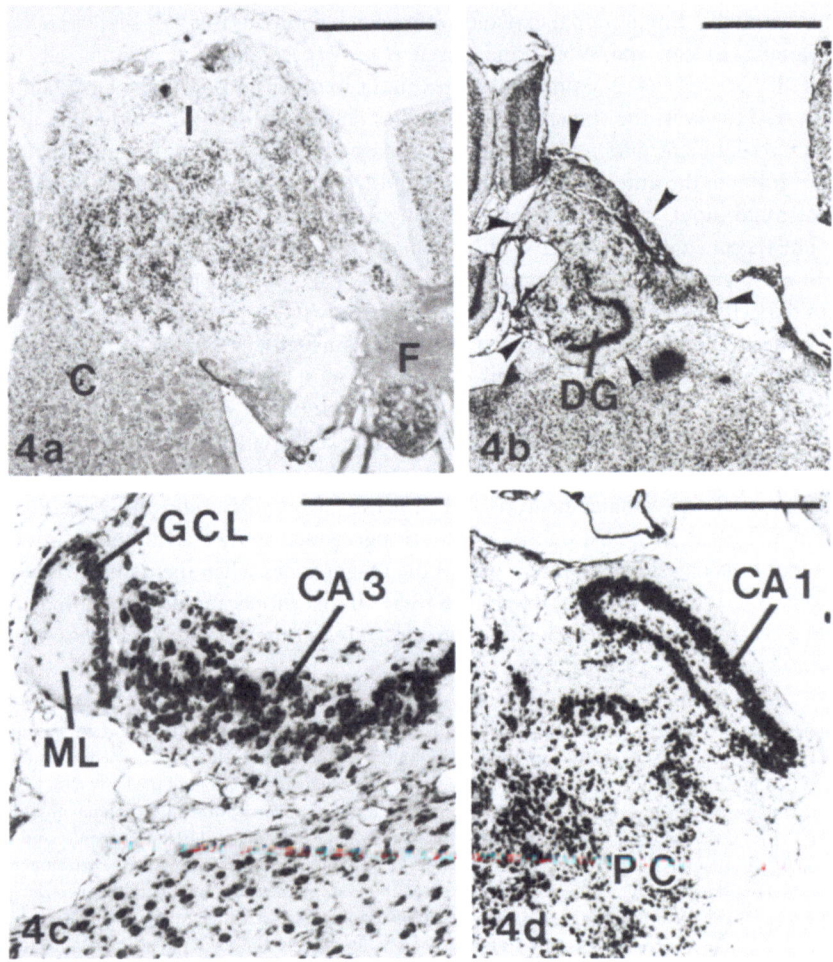

FIGURE 4. (a) Coronal section through an early gestational hippocampal implant (E14; survival: 3½ months) at the level of the host caudate nucleus. This implant is not highly organized at this level, but contains clusters of medium and large pyramidal neurons. The implant has expanded to fill the entire available space within the transplantation cavity. Though the implant is quite large and fuses with the host caudate nucleus (C) and fimbria (F), it does not encroach upon any structures in the host CNS. Bar = 1 mm. (b) Coronal section at the level of the rostral host hippocampus in a specimen with a late gestational hippocampal implant (arrowheads). Though this implant is well developed and contains a dentate-like region (DG), it is relatively small and occupies only a portion of the available space within the transplantation cavity. Donor: E19; survival: 4 months. Bar = 1 mm. (c) A laminated granule cell region (GCL) with accompanying molecular layer (ML) is present in this small late gestational hippocampal implant. In this specimen, the granule cells form a short straight cell layer between the ML and an area that contains a loose lamina of large pyramidal neurons (CA3). Bar = 200 μm. (d) Portion of an early gestational hippocampal implant. Though much of the interior of this implant is occupied by a diffuse distribution of pyramidal cells (PC), there is one subregion that contains a densely packed ovoid lamina of medium neurons (CA1). The apical processes of these cells extend into the cell-sparse area along the surface of the implant, and the basal processes and myelinated axons are concentrated within the interior of this tubular cell lamina. Bar = 500 μ m. From Ref. 40.

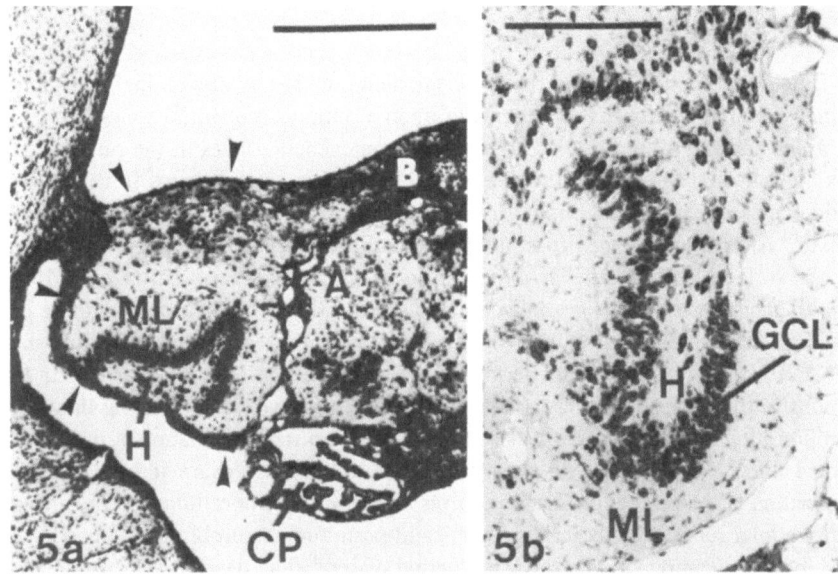

FIGURE 5. (a) This late gestational hippocampal implant contains a well-developed "dentate" sublobule (arrowheads) consisting of a hilus (H) curved granule cell layer and a molecular layer (ML). In this coronal section, the orientation of the crescent-shaped granule cell lamina is rotated 180° compared with the orientation of the host dentate gyrus. The adjacent sublobule (A) that contains large and medium pyramidal neurons possesses axonal connections with the dentate sublobule in subsequent sections. The dark areas of the implant (B) are due to the presence of silver grains, which represent anterograde axonal transport of isotope from another implant region that was injected with [³H] proline. A small choroid plexus (CP) also is present in this specimen. Bar = 500 μm. (b) A well-developed but small dentate-like region has formed in a subregion of this early gestational implant. The implant granule cells form a typical crescent-shaped lamina (GCL) with an external molecular layer (ML) and an internal hilus (H). Bar = 200 μm. From Ref. 40.

Though early gestational hippocampal transplants also contain granule cells, in the majority of specimens these cells are arranged in dense cell clusters of varying sizes. The most developed dentate-like regions are observed in areas of the transplants that fuse with the host hippocampal formation (Fig. 5). From this study, it is uncertain whether an input from the entorhinal cortex is required for the initial formation of a well-structured crescent-shaped granule cell lamina from neuroependymal cells or whether other "extrinsic" or "intrinsic" factors may influence the development of a completely formed dentate gyrus. However, these initial results indicate that future transplantation studies that obtain hippocampal tissue from a variety of embryonic donor ages and utilize various transplantation sites should provide excellent preparations in which to further investigate the influences of a number of genetic and epigenetic factors on the formation of the characteristic three-dimensional cytoarchitecture of Ammon's horn and the dentate gyrus.

3.3. Cerebellar Transplants

The cerebellum is another region of the neuraxis that exhibits a highly organized laminar cytoarchitecture that is dependent upon a complex sequential series of cell proliferations and migrations.[43,63-66] During early ontogenetic stages in the rat (E12–14), Purkinje cells and neurons in the deep cerebellar nuclei are generated from two separated anlagen of neuroepithelium located along the dorsolateral aspect of the fourth ventricle.[43,65] After these anlagen fuse along the midline, additional germinal cells migrate over the surface of the cerebellar plate and form an external germinal layer (E16–19) that will generate cerebellar granule cells and other cortical interneurons. As these postmitotic interneurons are produced during late embryonic and early neonatal stages (E17–P30), they migrate along Bergman glial processes in order to reach their final locations in the molecular layer or internal granule cell layer.[43,65,66] During this period of proliferation of cerebellar cortical neurons, there also is the formation of transversely oriented folia.[43,65] Thus, the development of the unique and characteristic trilaminar organization of the cerebellar cortex requires a precise series of cellular interactions and extensive migrations of both germinal cells and postmitotic neuroblasts.

Given the complex and protracted period of cerebellar development *in vivo*, it is particularly striking that cellular transplants dissected from both early and late gestational fetuses develop a trilaminar cortical organization (consisting of molecular, Purkinje cell, and granule cell layers) regardless of whether they are transplanted to the brain of neonatal[15,29] or adult rats,[36-38,40,] or to the anterior eye chamber.[6-10] These trilaminar cortical areas are separated from other regions of the implants that contain multipolar neurons, presumably from the deep cerebellar nuclei, that are embedded within a plexus of myelinated axons (Figs. 6 and 7). Thus, considerable neurogenesis and cell migration must occur subsequent to transplantation, especially in the early gestational transplants.[38,40]

The cytological appearance of early and late gestational cerebellar implants suggests that a full range of cerebellar neuronal types is present,[6,10,36,38,40] even in late gestational implants, which exhibit considerable posttransplantation cell death.[40] Moreover, many of the normal intrinsic cerebellar connections are maintained within the transplants, such as the basket cell and granule cell-parallel fiber inputs to Purkinje cells, the Purkinje cell projections to the deep nuclei, and the mossy fiber projections from the deep nuclei to the granule cell layer.[38,40] Though these projections are present within the cerebellar transplants, recent electron microscopic data indicate that the normal three-dimensional orientation of the parallel fibers within the molecular layer of the transplants is misaligned.[38] Observations of Golgi-impregnated cerebellar transplants to the intraocular (10) or intracephalic site (Fig. 8) indicate that the morphology of the granule cells, molecular layer interneurons, and neurons of the deep cerebellar nuclei is similar to that observed for these cell types in the normal adult cerebellum. However, the morphological appearance of the Purkinje cells in these two preparations differs. Kromer et al.[40] report that Purkinje cells in their early gestational intracephalic cerebellar transplants (E13) exhibit a well-developed dendritic arborization with spine-studded secondary and tertiary dendrites and smooth pear-shaped perikarya (Fig. 8), whereas Woodward et al.[10] report that Purkinje cells in their E15–16 intraocular cer-

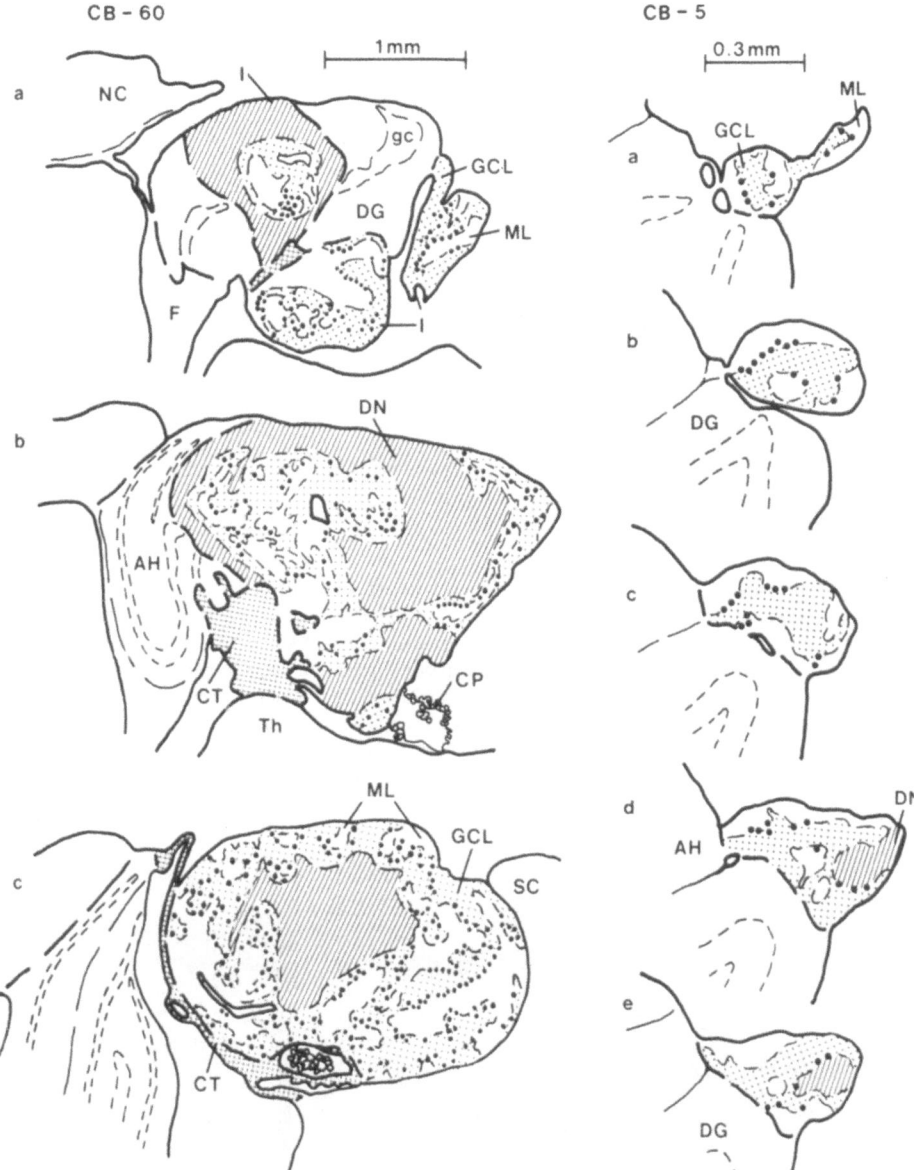

FIGURE 6. Schematic representations of camera lucida drawings from a series of frontal sections through an early gestational (CB-60) and a late gestational (CB-5) cerebellar implant. Striped areas represent regions containing neurons from the deep cerebellar nuclei (DN), finely dotted areas illustrate regions with granule cells (GCL), large dots indicate individual Purkinje cells, clear areas show the locations of the molecular layer (ML), and cross-hatched regions indicate the presence of nonneural connective tissue (CT). CB-60 (E12; survival: 3 months) is fused rostrally with the host dentate gyrus (DG) and is attached to the host superior colliculus (SC) caudally. This implant contains a well-defined choroid plexus (CP) and large invaginated and laminated cortical areas connected via myelinated axons with an extensive region containing deep cerebellar neurons. a–c, rostral-caudal. CB-5 (E17; survival: 4½ months) has a very small volume (0.11 mm^3) but still possesses a laminated cortex and a region with deep nuclear cells. The implant is attached to the lesioned surface of the host dentate gyrus. a–e, rostral-caudal. AH, Ammon's horn; CP, choroid plexus; DG, dentate gyrus; F, Fimbria; NC, neocortex; SC, superior colliculus; Th, thalamus. Modified from Ref. 40.

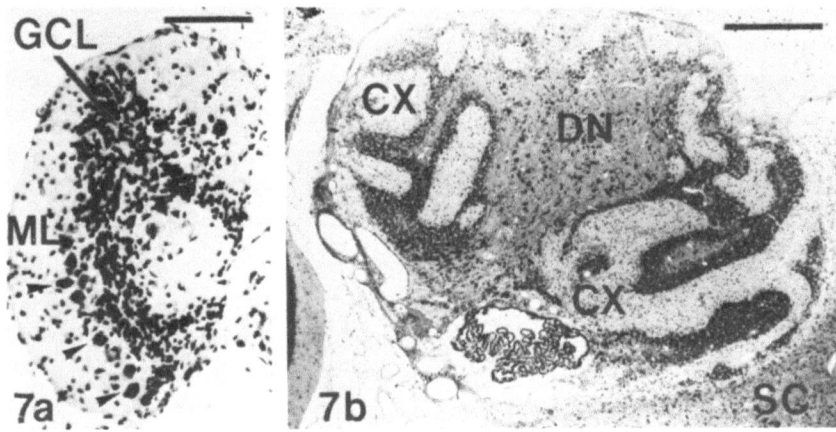

FIGURE 7. (a) Despite its small size, this late gestational cerebellar implant possesses a laminar organization with a well-defined outer molecular layer (ML), an intervening row of Purkinje cells (arrowheads), and a densely packed granule cell layer (GCL). Cresyl violet stain, bar = 100 μm. (b) Coronal section through an early gestational cerebellar implant. Regions composed of cortical tissue (CX) are readily identified by their trilaminar appearance and foliation, which distinguishes them from the densely myelinated areas containing neurons presumably from the deep cerebellar nuclei (DN). As for most cerebellar implants, a well-defined choroid plexus also is present. SC, host superior colliculus, Klüver–Barrera stain, bar = 0.5 mm. From Ref. 40.

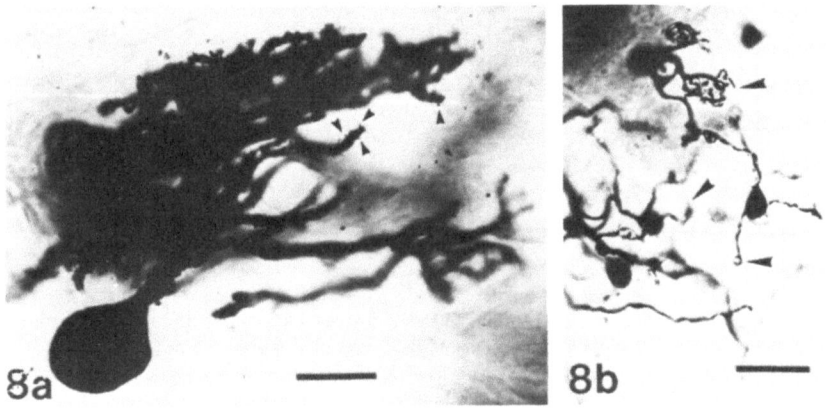

FIGURE 8. (a) Golgi-impregnated Purkinje cell in implant CB-32 (E13; survival: 2½ months). This cell, like most within the implant, possesses a single primary apical dendrite with a highly branched dendritic arborization. Both soma and proximal dendrite have smooth surface membranes whereas the secondary and tertiary dendrites contain spines (arrowheads). Bar = 20 μm. (b) Granule cells with typical claw-shaped termination (arrowheads) are located below the basal end of Purkinje cells where they form an internal granule cell layer within cortical regions of the implant, CB-32. Bar = 20 μm. From Ref. 40.

ebellar grafts exhibit incomplete maturation, for they retain their perisomatic thorns and possess poorly developed dendrites. The only major atypical features of the Purkinje cells in the *in cerebro* cerebellar transplants appear to be the lack of any preferential three-dimensional orientation of their dendritic branches,[40] and the presence of thin-necked spines on at least some thick (presumably proximal) dendrites.[38] Observations of *in vitro* cerebellar explants[67–72] indicate that these tissue preparations possess an appearance similar to that described for the *in oculo* grafts. In the cerebellar explants from early neonatal donors (P0-2), neurons of the deep cerebellar nuclei and cortical interneurons have a fairly normal appearance and perisomatic basket fibers are present around Purkinje cells.[67–70] Though Purkinje cells are present in these preparations, the vast majority exhibit only rudimentary development, as they still possess perisomatic filopodia and a sparse dendritic aborization.[67,69–73]

Kromer *et al.*[40] present several explanations to account for the extensive morphological differentiation of the Purkinje cells that is observed in their intracephalic cerebellar transplants to adult recipients. As the Golgi impregnations in this study were performed on large transplants from early gestational fetuses (in contrast to the smaller late gestational intraocular grafts and neonatal explants), differences in the gestational age of the donor, the density of granule cells, and the granule cell–Purkinje cell ratio could influence Purkinje cell development in the various preparations. Alternatively, the intracephalic transplants could possess a climbing fiber input, which is reported to be important for the maturation of Purkinje cells *in vivo*.[74] Due to the location of the intracephalic implants adjacent to the host hippocampus, there may be a climbing fiber input that originates from locus coeruleus neurons in the host CNS, which are known to have collateral axons that innervate both the telencephalon and the cerebellum.[75] A second source of climbing fibers could be intrinsic to the implant. Because these transplants are dissected at such an early embryonic age, there could be some contamination of the specimens with vestibular or inferior olivary neurons that arise from adjacent regions at the dorsolateral edge of the fourth ventricle.[76,77] However, since climbing fibers were not observed in those transplants prepared for electron microscopy in the study by Alvarado-Mallart and Sotelo,[38] this further suggests that when these afferents are absent from the transplants, Purkinje cells are still capable of developing normal, smooth pear-shaped perikarya with basket cell synapses but they appear to possess aberrant proximal dendritic spines.

Though *in vitro* cerebellar cultures fail to form folia after explantation,[68,72] several transplantation studies have reported varying degrees of fissure formation in intraocular grafts[6,10] and in intracephalic transplants to neonatal[15,29] and adult recipients.[36,38,40] In small late gestational cerebellar transplants, there are only minor cortical invaginations.[6,10,40] However, Wells and McAllister[29] describe several late gestational cerebellar transplants (injected into the cerebral cortex of neonatal hosts), that are somewhat larger in size and form more extensive folia along their superficial regions, which are located directly beneath the host cranium and dural covering. In deeper areas, there is invagination of the tissue along large blood vessels. Early gestational transplants that are placed in large intracephalic cavities[38,40] exhibit the most extensive fissure development. These specimens undergo considerable expansion within the transplantation cavity and develop numerous randomly oriented fissures that are located along large blood vessels that vascularize the transplant from the adjacent choroid plexus or host neuropil. These observations suggest that the development of fissures within the cerebellar cortex is not dependent upon physical forces associated with spatial constraints imposed by surrounding tissues. Instead, it appears that there is an intrinsic interaction between the external germinal cells in the transplanted cerebellar tissue and the vascular network arising from the host that probably initiates fissure formation and produces the nonoriented folia that are observed. Thus, these descriptions of cerebellar organization *in cerebro* are consistent with the hypothesis proposed by Mares and Lodin[78] that differential proliferation rates in various regions of the external germinal layer could account for folia formation. However, the lack of any preferential orientation of the fissures and folia and the Purkinje cell dendrites in the transplants suggests that the development of the precise planar orientation of the Purkinje cell dendrites and folia, which is observed *in vivo* but is absent in the cerebellar transplants,[38,40] is dependent upon more extensive, extrinsic influences not present within the ectopic cerebellar transplants. Future studies that utilize co-transplants of cerebellar tissue and other neural or non-neural tissues may prove beneficial for addressing this question.

4. Conclusions

Though techniques for successfully and consistently transplanting developing mammalian CNS tissue have been developed only recently, the experimental studies presented here already indicate that this procedure holds great potential for addressing questions concerning the development of the mammalian CNS. In particular, the intracephalic transplantation techniques provide an extremely valuable experimental approach for analyzing genetic and epigenetic factors operating during development that influence the morphology of individual neurons, guide growing axons, regulate synaptic specificity, and determine the unique three-dimensional organization in specific regions of the mammalian CNS. Thus, the developmental questions that can be explored with the *in cerebro* preparation should complement those that can be analyzed with *in vitro* explant or reaggregation procedures or with *in oculo* grafting techniques.

ACKNOWLEDGMENTS. The author wishes to thank Catherine LaRose, Steven Samuelsson, and Mary D. Anderson for their assistance in preparing the manuscript. Work in the author's laboratory was supported by grants from the National Institutes of Health (NS 18126) and the March of Dimes Birth Defects Foundation and by fellowships from the Alfred P. Sloan Foundation and the Charles E. Culpepper Foundation.

References

1. May, R. M., 1930, La greffe dans l'oeil de rat blanc adulte du tissu cerebral de rat nouveau-né, *Arch. Anat. Microsc.* **26:** 443.
2. Greene, H. S. M., and Arnold H., 1945, The homologous and heterologous transplantation of brain and brain tumors, *J. Neurosurg.* **2:**315.

3. Chatagnon, P.-A., 1952, Recherches sur la differenciation du neurone dans la greffe bréphoplastique endoculaire chez le rat blanc, *Arch. Biol.* **63**:199.

4. Olson, L., and Seiger, A., 1972, Brain tissue transplanted to the anterior chamber of the eye. 1. Fluorescence histochemistry of immature catecholamine and 5-hydroxytryptamine neurons reinnervating the rat iris, *Z. Zellforsch Mikrosk, Anat.* **135**:175.

5. Hoffer, B., Seiger, Å., Ljungberg, T., and Olson, L. 1974, Electrophysiological and cytological studies of brain in the anterior chamber of the eye: Maturation of cerebellar cortex *in oculo, Brain Res.* **79**:165.

6. Hoffer, B., Olson, L., Seiger, Å., and Bloom, F., 1975, Formation of a functional adrenergic input to intraocular cerebellar grafts: Ingrowth of inhibitory sympathetic fibers, *J. Neurobiol.* **6**:565.

7. Seiger, Å., and Olson, L., 1975, Brain tissue transplanted to the anterior chamber of the eye. 3. Substitution of lacking central noradrenaline input by host iris sympathetic fibers in the isolated cerebral cortex developed *in oculo, Cell Tissue Res.* **159**:325.

8. Seiger, Å., and Olson, L., 1977, Quantitation of fiber growth in transplanted central monoamine neurons, *Cell Tissue Res.* **179**:285.

9. Yellin, H., 1976, Survival and possible trophic function of neonatal spinal cord grafts in the anterior chamber of the eye, *Exp. Neurol.* **51**:579

10. Woodward, D. J., Seiger, Å., Olson, L., and Hoffer, B. J., 1977, Intrinsic and extrinsic determinants of dendritic development as revealed by Golgi studies of cerebellar and hippocampal transplants *in oculo, Exp. Neurol.* **57**:984.

11. Olson, L., Freedman, H., Seiger, Å., and Hoffer, B., 1977, Electrophysiology and cytology of hippocampal formation transplants in the anterior chamber of the eye. I. Intrinsic organization, *Brain Res.* **119**:87.

12. Goldowitz, D., Seiger, Å., and Olson, L., 1982, Anatomy of the isolated area dentata grown in the rat anterior eye chamber, *J. Comp. Neurol.* **208**:382.

13. Dunn, E., 1917, Primary and secondary findings in a series of attempts to transplant cerebral cortex in the albino rat, *J. Comp. Neurol.* **27**:565.

14. Le Gros Clark, W. E., 1940, Neuronal differentiation in implanted foetal cortical tissue, *J. Neurol. Psychiatry* **3**:263.

15. Das, G. D., 1974, Transplantation of embryonic neural tissue in the mammalian brain. I. Growth and differentiation of neuroblasts from various regions of the embryonic brain in the cerebellum of neonate rats, *TIT J. Life Sci.* **4**:93.

16. Das, G. D., 1975, Differentiation of dendrites in the transplanted neuroblasts in the mammalian brain, *Adv. Neurol.* **12**:181.

17. Lund, R. D., and Hauschka, S. D., 1976, Transplanted neural tissue develops connections with host rat brain, *Science* **193**:582.

18. Zimmer, J., 1978, Development of the hippocampus and fascia dentata: Morphological and histochemical aspects, *Prog. Brain Res.* **48**:171.

19. Jaeger, C. B., and Lund, R. D., 1979, Efferent fibers from transplanted cerebral cortex of rats, *Brain Res.* **165**:338.

20. Jaeger, C. B., and Lund, R. D., 1980, Transplantation of embryonic occipital cortex to the brain of newborn rats: An autoradiographic study of transplant histogenesis, *Exp. Brain Res.* **40**:265.

21. Jaeger, C. B., and Lund, R. D., 1980, Transplantation of embryonic occipital cortex to the tectal region of newborn rats: A light microscopic study of organization and connectivity of the transplants, *J. Comp. Neurol.* **194**:571.

22. Das, G. D., Hallas, B. H., and Das, K. G., 1980, Transplantation of brain tissue in the brain of rat. I. Growth characteristics of neocortical transplants from embryos of different ages, *Am. J. Anat.* **158**:135.

23. McLoon, S. C., and Lund, R. D., 1980, Identification of cells in retinal transplants which project to host visual centers: A horseradish peroxidase study in rats, *Brain Res.* **197**:491.

24. McLoon, L. K., Lund, R. D., and McLoon, S. C., 1982, Transplantation of reaggregates of embryonic neural retinae to neonatal rat brain: Differentiation and formation of connections, *J. Comp. Neurol.* **205**:179.

25. Freed, W. J., and Wyatt, R. J., 1980, Transplantation of eyes to the adult rat brain: Histological findings and light-evoked potential response. *Life Sci.* **27**:503.

26. Graziadei, P. P. C., and Kaplan, M. S., 1980, Regrowth of olfactory sensory axons into transplanted neural tissue. 1. Development of connections with the occipital cortex, *Brain Res.* **201**:39.

27. Lund, R. D., and Harvey, A. R., 1981, Transplantation of tectal tissue in rats. I. Organization of transplants and pattern of distribution of host afferents within them, *J. Comp. Neurol.* **201**:191.

28. Sunde, N., and Zimmer, J., 1981, Transplantation of central nervous tissue: An introduction with results and implications, *Acta Neurol. Scand.* **63**:323.

29. Wells, J., and McAllister, J. P., II, 1982, The development of cerebellar primordia transplanted to the neocortex of the rat, *Dev. Brain Res.* **4**:167.

30. Stenevi, U., Björklund, A., Kromer, L. F., Paden, C. M., Gerlach, J. L., McEwen, B. S., and Silverman, A. J., 1980, Differentiation of embryonic hypothalamic transplants cultured on the choroidal pia in brains of adult rats, *Cell Tissue Res.* **205**:217.

31. Stenevi, U., Björklund, A., and Svendgaard, N.-A., 1976, Transplantation of central and peripheral monoamine neurons to the adult rat brain: Techniques and conditions for survival, *Brain Res.* **114**:1.

32. Björklund, A., Stenevi, U., and Svendgaard, N.-A., 1976, Growth of transplanted monoaminergic neurons into the adult hippocampus along the perforant path, *Nature (London)* **262**:787.

33. Björklund, A., Segal, M., and Stenevi, U., 1979, Functional reinnervation of the rat hippocampus by locus coeruleus implants, *Brain Res.* **170**:409.

34. Björklund, A., and Stenevi, U., 1977, Reformation of the severed septohippocampal cholinergic pathway in the adult rat by transplanted septal neurons, *Cell Tissue Res.* **185**:289.

35. Møllgård, K., Lundberg, J. J., Beebe, B. K., Björklund, A., and Stenevi, U., 1978, The intracerebrally cultured 'microbrain': A new tool in developmental neurobiology, *Neurosci. Lett.* **8**:295.

36. Kromer, L. F., and Björklund, A., 1980, Embryonic neural transplants provide model systems for studying development and regeneration in the mammalian CNS, in: *Multidisciplinary Approach to Brain Development* (C. DiBenedetta, R. Balazs, G. Gombos, and G. Porcellati, eds.), pp. 409–426, Elsevier/ North-Holland, Amsterdam.

37. Kromer, L.F., Björklund, A., and Stenevi, U., 1979, Intracephalic implants: A technique for studying neuronal interactions, *Science* **204**:1117.

38. Alvarado-Mallart, R. M., and Sotelo, C., 1982, Differentiation of cerebellar anlage heterotopically transplanted to adult rat brain: A light and electron microscopic study, *J. Comp. Neurol.* **212**:247.

39. Kromer, L. F., Björklund, A., and Stenevi, U., 1981, Innervation of embryonic hippocampal implants by regenerating axons of cholinergic septal neurons in the adult rat, *Brain Res.* **210**:153.

40. Kromer, L. F., Björklund, A., and Stenevi, U., 1983, Intracephalic embryonic neural implants in the adult rat brain. I. Growth and mature organization of brainstem, cerebellar and hippocampal implants, *J. Comp. Neurol.* **218**:433.

41. Hallas, B. H., Das, G. D., and Das, K. G., 1980, Transplantation of brain tissue in the brain of rat. II. Growth characteristics of neocortical transplants in hosts of different ages, *Am. J. Anat.* **158**:147.

42. Gash, D., Sladek, J. R., Jr., and Sladek, C. D., 1980, Functional development of grafted vasopressin neurons, *Science* **210**:1367.

43. Altman, J., and Bayer, S. A., 1978, Prenatal development of the cerebellar system in the rat. I. Cytogenesis and histogenesis of the deep nuclei and the cortex of the cerebellum, *J. Comp. Neurol.* **179**:23.

44. Altman, J., and Bayer, S. A., 1980, Development of the brainstem in the rat. I. Thymidine-radiographic study of the time of origin of neurons of the lower medulla, *J. Comp. Neurol.* **194**:1.

45. Altman, J., and Bayer, S. A., 1980, Development of the brainstem in the rat. II. Thymidine-radiographic study of the time of origin of neurons of the upper medulla, excluding the vestibular and auditory nuclei, *J. Comp. Neurol.* **194**:37.

46. Altman, J., and Bayer, S. A., 1980, Development of the brainstem in the rat. III. Thymidine-radiographic study of the time of origin of neurons of the vestibular and auditory nuclei of the upper medulla, *J. Comp. Neurol.* **194**:877.

47. Altman, J., and Bayer, S. A., 1980, Development of the brainstem in the rat. IV. Thymidine-radiographic study of the time of origin of neurons in the pontine region, *J. Comp. Neurol.* **194**:905.

48. Kromer, L. F., 1982, Development of embryonic hippocampal transplants in the adult rodent CNS, *Soc. Neurosci. Abstr.* **8**:327.

49. Das, G. D., 1982, Extrapharenchymal neural transplants: Their cytology and survivability, *Brain Res.* **241**:182.

50. Seiger, Å., and Olson, L., 1973, Late prenatal ontogeny of central monoamine neurons in the rat· Fluorescence histochemical observations, *Z. Anat. Entwicklungsgesch.* **140**:281.

51. Lauder, J. M., and Bloom, F. E., 1974, Ontogeny of monoamine neurons in the locus coeruleus, raphe nuclei and substantia nigra of the rat. I. Cell differentiation, *J. Comp. Neurol.* **155**:469.

52. Sievers, J., Lolova, I., Jenner, S., Klemm, H. P., and Sievers, H., 1981, Morphological and biochemical studies on the ontogenesis of the nucleus locus coeruleus, *Bibl. Anat.* **19**:52.

53. Levitt, P., Moore, R. Y., and Garber, B. B., 1976, Selective cell association of catecholamine neurons in brain aggregates *in vitro*, *Brain Res.* **111**:311.

54. Hemmendinger, L. M., Garber, B. B., Hoffman, P. C., and Heller, A., 1981, Selective association of embryonic murine mesencephalic dopamine neurons *in vitro*, *Brain Res.* **222**:417.

55. Crain, S. A., Peterson, E. R., and Bornstein, M. B., 1968, Formation of functional interneuronal connections between explants of various mammalian central nervous tissues during development *in vitro*, in: *Ciba Foundation Symposium, Growth of the Nervous System* (G. E. W. Wolstenholme and M. O'Connor, eds.), pp. 13–31, Churchill, London.

56. Bayer, S. A., 1980, Development of the hippocampal region in the rat. I. Neurogenesis examined with ³H-thymidine autoradiography, *J. Comp. Neurol.* **190**:87.

57. Bayer, S. A., 1980, Development of the hippocampal region in the rat. II. Morphogenesis during embryonic and early postnatal life, *J. Comp. Neurol.* **190**:115.

58. Schlessinger, A. R., Cowan, W. M., and Gottlieb, D. I., 1975, An autoradiographic study of the time of origin and the pattern of granule cell migration in the dentate gyrus of the rat, *J. Comp. Neurol.* **159**:149.

59. LaVail, J. H., and Wolf, M. K., 1973, Postnatal development of the mouse dentate gyrus in organotypic cultures of the hippocampal formation, *Am. J. Anat.* **137**:47.

60. Beach, R. L., Bathgate, S. L., and Cotman, C. W., 1982, Identification of cell types in rat hippocampal slices maintained in organotypic cultures, *Dev. Brain Res.* **3**:3.

61. Levitt, P., and Rakic, P., 1980, Immunoperoxidase localization of glial fibrillary acidic protein in radial glial cells and astrocytes of the developing rhesus monkey brain, *J. Comp. Neurol.* **193**:815.

62. DeLong, G. R., 1970, Histogenesis of fetal mouse isocortex and hippocampus in reaggregating cell cultures, *Dev. Biol.* **22**: 563.

63. Miale, I. L., and Sidman, R. S., 1961, An autoradiographic analysis of histogenesis in the mouse cerebellum, *Exp. Neurol.* **4**:277.

64. Altman, J., 1966, Autoradiographic and histological studies of postnatal neurogenesis. II. A longitudinal investigation of the kinetics, migration and transformation of cells incorporating tritiated thymidine in infant rats, with special reference to postnatal neurogenesis in some brain regions, *J. Comp. Neurol.* **128**:431.

65. Korneliussen, H. K., 1968, On the ontogenetic development of the cerebellum (nuclei, fissures and cortex) in the rat, with special reference to regional variations in corticogenesis, *J. Hirnforsch.* **10**:379.

66. Rakic, P., 1971, Neuron–glia relationship during granule cell migration in developing cerebellar cortex: A Golgi and electron microscopic study in macacus rhesus, *J. Comp. Neurol.* **141**:283.

67. Wolf, M. K., 1964, Differentiation of neuronal types and synapses in myelinating cultures of mouse cerebellum, *J. Cell Biol.* **22**: 259.

68. Wolf, M. K., 1970, Anatomy of cultured mouse cerebellum. II. Organotypic migration of granule cells demonstrated by silver impregnation of normal and mutant cultures, *J. Comp. Neurol.* **140**:281.

69. Wolf, M. K., and Dubois-Dalcq, M., 1970, Anatomy of cultured mouse cerebellum. I. Golgi and electron microscopic demonstration of granule cells, their afferent and efferent synapses, *J.Comp. Neurol.* **140**:261.

70. Allerand, C. D., 1971, Patterns of neuronal differentiation in developing cultures of neonatal mouse cerebellum: A living and silver impregnation study, *J. Comp. Neurol.* **142**:167.

71. Seil, F. J., 1972, Neuronal groups and fiber patterns in cerebellar tissue cultures, *Brain Res.* **42**:33.

72. Hendelman, W. J., and Aggerwal, A. S., 1980, The Purkinje neuron. I. A Golgi study of its development in the mouse and in culture, *J. Comp. Neurol.* **193**:1063.

73. Calvet, M.-C., Lepault, A.-M., and Calvet, J., 1976, A procion yellow study of cultured Purkinje cells, *Brain Res.* **111**:399.

74. Sotelo, C., and Arsenio-Nunes, M. L., 1976, Development of Purkinje cells in absence of climbing fibers, *Brain Res.* **111**:389.

75. Nagai, T., Satoh, K., Imamoto, K., and Maeda, T., 1981, Divergent projections of catecholamine neurons of the locus coeruleus as revealed by fluorescent retrograde double labeling technique, *Neurosci. Lett.* **23**:117.

76. Harkmark, W., 1954, Cell migrations from the rhombic lip to the inferior olive, the nucleus raphe and the pons: A morphological and experimental investigation of chick embryos, *J. Comp. Neurol.* **100**:115.

77. Altman, J., and Bayer, S. A., 1978, Prenatal development of the cerebellar system in the rat. II. Cytogenesis and histogenesis of the inferior olive, pontine gray, and the precerebellar reticular nuclei, *J. Comp. Neurol.* **179**:49.

78. Mares, V., and Lodin, Z., 1970, The cellular kinetics of the developing mouse cerebellum. II. The function of the external granular layer in the process of gyrification, *Brain Res.* **23**:343.

15

Transplantation of Newborn Brain Tissue into Adult Kainic-Acid-Lesioned Neostriatum

P. L. McGeer, H. Kimura, and E. G. McGeer

1. Introduction

Other chapters in this volume detail many of the studies that have so far been performed indicating that transplanted tissue can survive in host brain and that it is immunologically privileged, apparently because it lies behind the blood–brain barrier.

As methods do not presently exist for treating diseases such as Huntington's disease, which involve progressive degeneration of neurons, the possibility of transplantation of viable brain tissue is of potential long-term interest to the treatment of this and other neurological disorders. The reasons for selective neuronal degeneration in the vast majority of neurological disorders are completely unknown. In the case of Huntington's disease, however, where an autosomal dominant gene is responsible, there might be reason to hope that cells free of the genetic fault might not be vulnerable to the degenerative process even if transplanted into an affected host. It is important to determine the breadth of conditions that can be tolerated for survival of brain tissue transplants and the degree to which transplanted tissue can make functional connections within the host. This chapter describes some preliminary results with respect to the transplantation of newborn brain tissue into neurotoxin-damaged host brain.

Kainic acid (KA) injections into rat neostriatum will reproduce many of the biochemical conditions encountered in Huntington's disease (see Table I). We describe here the survival of newborn neostriatal transplants in KA-lesioned rats as indicated by his-

P. L. McGeer, H. Kimura, and E. G. McGeer • Kinsmen Laboratory of Neurological Research, Department of Psychiatry, University of British Columbia, Vancouver, British Columbia V6T 1W5, Canada

TABLE I

Biochemical Similarities between Huntington's Disease and the Kainic Acid "Model"

Biochemical changes	Huntington's disease	KA "model"	Ref.
In neostriatum			
Indices of indigenous neurons			
Presynaptic GABA indices[a]	Markedly decreased		2
Binding sites for GABA	Decreased[b]		2, 3
Binding sites for benzodiazepines	Decreased[b]		4, 5
Presynaptic acetylcholine indices[a]	Markedly decreased		2
Binding sites for QNB (muscarinic)	Decreased		2, 6
Enkephalin levels	Decreased		2, 7
Binding sites for dopamine	Decreased		2, 6, 8
Binding sites for serotonin	Decreased		2, 9
Binding sites for kainic acid	Decreased		10–13
Indices of afferent neurons			
Presynaptic dopamine indices[a]	Normal or elevated		2
Presynaptic serotonin indices[a]	Normal or elevated		2, 14
Presynaptic noradrenaline indices[a]	Normal or elevated		2
β-Adrenergic binding sites	Normal		15, 16
Unassigned indices			
Angiotensin-converting enzyme	Decreased		2
TRH levels			17, 18
Ganglioside levels	Increased	Decreased	19
	Decreased		
DNA levels	Increased		19
γ-Hydroxybutyrate levels	Increased		20, 21
GABA transaminase activity			2
Ornithine δ-aminotransferase	Normal	Decreased	22, 23
	Decreased		
Aspartate transaminase	Decreased		22
In substantia nigra			
Of striatonigral neurons			
Presynaptic GABA indices[a]	Decreased		2
Substance P levels	Decreased		2
Other or unassigned			
Presynaptic dopamine indices[a]	Normal		2
GABA-binding sites	Increased		24, 25
Cholecystokinin levels	Decreased		26

[a] Including levels, uptake, release, turnover, and/or activity of synthetic enzymes; uptake in the case of cholinergic neurons is uptake of choline.
[b] At > 1 month following KA injections; no decrease in acute preparations.

tological staining and metabolic measurements with deoxyglucose. There was evidence of the migration of some transplanted neurons into the host neostriatal matrix. The model also demonstrated some potential problems associated with transplantation.

Injection of 5 nmoles of KA into the striatum of a Sprague–Dawley rat weighing approximately 300 g creates an area of neuronal destruction approximately 2.5 mm in diameter.[1] Figures 1–4 illustrate the effect, with cresyl violet staining, where normal (Figs. 1,2) and lesioned (Figs. 3,4) striatal tissue are compared. Figure 3 shows how neuronal tissue is replaced by glial tissue, just as in Huntington's disease. The reported biochemical changes following such lesions are summarized in Table I, compared with changes that have been noted in various studies on Huntington's disease patients. In both Huntington's disease and striatum-lesioned animals, indices associated with neurons that are indigenous to the striatum are decreased. Examples include glutamic acid decarboxylase (GAD), the enzyme associated with the synthesis of GABA, choline acetyltransferase (ChAT), the enzyme associated with the synthesis of acetylcholine, as well as postsynaptic binding sites for transmitters released in the striatum. Thus, [^{3}H]-QNB binding, which is reflective of cholinergic binding sites, [^{3}H]-GABA binding, which is reflective of GABA-binding sites, and [^{3}H]spiroperidol binding, which is associated with dopamine-binding sites, are decreased. On the other hand, indices associated with nerve endings of neuronal somata located outside the striatum are relatively unaffected. Thus, there are no such sharp decreases either in tyrosine hydroxylase, the enzyme that is associated with dopamine synthesis and that originates in neurons of the substantia nigra, or in high-affinity glutamate uptake, which is associated with glutamate nerve endings from the cortex. In the latter case, data are not available from Huntington's disease cases because high-affinity synaptosomal uptake cannot be measured on human postmortem material due to rapid postmortem deterioration. In the substantia nigra, where indigenous neurons are unaffected in both Huntington's disease and animals given a striatal injection of KA, the biochemical changes are associated with descending pathways to the nigra.

2. Methods

A series of rats were prepared with such lesions. Approximately 3 weeks later, half the animals were given a transplant of neonatal rat striatal tissue from the same species. Neonatal rats, less than 6 hr of age, were decapitated. The brain was quickly exposed and the jelly-like striatal tissue was identified visually. Tissue from approximately 10 neonatal rats was pooled in a sterile hypodermic syringe for transplantation to 8 rats consecutively. The pooled tissue was passed through a 26-gauge needle into the head of a second syringe to break up the clumps and produce a homogenized cell mass suitable for injection. Under stereotaxic guidance, the hypodermic syringe was lowered through the burrhole in the rat's skull previously produced during the KA lesion to a depth equal to the midpoint of the lesioned area. By depressing the syringe with a micrometer drive, an appropriate amount of tissue was deposited into the lesioned area. Two revolutions of the micrometer deposited approximately 100 μl of neonatal, neostriatal tissue into the adult host's brain. The injection time was approximately 2 min and the hypodermic needle was left in for a further 5 min to allow diffusion of the injected tissue. The needle then was withdrawn and the next rat injected until the series was completed. The rats showed no ill-effects from the surgical procedure.

The transplanted rats were sacrificed after varying periods of time. The three principal periods chosen were 3, 6, and 8 weeks. Different methods of sacrifice were employed depending on the type of test to be carried out on the transplanted brain tissue. Ten rats were sacrificed by perfusion through the heart with a 4% paraformaldehyde–1% glutaraldehyde mixture and prepared for cutting on a cryostat at 20-μm thickness. The cryostat sections were stained with cresyl violet by standard techniques.

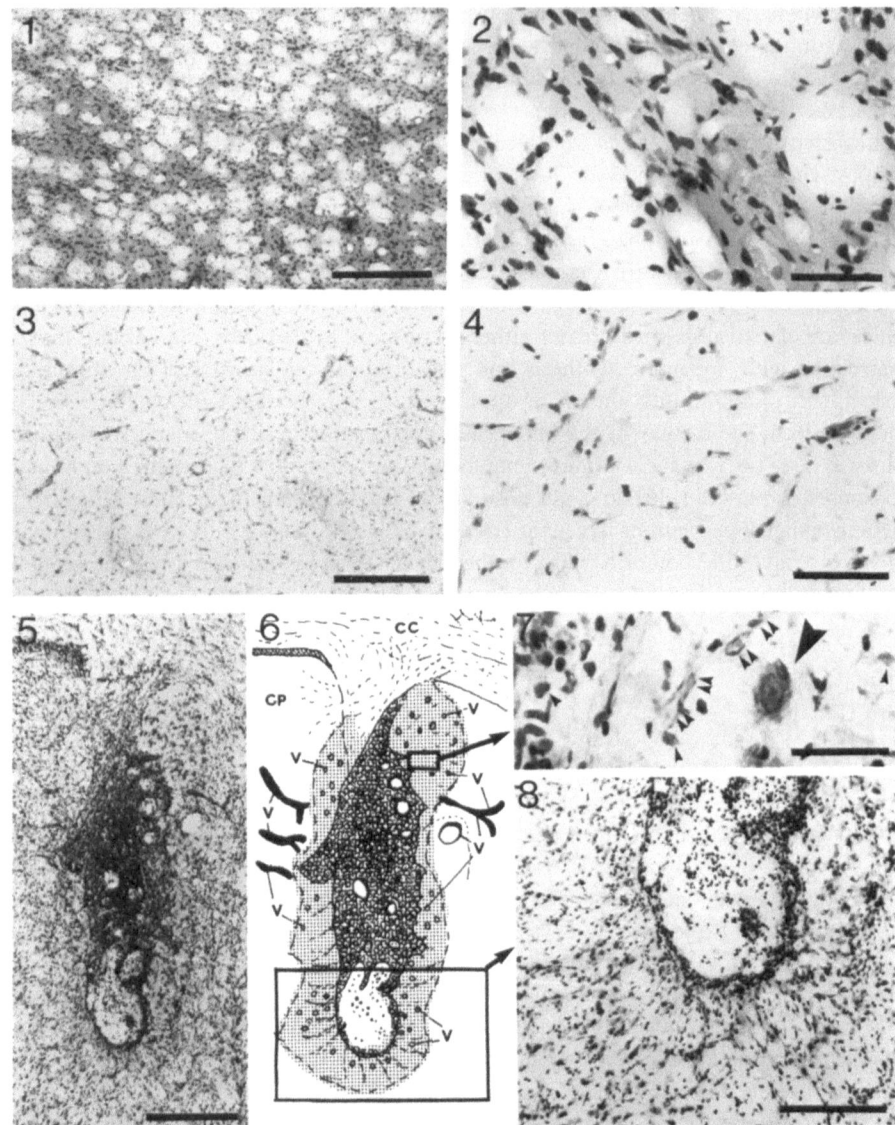

Figures 1–8. Histological examination of rat striata with cresyl violet staining.

Figure 1. Low-power magnification of the normal striatum. Bar = 500 μm.

Figure 2. High-power magnification of Fig. 1. Mostly medium-size cells (12–15 μm) are seen in the gray matter surrounding fiber bundles. Bar = 50 μm.

Figure 3. Low-power magnification of kainic acid-lesioned striatum. (Three months after the kainic acid injection.) Bar = 500 μm.

Figure 4. High-power magnification of Fig. 3. No neuronal somata are seen, while many glial cells are distributed. The arrangement of fiber bundles is indistinct due to the disappearance of neuronal cell bodies and their dendrites. Bar = 50 μm.

Figure 5. Striatum following kainic acid lesioning and transplantation (see text). In the center of the field, a densely staining column of transplanted neuronal tissue is seen. Bar = 500 μm.

Four rats were sacrificed by perfusing through the heart with 250 ml of chilled glyoxylic acid solution. Cross-sectional slabs approximately 5 mm thick were frozen under liquid propane chilled with liquid nitrogen and freeze-dried at $-43°C$ for 6 days. The tissue slabs were suspended in a jar of paraformaldehyde (40% humidity) and heated to 80°C for 1 hr. Following paraformaldehyde treatment, the slabs were infiltrated with paraffin under low vacuum. The slabs were then cut at 14-μm thickness for observation of catecholamine fluorescence.

A number of rats were implanted with a catheter into the femoral vein 1 day before sacrifice. Forty-five minutes before sacrifice, each rat was injected with 100 μCi/kg [U-^{14}C]deoxyglucose (282 mCi/mmole, Amersham) in saline. The rats were sacrificed by decapitation, the brains quickly removed, frozen, and 20-μm sections cut through the striatum on a microtome. The sections were mounted on chrome-alum-coated glass slides and placed on a hot plate (45°C) for 20 min. The dried tissues, still on the slide, were placed against X-ray film (X-mat R) for a period of 2 weeks.[27] The film was then developed with a dental X-ray film developer, and the slides were stained with cresyl violet.

3. Results

Six of the ten rats sacrificed for cresyl violet histochemistry had successful transplants while the remaining four had cavities in the striatal area, probably indicating an infection either from the KA lesion or, more probably, from the transplant and emphasizing the necessity for careful surgical precautions in experiments of this kind.

The appearance of transplanted tissue with cresyl violet staining is illustrated in Figures 5–8. A clear distinction between the area destroyed by KA prior to the transplant and the transplanted area can be seen in Fig. 5. The neuronal tissue can readily be identified by the presence of Nissl bodies in the cytoplasm of the large cells (Fig. 6). The neurons demonstrate their migratory tendencies by virtue of dendrites extending into the KA-lesioned field, which is characterized by an absence of neurons and the presence of only glial remnants. The transplanted neurons can even be distinguished from endogenous neurons outside the field because of their somewhat lighter-staining cytoplasm and their apparent migratory orientation. Some transplanted neurons can be detected a considerable distance from the injection site, obviously indicating a considerable capacity to undergo migration. Six weeks following injection, the migration can appear as great as 1 mm. Such long-migrating neurons are irregularly scattered throughout the lesioned field. The size of these neurons appears to be, on the average,

---

FIGURE 6. Schematic drawing illustrating special features of the neuronal region (dotted area) including the arrangement of reconstructed blood vessels. cc, corpus callosum; cp, striatum; v, blood vessels. Areas in the diagram enclosed in boxes are shown enlarged in the figures to the right.

FIGURE 7. A migrating neuron of large size (approx. 35 μm) is seen (large arrowhead). Beside the large neuron, capillaries (double small arrowheads) are situated. Lightly stained monocytes are also scattered in the field (single small arrowheads). Bar = 50 μm.

FIGURE 8. Neuronal cells are seen radiating to the periphery from the transplanted neuronal column. Bar = 200 μm.

somewhat larger than comparable neurons in unlesioned striata. Alternatively, it could be that the neurons with the greatest tendency to migrate are those that are of the large category (which make up only 1–2% of the population of the striatum), as opposed to the great majority of striatal cells, which are of medium to small size.

A prominent feature of the transplanted tissue is its tendency to stimulate revascularization. Following a simple KA lesion, the vascularity appears to atrophy. However, the transplants apparently stimulate endothelial tissue to rebuild the blood supply into the striatum. Numerous new capillaries can easily be detected in the migratory neuronal area (Figs. 5, 6). Under high power, neurons can be seen organizing themselves in the area close to the new vascular tissue. The vascularization is somewhat irregular, and it was not possible for us to determine whether the large migrating neurons carry with them new capillary tissues as they move.

The question of an immunological response to the transplanted tissue needs to be addressed. It should be noted that we have always observed monocytes in the transplanted area, but never in the KA-lesioned area. The significance of the monocyte infiltration, which is of modest proportions, is unknown but it may indicate an immunological response, either to infection or to neurons or glia. The size of the monocytes (15–20 μm) is indicated in Fig. 7.

The most noteworthy development in the different time periods after transplantation is that the migration at later times is substantially more than during the early times. Otherwise, the transplants tend to be similar in histological appearance.

Catecholamine fluorescence of KA-lesioned striata shows no change from normal beyond a possible slight decrease in intensity. Biochemical results have consistently shown, however, that such KA lesioning does not cause a decrease in either catecholamine or tyrosine hydroxylase levels,[2] at least until many months after the KA injection.[28] Thus, the nerve endings of the nigrostriatal dopaminergic neuronal system are capable of surviving for some time in the absence of postsynaptic target neurons. Little lipofuscin-like material can be detected in the KA-lesioned striatum (Fig. 9).

Following transplantation, a number of changes can be noted in the catecholaminergic fluorescence. The nerve endings organize themselves into rosettes or glomeruli around transplanted neurons. Lipofuscin-like material is noted irregularly scattered throughout the transplanted field, but does not extend into the KA-lesioned areas (Fig. 10). In Fig. 10, the lipofuscin-like particles are indicated with arrows, but under the fluorescence microscope, they can easily be distinguished from the catecholamines because the lipofuscin-like fluorescence is a bright yellow instead of green. When the glyoxylic acid-treated sections are subsequently stained with cresyl violet, the endothelium can be visualized. Thus, the rosettes are forming around neurons that have migrated away from the main transplanted area. If the main transplanted area itself is examined, fibers and terminals positively staining for catecholamines can be seen invading the area. Therefore, it is evident that existing fibers can expand and cluster around migrating cells, as well as penetrate the transplanted mass itself.

Figures 11–14 offer a comparison of deoxyglucose (DG) studies in a rat with a unilateral injection of KA in the left striatum (Fig. 11) with one similarly lesioned but later given a transplantation of neonatal striatal tissue into the lesioned area (Fig. 12). The animals shown in Figs. 11 and 12 were sacrificed 3 weeks after the transplant (6

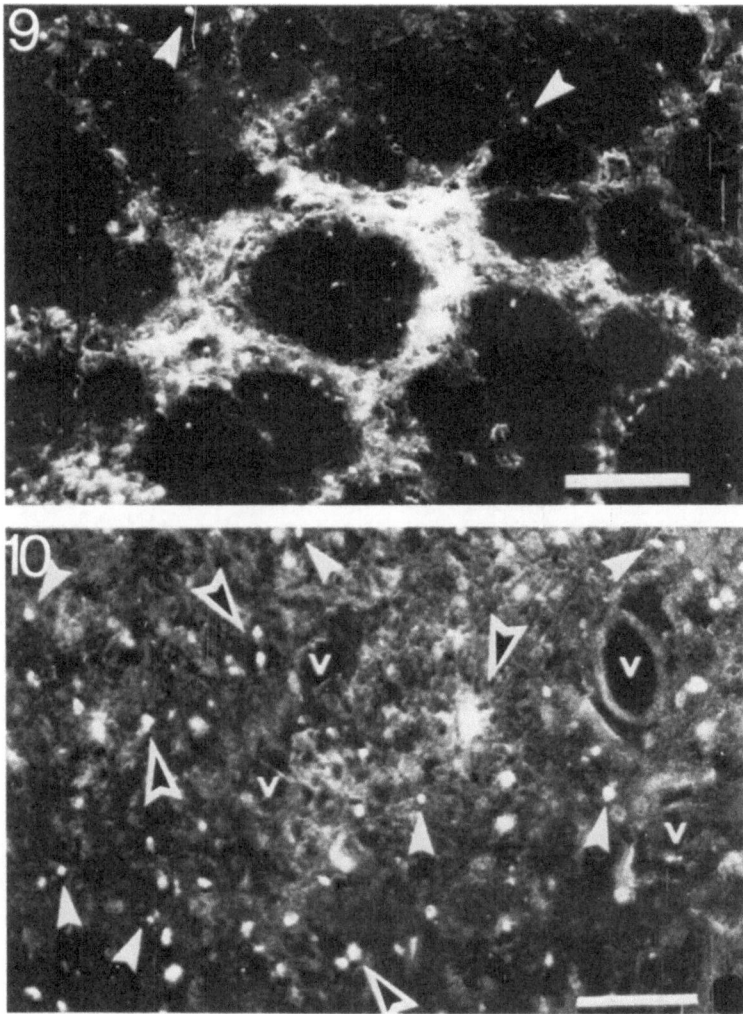

FIGURES 9 AND 10. Catecholamine fluorescence histochemistry in kainic acid-lesioned rat striatum without (Fig. 9) and with (Fig. 10) a transplant. The dopaminergic fluorescent fibers appear to be intact in Fig. 9, while they distribute irregularly through the transplanted tissue (Fig. 10). In addition, dopaminergic fluorescent rosettes or granules (large arrowheads with black insets) are only seen in Fig. 10. Lipofuscin-like fluorescent granules (small arrowheads) are significantly more abundant in Fig. 10 than in Fig. 9. v, blood vessels.

weeks after the KA injections). A high-power field of another transplanted rat is shown in Figs. 13 and 14. Figure 13 is the DG autoradiograph and Fig. 14 is the same area at still higher power stained by cresyl violet, showing the higher concentration of neurons in the metabolically active transplanted area, compared with the neuron-poor, KA-lesioned surround. As might be anticipated and as previously reported,[29] KA-lesioned

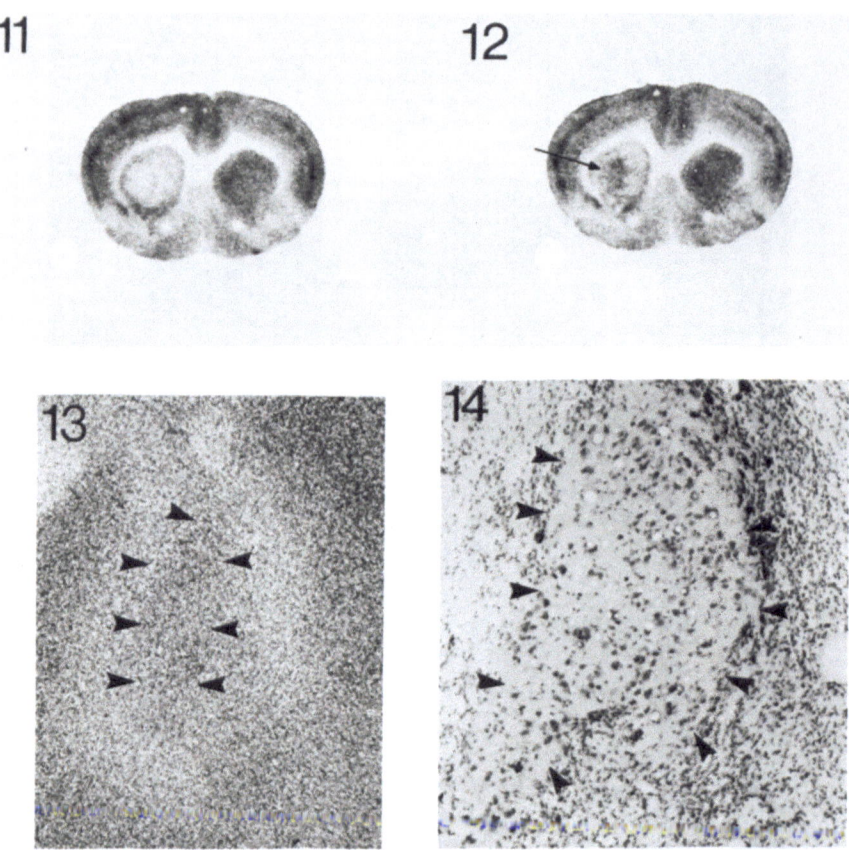

Figures 11–14. [^{14}C]-Deoxyglucose (DG) autoradiography.

Figure 11. Low power showing slice from a rat lesioned with kainic acid in the left striatum. Note its paleness compared to the normal, right striatum.

Figure 12. Similar view from a rat having a kainic acid lesion followed by a transplant. Note the relatively dark island of transplanted tissue (arrow) within the pale lesioned area.

Figure 13. Higher power of the striatum from another lesioned and transplanted rat. The transplanted tissue (outlined by arrowheads) shows a darker color than its immediate surround, indicating higher DG utilization.

Figure 14. Still higher-power cresyl violet stain of the area shown in Fig. 13. The transplanted tissue (arrowheads) is surrounded by the neuron-poor lesioned area.

areas, being virtually devoid of neurons, have a greatly decreased utilization of DG as compared to the normal, contralateral striatum (Fig. 11). In transplanted animals, the transplanted tissues can be detected as darkened areas within the clearly demarcated boundaries of the KA lesions (Figs. 12 and 13, arrows). The clear demonstration of a metabolically active zone in the center of a KA-treated field has never been seen in DG studies on a large number of rats that received no transplant, but is always observed in successfully transplanted animals.

4. Discussion

These results are a further demonstration that immature brain tissue can be successfully transplanted into damaged or deficient adult host brain. In these experiments, neostriatal tissue taken from newborn rats, as opposed to embryos, was transplanted directly into the damaged area. The immature neuronal tissue somehow stimulated the host brain to provide revascularization. This capacity to induce a revascularization is probably essential to the survival of the transplanted tissue. It is noteworthy that a simple KA lesion, which destroys neurons but nevertheless stimulates protein synthesis, presumably as a result of glial response,[30] induces no stimulation of revascularization even after long-term survival. It is uncertain from the specimens we have examined to date whether the large neurons that migrate out from the area of the transplant into the area depopulated of neurons stimulate vascularization to follow them.

The reorganization of the catecholamine system of the host to embrace the transplanted neurons with rosette-like collections of nerve endings would suggest that some functional connections have at least been commenced. The significance of the monocyte infiltration is not known, although it could signify an ultimate rejection of the transplant. Further work will be required to decide this point.

Schmidt et al.[31] have published a picture similar to that in Fig. 5 showing, by cresyl violet staining, the appearance of a core of neurons within the lesioned area in rats that had received a striatal injection of KA followed 5 days later by transplantation of embryonic striatal cells. In animals sacrificed 7 weeks after transplantation, they obtained biochemical evidence that these transplanted cells possessed considerable ChAT and near-normal GAD activity. The one discrepancy between their results and ours is that, in catecholamine histofluorescence studies, they found no evidence of appreciable penetration of dopamine fibers into the transplanted mass, although the surrounding gliotic, neuron-poor region was richly supplied with fluorescent dopamine terminals. The difference may be due to such factors as the varying time sequences employed or the use of embryonic as opposed to neonatal cells.

The neostriatum is a complex structure with considerable internal organization. The degree to which the transplanted tissue can organize and develop useful connections with host tissues is critical to determining whether this approach is of potential usefulness to such human disorders as Huntington's disease or whether it is merely an interesting example of biological symbiosis.

ACKNOWLEDGMENTS. We thank E. Singh, U. Scherer-Singler, P. Robertson, R. Gelman and J. Suzuki for excellent technical assistance and gratefully acknowledge the financial support of the Garfield Weston Foundation, the Huntington Disease Society of Canada, and the Medical Research Council of Canada.

References

1. McGeer, P. L., McGeer, E. G., and Hattori, T., 1978, Kainic acid as a tool in neurobiology, in: *Kainic Acid as a Tool in Neurobiology* (E. G. McGeer, J. W. Olney, and P. L. McGeer, eds.), pp. 123–138, Raven Press, New York.
2. Coyle, J. T., McGeer, E. G., McGeer, P. L., and Schwarcz, R., 1978, Neostriatal injections: A model for Huntington's chorea, in: *Kainic Acid as a Tool in Neurobiology* (E. G. McGeer, J. W. Olney, and P. L. McGeer, eds.), pp. 139–160, Raven Press, New York.
3. Zaczek, R., Schwarcz, R., and Coyle, J. T., 1978, Long-term sequelae of striatal kainate lesion, *Brain Res.* **152:**626.
4. Sperk, G., and Schlogl, E., 1979, Reduction of number of benzodiazepine binding sites in the caudate nucleus of the rat after kainic acid injections, *Brain Res.* **170:**563.
5. Mohler, H., and Okada, T., 1978, The benzodiazepine receptor in normal and pathological human brain, *Br. J. Psychiatry* **133:** 261.
6. Fields, J. Z., Reisine, T. D., and Yamamura, H. I., 1978, Loss of striatal dopaminergic receptors after intrastriatal kainic acid, *Life Sci.* **23:**569.
7. Schwarcz, R., Fuxe, K., Hokfelt, T., Terenius, L., and Goldstein, M., 1980, Effects of chronic striatal kainate lesions on some dopaminergic parameters and enkephalin immunoreactive neurons in the basal ganglia, *J. Neurochem.* **34:**772.
8. Weinreich, P., and Seeman, P., 1980, Effect of kainic acid on striatal dopamine receptors, *Brain Res.* **198:**491.
9. Fillion, G., Beaudoin, D., Rousselle, J. C., Deniau, J. M., Fillion, M. P., Dray, F., and Jacob, J., 1979, Decrease of [^{3}H]-5-HT high affinity binding and 5-HT adenylate cyclase activation after kainic acid lesion in rat brain striatum, *J. Neurochem.* **33:**567.
10. Vincent, S. R., and McGeer, E. G., 1979, Kainic acid binding to membranes of striatal neurons, *Life Sci.* **24:**265.
11. Henke, H., 1979, Kainic acid binding in human caudate nucleus: Effect of Huntington's disease, *Neurosci. Lett.* **14:**247.
12. Beaumont, K., Maurin, Y., Reisine, T. D., Fields, J. Z., Spokes, E., Bird, E. D., and Yamamura, H. I., 1979, Huntington's disease and its animal model: Alterations in kainic acid binding, *Life Sci.* **24:**809.
13. London, E. D., Yamamura, H. I., Bird, E. D., and Coyle, J. T., 1981, Decreased receptor-binding sites for kainic acid in brains of patients with Huntington's disease, *Biol. Psychiatry* **16:**155.
14. Neckers, L. M., Neff, N. H., and Wyatt, R. J., 1979, Increased serotonin turnover in corpus striatum following an injection of kainic acid: Evidence for neuronal feedback regulation of synthesis, *Naunyn-Schmiedebergs Arch. Pharmacol.* **306:**173.
15. Zahniser, N. R., Minneman, K. P., and Molinoff, P. B., 1979, Persistence of β-adrenergic receptors in rat striatum following kainic acid administration, *Brain Res.* **178:**589.
16. Enna, S. J., Bird, E. D., Bennett, J. P., Bylund, D. B., Yamamura, H. I., Iversen, L. L., and Snyder, S., 1976, Huntington's chorea: Changes in neurotransmitter receptors in the brain, *N. Engl. J. Med.* **294:**1305.
17. Spindel, E. R., Wurtman, R. J., and Bird, E. D., 1980, Increased TRH content of the basal ganglia in Huntington's disease, *N. Engl. J. Med.* **303:**1235.
18. Spindel, E. R., Pettibone, D. J., and Wurtman, R. J., 1981, Thyrotropin-releasing hormones (TRH) content of rat striatum: Modification by drugs and lesions, *Brain Res.* **216:**323.
19. Higatsberger, M. R., Sperk, G., Bernheimer, H., Shannak, K. S., and Hornykiewicz, O., 1981, Striatal ganglioside levels in the rat following kainic acid lesions: Comparison with Huntington's disease, *Exp. Brain Res.* **44:**93.
20. Ando, N., Gold, B. I., Bird, E. D., and Roth, R. H., 1979, Regional brain levels of γ-hydroxybutyrate in Huntington's disease, *J. Neurochem.* **32:**617.
21. Ando, N., Simon, J. R., and Roth, R. H., 1979, Inverse relationship between GABA and γ-hydroxybutyrate levels in striatum of rat injected with kainic acid, *J. Neurochem.* **32:**623.
22. Wong, P. T.-H., McGeer, E. G., and McGeer, P. L., 1982, Effects of kainic acid injection and cortical lesion on ornithine and aspartate aminotransferases in rat striatum, *J. Neurosci. Res.* **8:** 643–650.

23. Wong, P., T.-H., McGeer, P. L., Rossor, M., and McGeer, E. G., 1982, Ornithine aminotransferase in Huntington's disease, *Brain Res.* **231**:466.
24. Cross, A. J., and Waddington, J. L., 1981, Substantia nigra γ-aminobutyric acid receptors in Huntington's disease, *J. Neurochem.* **37**:321.
25. Waddington, J. L., and Cross, A. J., 1980, The striatonigral GABA pathway: Functional and neurochemical characteristics in rats with unilateral striatal kainic acid lesions, *Eur. J. Pharmacol.* **67**:27.
26. Emson, D. C., Rehfeld, J. F., Langvin, H., and Rossor, M., 1980, Reduction in cholecystokinin-like immunoreactivity in the basal ganglia in Huntington's disease, *Brain Res.* **198**:497.
27. Sokoloff, L., Reivich, M., Kennedy, C., Des Rosiers, M. H., Patlak, C. S., Pettigrew, K. D., Sakurada, O., and Shinohara, M., 1977, The [^{14}C]deoxyglucose method for the measurement of local cerebral glucose utilization: Theory, procedure and normal values in the conscious and anesthetized albino rat, *J. Neurochem.* **28**:897.
28. McGeer, E. G., McGeer, P. L., Hattori, T., and Vincent, S. R., 1979, Kainic acid neurotoxicity and Huntington's disease, *Adv. Neurol.* **23**:577.
29. Kimura, H., McGeer, E. G., and McGeer, P. L., 1980, Metabolic alterations in an animal model of Huntington's disease using the ^{14}C-deoxyglucose method, *J. Neurol. Transmiss. Suppl.* **16**: 103.
30. Jacubovic, A., Lin, D., and McGeer, E. G., 1979, Protein and RNA synthesis in kainic acid injected striata, *Brain Res.* **163**:289.
31. Schmidt, R. H., Björklund, A., and Stenevi, U., 1981, Intracerebral grafting of dissociated CNS tissue suspensions: A new approach for neuronal transplantation to deep brain sites, *Brain Res.* **218**:347.

16
Transplantation of Catecholamine-Containing Tissues to Restore the Functional Capacity of the Damaged Nigrostriatal System

WILLIAM J. FREED, BARRY J. HOFFER, LARS OLSON, AND RICHARD JED WYATT

1. Introduction

One of the ultimate goals of neurobiological research is to develop methods of reconstituting functional capacity following damage to the CNS. Most past efforts have approached this problem with attempts to enhance the function of the remaining, undamaged tissues by providing tropic influences or cellular "drive."[1] Drugs, chemical factors, or behavioral and electrical stimulation have been used. Such treatments might potentially stimulate the functional activity of spared tissues or increase the regrowth of fiber tracts.[2,3] For example, there have been encouraging results in reversing the effects of striatal lesions with the intracerebral administration of nerve growth factor.[4]

The success of these studies, however, appears limited, because such manipulations can at best only stimulate the growth of fibers from intact cells. When neurons are lost through retrograde degeneration or through damage directly to the parenchyma, tropic influences can only succeed if the cells can be made to proliferate. Unless and until this can be accomplished, providing the brain with new tissues may be the only ultimately

WILLIAM J. FREED AND RICHARD JED WYATT ● Preclinical Neurosciences Section, Adult Psychiatry Branch, National Institute of Mental Health, Saint Elizabeths Hospital, Washington, D.C. 20032.   BARRY J. HOFFER ● Department of Pharmacology, University of Colorado Medical Center, Denver, Colorado 80262.   LARS OLSON ● Department of Histology, Karolinska Institute, Stockholm, Sweden.

viable method of restoring function following damage to brain cell parenchyma. And, in fact, during the past few years there have been reports that some effects of brain lesions can be reversed by transplantation of new tissues to the damaged host brain.

Transplantation of brain tissue is, in many respects, a process entirely different from transplantation of peripheral organs. First, in peripheral organ transplantation, the primary obstacle is tissue rejection. The brain, however, is a "privileged site," where graft rejection is unlikely to occur (see Section 2), and is not a major obstacle. The primary obstacles to functionally significant brain grafts are (1) the complex interconnections of the transplanted cells with other neurons, which make the cells difficult to transplant without disturbing their physical integrity and which are not readily restored after transplantation, and (2) maintenance of nutritional requirements of grafted cells in their new environment.

Neurons are not readily removed from the brain without disturbing their processes. Stripping of axonal and dendritic processes from a fully differentiated neuron is almost invariably lethal. Thus, only neurons that have not yet developed processes can be transplanted. Processes are absent or insignificant for only the first few days after histogenesis. Undifferentiated neurons not only survive, but also retain the capacity to grow processes.[5,6] In contrast, a fully differentiated CNS neuron has, for the most part, exhausted its capacity to develop new functional connections.[7-9] The typical neuron, however, has many complex interconnections, which under normal circumstances develop early in ontogeny while other neurons are also developing their connections. A single brain nucleus, transplanted in isolation to a fully differentiated host brain, cannot be expected to become fully integrated with the host.

Simple survival of brain grafts is almost always achieved. Rather, the major challenge of intracerebral neural transplantation is to develop techniques through which neural grafts can be integrated into the structure of the host brain. Our approach to this problem has been twofold: first, to make use of a system where the connections are fairly simple, in order to characterize and understand the techniques and principles that are involved, and second, to employ this system to develop means of increasing the functional integration between graft and host brain.

It has now become apparent that small groups of only a few thousand neurons can have very significant functional consequences. One nucleus that has very great functional significance is the substantia nigra (SN), the source of much of the brain's dopamine. This nucleus contains about 3500 neurons,[10] which are the primary sources of the neurotransmitter dopamine in the brain. Loss of these cells in humans produces the clinical syndrome of Parkinson's disease, and their experimental destruction in rats produces a loss of eating, drinking, grooming, general motor behavior, and pervasive debilitation. The terminals of the dopamine-producing neurons are primarily located in the striatum, nucleus accumbens, and olfactory tubercule, and the dopaminergic pathways of the brain have been precisely mapped.[11-13]

We have, therefore, chosen the dopamine system for the development of brain grafting methodology. Our initial studies of brain grafts involved the dopamine-producing cells of the SN. These cells give rise to an ipsilateral projection to the corpus striatum and other areas of the forebrain. Denervation of this system gives rise to a well-characterized and easily quantifiable behavioral syndrome. Moreover, the nigrostriatal and mesolimbic dopaminergic systems are definitely or hypothetically involved in aging and a number of clinical disorders, including Parkinson's disease, schizophrenia, Huntington's disease, Lesch–Nyhan syndrome, and Gilles de la Tourette syndrome.[14–20] Thus, experimental application of brain grafting to the nigrostriatal system has the potential for clinical application, in addition to the possibility of yielding improvements in the understanding of the processes involved in these disorders.

2. Methods

2.1. Terminology

The scientific literature on transplantation has a terminology that may not be familiar to neuroscientists. It will be useful to summarize this terminology here.

Grafts are most often characterized in terms of their genetic relationship between donor and host. Thus, grafts within an individual have the greatest possible degree of similarity, and are called *autografts*. Grafts between different individuals that are genetically identical, i.e., between *syngeneic* or *isogeneic* individuals, are called *syngrafts* or *isografts*. Autografts and syn- or isografts are usually not rejected.

Grafts made between different individuals of the same species that are presumed, but not specifically known, to be genetically dissimilar are called *homografts*. But, when donor and recipient are from the same species and are definitely known to be genetically dissimilar, the grafts are termed *allografts*. When donor and recipient are of different species, the grafts are called *heterografts*, and when donor and graft species are widely disparate, such as of different orders, the grafts are called *xenografts*.

Under normal circumstances, homografts, allografts, heterografts, and xenografts will be rejected, while autografts and syn- or isografts survive transplantation on a long-term basis. There are, however, certain locations in the body where allografts and even xenografts may not be rejected. These locations, e.g., the anterior chamber of the eye, the hamster cheek pouch, and the brain, are endowed with *immunological privilege,* and are termed *privileged sites*.

Grafts are also sometimes classified in terms of location. A graft made to its normal bodily location is called a *homotopic* graft, while a graft placed in a location other than where it is normally found is a *heterotopic* graft. The terms *autoplastic, isoplastic, homoplastic, alloplastic, heteroplastic,* and *xenoplastic* are also sometimes used and refer to the genetic properties of the grafts. The default term homograft, for example, means *homoplastic*, not *homotopic*.

2.2. Dissection and Tissue Preparation

2.2.1. Fetal Brain Grafts

Dopamine-containing neurons are removed from the ventral mesencephalon essentially as described by Olson and Seiger[21] and Seiger and Olson.[5] In our studies, donors have typically been 17-day gestational rat embryos of the Sprague–Dawley strain. Crown–rump length (CRL) for these animals should be from 19 to 20 mm. The donors are deeply anesthetized with ether and both horns of the uterus are removed by cesarian section, using sterile instruments and materials. Clean, but not sterile procedures are used throughout. The intact uterus is rinsed under cool running tap water for about 30 sec and placed into a sterile petri dish. As a general convention, 2 hr is allowed for utilization of the tissue after sacrifice of the donor, although in practice the tissue is generally used within 1 hr. Each fetus is then removed and placed in a separate petri dish in sterile lactated Ringer's solution. The scalp is peeled back and the skull removed. The brain is removed and placed in a clean petri dish. The brain is arranged with the dorsal surface downward and manipulated to lie straight with the vertex of the mesencephalic flexure visible. With a pair of microdissecting scissors, a small wedge-shaped piece is removed from the vertex of the mesencephalic flexure. This piece includes the ventral surface of the brain and extends at least halfway to the cerebral aqueduct. It is about 0.5 mm thick in the rostral–caudal direction, about 2 mm wide after removing the lateral edges, and about 1 mm deep in the ventral–dorsal dimension. After removal, the piece is divided midsagittally, creating two pieces each containing more-or-less one individual SN. The maximum time allowed for use of the tissue after dissection is 30 min, although the tissue is usually used within 5 min. Other tissues, e.g., tectum or frontal cortex, are easily identified and can be removed as slabs and cut into appropriate-sized pieces.

2.2.2. Adult Adrenal Medulla Grafts

We have employed the adrenal medulla from animals weighing between 75 and 950 g. The donors are deeply anesthetized with ether or ketamine. The chest cavity and abdomen are opened, the aorta severed, and both adrenal glands are removed by grasping them with sterile curved forceps. Care is taken to avoid decapsulating the glands at this point. The whole adrenal glands are then placed into a sterile petri dish containing sterile lactated Ringer's solution. Under the dissecting microscope, a single longitudinal cut is made through the adrenal capsule and cortex with a pair of small curved dissecting scissors. About one-third of the width of the gland is removed in this first cut. The adrenal medulla should then be visible, and is removed from the adrenal cortex by cutting around its borders with micro dissecting scissors. It is important to avoid including adrenal cortex with the grafts, because adrenal corticosteriods are thought to play a role in maintaining the chromaffin cell phenotype. In grafting studies, we are usually interested in the neuronal phenotype that is epxressed only in the absence of corticosteriods.[22–27]

In our early experiments, we made a cut into the surface of the adrenal medulla, spread the edges of the cut, and removed small pieces from the interior of the medulla. More recently, we have been able to obtain preparations relatively free of adrenal cortex simply by carefully trimming the cortex away from the medulla. In our experiments, each adrenal medulla is cut into three to six pieces prior to transplantation. A maximum of 1 hr is allowed for use of the tissue after sacrifice of the donor.

2.2.3. Dissociated Cells

If individual dissociated cells could be transplanted, it would become possible to inject a few cells through small-gauge needles into the brain. If successful, this would allow for transplantation into multiple sites, resulting in a wider anatomical distribution of the grafted tissue. This would be an important advantage when attempting to transplant tissue for the purpose of influencing a large structure, such as the striatum of higher order animals, with functional consequences. In addition, grafts produced in this manner might produce less tissue damage and become better integrated with the host brain.

Björklund et al.[28] have developed a technique of this type, which was successful for 15-day (13–14 mm CRL) rat embryos. Older (19–20 mm CRL, or 17 day gestation) tissue generally yielded poor results, presumably because of the severe disruption of the cell processes caused by the dissociation procedure. Their procedure involved pooling of 9–11 pieces of brain tissue followed by incubation for 20 min in 200 μl of sterile saline containing 0.1% trypsin at 37°C. The trypsin was removed by several washings in saline at room temperature, and the cells were suspended by trituration with a fire-polished Pasteur pipette. The suspended tissue was used for as long as 90 min after preparation. Stereotaxic injections were made at a rate of 5 μl/1–2 min through beveled glass or stainless steel needles.

Perlow et al.[29] have transplanted dissociated, but subsequently cultured and reaggregated, bovine adrenal chromaffin cells to the rat ventricular system. The authors reported that these grafts survived. In our laboratory (A. Fine, unpublished), we have attempted to transplant dissociated adrenal chromaffin cells to the rat brain, with a limited degree of success. The adrenal medulla was dissociated by enzymatic treatment using hyaluronidase, DNase, and collagenase, followed by trituration. Histochemical fluorescence studies showed that yields of surviving cells were quite low, usually less than 1% of the tissue implanted. It is probable that survival of these cells can be substantially increased by technical improvements.

2.3. Implantation Techniques

Three categories of intracerebral graft implantation techniques have been employed in the nigrostriatal system. These include implantation into the cerebral ventricles, the brain parenchyma, and previously prepared cavities in the cerebral cortex.

2.3.1. Intraventricular Grafts

It is now generally conceded that the ventricular system is a favorable site for tissue transplantation. This was not always the case. In fact, early investigators who recognized that the brain was a privileged site for transplantation felt that immunological privilege did not extend to the ventricles. Murphy and Sturm,[30] who were the first to report favored homograft survival in the brain, reported that grafts contacting the ependymal lining of the ventricles did not survive. Similar conclusions have been expressed by Das et al.[31] Rosenstein and Brightman,[32] however, reported excellent survival of homografts in the ventricular system. We also have consistently obtained survival of grafts of many types of tissue in the lateral ventricles, including fetal SN,[33,34] other fetal brain tissues, adult adrenal medulla,[35] and kidney cortex pieces.[36] In virtually 100% of animals receiving grafts, surviving graft tissue is found when the animals are sacrificed 1 to 18 months after implantation of the tissues (Table I). Even allografts of fetal SN or adult adrenal medulla between inbred rat strains with differing major histocompatibility antigens survived consistently in the ventricles.[37]

The lateral ventricles are immediately adjacent to the striatum, nucleus accumbens, and lateral septum and are therefore an ideal location for transplantation of tissues intended to restore dopaminergic innervation to these structures. Additionally, grafts placed in the ventricles probably receive nourishment via the CSF immediately after implantation. The ventricular system also may provide a vehicle for conveying hormonal and chemical substances secreted by grafts to other periventricular structures.

2.3.2. Intraparenchymal Grafts

There have been many reports that homografts survive well when placed entirely within the brain parenchyma, not contacting the ventricular ependyma.[30,31,38,39] In a few animals, we discovered that parts of grafts aimed at the ventricles had become well integrated into the host brain tissue, although they were generally small as compared with ventricular grafts, containing 20–50 cells. This has prompted us to study the effects of both SN and adrenal medulla grafts intentionally placed into the substance of the striatum. Larger grafts placed by stereotaxic injection into the striatum, through relatively large-bore (19–21 gauge) needles, present at least two difficulties. First, a significant area of tissue damage is often found around or near the grafts. In contrast, ventricular grafts usually produce tissue damage only in the cortex, at the site of needle penetration. Also, some of the graft tissue is frequently displaced into the lateral ventricle. A second method of placing grafts into the substance of the brain with minimal disruption is through the use of dispersed cells as described by Björklund et al.[28] Thus, the use of

TABLE I
Frequency of Survival of Ventricular Homografts in Rats

Tissue	Duration of survival (months)	Reference	Number of rats examined histologically	Percent with surviving graft tissue
Embryonic substantia nigra	3	33; also unpublished data	13	100%
Embryonic substantia nigra	8–10	34; also unpublished data	16	100%
Embryonic substantia nigra	14–20	Unpublished data	6	100%
Embryonic cortex	9–17	Unpublished data	7	100%
Embryonic tectum	9–17	Unpublished data	6	100%
Adrenal medulla	5	35; also unpublished data	10	90%
Kidney cortex	6	36	9	100%
Total			67	98.5%

small grafts containing between 1 and about 50 cells is the most promising possibility for intraparenchymal transplantation.

2.3.3. Transplantation Cavities

A third method of placing solid tissue grafts into the brain is to insert the tissue into previously prepared "transplantation cavities." This method has been employed extensively by Björklund, Stenevi, and their colleagues.[40] These investigators remove an area of cerebral cortex, and the grafts are then placed upon the vascular bed that is formed on the walls of the cavity. The purpose of this maneuver is to provide better nourishment for the grafts. Again, early studies of brain grafting suggested that immunological privilege would not hold under these conditions. Medawar[41] concluded that homografts in the brain or anterior eye chamber would be rejected if penetrated by blood vessels. This suggested that heavily vascularized areas of the brain, e.g., transplantation cavities or the meninges, would not be favorable areas for survival of homografts. We also have found that fetal eye homografts survive poorly in previously prepared cortical transplantation cavities (Freed, Perlow, and Wyatt, unpublished manuscript). Björklund et al.[40] have nonetheless obtained excellent survival of fetal brain grafts in cortical transplantation cavities. Recently, Stenevi et al.[42] have transplanted fetal mouse SN to transplantation cavities in the rat. These grafts survived for at least several months. Therefore, immunological privilege may hold for cortical transplantation cavities in the same way that it does for the ventricular spaces and for the substance of the brain.

3. Histology and Catecholamine Production

3.1. Substantia Nigra Grafts

3.1.1. Nissl-Stained Material

When obtained from 17-day gestational embryonic rat donors, these grafts generally grow to fill a substantial part of the lateral ventricle when viewed in a coronal section. In the rostrocaudal dimension, the grafts usually extend for at least 1 mm, and sometimes as far as 3 mm. The grafts are generally fused with the ependymal lining on the lateral (striatal) side, and frequently on the medial (septum) side as well (Fig. 1). There are usually one or more locations where the ependymal lining is broken, and the graft tissue comes into direct contact with host brain. The remaining surfaces of the grafts are smooth and rounded and in contact with cerebrospinal fluid . Both glia and neurons are distributed throughout the grafts. The grafts frequently contain bands of large neurons, usually extending across the grafts oriented roughly in a mediolateral direction.

3.1.2. Histochemical Fluorescence

Catecholamines in grafted SN and adjacent brain regions were observed with histochemical fluorescence using the Falck–Hillarp method[43] or the method of de la Torre[44]

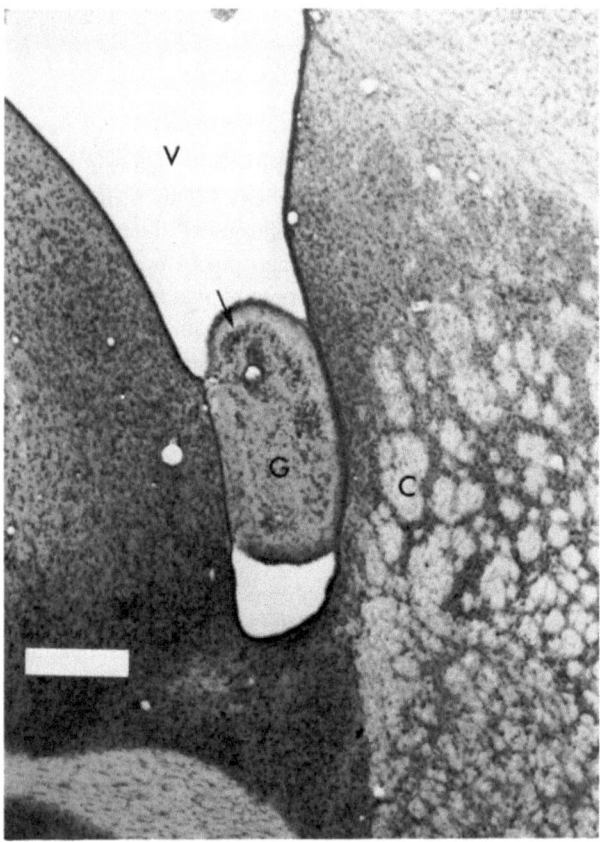

FIGURE 1. General appearance of a substantia nigra graft (G) in the rat lateral ventricle (cresyl violet stain).
Note the band of large neurons (arrow) across the top of the graft. V, lateral ventricle; C, caudate-putamen.
Bar = 0.5 mm.

with the addition of a perfusion as described by Church *et al.*[45] Virtually all animals
studied have been found to have grafts with numerous catecholamine-containing cells in
the range of about 2000 per animal.[33]

In grafts that have been studied up to 3 months after transplantation, the grafted
cells tend to be round in shape, having the appearance of immature dopaminergic cells.
Cells studied 9–20 months after grafting have been polygonal or irregular in shape,
similar to mature dopaminergic cells. Some cells had fine processes; however, the catecholaminergic cells often are too closely packed, or the fibers too dense to identify cells
with particular processes. Examples are seen in the lower right of Fig. 2 and in the two
groups of densely packed cells seen in the graft in Fig. 3A. Most grafts also contain
dense networks of fine catecholaminergic fibers, located in clumps in various regions
within the grafts and along the ventricular lining, especially around areas of contact
between graft and host brain.

The fluorescent cells are frequently grouped in bands with roughly the same configuration and orientation as the normal SN pars compacta. Björklund *et al.*[40] also have observed this, and in addition have described areas in the grafts, ventral to catecholamine cells, that contain fluorescent dendritic processes similar to the SN pars reticulata. In other grafts, catecholamine-containing cells are grouped or clumped irregularly.

The SN grafts reinnervate parts of the striatum of the host animals in every instance. This reinnervation is limited to parts of the striatum adjacent to graft tissue, and never extends for more than about 2 mm away from the borders of a graft in any direction. Close to the lateral ventricle, the reinnervation is frequently equal to, and sometimes even more intensely fluorescent than the normal, unlesioned striatum. The grafts themselves and the ependymal lining also tend to be hyperinnervated. With increasing distance from the graft, the density of reinnervation gradually decreases, although sometimes the extent of the reinnervation is fairly sharply demarcated (Fig. 3A). The total extent of reinnervation was never more than about 2 mm in the mediolateral dimension, 2 mm in the dorsoventral dimension, and about 2.5 mm in the rostrocaudal dimension. No individual animal exhibited reinnervation to that extent in all three dimensions; thus, the maximum total reinnervated volume was about 6 mm^3.

In some animals, individual catecholaminergic fibers reinnervating the caudate could be discerned, while the reinnervation in other rats had a diffuse appearance similar to the normal striatal dopaminergic innervation. Olson *et al.*[46] have observed both diffuse and discrete-fiber innervation of striatal grafts in the anterior eye chamber by SN grafts. Their interpretation was that more densely innervated tissue showed the

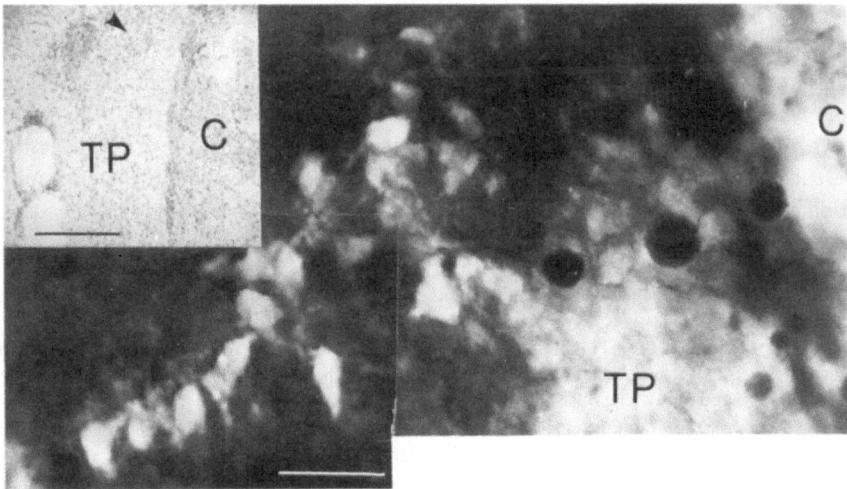

FIGURE 2. Location of catecholamine-containing cells in a substantia nigra graft. Upper left: A Nissl-stained section of a substantia nigra graft in the lateral ventricle of a rat. Lower right: An adjacent section revealing catecholamines by the SPG histochemical fluorescence technique,[44,45] shown at higher magnification. The arrowhead in the upper left photograph indicates the location of the cells shown in the lower right. TP, transplant; C, caudate-putamen. Bars: upper, 0.5 mm; lower, 0.05 mm.

diffuse fluorescence pattern, while individual fibers were visible where there was a sparser reinnervation.

Catecholaminergic fibers in some animals enter the host brains in bundles at small, circumscribed points associated with disturbances of the ependymal lining. In other animals, the points of entry of fibers into the grafts could not be precisely located, and seemed to be more diffuse. Björklund *et al.*[40] also describe a somewhat more diffuse entry of catecholaminergic fibers into the caudate-putamen from SN grafts in the cortex.

3.1.3. Biochemical Studies

We have also measured concentrations of dopamine in the grafts and the denervated striatum, to confirm the presence of dopamine in the grafts and to study the degree of reinnervation in quantitative terms. Stenevi *et al.*[47] have studied concentrations of dopamine in the hippocampus reinnervated by fibers from SN grafts. This reinnervation increased concentrations of dopamine in the hippocampus from about 0.002 μg/g wet wt to between 0.1 and 2.0 μg/g (mean 0.8 μg/g, corresponding to about 8 ng/mg protein).

For our experiments,[34] we have measured concentrations of dopamine in small circular needle punches taken from 0.4-mm-thick frozen brain sections by a sensitive gas-chromatographic mass-fragmentography assay.[48,49] The range of concentrations of dopamine in punch samples of SN grafts in the lateral ventricles of lesioned rats was from 0 to 298 ng/mg protein, averaging 96 ng/mg protein per sample or 88 ng/mg protein per rat. This is substantially greater than the average concentration of dopamine in the normal SN (31 ng/mg protein). The high concentrations of dopamine in the grafted SN may have been due to the hyperinnervation of the grafts by catecholaminergic fibers, in comparison with the normal SN, which contains few catecholamine fibers. In punches taken from parts of the caudate-putamen adjacent to SN grafts, concentrations of dopamine were increased as compared with either more ventrolateral parts of the striatum or with the striatum of control rats with sciatic nerve grafts (Fig. 4). This suggests the presence of a quantitatively substantial dopaminergic reinnervation of the dorsomodial caudate-putamen.

---→

FIGURE 3. Color-coded images of catecholamine fluorescence in substantia nigra and adrenal chromaffin cell grafts. Photographic negatives were processed by computer-assisted densitometry and color-coding as described by Goochee *et al.*[51] The most intense fluorescence is shown as white, followed by red, orange, yellow, and green. Deep blue and black indicate background levels of fluorescence. An arbitrary scale, showing internal computer gray levels, is shown in the upper right.

(A) A substantia nigra graft (G) in the ventricle that has partially reinnervated the host caudate-putamen. This graft was examined 18 months after transplantation. The reinnervation appears as the red, yellow, and green areas in the caudate-putamen (C). Fluorescence was induced by glyoxylic acid.[44] Bar = 1 mm. (B) An adrenal medulla graft showing diffusion of catecholamines around the graft, which is especially prominent to the right of the graft in the host striatum. The area occupied by grafted cells appears white. Falck–Hillarp fluorescence.[43] Bar = 0.5 mm. (C) A single grafted adrenal chromaffin cell, showing diffusion of catecholamines into adjoining fibrous tissue. The cell appears yellow, while areas of diffusion are green or light blue. Other areas around the cell were open ventricle, where any catecholamines that were present would have been washed away during the staining procedure. Glyoxylic acid-induced fluorescence.[44,45] Bar ~ 20 μm.

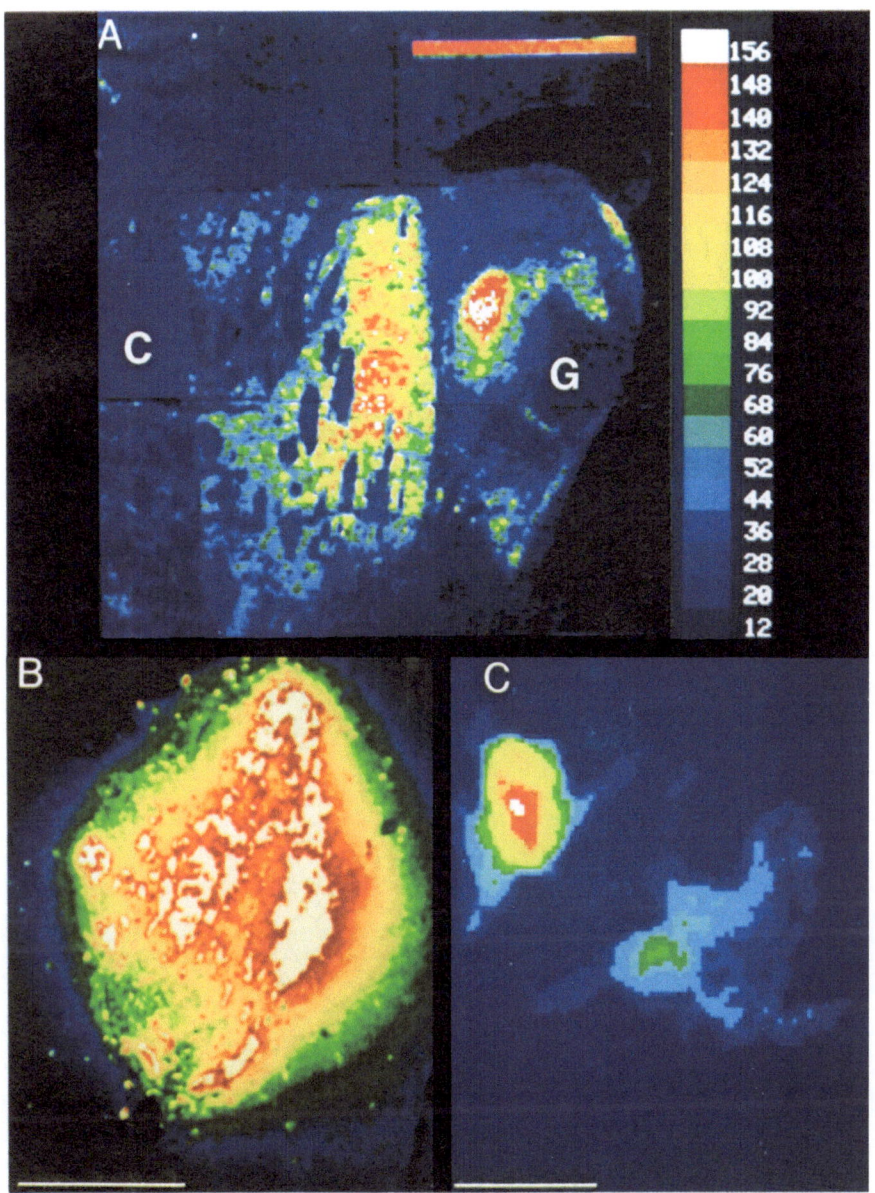

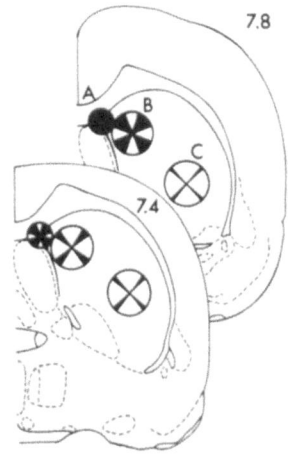

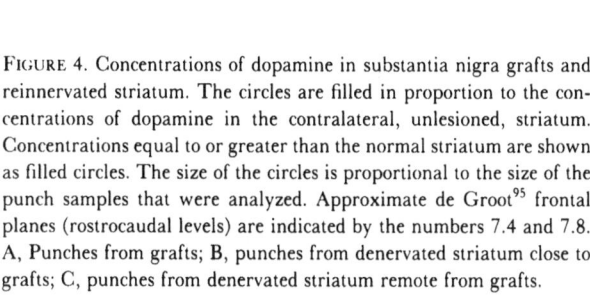

FIGURE 4. Concentrations of dopamine in substantia nigra grafts and reinnervated striatum. The circles are filled in proportion to the concentrations of dopamine in the contralateral, unlesioned, striatum. Concentrations equal to or greater than the normal striatum are shown as filled circles. The size of the circles is proportional to the size of the punch samples that were analyzed. Approximate de Groot[95] frontal planes (rostrocaudal levels) are indicated by the numbers 7.4 and 7.8. A, Punches from grafts; B, punches from denervated striatum close to grafts; C, punches from denervated striatum remote from grafts.

3.2. Adrenal Medulla Grafts

3.2.1. Histochemical Fluorescence

The adrenal chromaffin cell displays a remarkable degree of phenotypic plasticity. Under normal circumstances, when located in the adrenal medulla and surrounded by the adrenal cortex, the adrenal chromaffin cell is smooth and rounded, and contains large amounts of epinephrine. When removed from this environment, the rat chromaffin cell changes morphologically as well as biochemically, so that it begins to resemble a noradrenergic neuron; these changes occur when glucocorticoids are reduced and can be further promoted by nerve growth factor. The changes can be seen when the adrenal chromaffin cell is removed from the adrenal cortex and grown in tissue culture[24-27,50] or when chromaffin cells are transplanted to the anterior chamber of the eye.[22,23]

When transplanted to the anterior eye chamber, the adrenal chromaffin cell of the rat becomes elongated or irregular in shape and produces catecholaminergic fibers that sometimes innervate the host iris.[22] Such intraocular chromaffin cell grafts also will innervate grafts of the cerebral cortex in the eye chamber.[23] This suggested the adrenal medulla as a potential catecholamine-producing tissue for transplantation to the lateral ventricles. Use of the adrenal medulla would allow for the possibility of autotransplants in the denervated nigrostriatal system.

Adrenal chromaffin cells from 5- to 7-week-old donors grafted to the lateral ventricles were similar in appearance to those grafted to the anterior eye chamber.[22] They attained an intermediate degree of differentiation: some cells retained the rounded smooth appearance typical of normal chromaffin cells, while other cells had undergone varying degrees of transformation toward the adrenergic-neuron phenotype. Most of these cells had become somewhat elongated or developed short, coarse processes (Fig. 5). There also were a few cells with fine fibers and branching processes. The density of fine fibers, however, was very sparse in comparison to SN grafts. In particular, there was little or no evidence of reinnervation of the host brain by the grafts.

Many of these grafts, however, were surrounded by a halo of diffuse fluorescence,

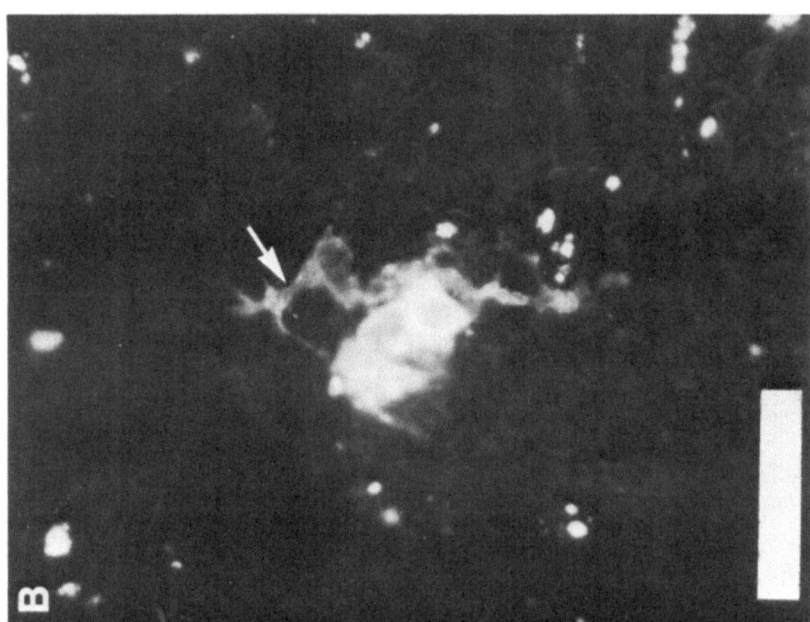

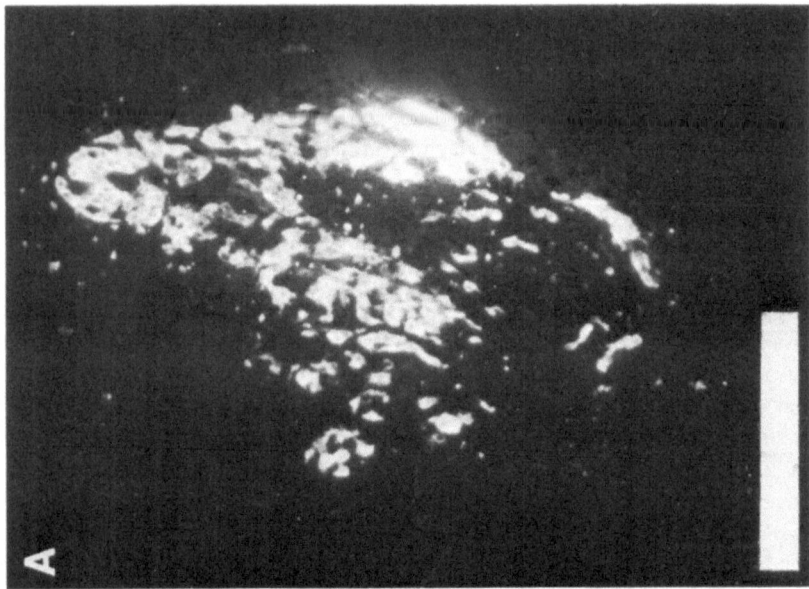

FIGURE 5. Adrenal medulla grafts in the lateral ventricle of the rat brain processed for Falck–Hillarp[43] histochemical fluorescence. (A) A graft that contained both rounded cells, similar to normal chromaffin cells, and cells that had become elongated. This graft is shown in Fig. 3B after color-coding for fluorescence intensity. Bar = 0.5 mm. (B) A small group of four grafted adrenal chromaffin cells that had developed short branching processes (arrow). Bar = 50 μm.

which was taken to be indicative of secretion of catecholamines from the grafts. This secretion was more easily seen in the form of color-coded images processed by computerized densitometry and image enhancement.[51] The graft in Fig. 5 is shown as a color-coded image in Fig. 3B. This clearly shows the halo of fluorescent staining surrounding the graft. Another grafted chromaffin cell, which was found isolated in an area of fibrous tissue, is shown after color-coding in Fig. 3C. The secretion of catecholamines around this cell could be seen in the fibrous tissue. In areas of open ventricular space, no fluorescence was seen, presumably because the fluorophore was washed away in the staining process. The diffuse fluorescent cloud that surrounded this cell was markedly different from the sharp borders of normal or grafted SN dopaminergic neurons or from the crisp appearance of catecholaminergic fibers in other parts of the brain.

The adrenal medulla grafts frequently contained significant numbers of autofluorescent macrophages, indicative of resorption or death of some of the grafted tissue. These macrophages were not intermixed with catecholaminergic cells, but rather in adjoining areas. There also were substantial amounts of fibrous tissue associated with the grafts, and in a few rats, areas of autofluorescent fibrous tissue (in one case, about 2 mm^3) were found which were almost devoid of cells with specific catacholamine fluorescence.

3.2.2. Catecholamines

Phenylethanolamine N-methyltransferase (PNMT), the enzyme catalyzing the conversion of norepinephrine to epinephrine, is present in the adrenal medulla. PNMT activity is increased by high concentrations of glucocorticoids and diminished in their absence.[27] The activity of dopamine β-hydroxylase, which catalyzes that conversion of dopamine to norepinephrine, is primarily under control of the neuronal innervation of the adrenal medulla,[52] which is cholinergic in nature.[53] The relative concentrations of dopamine, norepinephrine, and epinephrine therefore can be used as an indication of the state of differentiation of the adrenal medulla.

Concentrations of catecholamines were studied by a gas-chromatographic mass-fragmentography assay[48,49] using circular punch samples from 0.4-mm-thick coronal sections. Grafts in animals with and without SN lesions were studied, in addition to normal adrenal medulla.[54]

In general, concentrations of dopamine in the grafts were high but extermely variable (Fig. 6). Five of the fifteen grafts from lesioned hosts and nine of the eleven grafts from intact hosts had concentrations of dopamine greater than the highest value measured for intact striatum. Dopamine concentrations in some grafts were extremely high (Fig. 6 and Table II).

There was a substantial increase in the relative proportion of the total catecholamines found as dopamine in both types of grafts, and particularly in the grafts from nonlesioned hosts, as compared with the intact adrenal medulla. Also, the amount of epinephrine was greater in the grafts from lesioned as compared with nonlesioned hosts (medians 74 vs. 20, $p < 0.01$, two-tailed Mann–Whitney U test).

One possible explanation for the higher concentrations of dopamine in the grafts from nonlesioned hosts is that the grafts were taking up dopamine from the medium.

This would not, however, explain the higher concentrations of epinephrine in the grafts from lesioned rats. Another possibility is that increased dopamine β-hydroxylase activity occurs in the grafts from lesioned rats, perhaps combined with changes in tyrosine hydroxylase activity. In the normal adrenal medulla, dopamine β-hydroxylase and tyrosine hydroxylase are primarily under neuronal control[52] and the neuronal innervation is primarily cholinergic.[53] It is conceivable that the host striatum exerts a cholinergic influence on the grafted adrenal medulla, and that this is altered by SN lesions. Treatment with some drugs, particularly insulin, and cholinergic agents such as atropine and the ganglionic blocking agent chlorisondamine, produce alterations in adrenal catechol-

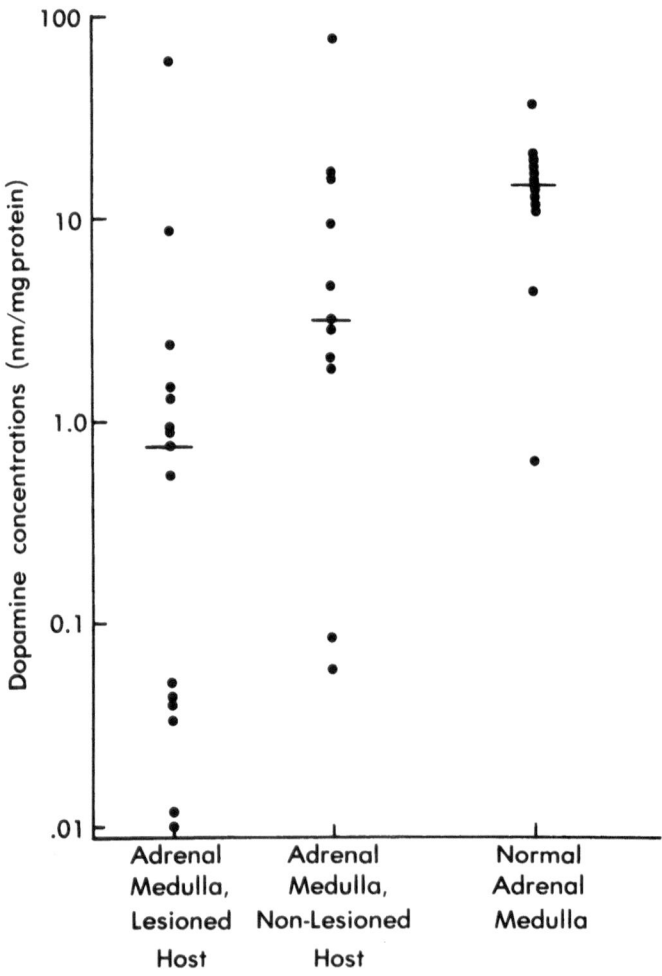

FIGURE 6. Concentrations of dopamine in adrenal medulla grafts in hosts with and without substantia nigra lesions. Although highly variable, the concentrations of dopamine were significantly greater in the grafts from nonlesioned hosts ($p < 0.01$, Mann–Whitney U test). Dopamine concentrations in normal intact adrenal medulla are shown for comparison. Horizontal bars indicate medians.

TABLE II
Dopamine Concentrations (ng/mg Protein) in Grafts and Host Brain Tissue

Tissue	Condition of host substantia nigra	Dopamine (mean ± S.E.M.)[a]	
Adrenal medulla grafts	Intact	1801	± 981
Adrenal medulla grafts	Lesioned	796	± 612
Normal adrenal medulla	—	2278	± 365
Substantia nigra grafts	Lesioned	88	± 33
Normal substantia nigra	Intact	32	± 4
Striatum (whole)	Intact	122	± 13
Striatum (punches)	Intact	130	± 6.8
Striatum (punches)	Lesioned, screened for rotational behavior	9.2 ±	4.0
Striatum (whole)	Lesioned, not screened	24 ±	7.8

[a]Note that the original data published in Freed et al.[54] were medians ± S I Q.R , rather than means.

amines and metabolizing enzymes somewhat similar to the lesion-related changes observed here.[52,55,56]

3.3. Comparison of Substantia Nigra and Adrenal Medulla Grafts

First, the adrenal chromaffin cell grafts were rarely integrated into the host brain as well as SN grafts. The borders of the grafts could always be located precisely, even in sections of intraparenchymal grafts. In contrast, typical SN grafts are free of macrophages and autofluorescent pigments and are imperceptibly fused with the host brain.

Fetal SN grafts produce numerous fine fibers that are found in dense clusters within the grafts, along the ventricular walls, and in the dorsomedial striatum. The host striatum is reinnervated by these fibers in such a way that with increasing distance into the striatum, the density of fibers decreases. The total amount of specific fluorescence in the reinnervated striatum was substantial and sometimes clearly exceeded that seen within the grafts themselves. Thus, a principal site of catecholamine secretion by SN grafts is in the striatum, rather than within the grafts. This suggests that these grafts release dopamine from terminals, and possibly from synaptic contacts in the reinnervated striatum.

In contrast, the adrenal medulla grafts produced few fibers and very little or no reinnervation of the striatum. In animals with adrenal medulla grafts, by far the largest fraction of specific fluorescence was located within the grafts. Nonetheless, these grafts contain large amounts of dopamine and were capable of reducing apomorphine-induced rotational behavior.[35] The grafts as well as individual grafted cells were sometimes surrounded by a fluorescent "cloud" when stained for catecholamines, suggesting active release of catecholamines from the grafted cells (Figs. 3B, C). The isolated sheep adrenal is also capable of secreting dopamine.[57] This indicates that the grafted adrenal medulla cells are acting in relative isolation from the function of the host brain, and appear to be secreting catecholamines that influence the striatum through passive diffusion to receptor sites.

Thus, SN grafts reinnervate the host striatum, where dopamine is presumably released from nerve terminals in close association with the denervated areas. The grafted cells themselves are sharply defined and there is no appearance of catecholamine release from the perikarya. This is illustrated by the patterns of reinnervation shown in Fig. 3A. In contrast, catecholamines are apparently released primarily from the cell perikarya of the adrenal medulla grafts. These grafts therefore can be conceptualized as chemical reservoirs, which release active neurotransmitter chronically in close juxtaposition to the site of action.

4. Rotational Behavior

4.1. The Rotational Behavior Model and Supersensitivity Phenomena

The caudate-putamen of the rat, as well as the olfactory tubercle, the nucleus accumbens, and other areas, are innervated by dopamine-containing axonal fibers that originate in the ipsilateral SN.[13,58] These dopaminergic projections are crucially important for the control of many behaviors and movements, including simple behaviors such as drinking, eating, and grooming, as well as more complex learned activities. Bilateral destruction of the SN in rats results in a generalized akinetic syndrome, including a loss of eating, drinking, and grooming behavior, as well as lack of movements in general.[59-63]

In light of the fact that this dopaminergic system contains a relatively small fraction of the brain's total neurons and synapses, its functional importance is all the more surprising. One interpretation is that the dopaminergic system performs a motivating or triggering role for all behaviors, while playing at most a minor role in controlling the occurrence of particular behaviors or their form. A second interpretation is that the dopaminergic system is involved in sensorimotor integration, so that animals with SN lesions become unresponsive to environmental stimulation.[62,64]

The rotational behavior model for the study of nigrostriatal function was first described by Ungerstedt.[65,66] When the SN is lesioned on one side of the brain of a rat, the animal will show turning movements in the direction of the lesion, which persist for only about 1 or 2 days. The posture and movements of the animal then adapt so that they appear to be essentially normal. Dopaminergic denervation supersensitivity was originally detected when it was found that these adapted animals rotate away from the lesioned side following administration of apomorphine, a postsynaptic dopamine agonist.[66] This was taken to indicate that postsynaptic denervation supersensitivity to dopamine had developed in the striatum ipsilateral to the lesion. The presence of this supersensitivity has since been confirmed by other techniques, including binding studies,[67] adenylate cyclase measurement,[68] and electrophysiological studies.[69]

Following the development of postsynaptic denervation supersensitivity in the striatum, the dopamine agonist apomorphine will stimulate the lesioned striatum to a greater degree than the normal, unlesioned striatum. Conversely, unilaterally lesioned rats rotate toward the side of the lesion when given amphetamine, which stimulates dopamine release.[65] This latter phenomenon has been interpreted as a measure of the relative excess of dopamine release from the normal, as compared with the lesioned side. Because dopaminergic terminals are not present on the side of the lesion, amphetamine induces greater dopaminergic activity on the normal than on the lesioned side.

These two techniques, apomorphine- and amphetamine-induced behavioral rotation, provide an ideal method for measuring the functional status of the nigrostriatal dopamine system (Fig. 7). Using rotational behavior, the functional status of the postsynaptic dopamine receptors of the lesioned side can be quantified by measurements of contralateral rotation following administration of apomorphine. Rotation in the ipsilateral direction following amphetamine can be used to measure the relative excess of releasable dopamine in the striatum that has not been denervated. Of course, the effectiveness of any dopamine remaining on the lesioned side may be magnified by the presence of denervation supersensitivity.

Rotation, or locomotion in circles, is easily quantified. To measure this behavior, rats are connected to a flexible wire spring by a rubber-band harness. The wire spring is used to turn a slotted disc, triggering three light-activated silicon-controlled rectifiers (Fig. 8). The apparatus and algorithm, described by Walsh and Silbergeld[70] records only full 360° turns in either direction.

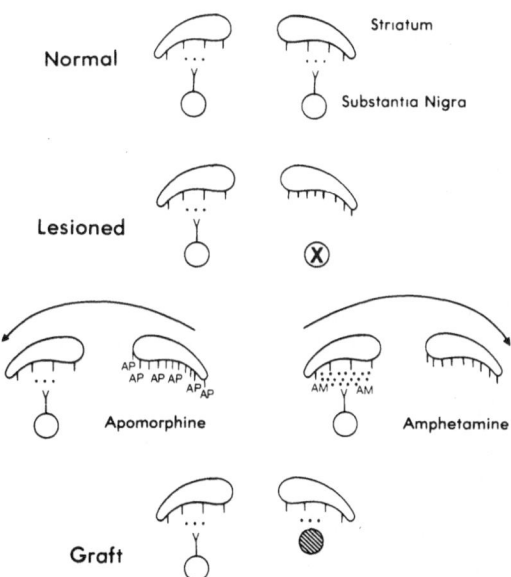

FIGURE 7. Presumptive mechanisms of lesion-induced rotational behavior and restoration by grafts. *Normal:* Dopaminergic neurons in both left and right substantia nigra (SN) tonically release dopamine (dots), maintaining normal striatal receptor sensitivity (hatches). *Lesioned:* When the right SN is destroyed, dopamine is no longer released into the right striatum (no dots shown). In response, striatal dopamine receptor sensitivity increases (more hatches). In this state, apomorphine (AP) will hyperstimulate the right striatum, causing rotation to the left. Amphetamine, however, stimulates the release of dopamine (more dots) on the left only, causing rotation to the right. *Graft:* When an SN graft is placed adjacent to the denervated striatum, dopamine is once again released tonically into the right striatum (dots). Striatal dopamine receptor sensitivity is partially normalized (normal number of hatches). The asymmetrical response to both amphetamine and apomorphine is therefore diminished.

Rat Rotometer

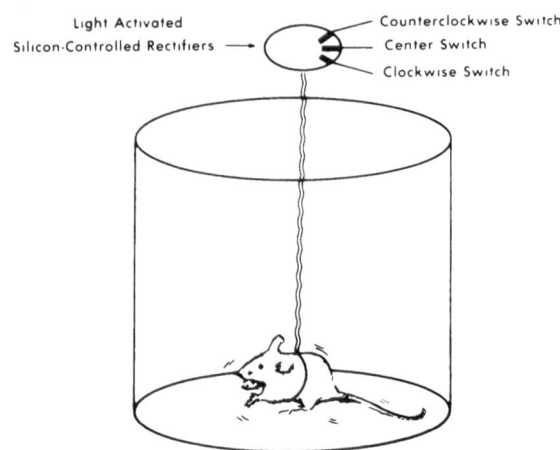

FIGURE 8. Arrangement of the apparatus for recording rotational behavior. Light-activated silicon-controlled rectifiers are employed as switches. When the center switch is crossed, a cycle is initiated. The next switch crossed (clockwise switch or counterclockwise switch) determines the direction of rotation. If the remaining switch is crossed, the cycle is completed and a turn is recorded. If the center switch is again crossed before the cycle is completed, a new cycle is initialized and no count is recorded. This algorithm records only full turns in either direction but rejects partial rotations.[70]

4.2. Embryonic Brain Grafts

4.2.1. Apomorphine-Induced Rotation

Fetal SN grafts have been found to reliably decrease apomorphine-induced rotational behavior. In our initial experiments[33,34] (also subsequent unpublished data), animals with embryonic SN grafts were compared with controls that had received grafts of adult sciatic nerve. Figure 9 shows the combined results from all animals tested with this procedure through the end of 1980. Overall, the reductions in rotation in the animals with SN grafts were greater than those of the controls; however, smaller but nonsignificant reductions in rotation were frequently observed in the animals with sciatic nerve grafts as well. Once established by SN grafts, reductions in apomorphine-induced rotation are stable for at least 6 months, while rotation in controls with sciatic nerve grafts is maintained at the slightly reduced levels seen just after grafting.[34]

In a recent experiment, we have investigated the effects of embryonic SN grafts in comparison to grafts of other embryonic brain areas. One of the control tissues used was the embryonic cortex, because the cortex normally provides a substantial innervation to the striatum.[71] The other was the tectum, because the tectum is physically close to the SN but does not innervate the striatum. Moreover, tectal grafts in the anterior eye chamber have tropic influences on cerebral cortex grafts similar to those of locus coeruleus grafts and grafts of other brain stem nuclei.[72]

All three types of grafts survived and contained neurons and glia. At least some of the cortex grafts had innervated the striatum, as determined by retrograde transport of Evan's blue and wheat germ agglutinin–horseradish peroxidase conjugate (U. Patel, H. E. Cannon-Spoor, and W. J. Freed, unpublished data). Apomorphine-induced rotation, however, was decreased significantly only by the SN grafts (Fig. 10).

Björklund and Stenevi[73] and Björklund et al.[40] initially reported that apomorphine-induced rotation was not decreased by embryonic SN grafts in cortical cavities. More recently, however, Dunnett et al.[74] found that apomorphine-induced rotation was decreased when a dose of 0.05 mg/kg was used, but not with a higher (0.25 mg/kg) dose. No data on rotation rates prior to transplantation were obtained.

We have obtained somewhat different dose–response relationships. First, we have observed similar reductions in rotation in animals tested with 0.1 or 0.25 mg/kg, even in our earliest experiments.[33] More recently, we have obtained a dose–response curve for apomorphine-induced rotation in 18 rats before and after receiving SN grafts (Fig. 11). There were significant reductions in the SN-grafted group for the 0.05 and 0.25 mg/kg doses, while the reductions at 0.1 mg/kg did not reach statistical significance ($p > 0.2$). We feel that it would be premature to attach any particular significance to the failure of the difference at 0.1 mg/kg to reach statistical significance. It should be pointed out that there are substantial differences between the experiments of Dunnett et al.[74] and ours, such as the difference in graft location and the substantially lower baseline rotation rates reported in their experiments (about 4/min) as compared with our studies (an average of about 8/min). It would seem, however, that reductions in apomorphine-induced rotation are not only restricted to the lower dosage range.

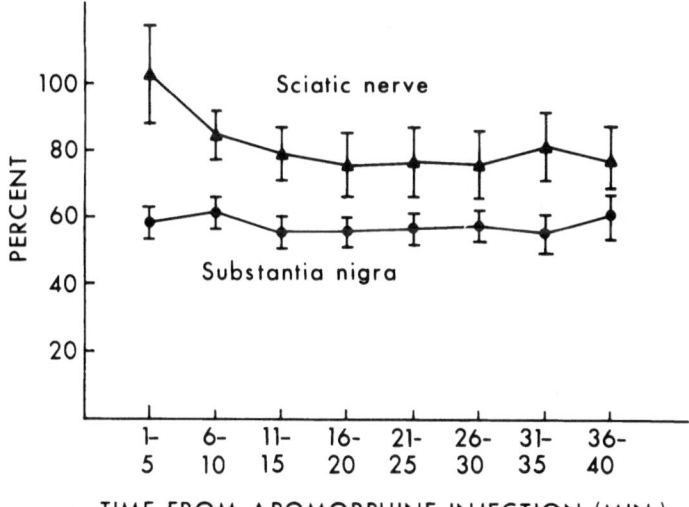

TIME FROM APOMORPHINE INJECTION (MIN.)

FIGURE 9. Reductions in apomorphine-induced rotation in animals with substantia nigra ($n = 50$ or sciatic nerve ($n = 18$) grafts. Data shown are mean percentages ($\pm$ S.E.M.) of baseline rotation rates for each 5-min segment of the 40-min testing sessions. The difference between the groups was significant by a two-way analysis of variance ($p = 0.008$).

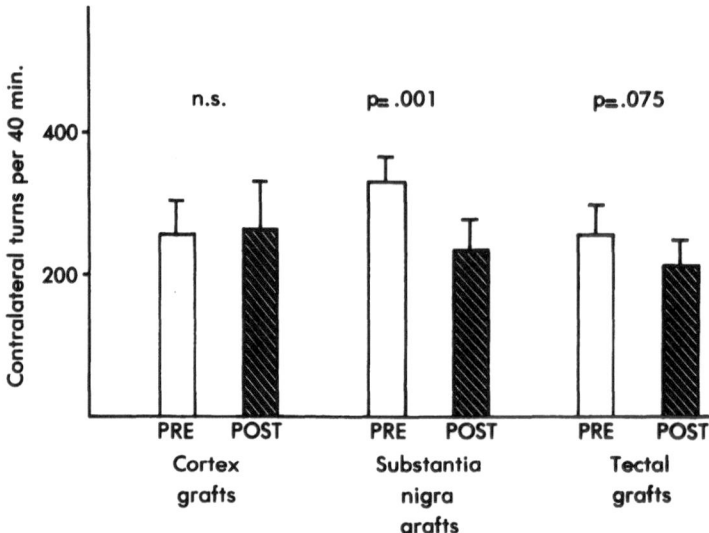

FIGURE 10. Apomorphine-induced rotation before (pre) and after (post) transplantation of fetal cortex, substantia nigra, or tectum to the right lateral ventricle. Probability levels shown are for the difference from pre- to post-transplantation (Scheffé), following a significant interaction effect on a two-way analysis of variance for one repeated measure. As is shown on the graph, only substantia nigra grafts significantly decreased apomorphine-induced rotation.

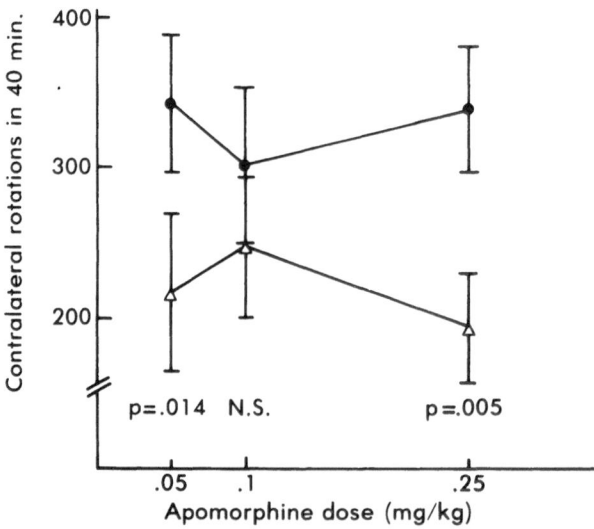

FIGURE 11. Dose–response curve for apomorphine-induced rotation before (●) and after (△) transplantation of substantia nigra to the lateral ventricle (means ± S.E.M.). The reductions were statistically significant for the 0.05 and 0.25 mg/kg dosages (Scheffé test). Note that the ordinate is discontinous.

4.2.2. Amphetamine-Induced Rotation

Björklund and Stenevi,[73] Björklund et al.,[28,40,75] and Dunnett et al.[74] have reported that amphetamine-induced rotation is decreased by embryonic SN grafts. In their first experiment,[73] four rats were examined for amphetamine-induced rotation 1 week after unilateral SN lesions, and on three occasions subsequent to grafting embryonic SN to the dorsal striatum. Turning was decreased or reversed in two of the four rats, and unaffected in the other two.

In a subsequent experiment, Björklund et al.[40] examined amphetamine-induced rotation in a larger group of rats ($n = 25$) according to essentially the same paradigm. Rotation was decreased 2 months after grafting, and tended to decrease somewhat more over the following 3 months. By 5 months after grafting, rotation was on the average about one-third of baseline. The most interesting aspect of this experiment is that when the grafts were surgically removed from five of these animals, their rotational behavior returned to baseline. This group[74,75] has also reported reductions of amphetamine-induced rotation as compared with unoperated controls in two subsequent publications.

Björklund et al.[28] also have studied amphetamine-induced rotation in animals that have received dispersed embryonic SN grafts into the body of the striatum.[28] Grafted cells, particularly when obtained from young (15-day gestational) donors, survived and were found distributed around the needle tract, producing a substantial reinnervation of surrounding areas. Reductions and even reversals in the direction of amphetamine-induced ipsilateral rotation were seen in five of six rats receiving grafts from 15-day embryos. Tissue from older (17-day gestational) donors was less effective, decreasing rotation to some degree in about half of the nine animals tested.

4.2.3. Relationship between Histochemistry and Behavior

Björklund et al.[40] have found that reductions of amphetamine-induced rotation are correlated with the degree of reinnervation of the striatum by SN grafts. In our experiments, reductions in apomorphine-induced rotation have not been related to features of the grafts, and in fact rotational behavior is not reduced in some cases even where striatal reinnervation is quite pronounced.

The most clear-cut feature of our animals in which SN grafts had decreased apomorphine-induced rotation has been the extent of the 6-OH-DA-induced striatal denervation. In an initial double-blind evaluation of five rats with reduced rotation (at least 70%) and of four others with no reductions, the animals with large reductions in turning were all found to have complete dopaminergic denervations of the striatum, while the denervations were incomplete in each of the four other rats.[33]

Although this difference could reflect some specific interaction between lesioned striatum and the grafts, the simplest explanation is that in animals with complete denervations, presynaptic dopamine reuptake and consequent metabolism is removed. This could then extend, both spatially and temporally, the action of dopamine released from graft-derived terminals in the striatum. This phenomenon would be analogous to that of presynaptic or prejunctional supersensitivity in the peripheral nervous system. In addition to the proliferation of postsynaptic receptors, the action of the natural transmitter can be accentuated by the absence of presynaptic terminals, because reuptake and metabolizing systems are absent.[76–78] This type of supersensitivity may be caused by factors such as decreased adrenergic neuronal reuptake, decreased cholinesterase activity, and loss of catechol-*O*-methyltransferase activity.[78] Although the above hypothesis provides a parsimonious explanation for the variable effects of the grafts on apomorphine-induced rotation, it is difficult to test experimentally. Because the reinnervation of the striatum by the grafts is limited, it may be that this hypothetical "presynaptic supersensitivity" is required for the grafts to be behaviorally effective.

4.3. Adrenal Medulla Grafts

Grafts of adrenal medullary tissue in the lateral ventricle reduce apomorphine-induced rotational behavior.[35] In this experiment, the grafts were obtained from 5–7-week-old donors. The reductions were similar in degree to those produced by ventricular SN grafts (Fig. 12). As these grafts did not reinnervate the host brain to any appreciable degree, the effects on rotation must be attributed to diffusion of catecholamines released from the cell soma (see Section 3.2).

4.4. Other Tissues

We also have grafted superior cervical ganglia to the lateral ventricles of rats with unilateral SN lesions[79] (also M. J. Perlow, W. J. Freed, and R. J. Wyatt, unpublished

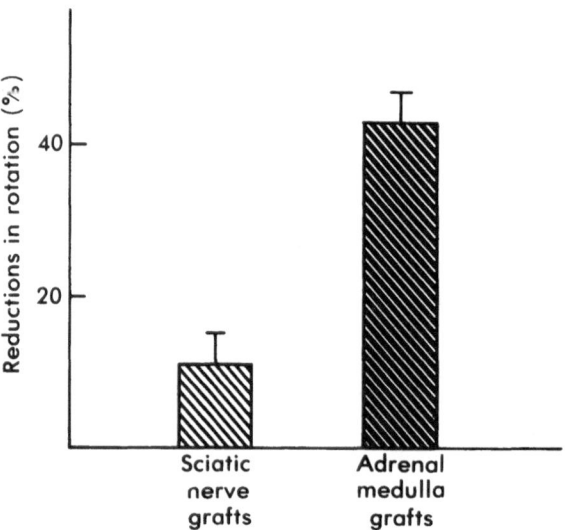

FIGURE 12. Effect of adrenal medulla grafts on apomorphine-induced rotation. Animals with substantia nigra lesions received either adrenal medulla or sciatic nerve grafts from young adult donors in the lateral ventricle adjacent to the denervated striatum. Data shown are percentage decreases from before to after transplantation (means = ± S.E.M.).

data). These grafts did not decrease apomorphine-induced rotation and contained few catecholaminergic cells.

4.5. The Issue of Controls

One question that frequently arises is the appropriate control tissue to be employed for behavioral studies of brain tissue grafting. We have always considered it essential, particularly in studies where we were attempting to decrease a behavior, to employ tissue in control experiments. As a first type of control, for both mechanical factors and the inevitable release of substances from dying or degenerating tissues, we felt that the tissue should be inactive in the sense that it would be incapable of either reinnervating brain tissue or secreting neuroactive substances. For this purpose, we usually have employed adult sciatic nerve, which is a neuronal tissue that contains supporting cells and nerve fibers that will degenerate *in situ*. Yet sciatic nerve is unlikely to restore functional capacity to damaged brain tissue in any meaningful sense. Strictly, however, the use of this tissue as a control would not justify any claims to the specificity of the effect of any particular active tissue or brain region in restoring functional capacity. This latter issue requires the use of fetal brain tissue, from other brain regions, for control grafts. Because it is conceivable that brain nuclei other than the SN could reverse effects of SN lesions, fetal brain tissue controls would, however, be a poor choice for any initial, exploratory experiment.

5. Other Behaviors

Although amphetamine- and apomorphine-induced rotational behaviors provide excellent methods of assessing the functional status of the nigrostriatal system, recovery of these behaviors does not necessarily imply that all behavioral deficits caused by SN lesions will also recover. This is particularly true in view of the fact that some dopamine-medicated behavioral functions may involve reciprocal connections between the striatum and dopamine-secreting nigral neurons. Moreover, some behaviors such as exploration appear to involve primarily the mesolimbic dopamine system, particularly the nucleus accumbens,[80,81] rather than the striatum.

Björklund, Dunnett, Stenevi, and their colleagues (Table III) have examined sensorimotor responding, spontaneous asymmetry, aphagia and adipsia, and self-stimulation[82] in animals with SN grafts. These experiments will be briefly reviewed here.

5.1. Spontaneous Asymmetry

Animals with unilateral SN lesions show a strong tendency to turn spontaneously toward the side of the lesion, which can be further accentuated by activation via mild tail-pinch.[74] In addition, animals with unilateral SN lesions have a strong tendency to

TABLE III
Studies of Behavioral Effects of Substantia Nigra Grafts by Björklund, Dunnett, Stenevi, and Their Colleagues[a]

Reference	Amphetamine-induced activation	Apomorphine sensitivity	Spontaneous asymmetry	Sensory neglect	Aphagia, adipsia	Akinesia after bilateral lesions
Solid tissue grafts in cortical cavities near dorsal striatum						
73	+	NS	NS	NS	NS	NS
40	+	0	NS	0	0	0
74	+	+	+	0	NS	NS
83	+	+	+	NS	0	0
75	+	NS	+	0	NS	NS
Solid tissue grafts in cortical cavities near lateral striatum						
84	0	0	0	+	?	+
Dissociated cell grafts in striatum						
28	+	NS	NS	NS	NS	NS

[a] +, effective; 0, no effect; NS, not studied.

make contralateral arm-choices in a T-maze.[74] Non-drug-related motor asymmetry has been employed by Dunnett et al.[74,83,84] and Björklund et al.[75] to study spontaneous behaviors in rats with nigrostriatal damage.

These authors have reported that these spontaneously asymmetrical behaviors are also decreased by SN grafts placed adjacent to the dorsal striatum.[74] For example, only 1 of 19 rats with lesions turned contralateral to the lesion in a T-maze while 19 of 51 rats with SN grafts made at least one contralateral turn.[74,75] These findings indicate, therefore, that behavioral effects of grafts can be detected even without drug activation.

5.2. Sensory Neglect

Normal rats will respond to tactile, visual, and olfactory stimuli by orienting and sometimes biting or grasping directed toward the stimulus. Rats with unilateral lesions of the nigrostriatal system do not respond normally to sensory stimuli, and will ignore stimuli contralateral to the lesion.[85,86] Björklund et al.[40,75] and Dunnett et al.[74] initially reported that SN grafts in cortical cavities located adjacent to the *dorsal* part of the striatum did not decrease lesion-induced sensory neglect. Subsequently, Dunnett et al.[84] reported that SN grafts in cortical cavities adjacent to the *lateral* border of striatum were effective in reducing the tendency of animals with unilateral SN lesions to neglect stimuli contralateral to the lesioned side. These laterally placed grafts had no effect on spontaneous or drug-induced rotation. From these results, Dunnett et al.[84] concluded that sensorimotor performance is mediated by dopaminergic mechanisms in lateral parts of the striatum, and therefore would not be affected by grafts in the ventricles or dorsal striatum.

5.3. Aphagia and Adipsia

Bilateral lesions of the SN produce a syndrome of severe and persistent aphagia, adipsia, and rapid weight loss and death[60,61] that in many respects resembles the "lateral hypothalamic syndrome."[87] Björklund and his colleagues have studied this phenomenon in animals with SN grafts in three instances. In the first of these experiments,[40] some rats with unilateral SN lesions were given bilateral intracortical SN grafts; then the remaining host SN was lesioned. The animals with SN grafts died within a median of 6 days, while the animals without grafts died in a median of 14 days. This difference is not statistically significant (Mann–Whithey U = 13, $p > 0.10$, two-tailed); thus, the grafts did not appear to have any effect. Dunnett *et al.*[83] prepared additional animals using the same sequence of surgical procedures, but kept some of the rats alive by gastric intubation. The animals with SN grafts initially lost weight more slowly than the control rats, but stabilized at a lower mean body weight. The control rats also consumed more bread and water than the rats with grafts. Thus, if anything, the SN grafts tended to impair the recovery of aphagia and adipsia after SN lesions.

Dunnett *et al.*[84] also have studied aphagia using the same paradigm but with the grafts placed adjacent to the lateral, rather than the dorsal, striatum. This procedure did not substantially alter the course of aphagia, adipsia, or survival duration after bilateral lesions, but the animals with SN grafts ate a sweetened wet cereal mash significantly more frequently than the unoperated controls. Thus, although the grafts did not exert a major effect, there was some tendency for the animals with grafts to eat a preferred food more readily.

In a recent preliminary experiment, we have transplanted fetal SN to the right lateral ventricles of 1-day-old rats ($n = 7$). Controls received sciatic nerve grafts ($n = 7$). When the animals had grown to between 400 and 800 g, they were coded and received bilateral lesions of the SN by stereotaxic injection of 6-OH-DA. The animals had free access to standard laboratory food and water at all times. Following the lesions, all animals lost weight, but those with SN grafts tended to lose weight somewhat more slowly (Fig. 13). The difference in percentage weight loss did not reach statistical significance ($p = 0.054$, two-tailed Mann–Whitney U test). It is interesting that the var-

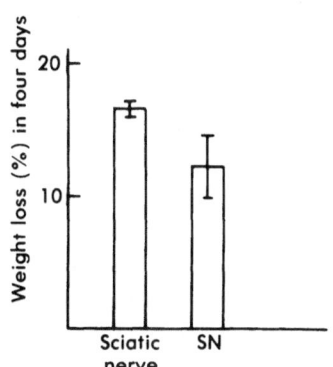

FIGURE 13. Effect of neonatal substantia nigra (SN) grafts on weight loss following bilateral SN lesions. Animals received either SN or sciatic nerve grafts 1 day after birth. When the animals matured, both host SN were destroyed. Data shown are percentage weight loss (means ± S.E.M.) during the 4 days following the lesions. The difference between the groups did not reach statistical significance ($p = 0.054$, Mann–Whitney U test).

iability was much greater in the group with SN grafts, suggesting that the grafts may have been effective only in some of the rats.

Our preliminary data are reminiscent of the short-term alleviation of weight loss in lesioned rats with SN grafts observed by Dunnett *et al.*[83] In their experiment, the slowed weight loss was reported only for rats being fed by gastric intubation, while the results of our experiment are also inconclusive because of the small number of subjects and short period of observation. We are, therefore, attempting to replicate this experiment with a larger number of rats and better long-term measures of eating and drinking.

5.4. Self-Stimulation

One of the most intriguing studies of behavioral properties of fetal brain grafts is the recent demonstration by Fray *et al.*[82] that rats will press a lever to initiate electrical stimulation of SN brain grafts. In this experiment, electrodes were implanted into SN (10 rats) or cortex (6 rats) grafts located in cortical cavities, or directly into the caudate-putamen (4 rats). At the end of 2 weeks of "training" (procedures were not specified), all animals were tested at the same current intensity. Rats with SN grafts responded at substantially higher rates than the controls. The effects of *d*-amphetamine and α-flupenthixol were similar to the effects of these drugs on self-stimulation in intact rats. The authors concluded that SN grafts can convey "specific, temporally-organized information axonally to the striatum."[82] Presumably, this means that, because the animals respond at rates as high as once every 6 sec, information must be transmitted from graft to striatum within a few seconds. This conclusion is supported by studies of stimulant drug self-administration, which have consistently found that rates of responding on continuous-reinforcement schedules increase as the dosage delivered per response is decreased. Responding ceases altogether, however, long before response rates reach one per 6 sec.[88,89] Therefore, high rates of responding on continuous-reinforcement schedules cannot be maintained simply by very low rates of drug delivery to remote, nonspecific sites. The finding of Fray *et al.*[82] therefore provides additional evidence that the SN grafts are not merely acting through nonspecific release of dopamine into the cerebrospinal fluid (see Section 3.3). More probably, these grafts release dopamine from terminals directly into the striatum, where the released dopamine can reach receptor sites within at most a few seconds, either through synaptic contacts or through short-distance diffusion.

6. Electrophysiology

An important consideration in terms of the capacity of SN grafts to function in place of the normal SN is whether the grafted cells are spontaneously active. We have therefore made extracellular unit recordings from SN neurons in ventricular grafts, and compared these neurons to SN zona compacta neurons *in situ*.[90]

The neurons of the normal SN zona compacta characteristically have spontaneous firing rates of between 0.5 and 10 Hz, and long-duration waveforms, in excess of 1.5 msec. Some cells that were not located in the zona compacta of the SN also have these properties. Therefore, neurons in the grafts were additionally characterized in anesthetized restrained rats in terms of their response to dopamine agonists and antagonists. Responses of transplanted and normal SN cells to dopaminergic drugs were measured from rate meter recordings of firing rates. Both normal and grafted cells in most cases were inhibited by dopamine agonists, such as dopamine, apomorphine, and epinine. Also, most normal and grafted cells were excited by the dopamine antagonists haloperidol and α-flupenthixol (Table IV).

In addition, activity was induced in nine "silent" cells by local application of glutamate. Each of these cells was also inhibited by epinine.

Therefore, SN grafts contain spontaneously active cells with electrophysiological properties and firing rates (4.0 $\pm$ 0.6 Hz) indistinguishable from normal nigral neurons. This suggests that the grafted dopaminergic neurons may continuously release dopamine from regenerated axon terminals. Thus, these neurons are probably not dependent upon amphetamine, apomorphine, or any environmental stimulus to express functional activity.

The fact that the rate of spontaneous activity in grafted neurons is similar to that of normal SN pars compacta neurons *in situ* suggests that the mechanisms regulating the activity of these neurons have not been grossly disturbed by transplantation. Under normal conditions, there appears to be an autoregulatory mechanism, whereby the activity of zona compacta neurons is inhibited by dopamine released from cell bodies or dendrites in the SN.[91] The grafted neurons respond normally to dopamine and dopamine agonists,[90] and dense dopaminergic fiber networks are frequently present within the grafts. Thus, the inhibitory autoregulatory mechanism is probably intact.

It also is possible that some excitatory influence on the cells may be present, which maintains the activity of the cells at nearly normal rates. One possibility is that excitation is maintained spontaneously, without an external input. A second possibility is that the activity is maintained by leakage of excitatory substances (e.g., glutamate) into the cerebrospinal fluid. The third, and most intriguing possibility is that a striatonigral feedback

TABLE IV

Properties of Spontaneously Active Substantia Nigra Cells *in Situ* and in Grafts[a]

	Cells in substantia nigra pars compacta	Cells in grafts
Firing rate 6.5 to 10 Hz and long-duration (> 2 msec) waveforms	28/28 cells[b]	47/82 cells
Inhibited by dopamine agonists (dopamine, apomorphine, epinine)	9/13 cells	9/11 cells
Excited by dopamine antagonists (haloperidol, α-flupenthixol)	12/15 cells	6/10 cells

[a] From Ref. 90.
[b] Eighteen cells with these properties but which were located outside the substantia nigra pars compacta were also found.

loop[92] has regenerated so that the grafts are regulated by the activity of excitatory substance P and inhibitory GABA neurons in the host striatum.

7. Future Directions

Ultimately, the ideal outcome of transplantation of dopamine-producing tissues to the caudate-putamen is to achieve complete reconstitution of nigrostriatal function. This outcome would take place in the event that the grafts become fully integrated into the host brain in such a way that all of the nigral efferents and afferents are restored. A complete restitution will not be a simple achievement, because complex multisynaptic connections, involving other brain areas, are probably formed during ontogeny in such a way that they are not easily duplicated in adulthood. In practice, therefore, the ideal outcome would involve (1) complete regrowth of the nigrostriatal dopaminergic system, so that the entire striatum becomes reinnervated, and (2) a complete functional reciprocal striatonigral reinnervation, so that the activity of the grafted cells is regulated in the same way as the normal SN.

In the experiments that we have described, the reinnervation of the striatum that is produced is not complete. Intraventricular grafts reinnervate a small but substantial fraction of the striatum within about 2 months, but even after 1½ years, the reinnervation is limited (see Fig. 3A). Intracortical and dispersed cell grafts also do not completely reinnervate the striatum.[28,40]

There are potentially at least two methods of increasing the degree of reinnervation of the striatum by SN grafts. The first might be simply to distribute grafts in many locations throughout the striatum. If a number of grafts were placed in various locations, each might reinnervate the surrounding area, and the sum of many small reinnervations could be equivalent to one complete reinnervation. The principal drawback to this technique is that a prohibitively large number of grafts might be required if the procedure is scaled up to larger species, e.g., humans.

The second possible approach is to determine why fiber regrowth is limited, and to try to promote the degree of reinnervation. For example, the degree of reinnervation may be limited because of competition from thalamic or cortical innervations[71,93] or interneurons for postsynaptic sites. Perhaps the hormonal environment of the adult brain may not be optimal for growth of dopaminergic axons, and administration of hormones could be beneficial. Or, growth factors present in immature striatum might be important; thus, combining grafts of embryonic striatum with SN grafts could be beneficial.

Corresponding to the limited dopaminergic reinnervation of the striatum, the grafts produce a limited behavioral reconstitution. Apomorphine-induced rotation is decreased only partially, while various other behavioral effects of SN lesions are influenced to varying degrees. Aphagia and adipsia produced by SN lesions, for example, are affected slightly if at all (see Section 5.3).

It would not be surprising to discover that some behavioral deficits produced by nigrostriatal lesions depend upon reciprocal connections between striatum and SN. For that reason, some behavioral deficits may be corrected only if the host brain innervates the grafts, or is otherwise able to regulate the activity of the grafted nigral neurons. At the present time, we do not know if this occurs, and it is possible that complete success of the grafts will not be achieved until advances in basic neurobiology make it possible to prompt the host brain to innervate grated tissues.[1,7]

8. Conclusions

One set of criteria for success of transplants has been given by Woodruff.[94] In his opinion, claims of permanent survival of a transplant can be made if (1) the grafts survive throughout the life of the recipient, (2) the recipient survives for at least 1 year, and (3) histological examination after death "gives no grounds for supposing that the transplant would have been destroyed had the recipient lived longer"[94] (p. 22).

In the case of embryonic brain grafts, these criteria have been met. In six animals with SN grafts and five with cortex or tectal grafts sacrificed 14 to 20 months after transplantation, all had surviving grafts that were indistinguishable from those examined a few months after transplantation. During this period, the recipients were about 2 years old and several rats from the same groups died of age-related diseases such as tumors. This would appear to be definitive evidence that the grafts survive permanently.

These grafts, in addition, meet criteria for survival based on functional activity.[94] In carefully controlled double-blind experiments, we have demonstrated that SN lesion-induced rotational behavior is significantly decreased by embryonic SN grafts, but not by grafts of fetal cortex or tectum (see Section 4.2.1). Björklund, Dunnett, Stenevi, and their colleagues have verified this conclusion[40,74,83] and have also found lesion-induced sensorimotor impairments to be favorably influenced by certain SN grafts.[84] These functional effects last for at least 6 months (see Section 4.2.1). Thus, fetal brain grafts become permanent functional constituents of the host brain and survive without loss of activity or deterioration until the death of the host.

While the functional effects of fetal SN grafts are, apparently, mediated by release of dopamine directly into the striatum from axon terminals, adrenal chromaffin cell grafts are, apparently, capable of decreasing rotational behavior by releasing dopamine directly from their cell bodies (see Section 4.3). In contrast to SN grafts, the principal site of dopamine release from adrenal medulla grafts is within the grafts, which are typically located at or close to the border between the striatum and the lateral ventricles. Catecholamine fiber production by these grafts is minimal. The functional effects of these grafts are therefore probably dependent upon diffusion of dopamine from the region of the grafts to striatal dopamine receptor. Thus, at the present time, adrenal medulla grafts act as living and self-sustaining localized drug-infusion pumps. The functional properties of these grafts might therefore be improved either by increasing the intimacy of contact between grafts and host striatum through the use of intraparenchymal or dissociated cell grafts, or by promoting the production of catecholaminergic fibers by the grafts. Thus, there are ample technical opportunities for improving the performance of adrenal chromaffin cell grafts. If the degree of fiber production and connectivity of fetal brain grafts can be increased, the functional performance of these grafts might also be improved. In the case of SN grafts, functional performance appears to be related either to reinnervation per se or at least to diffusion of catecholamines over very short distances from the terminals of grafted cells to striatal receptor sites. Whether actual synaptic contacts between graft and host are involved is, however, unknown. This, therefore, provides a simple model that allows the degree of success of these grafts to be measured and quantified in both functional and anatomical terms. Once the functional performance of these grafts has been optimized with this model, translation to more complex anatomical systems in the CNS may become possible.

9. Summary

When the SN is lesioned, the ipsilateral dopaminergic innervation to the corpus striatum is lost. In response to this loss, the denervated striatum becomes more sensitive to dopamine. In rats, this supersensitivity produces a functional deficit that can be measured by observing the number of rotations performed by the animals following administration of the dopamine agonist apomorphine. Two methods of treating this lesion-induced deficit with homologous tissue transplantation are described. In the first, embryonic SN is transplanted into the lateral ventricle, adjacent to the denervated striatum. These grafts partially reinnervate the striatum, increasing dopamine concentrations in the denervated regions. Presumably, rotational behavior is decreased because of this reinnervation and the consequent decrease in supersensitivity. In the second method, the adrenal medulla from young adult rats is transplanted to the lateral ventricles. These adrenal medulla grafts produce large amounts of dopamine, but do not reinnervate the striatum. Apparently, adrenal medulla grafts decrease rotational behavior by secreting dopamine into the lateral ventricles and striatum, which subsequently reaches striatal receptor sites by passive diffusion. Although these methods are only partially effective in correcting the functional deficit, they provide models for the reversal of the effects of lesions by brain grafting which may lead to improvements in the technique and possibly to eventual clinical application.

ACKNOWLEDGMENTS. The authors thank H. Eleanor Cannon-Spoor and Eleanor Krauthamer for their invaluable assistance in many of these experiments. We thank Theresa Hoffman for her expert preparation of the manuscript. We also acknowledge the NIMH technical development branch and the DSMR support staff for their many invaluable contributions.

References

1. Varon, S., 1977, Neural growth and regeneration: A cellular perspective, *Exp. Neurol.* **54:**1.
2. Finger, S., and Stein, D. G., 1982, *Brain Damage and Recovery: Research and Clinical Perspectives,* p. 384, Academic Press, New York.
3. Puchala, E., and Windle, W. F., 1977, The possibility of structural and functional restitution after spinal cord injury: A review, *Exp. Neurol.* **55:**1.
4. Hart, T., Chaimas, N., Moore, R. Y., and Stein, D. G., 1978, Effects of nerve growth factor on behavioral recovery following caudate nucleus lesions in rats, *Brain Res. Bull.* **3:**245.

5. Seiger, Å., and Olson, L , 1977, Quantitation of fiber growth in transplanted central monoamine neurons, *Cell Tissue Res.* **179**:285.
6. Varon, S., 1975, Neurons and glia in neural cultures, *Exp. Neurol.* **48**:93.
7. Cotman, C. W., Nieto-Sampedro, M., and Harris, E. W., 1981, Synapse replacement in the nervous system of adult vertebrates, *Physiol. Rev.* **61**:684.
8. Guth, L., 1974, Axonal regeneration and functional plasticity in the central nervous system, *Exp. Neurol.* **45**:606.
9. Steward, O., 1982, Assessing the functional significance of lesion-induced neuronal plasticity, *Int. Rev. Neurobiol.* **23**:197.
10. Anden, N.-E., Fuxe, K., Hamberger, B., and Hökfelt, T., 1966, A quantitative study of the nigro-neostriatal dopamine neuron system in the rat, *Acta Physiol. Scand.* **67**:306.
11. Anden, N.-E., Dahlström, A., Fuxe, K., Larsson, K., Olson, L., and Ungerstedt, U., 1966, Ascending monoamine neurons to the telencephalon and diencephalon, *Acta Physiol. Scand.* **67**:313.
12. Dahlström, A., and Fuxe, K., 1964, Evidence for the existence of monoamine-containing neurons in the central nervous system. I. Demonstration of monoamines in the cell bodies of brain stem neurons, *Acta Physiol. Scand.* **62**:1.
13. Ungerstedt, U., 1971, Stereotaxic mapping of the monoamine pathways in the rat brain, *Acta Physiol. Scand.* **367**:1.
14. Cotzias, G. C., Miller, S. T., Tang, L. C., and Papavasiliov, P. S., 1977, Levodopa, fertility, and longevity, *Science* **196**:549.
15. Finch, C. E., 1973, Catecholamine metabolism in the brains of ageing male mice, *Brain Res.* **52**:261.
16. Friedhoff, A. J., and Chase, T. N. (eds.), 1982, *Gilles de la Tourette Syndrome*, p. 454, Raven Press, New York.
17. Lloyd, K. G., Hornykeiwicz, O., Davidson, L., Schannak, K., Farley, I., Goldstein, M., Shibuya, M., Kelley, W. N., and Fox, I. N., 1981, Biochemical evidence of dysfunction of brain neurotransmitters in the Lesch–Nyhan syndrome, *N. Engl. J. Med.* **305**:110.
18. Meltzer, H. Y., and Stahl, S. M., 1976, The dopamine hypothesis of schizophrenia: A review, *Schizophr. Bull.* **2**:19.
19. Moskowitz, M. A., and Wurtman, R. J., 1975, Catecholamines and neurologic diseases, *N. Engl. J. Med.* **293**:332.
20. Wyatt, R. J., 1976, Biochemistry and schizophrenia. (IV). The neuroleptics—their mechanism of action: A review of the biochemical literature, *Psychopharmacol. Bull.* **12**:5.
21. Olson, L., and Seiger, Å., 1972, Brain tissue transplanted to the anterior chamber of the eye. I. Fluorescence histochemistry of immature catecholamine and 5-hydroxytryptamine neurons reinnervating the rat iris, *Z. Zellforsch. Mikrosk. Anat.* **135**:175.
22. Olson, L., 1970, Fluorescence histochemical evidence for axonal growth and secretion from transplanted adrenal medullary tissue, *Histochemie* **22**:1.
23. Olson, L., Seiger, Å., Freedman, R., and Hoffer, B., 1980, Chromaffin cells can innervate brain tissue: Evidence from intraocular double grafts, *Exp. Neurol.* **70**:414.
24. Tischler, A. S., and Greene, L. A., 1980, Phenotypic plasticity of pheochromocytoma and normal adrenal medullary cells, in: *Histochemistry and Cell Biology of Autonomic Neurons, SIF Cells, and Paraneurons* (O. Eranko, S. Soinila, and H. Paivarinta, eds.), pp. 61–68, Raven Press, New York.
25. Unsicker, K., Krisch, B., Otten, U., and Thoenen, H., 1978, Nerve growth factor-induced fiber outgrowth from isolated rat adrenal chromaffin cells: Impairment by glucocorticoids, *Proc. Natl. Acad. Sci. USA* **75**:3498.
26. Unsicker, K., Rieffert, B., and Ziegler, W., 1980, Effects of cell culture conditions, nerve growth factor, dexamethasone, and cyclic AMP on adrenal chromaffin cells *in vitro*, in: *Histochemistry and Cell Biology of Autonomic Neurons, SIF Cells, and Paraneurons* (O. Eranko, S. Sionila, and H. Paivarinta, eds.), pp. 51–59, Raven Press, New York.
27. Wurtman, R. J., Pohorecky, L. A., and Baliga, B. S., 1972, Adrenocortical control of the biosynthesis of epinephrine and proteins in the adrenal medulla, *Pharmacol. Rev.* **24**:411.
28. Björklund, A., Schmidt, R. H., and Stenevi, U., 1980b, Functional reinnervation of the neostriatum in the adult rat by use of intraparenchymal grafting of dissociated cell suspensions from the substantia nigra, *Cell Tissue Res.* **212**:39.

29. Perlow, M. J., Kumakura, K., and Guidotti, A., 1980, Prolonged survival of bovine adrenal chromaffin cells in rat cerebral ventricles, *Proc. Natl. Acad. Sci. USA* **77**:5278.

30. Murphy, J. E., and Sturm, E., 1923, Conditions determining the transplantability of tissues in the brain, *J. Exp. Med.* **38**:183.

31. Das, G. D., Hallas, B. H., and Das, K. G., 1979, Transplantation of neural tissues in the brains of laboratory mammals: Technical details and comments, *Experientia* **35**:143.

32. Rosenstein, J. M., and Brightman, M. W., 1978, Intact cerebral ventricle as a site for tissue transplantation, *Nature (London)* **276**:83.

33. Perlow, M. J., Freed, W. J., Hoffer, B. J., Seiger, A., Olson, L., and Wyatt, R. J., 1979, Brain grafts reduce motor abnormalities produced by destruction of nigrostriatal dopamine system, *Science* **204**:643.

34. Freed, W. J., Perlow, M. J., Karoum, F., Seiger, Å., Olson, L., Hoffer, B. J., and Wyatt, R. J., 1980, Restoration of dopaminergic function by grafting of fetal rat substantia nigra to the caudate nucleus: Long-term behavioral, biochemical, and histochemical studies, *Ann. Neurol.* **8**:510.

35. Freed, W. J., Morihisa, J. M., Spoor, E., Hoffer, B. J., Olson, L., Seiger, Å., and Wyatt, R. J., 1981, Transplanted adrenal chromaffin cells in rat brain reduce lesion-induced rotational behavior, *Nature (London)* **292**:351.

36. Morihisa, J. M., Cannon-Spoor, H. E., Wyatt, R. J., and Freed, W. J., 1981, Alteration of drinking behavior by transplantation of kidney cortex to the rat brain, *Soc. Neurosci. Abstr.* **7**:265.

37. Freed, W. J., Spoor, H. E., Sachs, D., and Wyatt, R. J., 1982, Immunological privilege in the brain with respect to functional brain grafts, *Soc. Neurosci. Abstr.* **8**:141.

38. Freed, W. J., and Wyatt, R. J., 1980, Transplantation of eyes to the adult rat brain: Histological findings and light-evoked potential response, *Life Sci.* **27**:503.

39. Lund, R. D., and Hauschka, S. D., 1976. Transplanted neural tissue develops connections with host rat brain, *Science* **193**:582.

40. Björklund, A., Dunnett, S. B., Stenevi, U., Lewis, M. E., and Iversen, S. D., 1980, Reinnervation of the denervated striatum by substantia nigra transplants: Functional consequences as revealed by pharmacological and sensorimotor testing, *Brain Res.* **199**:307.

41. Medawar, P. B., 1948, Immunity to homologous grafted skin. III. The fate of skin homografts transplanted to the brain, to subcutaneous tissue, and to the anterior chamber of the eye, *Br. J. Exp. Pathol.* **29**:58.

42. Stenevi, U., Björklund, A., Dunnett, S. B., and Gage, F. H., 1982, Cross-species neural grafting: Survival and function in an animal model of Parkinson's disease, *Soc. Neurosci. Abstr.* **8**:748.

43. Falck, B., Hillarp, N. -Å., Thieme, G., and Torp, A., 1962, Fluorescence of catecholamines and related compounds condensed with formaldehyde, *J. Histochem. Cytochem.* **10**:348.

44. de la Torre, J. C., 1980, An improved approach to histofluorescence using the SPG method for tissue monoamines, *J. Neurosci. Methods* **3**:1.

45. Church, A. C., Bunney, B. S., and Krieger, N. R., 1982, Neuronal localization of dopamine-sensitive adenylate cyclase within the rat olfactory tubercule, *Brain Res.* **234**:369.

46. Olson, L., Seiger, Å., Hoffer, B., and Taylor, D., 1979, Isolated catecholaminergic projections from substantia nigra and locus coeruleus to caudate, hippocampus, and cerebral cortex formed by intraocular sequential double brain grafts, *Exp. Brain Res.* **35**:47.

47. Stenevi, U., Emson, P., and Björklund, A., 1977, Development of dopamine sensitive adenylate cyclase in hippocampus reinnervated by transplanted dopamine neurons: Evidence for new functional contacts, *Acta Physiol. Scand. Suppl.* **452**:39.

48. Karoum, F., Garrison, C. K., Neff, N., and Wyatt, R. J., 1977, Trans-synaptic modulation of dopamine metabolism in the rat superior cervical ganglion, *J. Pharmacol. Exp. Ther.* **210**:654–661.

49. Karoum, F., and Neff, N. H., 1979, Analysis of dopamine and its metabolites in biological materials by mass fragmentography, in: *The Neurobiology of Dopamine* (A. S. Horn, J. Horf, and B. H. C. Westerink, eds.), pp. 63–76, Academic Press, New York.

50. Coupland, R. E., and MacDougall, I. D. B., 1966, Adrenaline formation in noradrenaline-storing chromaffin cells *in vitro* induced by corticosterone, *J. Endocrinol.* **36**:317.

51. Goochee, C., Rasband, W., and Sokoloff, L., 1980, Computerized densitometry and color coding of 14C deoxyglucose autoradiographs, *Ann. Neurol.* **7**:359.

52. Thoenen, H., 1975, Transsynaptic regulation of neuronal enzyme synthesis, in: *Handbook of Psycho-pharmacology,* Vol. 3 (L. L. Iversen, S. D. Iversen, and S. H. Synder, eds.), pp. 443–475, Plenum Press, New York.

53. Lewis, P. R., and Shute, C. C. D., 1969, An electron-microscopic study of cholinesterase distribution in rat adrenal medulla, *J. Mιcrosc. (Oxford)* **89:**181.

54. Freed, W. J., Karoum, F., Spoor, H. E., Morihisa, J. M., Olson, L., and Wyatt, R. J., 1983, Cate-cholamine content of intracerebral adrenal medulla grafts, *Braın Res.* **269:**184–189.

55. Patrick, R. L., and Kirshner, N., 1971, Effect of stimulation on the levels of tyrosine hydroxylase, dopamine β-hydroxylase, and catecholamines in intact and denervated rat adrenal glands, *Mol. Pharmacol.* **7:**87.

56. Snider, S. R., and Carlsson, A., 1972, The adrenal dopamine as an indicator of adrenomedullary hormone biosynthesis, *Naunyn-Schmιedebergs Arch. Pharmacol.* **275:**347

57. Lishajko, F., 1970, Dopamine secretion from the isolated perfused sheep adrenal, *Acta Physιol. Scand.* **79:**405.

58. Hökfelt, T., and Ungerstedt, U., 1969, Electron and fluorescence microscopical studies on the nucleus caudatus putamen of the rat after unilateral lesion of the ascending nigro-neostriatal dopamine neurons, *Acta Physιol. Scand.* **76:**415.

59. Heybach, J. P., Coover, G. D., and Lints, C. E., 1978, Behavioral effects of neurotoxic lesions of the ascending monoamine pathways in the rat brain, *J. Comp. Physιol. Psychol.* **92:**58.

60. Stricker, E. M., and Zigmond, M. J., 1976, Recovery of function after damage to central catecholamine-containing neurons: A neurochemical model for the lateral hypothalamic syndrome, *Prog. Psychobιol. Physιol. Psychol.* **6:**121

61. Ungerstedt, U., 1971, Adipsia and aphagia after 6-hydroxydopamine induced degeneration of the nigro-striatal dopamine system, *Acta Physιol. Scand.* **367:**95.

62. Ungerstedt, U., 1974, Brain dopamine neurons and behavior, in: *Neuroscιences: Thιrd Study Program* (F. O. Schmitt and F. G. Worden, eds.), pp. 695–703, MIT Press, Cambridge, Mass.

63. Ungerstedt, U., 1980, Behavioral pharmacology reflecting catecholamine neurotransmission, in: *Handbook of Experιmental Pharmacology,* Vol. 54 (L. Szekeres, ed.), pp. 499–519, Springer-Verlag, Berlin.

64. Kornhuber, H. H., 1974, Cerebral cortex, cerebellum, and basal ganglia: An introduction to their motor functions, in: *The Neuroscιences: Thιrd Study Program* (F. O. Schmitt and F. G. Worden, eds.), pp. 267–280, MIT Press, Cambridge, Mass.

65. Ungerstedt, U., 1971, Striatal dopamine release after amphetamine or nerve degeneration revealed by rotational behavior, *Acta Physιol. Scand.* **367:**49.

66. Ungerstedt, U., 1971, Postsynaptic supersensitivity after 6-hydroxydopamine induced degeneration of the nigro-striatal dopamine system, *Acta Physιol. Scand.* **367:**69.

67. Creese, I., and Synder, S. H., 1979, Nigrostriatal lesions enhance striatal ^{3}H-apomorphine and ^{3}H-spiroperidol binding, *Eur. J. Pharmacol.* **56:**277.

68. Kreuger, B. K., Forn, J., Walters, J. R., Roth, R. N., and Greengard, P., 1976, Stimulation by dopamine of adenosine cyclic 3′, 5′-monophosphate formation in rat caudate nucleus: Effect of lesions of the nigro-neostriatal pathway, *Mol. Pharmacol.* **12:**639.

69. Schultz, W., and Ungerstedt, U., 1978, Striatal cell supersensitivity to apomorphine in dopamine-lesioned rats correlated to behavior, *Neuropharmacology* **17:**349.

70. Walsh, M. J., and Silbergeld, E. K., 1979, Rat rotation monitoring for pharmacology research, *Pharmacol. Bιochem. Behav.* **10:**433.

71. Carman, J. B., Cowan, W. M., Powell, T. P. S., and Webster, K. E., 1965, A bilateral cortico-striate projection, *J. Neurol. Neurosurg. Psychιatry* **28:**71.

72. Björklund, H., Seiger, Å., Hoffer, B., and Olson, L., 1982, Trophic effects of brain areas on the developing cerebral cortex. I. Growth and histological organization of intraocular grafts, *Dev. Braın Res.* **6:**131–140

73. Björklund, A., and Stenevi, U., 1979, Reconstruction of the nigrostriatal dopamine pathway by intra-cerebral nigral transplants, *Braın Res.* **177:**555.

74. Dunnett, S. B., Björklund, A., Stenevi, U., and Iversen, S. D., 1981, Behavioral recovery following

transplantation of substantia nigra in rats subjected to 6-OHDA lesions of the nigrostriatal pathway. I. Unilateral lesions, *Brain Res.* **215**:147.

75. Björklund, A., Stenevi, U., Dunnett, S. B., and Iversen, S. D., 1981, Functional reactivation of the deafferented neostriatum by nigral transplants, *Nature (London)* **289**:497.

76. Trendelenburg, U., 1963, Supersensitivity and subsensitivity to sympathomimetic amines, *Pharmacol. Rev.* **15**:225.

77. Trendelenburg, U., 1966, Mechanisms of supersensitivity and subsensitivity to sympathomimetic amines, *Pharmacol. Rev.* **18**:629.

78. Fleming, W. W., 1976, Variable sensitivity of excitable cells: Possible mechanisms and biological significance, in: *Reviews of Neuroscience,* Vol. 2 (S. Ehrenpreis and I. J. Kopin, eds.), pp. 43–90, Raven Press, New York.

79. Wyatt, R. J., and Freed, W. J., Grafting dopamine-containing cells into the striatal region of substantia nigra lesioned rats, in: *Proceedings of World Health Organization's Study Group on Neuroplasticity and Repair in the Central Nervous System.* C. L. Bolis (ed)., Raven Press, N.Y., in press

80. Fink, J. S., and Smith, G. P., 1980, Mesolimbicortical dopamine terminal fields are necessary for normal locomotor and investigatory exploration in rats, *Brain Res.* **199**:359.

81. Kelly, P. H., Seviour, P. W., and Iversen, S. D., 1975, Amphetamine and apomorphine responses in the rat following 6-OHDA lesions of the nucleus accumbens septi and corpus striatum, *Brain Res.* **94**:507.

82. Fray, P. J., Dunnett, S. B., Iversen, S. D., Björklund, A., and Stenevi, U., Nigral transplants reinnervating the dopamine-depleted neostriatum can sustain intracranial self-stimulation, *Science* **219**:416–419.

83. Dunnett, S. B., Björklund, A., Stenevi, U., and Iversen, S. D., 1981. Behavioral recovery following transplantation of substantia nigra in rats subjected to 6-OHDA lesions of the nigrostriatal pathway. II. Bilateral lesions, *Brain Res.* **229**:457.

84. Dunnett, S. B., Björklund, A., Stenevi, U., and Iversen, S. D., 1981, Grafts of embryonic substantia nigra reinnervating the ventrolateral striatum ameliorate sensorimotor impairments and akinesia in rats with 6-OHDA lesions of the nigrostriatal pathway, *Brain Res.* **229**:209.

85. Ljungberg, T., and Ungerstedt, U., 1976, Sensory inattention produced by 6-hydroxydopamine-induced degeneration of ascending dopamine neurons in the brain, *Exp. Neurol.* **53**:585.

86. Marshall, J. F., and Teitelbaum, P., 1974, Further analysis of sensory inattention following lateral hypothalamic damage in rats, *J. Comp. Physiol. Psychol.* **86**:375.

87. Teitelbaum, P., and Epstein, A. N., 1962, The lateral hypothalamic syndrome: Recovery of feeding and drinking after lateral hypothalamic lesions, *Psychol. Rev.* **691**:74.

88. Goldberg, S. R., Hoffmeister, F., Schlichting, U. U., and Wuttke, W., 1971, A comparison of pentobarbital and cocaine self-administration in rhesus monkeys: Effects of dose and fixed-ratio parameter, *J. Pharmacol. Exp. Ther.* **179**:277.

89. Pickens, R., and Thompson, T., 1968, Cocaine-reinforced behavior in rats: Effects of reinforcement magnitude and fixed-ratio size, *J. Pharmacol. Exp. Ther.* **161**:122.

90. Wuerthele, S. M., Freed, W. J., Olson, L., Morihisa, J., Spoor, L., Wyatt, R. J., and Hoffer, B. J., 1981, Effect of dopamine agonists and antagonists on the electrical activity of substantia nigra neurons transplanted into the lateral ventricle of the rat, *Exp. Brain Res.* **44**:1.

91. Groves, P. M., Wilson, C. J., Young, S. J., and Rebec, G. V., 1975, Self-inhibition by dopaminergic neurons, *Science* **190**:522.

92. Gale, K., Costa, E., Toffano, G., Hong, J.-S., and Guidotti, A., 1978, Evidence for a role of nigral gamma-aminobutyric acid and substance P in the haloperidol-induced activation of striatal tyrosine hydroxylase, *J. Pharmacol. Exp. Ther.* **206**:29.

93. Buchwald, N. A., Price, D. D., Vernon, L., and Hull, C. D., 1973, Caudate intracellular response to thalamic and cortical inputs, *Exp. Neurol.* **38**:311.

94. Woodruff, M. F. A., 1960, *The Transplantation of Tissues and Organs,* Thomas, Springfield, Ill.

95. de Groot, J., 1959, *The Rat Forebrain in Stereotaxic Coordinates, Verh. K. Ned. Akad. Wet. Afd. Natuurkd.* **52**:1–40.

17

Transplantation Strategies in Spinal Cord Regeneration

HOWARD NORNES, ANDERS BJÖRKLUND, AND ULF STENEVI

1. Introduction

Transplantation models in the brain have proven successful under conditions in which transplants serve as a "bridge" for the regeneration of axons across a site of injury, or as "release" or "driving" units to replace missing inputs to a particular target area.[1-8] Similar models have been applied to the mammalian spinal cord including: (1) intraspinal transplants to form a bridge for the regeneration of spinal cord axons, (2) extraspinal transplants with only the end or ends of the transplants inserted into the cord to bypass the region of injury, and (3) intraspinal neural implants to replace missing supraspinal inputs. These models demonstrate that neurons of the spinal cord possess the ability to regenerate axons several millimeters into both intrinsic and extrinsic transplants; however, the growth of these axons into the tissue of the host spinal cord has been limited. By contrast, embryonic CNS neurons transplanted into the adult spinal cord, possess the ability to grow axons that penetrate several millimeters into spinal cord tissue. This review provides an overview of the attempts to promote regeneration of spinal cord connections by using various transplantation paradigms.

HOWARD NORNES ● Department of Anatomy, Colorado State University, Fort Collins, Colorado 80523.   ANDERS BJÖRKLUND AND ULF STENEVI ● Departments of Histology and Ophthalmology, University of Lund, Lund, Sweden.

2. Spinal Cord Transplantation Models

2.1. Intraspinal Transplants for Axonal Bridging

2.1.1. Adult Peripheral Tissue

The use of segments of peripheral nerves for making intraspinal transplants has a long history. Early in the century, Tello[9] and Ramón y Cajal[10] proposed the use of peripheral nerve grafts because of their prominent ability to promote axonal regeneration of peripheral neurons. Axons from the proximal stump of a cut peripheral nerve regenerate long distances within the Schwann cell environment of the distal nerve stump,[10] and studies, described in this section, similarly demonstrate that this environment also can support regeneration of CNS neurons. However, numerous studies in the earlier literature along these lines remain notoriously controversial and inconclusive, because of the inadequacy of the classical silver staining techniques to identify the origins of axons appearing in the grafts and difficulties inherent in the functional analysis of spinally transected animals.[11,12]

Sugar and Gerard[13] were the first to apply this strategy for the promotion of spinal cord regeneration. They aroused much attention by reporting functional recovery in rats with transected cords. Degenerated sciatic nerves from donor rats (homografts) were inserted into the 2- to 3-mm gap at the site of the spinal transection. Physiological recovery was reported in 13 of 32 implanted animals that survived more than 4 weeks. A return of coordinated movement and a hindlimb response to electrical stimulation of the brain stem were reported. They also claimed that the anatomical observations correlated well with the physiological results. In the 13 animals showing functional recovery, "new fiber bundles connected the cord stumps across the scar, roughly corresponding with the amount of functional return." However, the origins of these axons were never assessed.

Three subsequent studies did not support the interpretations of Sugar and Gerard. Barnard and Carpenter[14] repeated the sciatic nerve transplant experiments in rats using both auto- and homografts of fresh and degenerated segments of nerves. Sixty of their rats survived beyond 25 days and 40 beyond 35 days, and in no case was functional recovery observed beyond that of a spinal animal. In both transplant and control animals, a slight touch of the foot or adjacent tail region resulted in rapid kicking movements and flexion of the lower part of the body. During locomotion, the legs usually dragged and occasionally had alternating stepping movements. Electrical stimulation on the surface of the cord, cephalic to the transplant, did not produce a hind leg response across the transplant. Nerve fibers were reported to penetrate into the transplant, but none could be traced across the entire transplant. They occurred singly and in no instance were bundles of nerve fibers observed as described by Sugar and Gerard. With this limited regeneration of axons and lack of functional recovery, Bernard and Carpenter suggested that the positive results of Sugar and Gerard were due to incomplete transsections of the cords in their host animals.

Feigen *et al.*[15] repeated the sciatic nerve experiment in rats, using mainly autologous grafts of fresh and degenerated segments of nerve. One hundred and two rats received transplants and 46 survived 30 days, 34 for 60 days, and 40 for 120–400 days.

No difference in reflex activity was observed between the animals with and without transplants. A motor response was observed in the hind legs in nearly all of their spinal animals following crude stimulation of the brain with an electroshock apparatus. This response was observed as early as 4 days following transection, which was likely to be earlier than the time required for any axon to regenerate across the spinal gap. Animals that subsequently had large gaps and dense scar tissue at the site of the transection also showed this hind leg response. The interpretation was that this response was not due to regeneration, but to relay of the stimulus through the periphery. Feigen *et al.* suggested that this peripheral pathway also could explain the hind leg response observed by Sugar and Gerard.

Brown and McCouch[16] repeated the peripheral nerve transplant experiment in 17 dogs and 34 cats. Segments of both fresh and degenerated nerves, stripped of their sheaths, were implanted into the gap at the site of the transection. In other animals an emulsion of degenerated nerve was painted on the stumps of the cord. Electrical stimulation of the cerebral cortex or brain stem produced no detectable motor responses below the transplant, and their anatomical studies revealed a typical abortive regeneration into the scar-transplant tissue.

The studies repeating the Sugar–Gerard experiment thus neither provided evidence for the regeneration of spinal cord axons across the transplant to the opposite stump of the cord, nor did they show any recovery of motor and sensory functions greater than in nontransplanted, spine transected animals. The conclusion is that Sugar and Gerard misinterpreted their results.

Following these early experiments, a significant improvement was made in the surgical methods for making intraspinal transplants with the introduction of the subpial spinal transection. Kao *et al.*[17,18] made sciatic nerve, nodose ganglion, or brain tissue transplants into subpial transection sites. Although no improvement of function was reported using this surgical modification, the invasion of cells forming the collagenous scar was eliminated or significantly reduced. The method also preserved the pia–arachnoid membrane, which presumably provided better blood and CSF flow to the transplant. Nevertheless, the pathological changes intrinsic to the cord still occurred. Astroglial scarring and cavities formed at each cord–transplant junction. An attempt to eliminate this problem was made by using a delayed implantation method.[19,20] One week after transection, when the process of cavitation neared completion, the necrotic tissue was removed from the stumps of the cord and a sciatic nerve was placed into the transection gap in contact with the fresh stumps. The cavitation at the transplant–cord junction was eliminated by this second operation; however, the glial basement membrane (astroglial scar) still formed at each end of the spinal cord.

There was no doubt that axons had regenerated into intrinsic transplants made by these early investigators, but a common problem was knowing the origin of the regenerated fibers. With silver staining methods, it was impossible to know if they were from spinal cord, dorsal root ganglia, or autonomic neurons. To clarify this issue, Richardson *et al.*[21,22] performed dorsal root lesion and axonal tracer studies to confirm that many axons appearing in an intraspinal transplant of peripheral nerve were indeed from spinal neurons. Segments of autologous sciatic nerves were placed into the gaps of subpially transected cords in rats. Following survival times of 1–4 months, some axons in the grafts originated from cell bodies in the nearby segments of the spinal cord; others originated from dorsal root ganglia up to seven segments distal to the level of the transplant. No claim was made that any of these fibers crossed the entire transplant and into the opposite stump of the cord. There also was no evidence for the regeneration of long descending tracts into the transplants. Again, no functional recovery attributable to the transplant was reported.

Cultured Schwann cells also have been added to the peripheral nerve bridge preparation to further enhance regeneration of spinal cord axons. Previous work had shown that transplanted Schwann cells can remyelinate experimentally demyelinated axons of the spinal cord.[23,24] Nonneuronal cells cultured from dorsal root ganglia were added around the segments of peripheral nerve inserted at the site of transection.[25] Enhanced wound healing and axon growth were reported for the animals with the cultured cells, but, again, the identity of the axons (peripheral or central) was not ascertained.

Intraspinal transplants of smooth muscle have also been used to investigate the regenerative ability of the descending catecholamine fibers originating from the brain stem.[26] A mitral valve or iris was inserted into the cord with a glass rod. The catecholamine fibers, visualized with the histofluorescence method, regenerated through the necrotic zone and into the transplant. The terminal pattern formed in the grafts was similar to that of the normal innervation of the mitral valve or iris, again showing an affinity between the axons and their target tissue. Adult musculature,[13] cerebellum,[17,27] and cranial ganglia[18] tissues also have been transplanted into the cord with unnoteworthy results.

2.1.2. Embryonic CNS Tissue

Embryonic CNS tissue has the advantage of providing growing axons from the bridging tissue into the spinal cord of the host. This creates the potential for constructing a relay-type bridge if host spinal axons can regenerate and form synaptic connections in the embryonic bridging tissue.

Sugar and Gerard[13] initiated the intraspinal embryonic transplant experiments by using cerebral cortex tissue from 30- to 40-mm rat fetuses. In most cases, the embryonic tissue was displaced from the gap at the site of transection, presumably during the initial surgical procedure. When the embryonic tissue did remain in place, either a dense degenerated mass or a random cluster of cells formed that did little or nothing to promote regeneration.

Recent studies have extended this earlier work and demonstrated a remarkable growth potential of CNS embryonic tissue when placed in the adult spinal cord.[28-30] Embryonic tissue of rat hippocampus, spinal cord, and brain stem was placed into 1- to 3-mm gaps in subpially transected spinal cord of adult rats. Tissue taken from appropriate embryonic stages filled the gaps and formed regions of continuous tissue union with the host cords. The tissue bridge illustrated in Figure 1 was formed from a graft comprising the entire lower brain stem of an 11- to 12-day-old rat fetus (crown–rump length 7.5 mm) placed into the gap of a subpially transected cord (T_{10}–T_{11}) in an adult rat. Five months following transplantation, the graft had reached a length of about 3.5

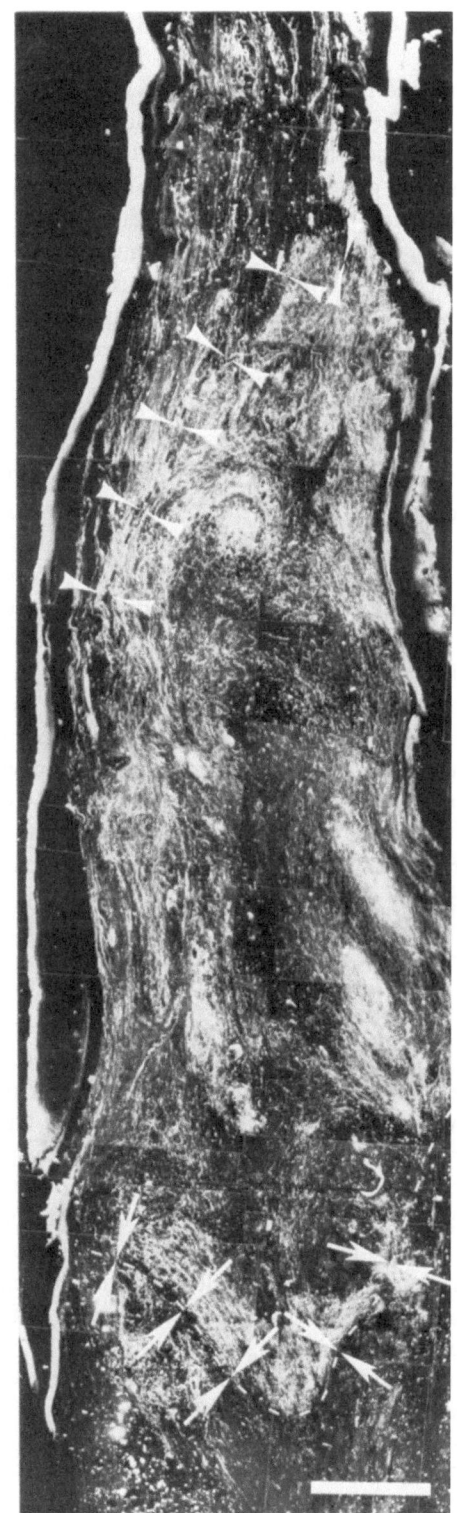

FIGURE 1. Rhombencephalic transplant from an 11- to 12-day-old fetus into the gap of a subpial transection of the spinal cord (T_{10}–T_{11}) in an adult rat (specimen 15 J, 5 months' survival). Fluorescence photomontage from a horizontal section through the most ventral part of the graft. At sacrifice, the transplant was about 3–3½ mm long and had fused well with both rostral (arrowheads) and caudal (arrows) stumps of the cord. The graft contains numerous clustered CA-containing cell bodies and rich CA-fluorescent terminal plexuses. At the rostral graft–cord junction, abundant CA fibers pass between the host cord and the graft. These fibers may in part represent supraspinal noradrenergic axons that have grown into the brain stem graft from the rostral stump of the severed cord. × 110. From Ref. 29 with permission.

mm and fused well with both rostral (arrowheads) and caudal (arrows) stumps of the cord. The transplant contained numerous clusters of catecholamine (CA) cell bodies and a rich fiber plexus. At the rostral cord junction, CA fibers passed between the host cord and the graft, which may represent supraspinal noradrenergic axons grown into the brain stem graft from the rostral stump of the severed cord. It was a common observation that CA fibers from the rostral stump of the cord had regenerated into the embryonic tissue bridge. Neurons in the transplant were reported to grow axons from distances of up to 16–25 mm into the host spinal cord tissue.[28–30] Functional recovery attributable to the transplants has not been observed.

2.1.3. Synthetic Material

Transplants of synthetic material have the advantage of providing readily available donor stock and also eliminating the potential immune-rejection problem. One method has been to wrap a 1-cm strip of Millipore filter around the site of transection to form a cuff for the regenerating axons.[31,32] This has been attempted in the adult feline cord in the upper thoracic region. The cuff was reported to eliminate constrictive meningeal scarring, and groups of linearly arranged axons of unknown origin were observed immediately deep to the pia–arachnoid complex in the Millipore-protected animals. No return of function was observed. A second method uses a cell-free bovine collagen material to establish a bridge at the site of transection in spinal cord of rats.[33] In these experiments, the cord was transected and the gap was filled with the semifluid collagen, which subsequently solidified. After 90 days, a few CA fibers had regenerated from the rostral stump of the cord into the collagen bridge, and it was claimed that some of them actually crossed the entire bridge to the distal stump of the cord. No clear evidence of functional recovery was observed.

2.2. Extraspinal Transplants for Axonal Bridging

Peripheral nerve transplants were made extraspinally to bridge regions of the cord or to bridge between cord and brain stem. This approach, which originated in the 1950s, made use of two surgical designs. In one type, a peripheral nerve above or below the transection was cut and inserted into the spinal stump on the opposite side[34–36]; the second method was to dissect out a segment of a peripheral nerve and insert one end into the brain and the other into the spinal cord.[37]

Two investigators used the first surgical method and reported promising results. An intercostal nerve originating from the cephalic side of the transection was inserted into the distal isolated segment of the cord. Turbes and Freeman[34] reported recovery of "reflex standing and walking" in 25 of 30 dogs with this type of bridging, a condition that rarely occurred in their laboratory following surgical transection of the cord. Interestingly, surgical resection of this inserted nerve was reported to result in loss of function. Galabov[36] obtained similar results with the same type of experiments using four dogs. The reciprocal experiment also was done where an intercostal nerve originating from the distal isolated segment of a transected cord was inserted into the proximal stump.[35]

Good functional recovery was reported in 6 of 19 animals, and electrical stimulation above or below the points of origin or insertion of the nerve bridge showed conduction of electrical potentials in both directions through the bridge.

The second surgical method for making extraspinal transplant bridges has been used more recently by David and Aguayo on rats.[37] A 35-mm segment of sciatic nerve was completely dissected free and placed in the subcutaneous tissue along the back of the animal. One end was inserted into the medulla oblongata and the other into the lower cervical/upper thoracic region of the spinal cord. Axons grew in both directions through this extradural bridge as revealed by HRP tracing studies. Cells from both sides of the bridge grew axons 35 mm through the peripheral nerve and approximately 2 mm beyond the end of the inserted nerve into the host cord. This is the first conclusive evidence that intrinsic spinal cord neurons have the ability to regenerate axons over a relatively long distance. This regeneration occurred within the Schwann cell environment of the peripheral nerve bridge and essentially stopped when entering the host CNS tissue.

2.3. Intraspinal Transplants for Replacing Supraspinal Inputs

This strategy involves implanting embryonic supraspinal neurons, which normally form descending spinal tracts, directly into the isolated distal portion of the transected spinal cord. Among the supraspinal systems lost following spinal transection are the norepinephrine- and serotonin-containing innervations whose cells of origin are located in the brain stem. These monoamine systems have strong effects on the activity of the mammalian spinal cord, including stepping and reflexes.[38-41]

Transplants of both embryonic locus coeruleus (norepinephrine-containing) and raphe (5-hydroxytryptamine-containing) tissue survived in the adult spinal cord of a rat when inserted by a glass pipette[42] or placed in a small subpial cavity made by aspiration.[29,30] Figure 2 is a camera lucida drawing of the distribution of CA fibers from an embryonic locus coeruleus implant into the adult spinal cord of a rat after 6 months' survival. The host spinal cord had been denervated of its normal adrenergic innervation with 6-hydroxydopamine prior to transplantation. The transplant (T) fused well with the host spinal cord, and growth of CA fibers extended bilaterally up to a distance of about 12 mm. The density of the terminal network was in the normal range for a distance up to 5.5 mm from the graft (level b). Non-adrenergic neurons in the transplant also grew axons into the host spinal cord. They grew even farther (at least 16 mm) from the transplant than the CA fibers as revealed by retrograde tracer studies (see Fig. 2). No gross functional changes were evident in these animals with monoamine transplants based on subjective observations (unpublished observations).

3. Discussion

The four different transplantation strategies currently being explored in the adult mammalian spinal cord are schematically summarized in Figure 3. Intraspinal (Fig.

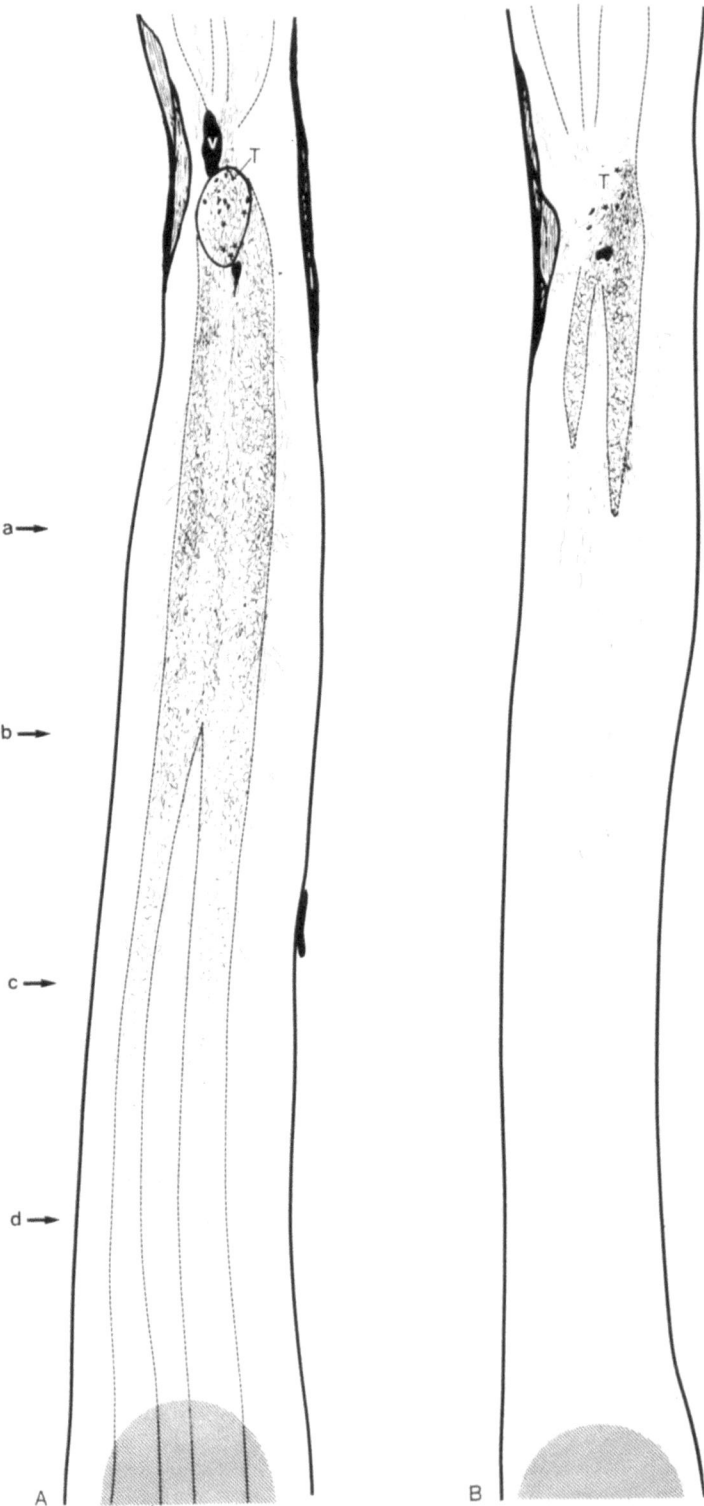

3A) and extraspinal (Fig. 3B) bridge grafts are used with the aim of promoting axonal regeneration across the transection. Intraspinal bridge grafts of embryonic CNS tissue (Fig. 3C) are used with the idea of creating a relay for synaptic transmission across the transection, and intraspinal implants of embryonic brain stem neurons (Fig. 3D) are explored for their ability to substitute, anatomically and functionally, for lost supraspinal inputs of the isolated cord. Each of these experimental approaches to spinal cord regeneration are discussed in further detail below.

3.1. Intraspinal Transplants for Axonal Bridging

Transplants were placed at the site of injury/transection with the objective of promoting the regeneration of the damaged spinal cord axons (Fig. 3A). Adult tissues (primarily peripheral nerve), embryonic tissues, and synthetic materials have been used. Regenerated fibers were observed in all the above transplant types, and recently it has been shown that some of the fibers originated from neurons in the spinal cord. A bridge made by a peripheral nerve graft contained spinal cord axons originating from the nearby segments, as well as from dorsal root fibers from up to seven segments away from the transplant.[21,22] A limiting factor in this model has been the lack of regenerating axons through the host transplant and into the opposite stump of the cord. In similar experiments recently performed in the septohippocampal system of adult rats, it has been shown, however, that the lesioned septal cholinergic neurons of the host regenerated through a bridge graft to reinnervate part of the denervated hippocampus.[6] In this case, a graft of either embryonic hippocampus or adult iris was used. Electrophysiological experiments[43] also showed that these axons formed synaptic connections with neurons in the host hippocampus.

←

FIGURE 2. Camera lucida drawings of the distribution of CA fibers from a locus coeruleus transplant into an adult spinal cord. The transplant was placed into a small subpial cavity in the lower thoracic region 6 months before sacrifice. The host spinal cord had been denervated of its normal adrenergic innervation with 6-OH-DA prior to transplantation. The shaded area in the most distal (caudal) region of the spinal cord in the drawings illustrates the area of intense fluorescence at the site of injection of the fluorescent tracer True Blue. Specimen 13 I; crown–rump length of donor fetus 16 mm. (A) Drawing from a horizontal section through the intermediate gray including the transplant (T) and the ventral gray more caudally. The transplant fuses well with the host tissue, and both CA-containing neurons (circles) and True Blue-labeled cell bodies (stars) are scattered throughout the transplant. An ependymal-lined cavity (V) is associated with the transplant. The cavity seems to be in continuity with the central canal. Note the growth of CA fibers from the transplant bilaterally in the intermediate and ventral gray matter. Roughly, normal density of CA fibers exist up to about 3–5.5 mm from the transplant (levels a, b). They gradually become more sparse 8.5 mm (level c) and 11 mm (level d) from the graft. True Blue fluroescing cell bodies in the transplant show that the axons from non-adrenergic cells in the transplant extend caudally at least 13–14 mm from the transplant. Only a few CA fibers have grown rostrally in this specimen. (B) Drawing of a section ventral to the one in (A), showing the more ventral aspect of the transplant in specimen 13 I. Scattered CA-containing (circles) and non-adrenergic True Blue-labeled cell bodies (stars) are in the transplant (T), and large numbers of CA fibers have grown out from the transplant bilaterally into the ventral gray. They continue in the white matter for at least 8–9 mm. From Ref. 29 with permission.

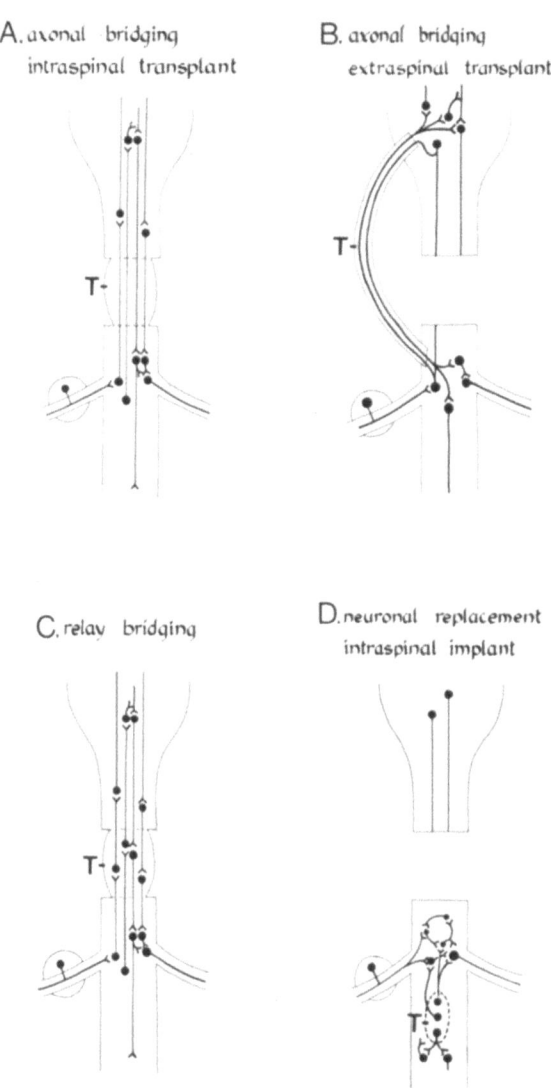

FIGURE 3. (A, B) Models designed to promote regeneration of host axons, either through or around the site of injury. The intraspinal transplant (T) in (A) forms a bridge across the site of injury for both ascending and descending fibers. The extraspinal transplant (T) in (B) forms a bypass pathway at the site of injury for both ascending and descending fibers. A modification of (B) is discussed in the text in which a peripheral nerve above or below the transection is cut and inserted into the opposite stump of the spinal cord to bypass the site of injury. (C, D) Models in which embryonic CNS tissues are used. In the relay bridge (C), the host axons grow and form synapses in the transplant (T), and the cells in the transplant grow into and form synapses in the opposite side of the site of transection. In the neuronal replacement model (D), embryonic brain stem neurons that normally form descending pathways into the cord are inserted directly into the isolated spinal cord to drive the intrinsic pattern generators.

3.2. Extraspinal Transplants for Axonal Bridging

The transplants in this case were positioned outside the vertebral column, with one or both ends of the peripheral nerve inserted into the spinal cord or brain stem tissue, in order to bypass the site of injury (Fig. 3B). Two bypass surgical designs have been used: (1) a segment of peripheral nerve was laid along the vertebral column with the two ends inserted into the cord to span a region, and (2) a peripheral nerve was cut, which originated immediately above or below the site of transection, and its proximal end was inserted into the opposite segment of the spinal cord. The significant finding in the first type of preparation was that CNS axons possess the capability of regenerating long distances (up to 35 mm) through the peripheral nerve bridges.[37] Unfortunately, when the axons again enter CNS tissue at the opposite end of the bridge, regeneration essentially stops; they extend only 1–2 mm in the host spinal cord. The latter model of using an intact peripheral nerve stump has not been rigorously tested, although the three published reports[34–36] claimed some functional recovery.

3.3. Intraspinal Transplants for Relay Bridging

Embryonic CNS tissue was used in these experiments. It has been shown that central neurons can regenerate into intraspinal transplants of embryonic CNS tissue, and that neurons in the transplant can grow axons for distances of at least 16–25 mm in the host spinal cord.[28–30] If functional synapses were formed in this anatomical relay bridge, it seems conceivable that a spinospinal pathway could be restored across the graft. Against the background of previous studies, the distances of axon growth from the embryonic neural transplants into the host cord are quite remarkable, particularly when realizing that adult neurons, both central and peripheral, have not been observed to grow more than 1–2 mm within the spinal cord tissue.[37,44–46] Even in lamprey, which has been reported to recover swimming and crawling behavior following transection of the cord, the majority of the regenerating spinal cord axons terminated within 300 μm from the center of the transection, and none were observed to extend beyond 3.6 mm.[48]

The relay type of bridge at present is a promising bridging strategy because of the apparent inability of the damaged spinal cord neurons to regenerate their axons (1) across the entire transplant to the opposite stump of the cord and (2) any significant distance within the denervated spinal cord tissue. The functional capability of the isolated cord also suggests that a simple spino-spinal relay circuit may result in functional restoration. Locomotion has been shown to be generated by neuronal networks intrinsic to the cord. These intraspinal "pattern generators" are controlled by supraspinal systems that act tonically to drive or modulate their activity.[40,49,50] It seems conceivable that a more general type of input, which could be restored by a spino-spinal relay circuit, might be enough to modulate or activate a gating circuit, or to drive the pattern generators of the isolated spinal cord.

3.4. Intraspinal Transplants for Replacing Missing Supraspinal Inputs

The rationale of this model is to provide driving or releasing units to the intrinsic circuitry, i.e., pattern generators and/or reflexes, of the isolated spinal cord (see Fig. 3D). This approach gains support from the studies in brain-lesioned animals in which intracerebral neuronal implants result in restoration of function. It has been applied to several CNS systems, including the nigrostriatal dopamine pathway,[1-3,7] hypothalamic vasopressin- and gonadotropin-releasing systems,[5,8] and hippocampal and CA systems,[7,51,52] all of which are reviewed elsewhere in this volume.

In the adult spinal cord of mammals, this model has the potential for being used as a functional assay to determine the degree and specificity of reinnervation of the spinal cord from particular types of transplants. The brain stem monoamine systems are among the supraspinal systems that are lost following transection of the cord, and this model is currently being applied to these systems. The descending monoamine systems are known to have prominent effects on the segmental activity of the spinal cord. Thus, when DOPA, the precursor of the CAs, is injected into acutely spinalized cats, dramatic changes occur in spinal activity. Stepping appears, which cannot be evoked in the acute cat without DOPA, and the animal can walk on a moving treadmill adjusting its speed and gait to the speed of the treadmill.[53,54] In addition to its effect on locomotion, DOPA causes pronounced reflex changes in the spinal cord.[38,39,41] 5-HTP, the precursor to serotonin, elicits reflex changes in cats similar to those seen with DOPA,[41] and 5-HTP initiates locomotion-like discharges in rabbits.[55] With this in mind, it has been found that both embryonic noradrenaline and serotonin neurons survive when implanted into the denervated spinal cords of rat.[29,30] The noradrenaline cells were found to grow CA fibers up to 12 mm into the host spinal cord with a normal density of terminal plexuses up to 5.5 mm. Considering the influence of monoamines on spinal cord function, this model will provide possibilities to explore the functional specificity and reinnervation of these particular intraspinal embryonic transplants. This could be an important first step to the application of transplantation models to clinical medicine.

4. Concluding Remarks

In general, it is remarkable how relatively little progress has been made in transplantation approaches to spinal cord damage since the early studies of Sugar and Gerard[13] and Turbes and Freeman.[34] It is obvious that the numerous attempts that were made in the 1940s and 1950s were hampered by the inadequacy of the methodology available. It is only during the last decade that modern neuronal tracing techniques, and the increased knowledge of the chemically defined systems in the CNS, have provided the tools for the further exploration of the potentials of the various transplantation strategies for the restoration of structure and function following spinal cord damage. The important progress made over the last few years has been to reestablish the transplantation strategies as valid and interesting experimental models in spinal cord regeneration research. Today we know that the different intra- and extraspinal transplant bridges and intraspinal neural implants can work in a basic sense—we know that both periph-

eral and central neural tissues can be grafted with survival into the cords of adult experimental animals and, most importantly, that a whole range of neuronal regenerative
phenomena take place in association with these grafts. The assumption of the early
workers, such as Tello,[9] Sugar and Gerard,[13] and Campbell and colleagues,[31-32] that
the intrinsic neurons of the damaged cord can regenerate into grafts has been substantiated. There is evidence, in particular from the studies on embryonic neural implants,
that ingrowing axons can extend for relatively long distances within the denervated
spinal cord. Thus, the question now is not whether neural grafting to the spinal cord is
possible, but how neural transplants can be used to obtain structural and functional
restoration of defined components of the spinal cord circuitry. The value of current
transplant strategies lies not so much in their immediate application to clinical problems
of spinal cord injury, but in their potential as experimental models for the exploration
of the fundamental issues of axonal regeneration and recovery of function in the damaged spinal cord.

ACKNOWLEDGMENTS. This work was supported by grants from the Paralyzed Veterans
of America (OBR-156), the Spinal Cord Society, and the Swedish MRC (04x-3874).

References

1. Björklund, A., and Stenevi, U., 1979, Reconstruction of the nigrostriatal dopamine pathway by intracerebral nigral transplants, *Brain Res.* **17**:555.
2. Björklund, A., Stenevi, U., Dunnett, S. B., and Iversen, S. D., 1981, Functional reactivation of the
   deafferented neostriatum by nigral transplants, *Nature (London)* **289**:497.
3. Perlow, M. J., Freed, W. J., Hoffer, B. J., Seiger, Å., Olson, L., and Wyatt, R. J., 1979, Brain grafts
   reduce motor abnormalities produced by destruction of neostriatal dopamine system, *Science* **204**:643.
4. Freed, W. J., Perlow, M. J., Karoum, F., Seiger, Å., Olson, L., Hoffer, B. J., and Wyatt, R. J., 1980,
   Restoration of dopaminergic functions by grafting of fetal rat substantia nigra to the caudate nucleus:
   Long-term behavioral, biochemical, and histochemical studies, *Ann. Neurol.* **8**:510.
5. Gash, D., Sladek, J. R., Jr., and Sladek, S. D., 1980, Functional development of grafted vasopressin
   neuron, *Science* **210**:1367.
6. Kromer, L. F., Björklund, A., and Stenevi, U., 1980, Regeneration of the septohippocampal pathways
   in adult rats is promoted by utilizing embryonic hippocampal implants as bridges, *Brain Res.* **210**:173.
7. Dunnett, S. B., Low, W. C., Iversen, S. D., Stenevi, U., and Björklund, A., 1982, Septal transplants
   restore maze learning in rats with fornix–fimbria lesion, *Brain Res.* **251**:335.
8. Krieger, D. T., Perlow, M. J., Gibson, M. J., Davies, T. F., Zimmerman, E. A., Ferin, M., and
   Charlton, H. M., 1982, Brain graft reverse hypogonadism of gonadotropin releasing hormone deficiency, *Nature (London)* **298**:468.
9. Tello, F., 1911, La influencia del neurotropismo en la regeneracio de los centros nervioses, Trab., *Lab.
   Invest. Biol. Madred* **9**:123.
10. Ramón y Cajal, S., 1928, Degeneration and regeneration of the nervous system (R. M. May, ed.,
    transl.), Hafner, New York.
11. Puchala, E., and Windle, W. F., 1977, The possibility of structural and functional restoration after
    spinal cord injury: A review, *Exp. Neurol.* **55**:1.
12. Guth, L., Brewer, C. R., Collins, W. F., Goldberger, M. E., and Perl, E. R., 1980, Criteria for eval-
    uating spinal cord regeneration experiment, *Exp. Neurol.* **69**:1.
13. Sugar, O., and Gerard, R. W., 1940, Spinal cord regeneration in the rat, *J. Neurophysiol.* **3**:1.

14. Barnard, J. W., and Carpenter, W., 1950, Lack of regeneration in spinal cord of rat, *J. Neurophysiol.* **13**:223.
15. Feigin, I., Geller, E. H., and Wolf, A., 1951, Absence of regeneration in the spinal cord of the young rat, *J. Neuropathol. Exp. Neurol.* **10**:42-.
16. Brown, J. O., and McCouch, G. P., 1947, Abortive regeneration of the transected spinal cord, *J. Comp. Neurol.* **87**:131.
17. Kao, C. C., Shimizu, Y., Perkins, L. C., and Freeman, L. W., 1970, Experimental use of cultured cerebellar cortical tissue to inhibit the collagenous scar following spinal cord transection, *J. Neurosurg.* **3**:127.
18. Kao, C. C., 1974, Comparison of wound healing process in transected spinal cords grafted with autologous brain tissue, sciatic nerve, and nodose ganglia, *Exp. Neurol.* **44**:424.
19. Kao, C. C., Chang, L. W., and Bloodworth, J. M. B., Jr., 1977, Successful axonal regeneration to bridge the gap of transected mammalian spinal cords: An electron microscopic study of the results of delayed microsurgical nerve grafting, *Exp. Neurol.* **54**:*591–615*.
20. Kao, C. C., Chang, L. W., and Bloodworth, J. M. B., Jr., 1977, The mechanism of spinal cord cavitation following spinal cord transection. Part 3. Delayed grafting with and without retransection, *J. Neurosurg.* **46**:757.
21. Richardson, P. M., McGuiness, U. M., and Aguayo, A. J., 1980, Axons from CNS neurons regenerate into PNS grafts, *Nature (London)* **284**:264.
22. Richardson, P. M., McGuiness, U. M., and Aguayo, A. J., 1982, Peripheral nerve autografts to the rat spinal cord: Studies with axonal tracing methods, *Brain Res.* **237**:147.
23. Blakemore, W. F., 1977, Remyelination of CNS axons by Schwann cells transplanted from the sciatic nerve, *Nature (London)* **266**:68.
24. Duncan, I. D., Aguayo, A. J., Bunge, R. P., and Wood, P. M., 1981, Transplantation of rat Schwann cells grown in tissue culture into mouse spinal cord. *J. Neurol. Sci.* **49**:241.
25. Wrathal, J. R., Rigamonte, D. D., Braford, M. R., and Kao, C. C., 1982, Reconstruction of the contused cap spinal cord by the delayed nerve graft technique and cultured peripheral nonneuronal cells, *Acta Neuropathol.* **57**:59.
26. Björklund, A., Katzman, R., Stenevi, U., and West, K. A., 1971, Development and growth of axonal sprouts from noradrenaline and 5-hydroxytryptamine neurons in the rat spinal cord, *Brain Res.* **31**:21.
27. Aihara, H., 1970, Autotransplantation of cultured cerebellar cortex for spinal cord reconstruction, *Brain Nerve* **22**:769 (in Japanese).
28. Bregman, B. S., and Reier, P. J., 1982, Transplantation of fetal spinal cord tissue to injured spinal cord in neonatal and adult rats, *Soc. Neurosci. Abstr.* **8**:870.
29. Nornes, H., Björklund, A., and Stenevi, U., 1983, Reinnervation of the denervated adult spinal cord of rats by intraspinal transplants of embryonic brainstem neurons, *Cell Tissue Res.* **230**:15.
30. Nornes, H., Björklund, A., and Stenevi, U., 1981, Embryonic CNS tissue implanted into adult spinal cords, *Soc. Neurosci. Abstr.* **7**:678.
31. Campbell, J. B., Andrew, C., Bassett, L., Husby, J., and Noback, C. R., 1958, Axonal regeneration in the transected adult feline spinal cord. *Surg. Forum* **8**:528.
32. Bassett, C. A. L., Campbell, J. B., and Husby, J., 1959, Peripheral nerve and spinal cord regenerative factors leading to success of a tubulation technique employing Millipore, *Exp. Neurol.* **1**:386.
33. de la Torre, J. C., 1982, Catecholamine fibre regeneration across a collagen bioimplant after spinal cord transection, *Brain Res. Bull.* **9**:545.
34. Turbes, C. C., and Freeman, L. W., 1958, Peripheral nerve–spinal cord anastomosis for experimental cord transection, *Neurology* **8**:857.
35. Perkins, L., Babbini, A., and Freeman, L. W., 1964, Distal–proximal nerve implants in spinal cord transection, *Neurology* **14**:949.
36. Galabov, G., 1966, Regeneration of sectioned spinal cord by implantation of a peripheral nerve, *Comp. Neurol. Acad. Bulg. Sci.* **19**:449.
37. David, S., and Aguayo, A. J., 1981, Axonal elongation into peripheral nervous system "bridges" after central nervous system injury in adult rats, *Science* **214**:931.
38. Andén, N.-E., Jukes, M. G. M., and Lundberg, A., 1966, The effect of DOPA on the spinal cord. 2. A pharmacological analysis, *Acta Physiol. Scand.* **67**:387.

39. Jankowska, E., Jukes, M. G. M., and Lundberg, A., 1967, The effect of DOPA on the spinal cord. 6. Half-centre organization of interneurons transmitting effects from flexor reflex afferents, *Acta Physiol. Scand.* **70**:389.

40. Grillner, S., 1973, Locomotion in the spinal cat, in: *Control of Posture and Locomotion* (R. B. Stein, K. G. Pearson, R. S. Smith, and J. B. Redford, eds), pp. 515–536, Plenum Press, New York.

41. Lundberg, A., 1982, Inhibitory control from brain stem of transmission from primary afferents to motorneurons, primary afferent terminals and ascending pathways, in: *Brain Stem Control of Spinal Mechanisms* (B. Sjölund and A. Björklund, eds.), Elsevier Biomedical Press, Amsterdam, New York, Oxford.

42. Nygren, L. G., Olson, L., and Seiger, Å., 1977, Monoaminergic reinnervation of transected spinal cord by homologous fetal brain grafts, *Brain Res.* **129**:227.

43. Segal, M., Stenevi, U., and Björklund, A., 1981, Reformation in adult rats of functional septo-hippocampal connection by septal neurons regenerating across an embryonic hippocampal connection by septal neurons regenerating across an embryonic hippocampal bridge, *Neurosci. Lett.* **27**:7.

44. Stensaas, L. J., Burgess, P. R., and Horch, K. W., 1979, Regenerating dorsal root axons blocked by spinal cord astrocytes, *Soc. Neurosci. Abstr.* **5**:684.

45. Nathaniel, E. J. H., and Nathaniel, D. R., 1973, Regeneration of dorsal root fibers into the adult rat spinal cord, *Exp. Neurol.* **40**:333.

46. Perkins, C. S., Carlstedt, T., Mizuno, K., and Aguayo, A. J., 1980, Failure of regenerating dorsal root axons to regrow into the spinal cord. *Can. J. Neurol. Sci.* **7**:323.

47. Rovainen, C. M., 1976, Regeneration of Müller and Mauthner axon after spinal transection in the larval lamprey, *J. Comp. Neurol.* **168**:545.

48. Selzer, M. E., 1978, Mechanism of functional recovery and regeneration after spinal cord transection in larval sea lamprey, *J. Physiol. (London)* **277**:395.

49. Shik, M. L., Severin, F. V., and Orlovsky, G. N., 1966, Control of walking and running by means of electrical stimulation of the midbrain, *Biophysics* **11**:756.

50. Stein, P. C., 1978, Motor systems with specific reference to control of locomotion, *Annu. Rev. Neurosci.* **1**:61.

51. Björklund, A., Segal, M., and Stenevi, U., 1979, Functional reinnervation of rat hippocampus by locus coeruleus implants, *Brain Res.* **170**:409.

52. Low, W. C., Lewis, P. R., Bunch, S. B., Dunnett, S. B., Thomas, S. R., Iversen, S. D., Björklund, A., and Stenevi, U., 1982, Functional recovery following neural transplants of embryonic septal nuclei into adult rats with septohippocampal lesions: The recovery of function, *Nature (London)* **300**:260.

53. Budakova, N. N., 1973, Stepping movements in the spinal cat due to DOPA administration, *Fiz. Zh. SSSR im. I.M. Sechenova* **59**:1190.

54. Forssberg, H., and Grillner, S., 1973, The locomotion of the acute spinal cat injected with clonidine i.v., *Brain Res.* **50**:184.

55. Viala, D., and Buser, P., 1969, The effects of DOPA and 5-HTP on rhythmic efferent discharges in hindlimb nerves in the rabbit, *Brain Res.* **12**:437.

18

Some Consequences of Grafting Autonomic Ganglia to Brain Surfaces

JEFFREY M. ROSENSTEIN AND MILTON W. BRIGHTMAN

1. Introduction

Neural transplantation has been concerned with both the survival potential of grafts and their incorporation or connections within the host brain substance. Recent investigations have shown that grafts can produce a relatively normal cellular pattern in the cerebellum[1] and can make functional connections in the rat visual system,[2,3] hippocampus,[4] basal ganglia,[5] or hypothalamic system.[6] A common thread throughout such experiments is that the neural tissues that survived and differentiated within the host brain were exclusively from fetal donors. Often, homologous fetal tissue was used as grafts that might ameliorate deficiencies in the host. For example, Gash et al.[6] implanted fetal hypothalamus in the vasopressin-deficient Brattleboro rat, and Björklund and Stenevi[4] and Perlow, Wyatt, and associates[5] inserted fetal substantia nigra grafts to supplant a damaged dopamine system. It is well known that the "younger" the CNS neural tissue graft, the better its chances of survival and rate of growth. Recently, cultured fetal CNS tissue has been utilized successfully as donor tissue.[7] In sharp contrast, grafts of the peripheral nervous system (PNS) must come from a more mature source in order to survive upon the surface of the CNS.[8,9]

1.1. Superior Cervical Ganglion as a Graft

The system of neural transplantation that we are pursuing differs from the other methods where brain tissue is necessarily damaged when the graft is inserted into it. In

JEFFREY M. ROSENSTEIN ● Department of Anatomy, George Washington University Medical Center, Washington, D.C. 20037.   MILTON W. BRIGHTMAN ● Laboratory of Neuropathology and Neuroanatomical Sciences, National Institutes of Health, Bethesda, Maryland 20205.

our system, the brain surface is left as intact as possible. This approach thus enables one to investigate structural interactions between a foreign neural graft and the relatively undamaged recipient brain. One of our original intentions was to study the growth and development of the superior cervical ganglion (SCG) as an *in vivo* organ culture upon the host brain.[8,9] The SCG has been extensively studied *in vitro* and comparisons of its development (in fact, regeneration) within the ambient milieu of the host ventricular system could be made. According to fluorescence studies, autografted SCG survives after mechanical and pharmacological lesions.[10] Thus, the allografted SCG, from a nonsibling donor, provided a well-defined and relatively uncomplicated neural tissue to ascertain both cellular alterations and factors necessary for foreign axons to enter the minimally damaged brain.[8,9]

Very recent analyses, however, show that the neuronal composition of the SCG is not really as simple as previously believed. Discovery of several different peptides within the ganglion[11] has prompted the notion that the classically noradrenergic ganglion cell bodies may alter their transmitter during stages of development or may even acquire additional peptide transmitters.[12] These findings will provide exciting avenues to follow, by immunocytochemical methods, possible changes in peptide content of transplanted ganglion cells.

An additional potential usefulness of the allografted SCG in studying neural development is that the ganglion is in the domain of the PNS; it lies outside of the blood–brain barrier (BBB). From its peripheral residence it has a direct and well-known projection to pial and choroidal blood vessels. Recent fluorescence preparations show that its intact noradrenergic projection may sprout into the hippocampal formation after a lesion to a cholinergic system.[13] These projections appear to lie outside the BBB. Might, then, a foreign allograft from outside the BBB affect the BBB of the host? The transplants' imposition into the host central blood vascular system could alter the fluid physiology at the boundary between graft and brain and provide for unique opportunities to follow changes in the BBB (see Section 5).

One of the more intriguing concepts that has emerged from our investigations of neural transplantation concerns apparent differences in the growth potential or regenerative properties between autonomic neurons and central neurons. Unlike CNS tissue, which must be fetal for it to grow in the host brain or *in vitro,* autonomic ganglia (and sensory ganglia, unpublished observations) must be *mature* to thrive after transplantation.[8,9] The fetal state of autonomic and sensory neural tissue is not compatible with survival upon the brain surfaces. One possibility for this survival requirement is that transplanted autonomic neurons seem to operate on a different level in that only after they have once made target contact *in situ,* receiving perhaps retrograde influence from nerve growth factor (NGF),[14] can they grow *again* after transplantation to find another target.[9] Another possibility is that fetal central neurons are, as yet, "uncommitted" to a specific neuronal interconnection and in their new host microenvironment, they may seek to find an appropriate target to ensure their survival. This possibility is supported by our finding that sympathectomy of the host markedly enhances the number of surviving transplanted ganglion cells (see Section 6). The concept does not hold, however, for central neurons, as transplantation of mature CNS tissue, presumably after appropriate contacts are made, is fruitless.[15] Finally, in addition to the requirement for an appropriate synaptic target, differences in nutritive requirements may have considerable bearing on survival of individual neurons.[9]

1.2. Fourth Ventricle as a Graft Site

Using the fourth ventricle as a transplantation site enables us to assess the interactions of foreign transplanted neurons and the undamaged brain (for detailed methodology see Refs. 8 and 9). The present system has advantages in that (1) the avoidance of injury allows for analysis of cellular and physiological interactions between brain and graft uncomplicated by excessive scar formation and inflammatory response, (2) the system can accommodate a wide variety of donor tissue, and (3) it is reliable and reproducible at any age of the host.

The following descriptions will show that the undamaged developing brain has a remarkable accomodative capacity to the presence of an autonomic graft. Examination of cellular changes in the relatively intact host brain is elemental for comparison of cellular responses that might take place in host brains that have been damaged by graft placement. Changes occur in host morphology, cellular migration and development at the internal and external brain surfaces.[16,17] We have yet to determine, however, whether the noninvasive transplant method will allow for transplanted axons to penetrate and become incorporated with the host's CNS or whether this invasion can only succeed after mechanical disruption of the brain surface.[9] Do the "alluring substances" postulated by Ramón y Cajal (1928)[18] need be liberated at a lesion site to induce the ingress of foreign axons? The growth characteristics of fetal neural graft coupled with a necessary lesion at the time of graft placement might provide the conditions necessary for successful transplantation. The response of the astrocyte might also play a key role (see Section 3). We have shown that an autonomic ganglion graft must provide some attractant for neural and glial tissue to migrate anomalously out of the developing brain.[16,17] Yet, movement in the opposite direction—grafted autonomic axons into the undamaged brain parenchyma—is meager.[9]

The composition of the cerebrospinal fluid (CSF) would seem an inadequate medium for the support and long-term growth of neural tissue. The use of the CSF as a medium for an *in vitro* system was attempted with some success by Martinovic[19] and then abandoned nearly half a century ago. The CSF has very low concentrations of amino acids and protein compared with serum[20] and only one-tenth of the 600 mg/100 ml of glucose mandatory to grow most neural tissues *in vitro*. Survival of neural grafts must then be supported by host blood vessels that anastomose with the graft's residual capillaries or with new vessels forming in the host and entering the graft. Formation of vascular connections in neural grafts is one of our active areas of study (see Section 5), yet even the glucose concentration of blood is only about one-sixth of that required for *in vitro* growth. Even though the acquisition of a target site might be necessary for the maintenance of the metabolic machinery in individual grafted autonomic or central neurons[8,9] their initial and sustained survival must depend on a receptive host vasculature.

2. Regeneration in the SCG Graft

When a portion of the mature rat SCG is transplanted to the host fourth ventricle, it regenerates vigorously. Because the ganglion graft has been severed from its afferent cholinergic connections, it initially undergoes, within the first few days after grafting, a period of degeneration. Simultaneously, the chromatolytic neurons sprout numerous neurites. These new axon sprouts in the mature ganglion graft have terminals recognizable as growth cones (Fig. 1) of a morphology similar to that of newborn SCG explanted to the culture dish.[21] The presence of both growth cones in the graft and attendant changes in Schwann cell morphology suggested that the SCG allograft underwent a process of redevelopment.[9] Over time, the numbers of both neurons and axons diminished and it was postulated that those that survived, some for at least 2½ years, had established synaptic contacts that assured their survival.[8,9] One sign of a neuron's functional survival is its ability to synthesize and transport its transmitter.

Histochemical Localization of Noradrenaline

It is well known that, *in situ*, cholinergic connections in the SCG are needed for the full expression of enzymes in the biosynthesis of noradrenaline.[22] Because the allografted SCG regenerated extensively when separated from both its cholinergic afferents and target sites, we sought to determine if these transplanted neurons might produce noradrenaline (NA). Dissociated autonomic neurons from newborn rats will produce NA,[23] and dependent on certain factors in the culture system, both NA and acetycholine can be expressed.[24,25]

To localize NA intracellularly, we utilized electron microscopic histochemical methods that selectively label this neurotransmitter only in its membrane-bound form. The methods used were· the classical potassium permanganate method of Richardson (1966),[26] the sodium chromate–potassium dichromate procedure for the electron microscopic visualization of the chromaffin reaction,[27] and the administration of the amine analog 5-hydroxydopamine (5-OH-DA) followed by aldehyde fixation. Each produces a membrane-bound, electron-dense reaction product, which, in the SCG, represents NA.[28] Ultrathin sections of the transplant were not stained so as not to mask the appearance of the reaction products. In the tissue stained by the chromate–dichromate method, which produced nearly identical results as the permanganate but with much improved fixation, many of the boutons contained granular vesicles. These vesicles are characteristic of NA visualization and ranged between 40 and 60 mm in size (Fig. 2). Reaction

---

→

FIGURE 1. Portions of growth cones (G) in SCG allograft contain much smooth endoplasmic reticulum, a few dense-core vesicles, and mitochondria. 3 days P.O. (postoperative age of the transplant). × 16,000.

FIGURE 2. Noradrenergic bouton (b) contains vesicles with dense reaction product after chromate–dichromate fixation. s, Schwann process. 6 weeks P.O. × 68,000.

FIGURE 3. Thin neurite (n) contains reaction product in both vesicles and tubular reticulum. 1 month P.O. × 58,000.

FIGURE 4. Noradrenergic storage site is located at the periphery of a 30-month-old transplanted ganglion cell. × 51,000.

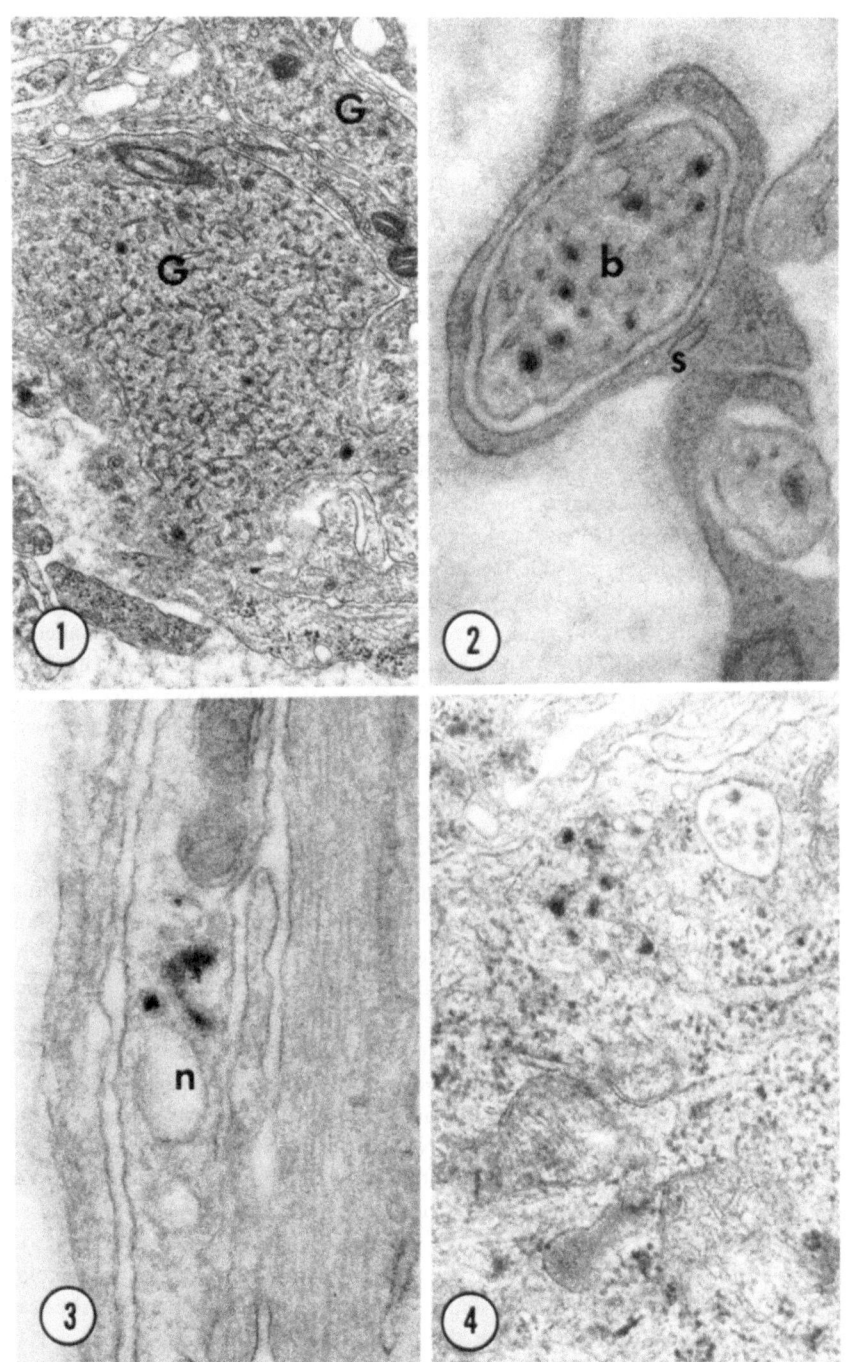

product was found in both vesicles and tubular reticulum suggesting a transport from the cell body into even the thinnest of neurites (Fig. 3). It should be noted that not all boutons contained reaction product and, in most growth cones, it was absent. Studies are in progress to determine if these structures might arise from cholinergic neurons[29] or peptidergic neurons, such as those with substance P-like immunoreactivity.[12]

NA storage sites in ganglion cell somata were easily identified as a cluster of granular vesicles near the periphery of the cell. These sites were labeled in neurons surviving for at least 30 months after an *intravenous* injection of 5-OH-DA, and fixation with aldehyde (Fig. 4). Of significance is that, within the brain, 5-OH-DA cannot cross the BBB[30] but the SCG graft residing within the host brain is readily labeled. Moreover, boutons in the graft, after 5-OH-DA administration, were considerably more intensely labeled; the labeled vesicles were larger and appeared to contain a greater "amount" of the reaction product (Fig. 5; cf. Fig. 2).

Interestingly, and depicted by all three fixation methods, was the presence of reaction product in Schwann cells particularly after 5-OH-DA administration. The product appeared in membrane-bound organelles morphologically reminiscent of lysosomes in close association with the Golgi complex (Fig. 6). Possibly, the lysosomal system in the Schwann cell might be responsible for taking up and degrading NA in the graft. The NA could become available to the Schwann cells after the death and breakdown of a neuron or its boutons. This suggestion is reinforced by the marked increase in label after the administration of 5-OH-DA. The catecholamine analog 5-OH-DA can, in the first instance, reach the ganglion cell and its processes because the blood vessels of the grafted ganglion are permeable to macromolecules as they are within the SCG residing *in situ*.

3. Tropic Effects

According to Jacobson,[31] the term *chemotropism* may be used to describe the movement of neural tissue toward some source of diffusible substance acting at a distance. Although recent *in vitro* studies have been suggestive of some tropic influences in chick sensory ganglia,[32] there is, however, scant information regarding chemotropic phenomena in the mammalian nervous system.[33]

Transplantation of the mature SCG to the undamaged cerebellar or medullary surfaces has led to unexpected cellular consequences, which are suggestive of postnatal, *in situ,* chemotaxis. Where the graft was placed, neurons and associated neuropil could migrate out of the confines of the intact CNS and freely enter the regenerating graft. The entry was unimpeded by either glia limitans or basal lamina. Within the developing cerebellum itself, neuronal migration and neuronal differentiation were changed at sites close to the SCG.

These findings demonstrated a unique phenomenon in mammalian neurobiology: neural and glial tissue migrates anomalously in response to an autonomic tissue transplant. The movement was not effected by other tissues or inert structures.[17] When migration did take place, it involved neuronal and nonneuronal cells; there was both a *neurotropic* and a *gliotropic* response. Conceivably, forces that induced neural tissue migration were separate and distinct from those that induced glial migration. Alterna-

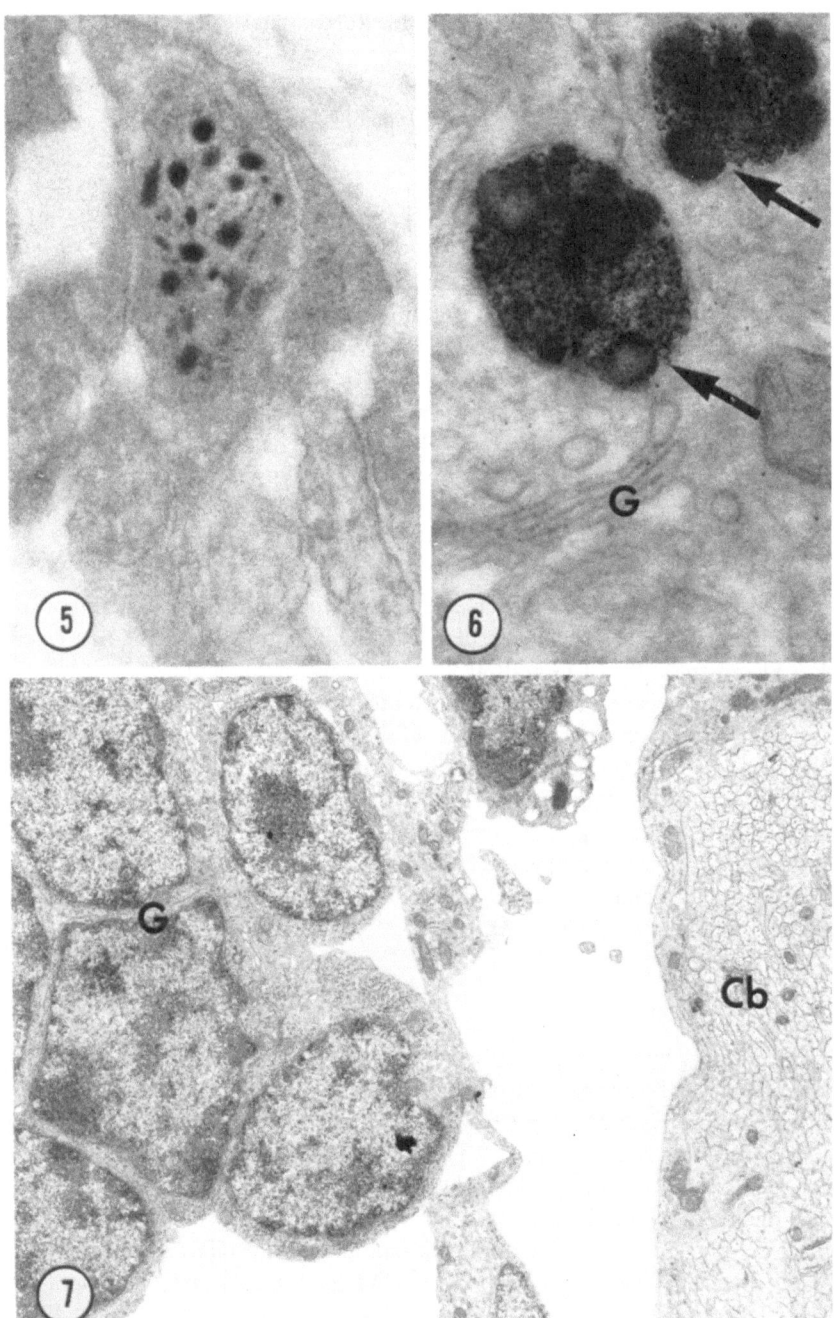

FIGURE 5. Vesicles in noradrenergic bouton are heavily labeled after intravenous administration of 5-hydroxydopamine. 2 weeks P.O. × 65,000.

FIGURE 6. Chromate reaction product is present in lysosome-like structures (arrows) near Golgi complex (G). 5 months P.O. × 70,000.

FIGURE 7. Cluster of granule cells (G) extends out of the cerebellum (Cb) in response to the placement of an SCG graft. 4 weeks P.O. × 9200.

tively, the neural and glial tissue might be coupled physically or chemically and anomalous migration of the two tissues was simultaneous and directed.

3.1. Neurotropic Effects

Placement of an SCG allograft induced, in some instances, neural tissues to remain arrested permanently at their site of germination from where they would normally embark on a patterned migration. These findings have been corroborated by other investigators using fetal CNS serotoninergic tissue.[34] In other instances, tissue bridges containing neurons and associated neuropil emanated the brain surface and extended into the SCG graft. These labile neurons in both systems we have so far studied, cerebellum and olfactory bulb, were predominantly granule neurons.

3.1.1. Cerebellar Neurons

Migration of external granule cells, precursors of granule neurons, and of basket and stellate interneurons of the molecular layer begins at the end of the first postnatal week.[35] Placement of the SCG graft at this time markedly alters these "programmed," stereotyped cellular patterns. Groups of chronologically and, presumably, synaptically mature granule neurons often extended out of the cerebellum for considerable distances. Not being covered by glia, some were in direct contact with CSF (Fig. 7). Mossy fiber afferent terminals ascending from the spinal cord as well as small Purkinje dendritic branchlets and associated spines formed part of the tissue bridges. This postnatal mobility suggested an extensive plasticity of these structures that could be expressed in the presence of the transplants. A complete account of morphological changes in the cerebellum may be found in previous reports.[16,17]

3.1.2. Olfactory Neurons

The olfactory bulb, like the cerebellar cortex, is composed of a laminated arrangement of neurons and neuropil. The olfactory neurons originate prenatally from the subependymal and matrix layers of the olfactory ventricle and, postnatally, undergo migration and subsequent proliferation.[36,37] The external granule cells of the olfactory bulb possess a migratory capacity estimated to be 50 μm/day[36]; the migration proceeds through the internal granule layer and mitral layer for at least 73 days after birth. We sought to determine if postnatal granule cell migration in the olfactory formation could be altered in some manner comparable to that of its analogous cell type in the cerebellum.

Accordingly, we placed a mature SCG allograft on the surface of the olfactory bulb in 6-day-old rat pups and allowed postoperative periods of up to 6 months. As in the experiment with the cerebellum, the olfactory bulb was essentially undisturbed; one difference was that the overlying skull had to be removed in order to reach the bulb surface. In some specimens, the results bore striking similarities to those in the cerebellum. Bridges of bulb tissue extended from the surface to reach the regenerating graft. The bridges were composed of longitudinally arranged columns of cells that appeared to arise from the periventricular zone (Pagnanelli and Brightman, unpublished observations). More often, there was no external bridge, but a long column of intrinsic cells extending from the bulb's interior to its subpial surface. The cell columns were comprised of a heterogeneous population including granule cells, astrocytes, microglia, and an occasional mitral cell (Fig. 8). The induced rearrangement of normal cellular patterns of granule cells in both cerebellum and olfactory bulb leads to speculation that this class of neuron, one that arises from a periventricular germinal zone and retains a postnatal migratory capability, may be responsive to some neurotropic factor secreted by the SCG transplant. It has been suggested that the postnatal origin of granule cells in the CNS may be important for interactions with environmental stimuli after birth.[36] Such stimuli, apparently provided by an autonomic graft, induce anomalous movement of the granule cells. We plan to determine if a third group of granule neurons might respond in the same generalized manner: those of the hippocampal formation.

3.2. Gliotropic Effects

The responses of astrocytes to a graft present a particularly intriguing and yet untapped area of study in neural transplantation. When a mature SCG graft was gently placed on the brain, a subtle gliosis took place and, over time, cell processes with astrocyte-like characteristics were identified within the graft.[9] As the brain was undamaged and thus without "space" to accomodate the graft, something of a paradox was created. The host offered the SCG a rich blood supply for its nourishment, yet the gliotic reaction in this system of minimal disturbance might have limited any axonal penetration. In aforementioned fluorescence microscopic studies where grafted axons entered the host brain, little description was made of astroglial activity in the host.[10] The observation of foreign axons from a fetal CNS graft entering the host is indeed an exciting one; further knowledge of both their mechanisms of entry and relationship with the astrocytes may serve to expand our approaches and application of neural grafting techniques. As described below, a mature ganglion graft produces an actively *directed* astrogliosis in the neonatal brain. It is conceivable that mature PNS tissue grafts may effect a gliotropic response to a greater degree than fetal CNS implants.

3.2.1. Cerebellum

In the tissue bridge between SCG explants and the cerebellum, astrocytes were a major component. Astrocytic processes were enmeshed amongst the neurons and their axons so as to form a lattice-like scaffold that the neural tissue might have utilized to migrate anomalously. In order to further explore the extent of gliosis and glial mobility at a transplant site, we have utilized the immunocytochemical marker for astrocytes in the CNS, glial fibrillary acidic protein (GFAP). This protein is the major component in brain astrocytes[38] but is found occasionally in the PNS.[39] Of particular importance to this study is that GFAP does not appear in the autonomic ganglion.[39] In order to follow the distribution of astrocytes, we used the elegant method of Trapp et al.[40] in which

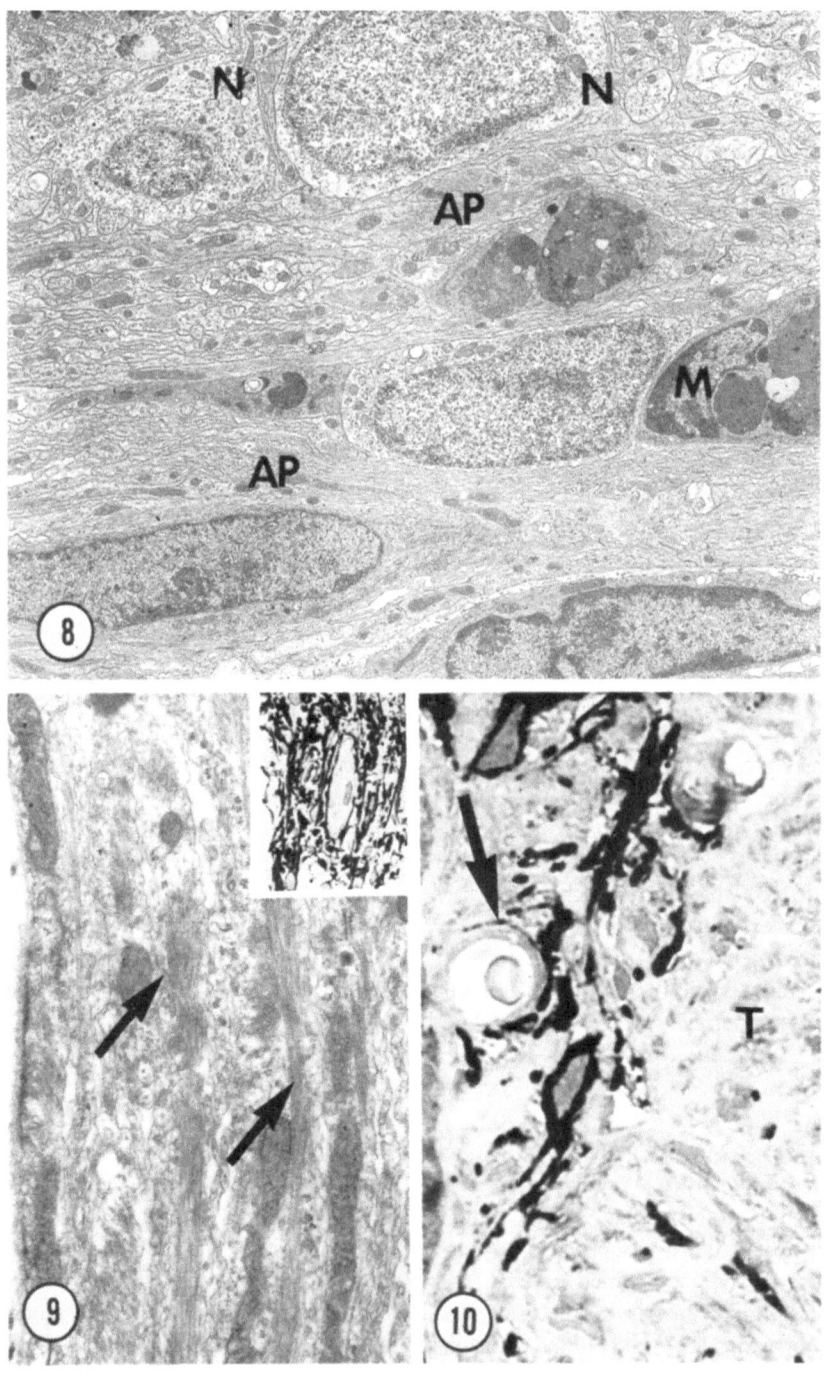

plastic-embedded, 1-μm sections are immunostained and adjacent tissue in the block can be subsequently thin-sectioned.

In areas where tissue bridges arose from the cerebellum, we could not identify normal Bergmann glia. These astrocytes typically have a lucent appearance and contain only a few filaments, although the cells stain for GFAP. Instead, the astrocytes of the bridge contained thick bundles of filaments oriented toward the cerebellar surface and SCG graft (Fig. 9). We considered these cells to be reactive Bergmann glia. Staining with GFAP antisera indicates that the tissue bridges, generally, were surrounded by an astroglial "shell" (Fig. 9, inset), but within the bridge the astrocyte processes appear to be randomly arranged. However, not all CNS neural tissue may be contained within the bridge (Fig. 7), and at the point of penetration of the cerebellar tissue into the SCG graft there was no intervening glia.[17]

3.2.2. Medulla

The dorsal medullary surface was another region in which glial movement was induced by the SCG graft. The response was primarily glial because most of the neural tissue in this portion of the CNS does not have the capacity to migrate at the time of transplantation[17] (1–2 weeks of age).

When the SCG graft was placed on the medulla, there was a progressive astroglial invasion of the graft. By 1 week, the glia limitans was thickened but invasion did not take place. For periods of up to 1 month, a gliosis was produced consisting of insinuating astrocytic processes, some of which were packed with filaments (Fig. 11). After this time, GFAP-positive astrocytic cell bodies and thickened astrocytic processes (Fig. 12) actively extended for a considerable distance of nearly 1 mm, into some grafts. When the graft lay on the ependyma, astrocytes might push them apart to make their entry. In an SCG autograft, where the animal's own ganglion was used and thereby leaving certain adrenergic sites denervated, there was a considerable reduction in astrocyte migration (unpublished observations). As the SCG does not stain for GFAP, it is relatively easy to identify the invading astrocytes.

In some cases, astrocytes lay close to capillaries that were contacted by typical end feet, suggesting that these were blood vessels from the CNS (Fig. 10). Processes that contained filaments and formed gap junctions and enveloped neurites that we previously termed *reactive Schwann cells*[9] could, in fact, be astrocytes. If they are astrocytes, then it may be possible that, instead of "walling off" the foreign graft, the astrocytes may serve either to guide or to *incorporate* the neural elements of the graft into the host brain. The factor(s) that initiates the directed gliotropic responses remains to be elucidated.

←——

FIGURE 8. Portion of a cellular bridge from the olfactory bulb that extends toward an SCG graft contains both neurons (N) and astrocytic processes (AP) and microglial cells (M). 2 months P.O. × 7000. Micrograph by Dr. D. M. Pagnanelli.

FIGURE 9. Astroglia laden with filament bundles (arrows) form the lateral edge of a cerebellar tissue bridge. × 24,000. Inset: GFAP-positive staining of lateral portion of a tissue bridge. × 180. 4 months P.O.

FIGURE 10. GFAP-positive stained process within SCG transplant (T). Note astroglia end feet on blood vessel (arrow). 6 weeks P.O. × 720.

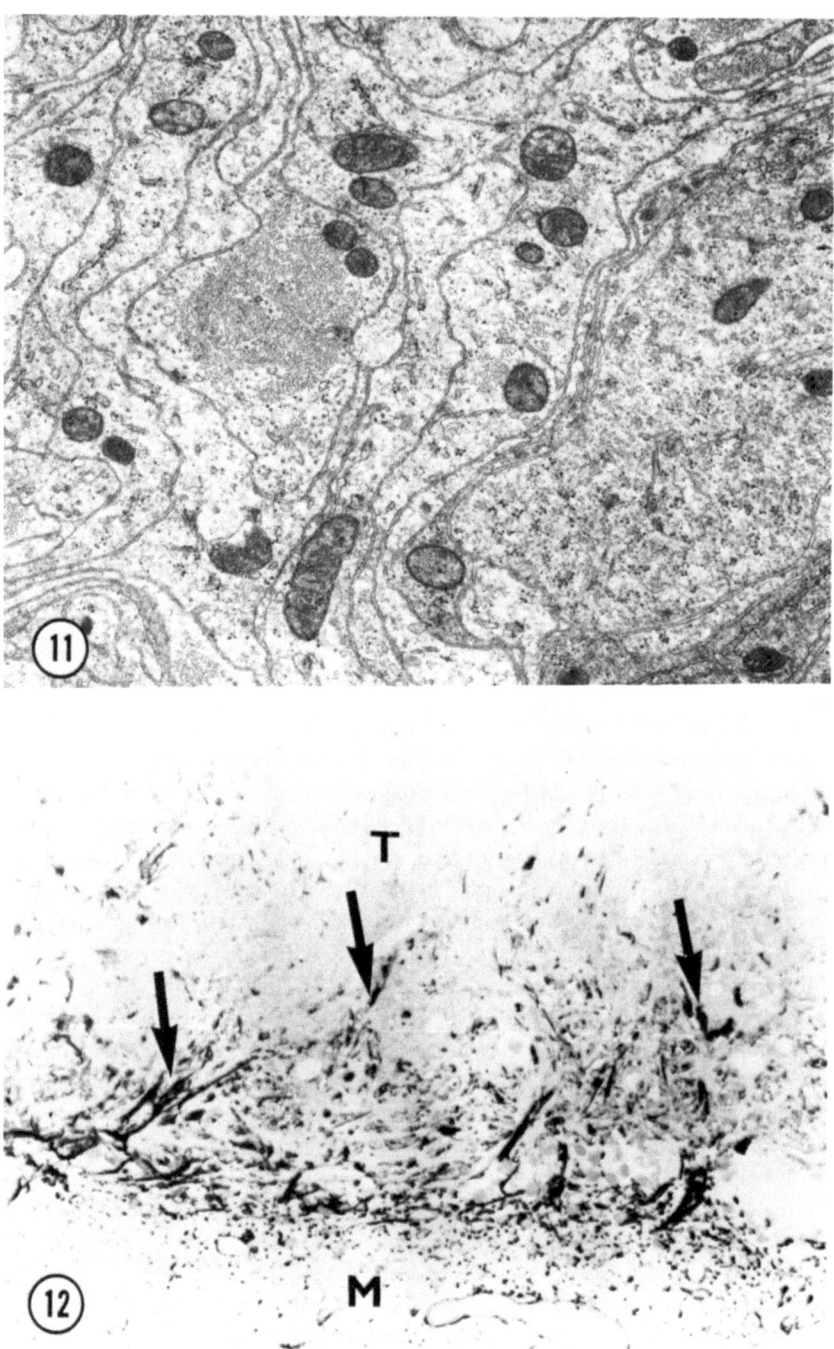

FIGURE 11. Astrocytic processes form interdigitating laminae on the medullary surface beneath SCG graft. 4 months P.O. × 28,000.

FIGURE 12. GFAP-positive processes (arrows) arise from the medullary surface (M) and extend into an SCG transplant (T). 6 weeks P.O. × 180.

3.2.3. Schwann Cells

In addition to astrocytes, Schwann cells were another nonneuronal component of the tissue bridge between cerebellum and graft. The Schwann cells were surrounded by a basal lamina, had typical lobulated nuclei, and did not contain filaments. The lack of any discernible filaments set them apart from the reactive astrocytes. The Schwann cells' basal lamina further indicated that they were not oligodendroglia. The Schwann cells occurred in clusters within the cerebellum; the spacing between processes approximated that in the PNS and so could be distinguished from the CNS tissue (Fig. 13). Although closely aligned with granule cells, parallel fibers, astrocytes, or oligodendroglia (Fig. 13), they appeared to be separated by thin laminae of filament-laden astrocytic processes. Often, the basal laminae of Schwann cells and astroglial abutted one another.

Schwann cells penetrate the CNS under a variety of conditions including X-irradiation[41] or direct transplantation into the brain.[42] In our system of minimal disturbance, the Schwann cells, which must have originated from the SCG graft, might have utilized the astroglial guidelines in a tissue bridge to enter the host cerebellum. The bridges, in some instances, might be considered a "two-way street," along which astrocytes migrated toward the PNS graft. Subsequently, Schwann cells moved toward the CNS. This sequential and reciprocal exchange of nonneuronal cells seems a reasonable suggestion as astroglial bridges have been previously implicated in the entry of Schwann cells into the CNS in both experimental encephalomyelitis[43] and where sciatic nerve is directly transplanted into the brain.[42]

4. Hypothalamic Grafts to the Brain Surface

When fragments of fetal hypothalamus that include neurosecretory cell groups and median eminence are transplanted to the floor of the third ventricle, they not only survive but, in about 25% of the recipients, secrete their antidiuretic hormone. The recipients, Brattleboro rats homozygous for the trait of diabetes insipidus, are able to concentrate their urine and decrease their water intake to near-normal levels.[6] The target of the neurosecretory axon terminals is the vessel wall of fenestrated capillaries permeable to peptides and proteins such as HRP.[48]

We have transplanted neurosecretory tissue from the fetuses of normal rats to the more accessible fenestrated capillaries of the choroid plexus and area postrema of the fourth ventricle in Brattleboro recipients. Do the fenestrated capillaries of these other circumventricular organs, which are not the normal targets of neurosecretory axons, make functional contact with the neurosecretory terminals of transplanted hypothalamus?

Although the hypothalamic fragment survives ectopically on the brain surface in about 20% of attempts, the main problem with this approach is the impassible choroid plexus epithelium, Even when both the recipient's SCG are removed so that ingrowing neurosecretory neurites have denervated pial and choroidal capillaries to innervate, they are confronted by the choroidal epithelium. Neurites intertwine with the epithelium's microvilli but cannot pass the tight junctions between epithelial cells to reach the stromal capillaries (see Fig. 17).

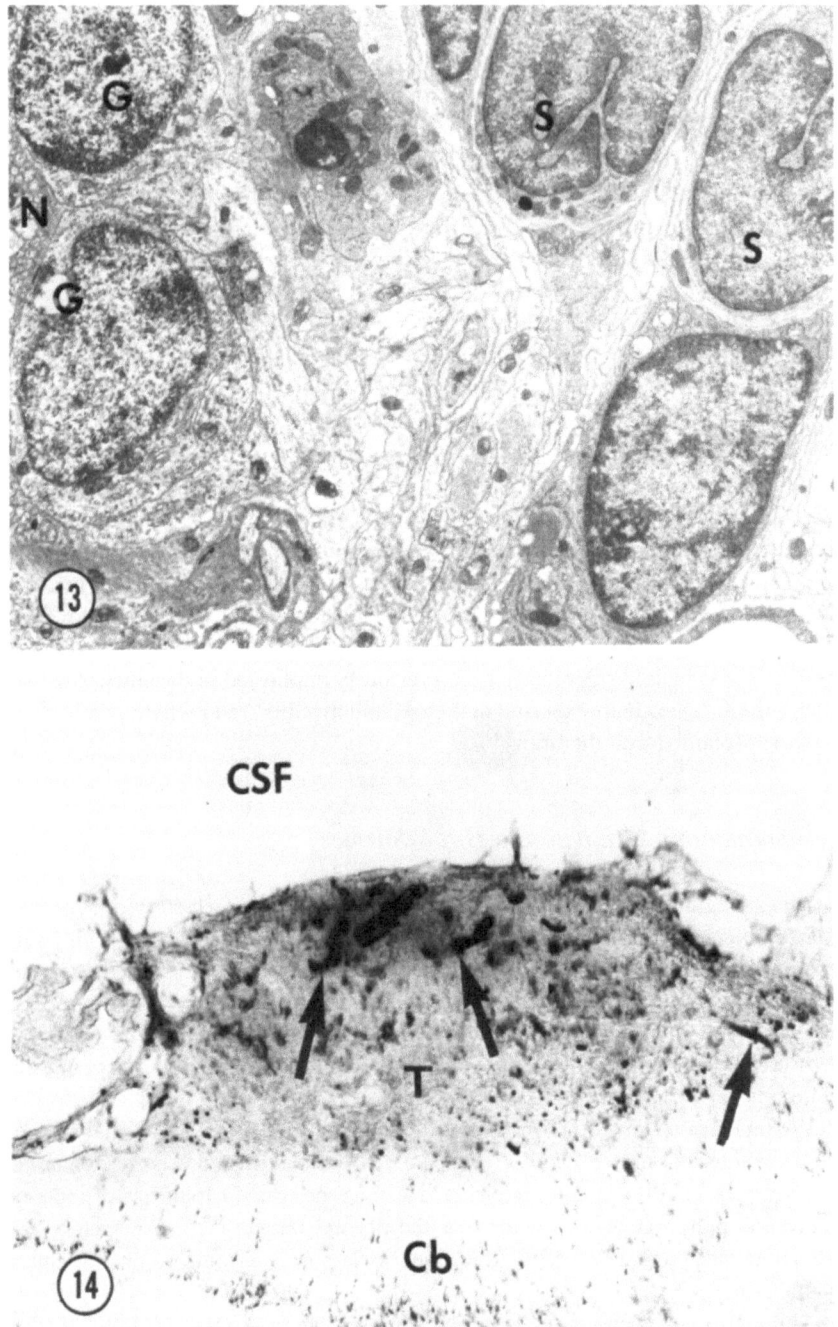

FIGURE 13. Schwann cells (S) lie within a tissue bridge and are in close association with cerebellar granule cells (G) and neuropil (N). 7 months P.O. × 7200.

FIGURE 14. SCG transplant (T) nestles on cerebellar surface (Cb) in ambient CSF. Two minutes after intravenous HRP administration reaction, product begins to fill the transplant. Arrows indicate vessels laden with HRP. 40-μm vibratome section. 3 months P.O. × 140.

Although many neurons survive for at least 3 months after transplantation and the hypothalamic neuropil appears normal, immunocytochemical tests for the presence of vasopressin and neurophysin have been negative (Korino and Brightman, unpublished observations). In many of the unmyelinated axons of the graft, large dense-cored vesicles, about 100 mm, are interspersed among synaptic vesicles and other organelles. The vesicles are probably not neurosecretory but might be analogous to those described recently in the primate's substantia gelatinosa.[49]

The vessels of the iris also are fenestrated and permeable to macromolecules. Grafts to the anterior chamber of the eye, however, have been likewise unsuccessful in establishing normal diuresis even though the hypothalamic tissue survives. It would appear that, in our system, the fenestrated vessels of the transplanted median eminence may not survive. In the absence of their normal target (the fenestrated capillaries of the median eminence and posterior lobe of the pituitary gland), the neurosecretory neurons, if they do persist, fail to synthesize their hormone.

5. Effects on the Blood–Brain Barrier

The imposition of the peripheral ganglion graft, with a type of vasculature distinct from that of the CNS, may be expected to interact with the brain in ways other than a tropic one. Neural transplants thrive when, presumably, supplied by host blood vessels. Within 12 to 24 h after transplantation, SCG grafts become vascularized by, initially, a few vessels that can be filled with India ink perfused through the aorta. It appears, therefore, that the vessels of the graft soon anastomose with those of the pia or parenchyma (Tsubaki and Brightman, unpublished observations). The vessels of the graft and host may be morphologically and functionally distinguishable if the graft's endothelium retains its *in situ* characteristics: fenestrae and permeability to macromolecules such as HRP.[44] If these endothelial features are retained, and if the graft is close to or fused with the host, blood-borne molecules might be able to enter the extracellular fluid of the brain.

Circumventing the Blood–Brain Barrier

The question may be restated: could an autonomic graft be used as an indwelling portal that would allow entry of blood-borne macromolecules to brain regions where, normally, they are excluded? To answer this question, we injected HRP into the femoral vein in animals that had SCG grafts in the fourth ventricle for several months. HRP has a molecular weight of 40,000 and intravascular administration of this glycoprotein should not allow its entry into either brain extracellular spaces or the CSF except at certain large arterioles.[45] It should be recalled from Section 2 that intravascular administration of the catecholamine analog 5-OH-DA, which is a relatively small molecule (MW 205), normally excluded from the brain by the BBB, gained access to the transplant and labeled adrenergic storage sites.

How might the glycoprotein HRP be distributed in the SCG graft and surrounding tissue after vascular administration? Thus far, we have allowed HRP to circulate for time periods between 2 and 30 min. As early as 2 min circulation and subsequent fixation with aldehydes, HRP had time to exude from vessels in the graft (Fig. 14). The perivascular region around a number of vessels at this early period was heavily laden with reaction product including some vessels that appeared to extend between cerebellum and the graft (Fig. 14). The perivascular clefts of the vessels in the underlying brain had no discernible reaction product at this early time, even though the SCG tissue appeared to fuse with the subjacent cerebellum.

By 15 min circulation time, the entire SCG graft was permeated by HRP, which spread to the extracellular clefts of the underlying brain. Individual ganglion cells and axon bundles within the graft were outlined where HRP had become trapped within extracellular spaces (Fig. 15). Occasionally, some neural processes took up the exogenous protein. This distribution of HRP was comparable to that within the *in situ* SCG[44]. Near the transplant, however, either the host CNS vessels anastomosed with the residual portions of permeable capillaries or, more unlikely, the impermeable CNS vessels, as they entered the domain of the peripheral graft, changed their permeability properties.

Possible Changes in Brain Vessels

After 30 min circulation time, the extracellular clefts of the host's medulla or cerebellum contiguous with the SCG graft contained HRP reaction product (Fig. 16). Our interpretation of this extracellular dissemination of protein is that the HRP has infiltrated the continuous intracellular spaces between graft and brain. If the graft tissue did not fuse with the brain, the protein probably entered the CSF before reaching the brain surface. Interestingly, in the areas where HRP flooded host medulla or cerebellum, endothelial linings of capillaries contained a far greater number of "transporting" vesicles and cisternae than a normal animal that has received intravascular HRP (Fig. 16). Possibly, the placement of the graft had stimulated the cell membrane of contiguous CNS endothelial cells to invaginate as pits which have been implicated in vesicular transport.[45] Alteration of homeostasis such as hyperosmolarity[45,46] or tumor induction[47] can change BBB permeability to protein. It would appear that after transplantation of a peripheral ganglion, there is an irreversible flow of blood-borne macromolecules from the extracellular space of the graft into those of the host brain.

FIGURE 15. HRP reaction product permeates SCG transplant 15 min after intravenous administration, surrounding neurons and axon bundles. 1-μm section counterstained with toluidine blue. 3 months P.O. $\times$ 400.

FIGURE 16. Extracellular spaces of the neuropil and around capillaries (Cap) within the medulla subjacent to an SCG are filled with HRP 30 min after intravenous injection. 6 weeks P.O. $\times$ 9500.

FIGURE 17. Hypothalamic graft from a normal, 18-day-old rat fetus contains many neurites (small arrows) and synaptic terminals (T), none of which penetrate beyond the microvilli (large arrows) of the choroid plexus epithelium (E) of a recipient, Brattleboro rat. 3 months P.O. $\times$ 26000. Kirino and Brightman (unpublished results).

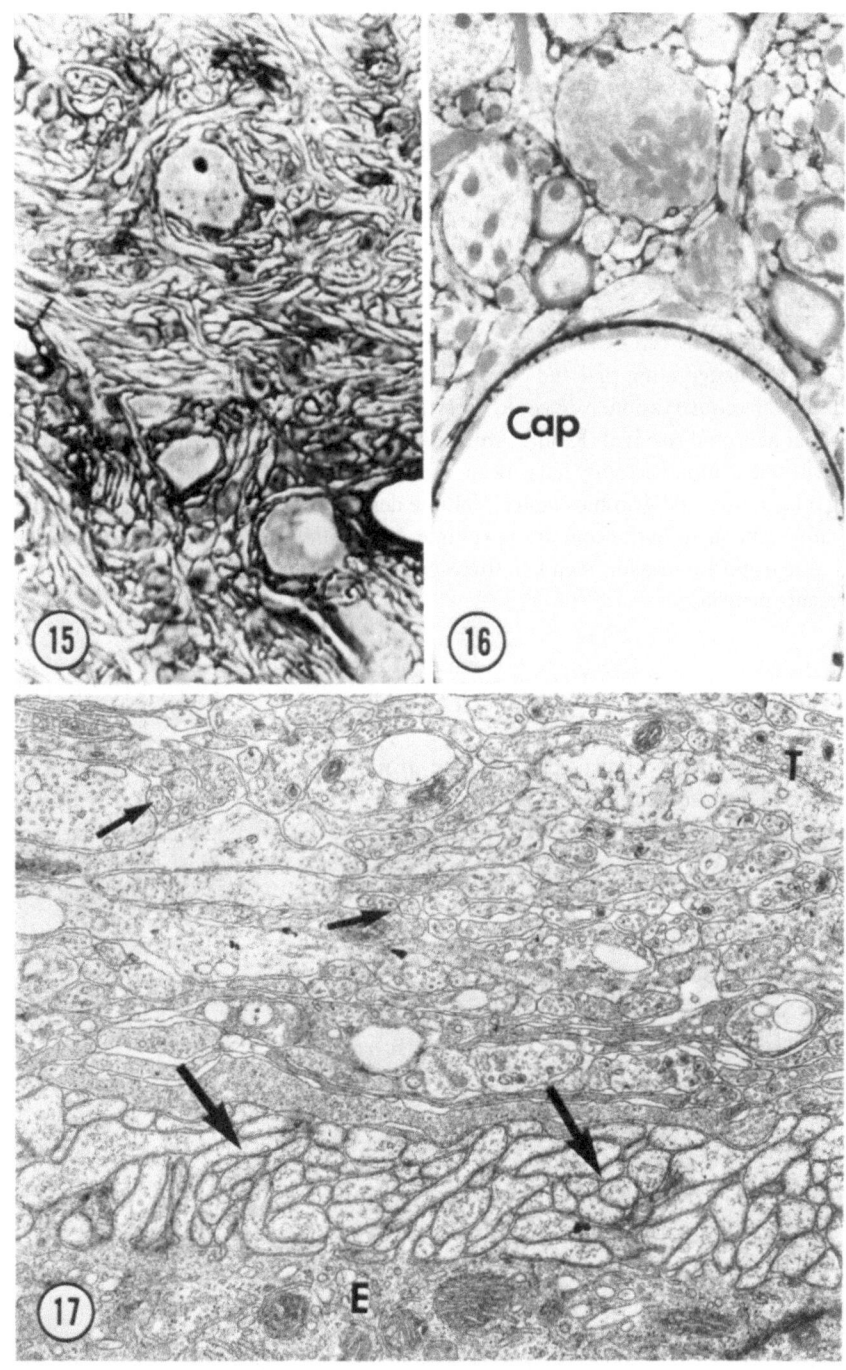

6. Enhanced Survival of Transplanted Autonomic Neurons

In previous reports,[8,9] we suggested that the acquisition and maintenance of target sites might be necessary for the long-term survival of transplanted neurons. This requirement is complementary to, but probably follows the rapid establishment (12–24 h) of a nutritive blood supply. When the SCG is allografted into the unaltered ventricle, there is an inevitable and constant demise of the ganglion cells. As the animal's own SCG occupied their normal sites in the host brain, none were available to the grafted neurons. Those grafted neurons that did survive probably formed contacts upon newly formed blood vessels or each other.

By allografting the SCG into a recipient that had been bilaterally sympathectomized, thereby denervating pial and choroidal vessels, natural targets for the SCG became available for reinnervation. We could ascertain if the denervated sites could lead to an increased neuronal survival. Results, thus far, are striking. Transplants in animals with and without sympathectomy have been measured for morphometric volume density using a computer and graphics tablet. Volume density (Vv) is the ratio of the volume of a certain component to the containing volume. Sequential series of 1-μm sections can be used to extrapolate measurements to three-dimensional structures. In this equation, Vv represents neurons:

$$Vv\% = \frac{\text{total neuronal area}}{\text{total transplant area}}$$

When compared to an undisturbed ganglion *in situ,* a 1-month-old graft in a host with its own SCG intact contains a 14% neuronal cell body area, whereas in a sympathectomized host, over 50% of the neurons are viable (Fig. 18). By 6 months, there is nearly

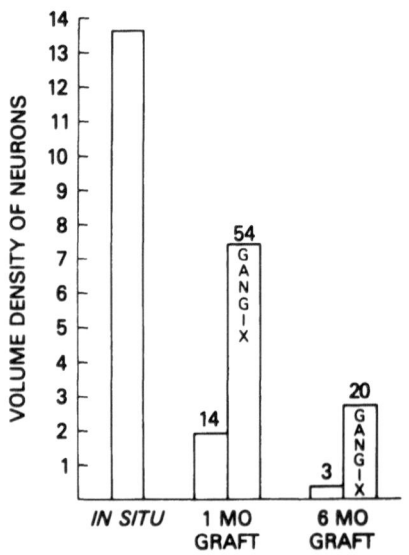

FIGURE 18. Based on morphometric volume density measurements, neuronal survival in SCG allografts is markedly enhanced when the host has received a sympathectomy (GANGLX) prior to graft placement.

a sevenfold increase in surviving neurons of the SCG grafts in the host that had been sympathectomized prior to receiving the graft. These results suggest a similar conclusion drawn from other studies where a transplant may supplant deficiencies in the host. Thus, a lesion to the host that denervates appropriate target sites enables transplanted neurons to survive to a greater degree than in a nondenervated host. This is probably due to the neurons' ability to form contacts with targets that have become available.

ACKNOWLEDGMENT. Supported in part by NIH Grant NS-17468 to J.M.R.

Note Added in Proof
    Additional data concerning experiments on the blood–brain barrier may be found in: Rosenstein, J.M., and Brightman, M.W., 1983, *Science* **221**:879.

References

1. Das, G. D., and Altman, J., 1972, Studies on the transplantation of developing neural tissue in the mammalian brain. I. Transplantation of cerebellar slabs into the cerebellum of neonate rats, *Brain Res.* **38**:233.
2. Jaeger, C. B., and Lund, R. D., 1980, Transplantation of embryonic occipital cortex to the tectal region of newborn rats: A light microscopic study of organization and connectivity of the transplants, *J. Comp. Neurol.* **194**:571.
3. Lund, R. D., and Harvey, A., 1981, Transplantation of tectal tissue in rats. I. Organization of transplants and pattern distribution of host afferents within them, *J. Comp. Neurol.* **201**:191.
4. Björklund, A., and Stenevi, U., 1979, Reconstruction of the nigrostriatal dopamine pathway by intercerebral implants, *Brain Res.* **177**:555.
5. Perlow, M. J., Freed, W. J., Hoffer, B. J., Seiger, Å., Olson, L., and Wyatt, R., 1979, Brain grafts reduce motor abnormalities produced by destruction of nigrostriatal dopamine system, *Science* **204**:643.
6. Gash, D., Sladek, C. D., and Sladek, J. R., Jr., 1980, A model system for analysing functional development of transplanted peptidergic neurons, *Peptides* **1** (Suppl. 1):125.
7. McLoon, L. K., Lund. R. D., and McLoon, S. C., 1982, Transplantation of reaggregates of embryonic neural retinae to neonatal rat brain: Differentiation and formation of connections, *J. Comp. Neurol.* **205**:179.
8. Rosenstein, J. M., and Brightman, M. W., 1978, Intact cerebral ventricle as a site for tissue transplantation, *Nature (London)* **275**:83.
9. Rosenstein, J. M., and Brightman, M. W., 1979, Regeneration and myelination in autonomic ganglia transplanted to intact brain surfaces, *J. Neurocytol.* **8**:359.
10. Stenevi, U., Björklund, A., and Svendgaard, N. -A., 1976, Transplantation of central and peripheral monoamine neurons to the adult rat brain: Techniques and conditions for survival, *Brain Res.* **114**:1.
11. Hökfelt, T., Johannson, O., Ljungdahl, A., Lundberg, J. M., and Schultzberg, M , 1980, Peptidergic neurons, *Nature (London)* **283**:515.
12. Kessler, J. A., and Black, I. B., 1982, Regulation of substance P in adult sympathetic ganglia, *Brain Res.* **234**:182.
13. Madison, R., Crutcher, K. A., and Davis, J. N., 1981, Sympathohippocampal neurons are inside the blood–brain barrier, *Brain Res.* **213**:183.
14. Hendry, I. A., 1975, The response of adrenergic neurons to axotomy and nerve growth factor, *Brain Res.* **94**:87.
15. Glees, P., 1955, Studies on cortical regeneration with special reference to central implants, in: *Regeneration in the Central Nervous System* (W. Windle, ed.), pp. 94–111, Academic Press, New York.

16. Rosenstein, J. M., and Brightman, M. W., 1980, Arrest and migration of cerebellar neurons towards grafts of autonomic ganglion, *Peptides* **1**:221.

17. Rosenstein, J. M., and Brightman, M. W., 1981, Anomalous migration of central nervous tissue to transplanted autonomic ganglia, *J. Neurocytol.* **10**:387.

18. Ramón y Cajal, S., 1928, *Degeneration and Regeneration of the Nervous System* (R. M. May, ed., transl.), Vol. 1, Oxford University Press, London.

19. Martinovic, P. N., 1930. Migration and survival of the nerve cells cultivated in the cerebrospinal fluid of the embryo and the young animal, *Arch. Exp. Zellforsch. Besonders Gewebezuech.* **10**:145.

20. Rappoport, S. I., 1976, *Blood–Brain Barrier in Physiology and Medicine,* Raven Press, New York.

21. Rees, R. P., and Bunge, R., 1974, Morphological and cytochemical study of synapses formed in culture between isolated superior cervical ganglion neurons, *J. Comp. Neurol.* **157**:1.

22. Black, I. B., and Mytilineou, C., 1976, Trans-synaptic regulation of the development of end organ innervation by sympathetic neurons, *Brain Res.* **101**:503.

23. Patterson, P. H., 1978, Environmental determination of autonomic neurotransmitter function, *Annu. Rev. Neurosci.* **1**:1.

24. Iacovitti, L., Joh, T. H., Park, D. H., and Bunge, R. P., 1981, Dual expression of neurotransmitter synthesis in cultured autonomic neurons, *J. Neurosci.* **1**:685.

25. Potter, D. D., Landis, S. C., and Furshpan, E. J., 1980, Dual function during development of rat sympathetic neurons in culture, *J. Exp. Biol.* **89**:57.

26. Richardson, K. C., 1966, Electron microscopic identification of autonomic nerve endings, *Nature (London)* **210**:765.

27. Tranzer, J. P., and Richards, J. G., 1976, Ultrastructural cytochemistry of biogenic amines in nervous tissue: Methodologic improvements, *J. Histochem. Cytochem.* **24**:1178.

28. Landis, S. C., 1977, Growth cones of cultured sympathetic neurons contain adrenergic vesicles, *J. Cell Biol.* **78**:8.

29. Yamauchi, A., Lever, J. D., and Kemp, K. W., 1973, Catecholamine loading and depletion in the rat superior cervical ganglion, *J. Anat.* **114**:272.

30. Molliver, M. E., and Kristt, D. A., 1975, The fine structural demonstration of monoaminergic synapses in immature rat neocortex, *Neurosci. Lett.* **1**:305.

31. Jacobson, M., 1978, *Developmental Neurobiology,* 2nd ed., Plenum Press, New York.

32. Liu, H. M., Balkovic, E. S., Sheff, M. F., and Zacks, S. I., 1979, Production in vitro of a neurotropic substance from proliferative neurolemma-like cells, *Exp. Neurol.* **64**:271.

33. Weiss, P., and Taylor, A. C., 1944, Further experimental evidence against "neurotropism" in nerve regeneration, *J. Exp. Zool.* **95**:233.

34. Yamamoto, M., Chan-Palay, V., Steinbusch, H., and Palay, S. L., 1980, Hyperinnervation of arrested granule cells produced by transplantation of monoamine containing neurons into the fourth ventricle of rat, *Anat. Embryol.* **159**:1.

35. Altman, J., 1969, Autoradiographic and histological studies of postnatal neurogenesis. III. Dating the time of production and onset of differentiation of cerebellar microneurons in rats, *J. Comp. Neurol.* **136**:269.

36. Altman, J., 1966, Autoradiographic and histological studies of postnatal neurogenesis. II. A longitudinal investigation of cells incorporating tritiated thymidine in neonate rats, with special reference to postnatal neurogenesis in some brain regions, *J. Comp. Neurol.* **128**:431.

37. Hinds, J. W., 1968, Autoradiographic study of histogenesis in the mouse olfactory bulb, *J. Comp. Neurol.* **134**:305.

38. Eng, L. F., 1980, The glial fibrillary acidic (GFA) protein, in: *Proteins of the Nervous System* (R. Bradshaw and D. Schneider, eds.), Raven Press, New York.

39. Jessen, K. R., and Mirsky, R., 1980, Glial cells in the enteric nervous system contain glial fibrillary acidic proteins, *Nature (London)* **286**:736.

40. Trapp, B. D., McIntyre, L. J., Quarles, R. H., Sternberger, N. H., and Webster, H. de F., 1979, Immunocytochemical localization of rat peripheral nervous system myelin proteins: P_2 protein is not a component of all peripheral nervous system myelin sheaths, *Proc. Natl. Acad. Sci. USA* **76**:3552.

41. Heard, J. K., and Gilmore, S. A., 1980, Intramedullar Schwann cell development following x-irradiation of mid-thoracic and lumbosacral spinal cord levels in immature rats, *Anat. Rec.* **197**:85.

42. Weinberg, E. L., and Raine, C. S., 1980, Reinnervation of peripheral nerve segments implanted into the rat central nervous system, *Brain Res.* **198**:1.

43. Raine, C. S., Traugott, V., and Stone, S. H., 1978, Glial bridges and Schwann cell migration during chronic demyelination in the CNS, *J. Neurocytol.* **7**:541.

44. Jacobs, J. M., 1977, Penetration of systematically injected horseradish peroxidase into ganglia and nerves of the autonomic nervous system, *J. Neurocytol.* **6**:607.

45. Brightman, M. W., Hori, M., Rappoport, S., Reese, T. S., and Westergaard, E., 1973, Osmotic opening of tight junctions in cerebral endothelium, *J. Comp. Neurol.* **152**:317.

46. Brightman, M. W., 1977, Morphology of blood–brain interfaces, in: *The Ocular and Cerebrospinal Fluids* (L. Z. Bito, H. Davson, and J. D. Fenstermacher, eds.), *Exp. Eye Res.* **25**:1.

47. Groothius, D. R., Fischer, J. M., Lapin, G., Bigner, D. D., and Vick, N. A., 1982, Permeability of different experimental brain tumor models to horseradish peroxidase, *J. Neuropathol. Exp. Neurol.* **41**:164.

48. Brightman, M. W., and Reese, T. S., 1969, Junctions between intimately apposed cell membranes in the vertebrate brain, *J. Cell Biol.* **40**:648.

49. Knyihar-Csillik, E., Csillik, B., and Rakic, P., 1982, Ultrastructure of normal and degenerating glomerular terminals of dorsal root axons in the substantia gelatinosa of the rhesus monkey, *J. Comp. Neurol.* **210**:357.

Index